ANCIENT TEXTILES SERIES VOL. 18

PREHISTORIC, ANCIENT NEAR EASTERN AND AEGEAN TEXTILES AND DRESS

an Interdisciplinary Anthology

edited by
Mary Harlow, Cécile Michel and Marie-Louise Nosch

Oxbow Books
Oxford & Philadelphia

Published in the United Kingdom in 2014 by
OXBOW BOOKS
10 Hythe Bridge Street, Oxford OX1 2EW

and in the United States by
OXBOW BOOKS
908 Darby Road, Havertown, PA 19083

Paperback Edition: ISBN 978-1-78297-719-3
Digital Edition: ISBN 978-1-78297-720-9

A CIP record for this book is available from the British Library

Printed in Malta by Gutenberg Press

For a complete list of Oxbow titles, please contact:

UNITED KINGDOM
Oxbow Books
Telephone (01865) 241249, Fax (01865) 794449
Email: oxbow@oxbowbooks.com
www.oxbowbooks.com

UNITED STATES OF AMERICA
Oxbow Books
Telephone (800) 791-9354, Fax (610) 853-9146
Email: queries@casemateacademic.com
www.casemateacademic.com/oxbow

Oxbow Books is part of the Casemate Group

Front cover:Detail of the skirt, showing the loose belt on ivory figurine NAM 6580, Prosymna.

Contents

Acknowledgements

This anthology forms part of the *Programme International de Collaboration Scientifique* (*PICS*) *TexOrMed = Textiles from the Orient to the Mediterranean*, between the Danish National Research Foundation's Centre for Textile Research and the CNRS Archéologies et Sciences de l'Antiquité – Histoire et Archéologie de l'Orient Cunéiforme research group (2012–2014). We thank the CNRS and the DNRF for their support.

This anthology, Mary Harlow, Cécile Michel and Marie-Louise Nosch (eds), *Prehistoric, Ancient Near Eastern and Aegean Textiles and Dress: an interdisciplinary anthology*, Ancient Textiles Series 18, Oxbow Books, Oxford (2014) is the first volume of two, which group interdisciplinary contributions to the field of textile research. The second volume is Mary Harlow and Marie-Louise Nosch (eds.), *Greek and Roman Textiles and Dress: an interdisciplinary anthology*, Ancient Textiles Series 19, Oxbow Books, Oxford (2014).

We thank our colleagues at the Centre for Textile Research for their valuable help and advice, especially Dr. Giovanni Fanfani. We also thank Clare Litt, editor in chief, and Sam McLeod at Oxbow Books, for the always smooth collaboration and professional help. Finally, we thank the authors for their excellent contributions, trust and patience.

Copenhagen, December 2013

The editors
Mary Harlow, Cécile Michel and Marie-Louise Nosch

Contributors

GIULIA BACCELLI is an archaeologist. She graduated at the University of Florence, Italy with a thesis on spinning and weaving tools found in Tell Barri, Syria. She completed her PhD in 2011 at the University of Tübingen, Germany. Her thesis dealt with the meaning and value of textiles in 2nd millennium BC Syria, particularly focusing on the Royal Grave of Qatna. She has participated in several archaeological excavations in Syria (Tell Barri, Tell Beydar and Qatna) as field archaeologist and she is currently scientific collaborator at the University of Tübingen, Germany.

BENEDETTA BELLUCCI is an art historian and archaeologist. She received her PhD in Ancient Near Eastern Archaeology and Art History at the University of Pavia (Italy) in 2009 with a thesis on the representation of composite creatures on Late Bronze Age seals and seal impressions in Northern Syria and Southern Anatolia. She has participated in several archaeological excavations in Italy, Syria, Turkey, Libya and Qatar as field archaeologist and finds registrar.

TINA BOLOTI is an archaeologist and a PhD candidate at the University of Crete, whose research is co-financed by the European Union (European Social Fund – ESF) and Greek national funds (Research Funding Program: Heraclitus II). Her thesis, which examines the functional and symbolic role of cloth and clothing in rituals in the Aegean Late Bronze Age, constitutes a combined study of the related iconography and the Linear B archives. She participates in archaeological research programs of *The Archaeological Society at Athens* (publication of the Greek excavations at Mycenae) and the *Academy of Athens* (research in the prehistoric settlement on Koukonisi, Lemnos), while she is collaborator of the *Centre for Research & Conservation of Archaeological Textiles*.

CHAIDO KOUKOULI-CHRYSANTHAKI is Honorary Ephor of the Greek Ministry of Culture. She has been Director of the 18th Ephoria of Classic and Prehistoric Antiquities of Kavala (North Greece) during which time she has lead major archaeological projects in the vicinity of Kavala, the island of Thassos and East Macedonia in general. Among her several excavation projects, prehistoric research is best represented by the excavation of the Bronze Age settlement of Skala Sotiros on Thassos, co-directorship of the Greek/Bulgarian investigation of Neolithic Promahon-Topolnica and co-directorship of the Greek/French excavation of Dikili Tash. Dr. Koukouli-Chrysnathaki has published extensively on the archaeology of Thassos and the Macedonian region of Northern Greece.

RICHARD FIRTH is a Research Associate of the University of Bristol, since 2005 a collaborator of the Danish National Research Foundation's Centre for Textile Research, University of Copenhagen, and more recently an editor for the Cuneiform Digital Library Initiative. He has a degree in Mathematics from Cambridge, a PhD in Elementary Particle Physics from Durham and has had a career in research and development in the UK nuclear power industry. He has written numerous papers on Linear B topics including extensive studies of the find-places of the Linear B tablets at Knossos. More recently he has written a number of papers related to the Ur III textile industry.

AGNÈS GARCIA-VENTURA is an ancient historian, post-doctoral researcher at "Sapienza" Università di Roma (Italy). She was awarded her PhD by the Universitat Pompeu Fabra (Barcelona, Spain) in 2012 with a thesis on the textile production in Ur III Mesopotamia. Her research focuses on textiles and gender in Mesopotamia with particular attention to visual imagery sources such as foundation figurines and to the organisation of work as reflected in the Ur III administrative texts. She is also carrying out research into the historiography of Ancient Near Eastern studies in Spain during the 20th century and into Phoenician and Punic musical performance.

VALENTINA GASBARRA received her PhD in *Linguistics* at the University of Rome "Sapienza". She is Postdoc research fellow at the Department of Document Studies, Linguistics and Geography of the University of Rome "Sapienza" within the activities of the Italian PRIN-Project "Linguistic representations of identity. Sociolinguistic models and historical linguistics". She has studied abroad (Oxford) and she has been part of different national Research projects. Her research interests mainly concern Mycenaean Greek, with particular focus on phonology (Mycenaean labiovelars), nominal morphology and morpho-syntax (the different typology of compounds in Linear B archives), linguistic contacts between Mycenaean and Semitic languages in the Eastern Mediterranean area during the 2nd millennium BC, and the role of Hittite as bridge-language.

SALVATORE GASPA is a historian specialized in Ancient Near Eastern studies. His interests focus on the Assyrian material culture and on the history of the Neo-Assyrian Empire. His methodological approach combines textual, iconographical, and archaeological investigation of the Assyrian *realia*. In 2003–2004 he was granted two Finnish Government Fellowships at the University of Helsinki for doctoral training and research in Neo-Assyrian texts and terminology. He obtained a PhD in Semitic Linguistics at the University of Florence in 2007 with a study on the terminology of vessels in the Neo-Assyrian texts and a second PhD in Ancient Near Eastern history at the University of Naples "L'Orientale" in 2011 with a research on foods and food practices in the Assyrian cult. In 2012 he published a monograph on foods and food practices in the state cult of the Assyrian Empire, while a book on the lexicon of Neo-Assyrian vessels appeared in 2014. As a Marie Curie Intra-European Fellow at the Danish National Research Foundation's Centre for Textile Research of the University of Copenhagen 2013–2015, he is carrying out a research project on textiles in the Neo-Assyrian Empire.

MARY HARLOW is an ancient historian, senior lecturer at the School of Archaeology and Ancient History, University of Leicester. In 2011–2013 she was guest professor at the Danish National Research Foundation's Centre for Textile Research at the University of Copenhagen. She works on Roman dress and the Roman life course. Her research combines literary studies, iconography and archaeology and methodologies derived from history, anthropology and sociology. She is publishing the Cambridge Key Themes volume on Roman dress.

ELENI KONSTANTINIDI-SYVRIDI is a curator at the Collection of Prehistoric, Egyptian and Oriental Antiquities at the National Archaeological Museum, Athens. She graduated from the Department of History and Archaeology at the University of Ioannina, Greece and received her PhD at the University of Birmingham, UK, Department of Ancient History and Archaeology. Her thesis entitled *Jewellery in the burial context of the Greek Bronze Age* was published in 2001 (BAR IS). Her research focuses on Late Bronze Age Aegean and Eastern Mediterranean, with particular interest in Mycenaean jewellery and dress. She has given seminars on the history and technology of jewellery, in Greece and abroad and she is currently working on a project for the reconstruction and terminology of ancient jewellery techniques, partly sponsored by INSTAP. She has participated in the edition of several books and written articles on the Late Bronze Age.

Paula Mazăre is an archaeologist. She was awarded her PhD by the University of Alba Iulia, Romania in 2012 with a thesis on Neolithic and Copper Age textile production in Transylvania (Romania). Her research is focused mainly on prehistoric textile tools and textile imprints/vestiges, but it combines computing and experimental archaeology, iconography, mythology and ethnographic data as sources for interpreting the practical and symbolic meaning of textile production. As team member of different archaeological excavations/research projects in the Roman city of Alba Iulia (Apulum) she published the first study on textile production in Apulum (Roman province of Dacia).

Cécile Michel is a historian and assyriologist, Director of Research at the National Centre of Scientific Research (CNRS) in the team Histoire et Archéologie de l'Orient Cunéiforme (Archéologies et Sciences de l'Antiquité) at Nanterre. She has collaborated with the Centre for Textile Research (CTR) since 2005. Working on the decipherment and study of cuneiform texts from the first half of the 2nd millennium BC (private archives of merchants, state administrative archives), her main research interests are Mesopotamian trade, Upper Mesopotamian and Anatolian societies, gender studies, daily life and material culture (fauna, food, metals, textiles), calendars and chronology, history of sciences, education, writing and computing. Heading an International Collaboration Scientific Programme (PICS) *Textile from Orient to the Mediterrenean* (*TexOrMed*) with Marie-Louise Nosch, she organized and published international conferences on textile terminologies and wool economy.

Marie-Louise Nosch is a historian and director of the Danish National Research Foundation's Centre for Textile Research (CTR) at the University of Copenhagen and the National Museum of Denmark. She is a professor in ancient history. She was awarded her PhD by the University of Salzburg in 2000 with a thesis on Mycenaean textile administration in Linear B but has subsequently merged Linear B studies with experimental archaeology and textile tool studies; as director of the Centre for Textile Research, she has launched research programmes combining archaeology and natural sciences. She is author and co-author of works on Aegean Late Bonze Age textile production in the Mycenaean palace economies.

Stratis Papadopoulos received his PhD from the Aristotle University of Thessaloniki. He has worked extensively with the 18th Ephorate of Classical and Prehistoric Antiquities of Kavala (North Greece), in several excavation projects on Thassos island. He has directed the excavations of the Early Bronze Age Settlement at Skala Sotiros, the Final Neolithic-Early Bronze Age settlement at Aghios Ioannis, the Early/Middle Bronze Age at Aghios Antonios, Potos and, finally, he has cooperated in the excavation of Neolithic Limenaria. Dr. Papadopoulos has taught prehistoric archaeology at the Democritus University of Thrace and the Kavala Technical Institute. He leads several interdisciplinary research projects related to prehistoric Thassos, Eastern Macedonia and Aegean Thrace.

Louise Quillien is a historian – she received her *Agrégation d'histoire* diploma in 2011 – and is currently a PhD candidate at the University of Paris I Panthéon-Sorbonne and a member of the team Histoire et Archéologie de l'Orient Cunéiforme (Archéologies et Sciences de l'Antiquité) at Nanterre. Her initial education is in general geography and history, with a specialization in Ancient Near-Eastern history and Akkadian language. In her PhD thesis on "Textiles in Mesopotamia, 750–500 BC: manufacturing techniques, trade and social meanings", she utilizes different sources from textual archives, archaeology and iconography, in order to understand the role of textiles in ancient Near Eastern society.

Caroline Sauvage is an archaeologist, assistant Professor at Loyola Marymount University. She is affiliated with the Danish National Research Foundation's Centre for Textile Research, Copenhagen through a Marie-Curie extra-European fellowship (2014). Her research interests include trade and maritime exchanges in the eastern Mediterranean, as well as the development and use of textile tools during the Late Bronze Age and early Iron Age. Her research focuses on exchanges, the status of objects, their representations and use as identity markers in the eastern Mediterranean area as a whole. Her work is based on the study of material artifacts and their interconnections, and aims to avoid the classic pitfalls of disciplinary partitioning in the study of eastern Mediterranean societies and group identities.

Orit Shamir is the Curator of Organic Materials, Israel Antiquities Authority. She holds a PhD in Archaeology and wrote her thesis on *Textiles in the Land of Israel from the Roman Period till the Early Islamic Period in the Light of the Archaeological Finds*. Her MA thesis in Archaeology was on "Textile Production in Eretz-Israel at the Iron Age in the Light of the Archaeological Finds". She has researched and published widely on textiles.

Ariane Thomas is curator of the Mesopotamian collections in the department of Ancient Near Eastern Antiquities at the Louvre Museum. In this capacity, she teaches Ancient Near Eastern archaeology at the Ecole du Louvre. She also participates in several archaeological excavations in the Middle East. She was awarded her PhD by the Sorbonne University (Paris IV) in 2012 with a thesis on Ancient Mesopotamian royal costumes from the 3rd to the 1st millennium BC. As a curator, she develops research programs that combine archaeology, history, epigraphy and science, particularly focused on textiles and clothing.

Sophia Vakirtzi is a PhD candidate at the University of Crete, Greece. Her research focuses on the production of yarn at Bronze Age settlements of the Aegean islands, on the basis of spindle whorls recovered from settlements and cemeteries. She has collaborated with the on-going Bronze Age excavations of Akrotiri on Thera, Skarkos on Ios, Koukonisi on Lemnos, and she has studied textile tools from Skala Sotiros and Aghios Ioannis on Thassos, Ayia Irini on Keos, Grotta and Aplomata on Naxos, Phylakopi on Melos, Kastri on Syros and the prehistoric Heraion on Samos.

MATTEO VIGO is a Hittitologist. He was trained at the University of Pavia (Italy) in ancient Anatolian languages and civilizations. As a PhD student he underwent research training activities in Germany at the Universities of Konstanz and Tübingen. He received a Post-Doc funded by Rotary International at the Oriental Institute of the University of Chicago. He collaborates with the Chicago Hittite Dictionary Project (CHD). His interests focus mainly on Hittite civilization, Hittite administration, Hittite international diplomacy but also Hittite lexicography and etymology. As an Intra-European Marie Curie Fellow 2013–2015, he is currently involved in a research project on Hittite Textile Terminology (TEXTHA) at the Centre for Textile Research. He is employed at the University of Copenhagen where he teaches Hittite language and civilization.

1. Investigating Neolithic and Copper Age Textile Production in Transylvania (Romania). Applied Methods and Results

Paula Mazăre

The functional, practical and symbolic importance of textiles in everyday life, and also during special events (ceremonies, celebrations, rituals, etc.) within human communities, has been highlighted by numerous studies in anthropology, history and archaeology.[1] The textile remains found during archaeological excavations are seen, according to researchers like Penelope Walton and Gillian Eastwood (1983), as being "the remains of one of man's more intimate artefacts". However, the importance of textile products and that of the activities devoted to textile production in prehistory was generally ignored by Romanian archaeologists. This is due to the scarcity of such archaeological remains.[2]

The significant advances made by researchers in the West compared with the sporadic and inconsistent efforts from Romania[3] now fully justifies the need for a systematic and scientific approach, intended towards aligning Romania with the Western European map of discoveries and research on prehistoric textiles. This is particularly true since the new trends for this field of study suggest a growing interest in this area of research.[4]

This paper aims to summarise the research performed as part of my PhD thesis, entitled "*The craft of textile production at Neolithic and Copper Age communities in Transylvania*" which was finalised in July 2012. The main focus of the research was to characterise the craft of textile

[1] Cordwell and Schwarz 1979; Schneider 1987; Barber 1991, 1994, 2007; Smith 2002; Larsson Lovén 2002; Bergerbrant 2007; Gleba *et al.* 2008.

[2] As far as is known, for Romania only one prehistoric textile product was found; more precisely, the remains of a burnt bedspread discovered in Sucidava-Celei. According to the strata in which it was found, it was chronologically dated at the beginning of 3rd millennium BC (Nica 1981).

[3] Gumă 1977; Zaharia and Cădariu 1979; Aghiţoaie and Draşovean 2004; Săvescu 2004; Marian 2006, 2008, 2009, 2010, 2012; Marian and Ciocoiu 2004a, 2004b, 2005; Marian and Anăstăsoaei 2007, 2008; Marian and Bigbaev 2008; Marian *et al.* 2004, 2005; Văleanu and Marian 2004; Mazăre 2008, 2010, 2011a, 2011b, 2011c, 2012, 2013; Mazăre *et al.* 2012; Prisecaru 2009a, 2009b.

[4] The author refers here to just a few papers and works, as follows: Good 2001; Bichler *et al.* 2005; Müller *et al.* 2006; Tiedemann and Jakes 2006; Médard 2006; 2010; Baldia and Jakes 2007; Cardon 2007; Gillis and Nosch 2007a, 2007b; Mårtensson *et al.* 2009; Breniquet 2008; Gleba 2008, 2012; Gleba and Mannering 2012; Rast-Eicher 2008; Frei *et al.* 2009a, 2009b; Vanden Berghe *et al.* 2009; Andersson Strand *et al.* 2010; Hurcombe 2010; Chmielewski and Gardyński 2010; Cybulska 2010; Cybulska *et al.* 2008; Mannering *et al.* 2010; Michel and Nosch 2010; Grömer 2010; Brandt *et al.* 2011; Leuzinger and Rast-Eicher 2011; Bergfjord and Holst 2010; Murphy *et al.* 2011; Bergfjord *et al.* 2012; Andersson Strand and Nosch 2013; Hopkins 2013; Rast-Eicher and Bender Jørgensen 2013.

production (with all its economic, social and symbolic implications) during the Neolithic and Copper Age, within the geographical context of Transylvania. This was archeived using the main evidence preserved in the local soil conditions (specific for the entire southeast of Europe): the pottery textile imprints and textile tools (spindle-whorls, loom-weights and spools).

The unusual character of this paper in the context of Romanian archaeological research justifies to a large extent the limitation of the research area to a confined geographical unit, represented here by Transylvania. Among the obstacles encountered during this research were the difficulty of finding and gathering the material necessary for such a study (over 15 museum collections were browsed and not always successfully), the absence of data for the context of discovery, and the difficulty of cultural and chronological affiliation/classification for some of the artefacts.[5]

Area of the research

Transylvania as an entity is defined as the Inner-Carpathian area of Romania, a historical region which was known during the Middle Ages as "The Voivodate of Transylvania" or "Voivodal Transylvania". Geographically, it corresponds to the Depression of Transylvania, bordered by segments of the Carpathian Mountains on the East, South and West .

Figure 1.1 shows that the sites examined for this study do not cover the entire surface of Transylvania evenly. This is partly due to the fact that only some specific Neolithic and Copper Age sites were archaeologically excavated and even these were not always systematically and exhaustively researched. In addition, access to museum collections were in some cases restricted. As a result there is an uneven distribution of the analysed material within the area of study. In order to compensate, the research and analysed materials belonging to some north-western settlements, situated beyond the geographical limits previously established, were used.

Materials originating from 54 sites, consisting of textile imprints, spindle-whorls, loom-weights and spools were studied. There is a clear disproportion among the three categories of materials. The most representative are the loom-weights (identified in 45 sites, including three sites with discoveries of spools), then the spindle-whorls (23 sites) and the textile imprints (identified only in 11 sites).

Cultural and chronological framework

Defining the Neolithic and Copper Age and establishing a chronology is somewhat of a difficult task if one considers the different periodisation systems proposed by the literature.[6] The terminology used is also a subject of interpretation and dispute. In Romanian archaeological literature the term *Aeneolithic or Chalcolithic*[7] *is found and* used to broadly designate the same period known in the Western literature as 'Copper Age' (or 'Kupferzeit' in German). The author preferred the term Copper Age instead of Aeneolithic in order to align with an older current[8]

[5] The absence of references and methodological models in the Romanian scientific literature was compensated mostly with a three month research internship (May–August 2009) at the Danish National Research Foundation's Centre for Textile Research (*CTR)*, University of Copenhagen.

[6] Ursulescu 1986–1987, 72; Ursulescu 1993, 22; Maxim 1999, 1; Schier and Draşovean 2004, 46; Lazarovici and Lazarovici 2006, 1–10; Petrescu-Dîmboviţa 2010, 113.

[7] Petrescu-Dîmboviţa and Vulpe 2010.

[8] Schier and Draşovean 2004; Lazarovici and Lazarovici 2006, 2007; Gogâltan 2008; Gogâltan and Ignat 2011; Diaconescu 2009.

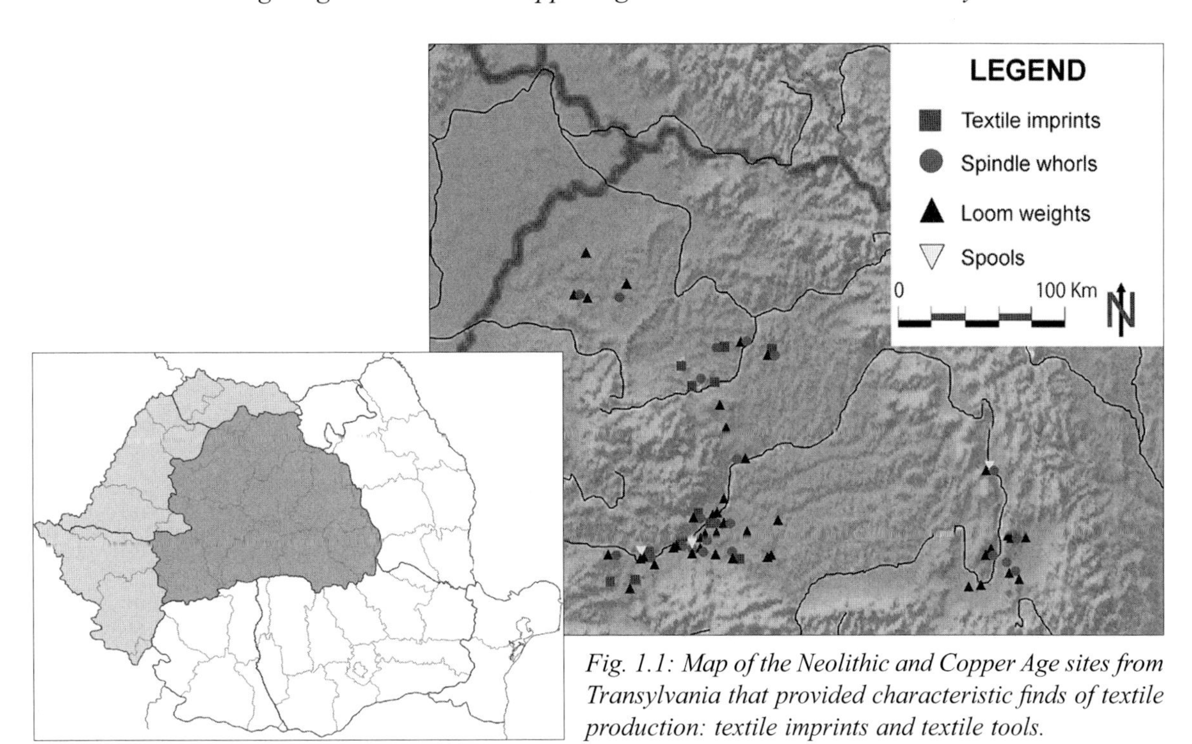

Fig. 1.1: Map of the Neolithic and Copper Age sites from Transylvania that provided characteristic finds of textile production: textile imprints and textile tools.

that aims at adapting the archaeological realities from present day Romania to the central and western European terminology.

Since there are many contradictory opinions regarding the final phase of the Copper Age (*Late Copper Age* or *Late Aeneolithic*),[9] that time period was not included as part of the research. Therefore, the research is carried out upon the following cultures/cultural groups which are chronologically situated between *c.* 6000–3500 BC: Starčevo-Criş (*c.* 6100–5300 BC), Vinča (*c.* 5400–4500 BC), Cluj-Cheile Turzii-Lumea Nouă cultural complex (*c.* 5200–4900 BC), Linear Pottery culture (*Notenkopf* horizon *c.* 5000–4950 BC), Iclod (*c.* 4900–4600 BC), Suplac (*c.* 5200–4650 BC), Oradea-Salca-Herpály (*c.* 5000–4500 BC), Turdaş (*c.* 5050–4930 BC), Foeni (*c.* 4750–4450 BC), Petreşti (*c.* 4600–4000 BC), Ariuşd (Cucuteni A1–A4 – *c.* 4600–4050 BC), Tiszapolgár (*c.* 4500–4000 BC) and Bodrogkeresztúr (*Scheibenhenkel* horizon – *c.* 4050–3500 BC).[10]

[9] Although quite disregarded lately there is still a theory that states that the realities of the *Late Aeneolithic/Late Copper Age* are better described by a "transitional" period, ranging from Neolithic (Eneolithic) to the Bronze Age. This theory was launched in Romania by M. Petrescu-Dîmboviţa (1950, 119) and was later adopted by most researchers. It is still being used, as shown by Fl. Gogâltan, by the followers of the 'Thracological School' (Gogâltan and Ignat 2011, 7). There is also no common ground among specialists on defining the end of the Copper Age and the beggining of the Bronze Age respectively. According to some authors, the Copper Age ends somewhere *c.* 2500 BC (Gogâltan 2008, 81; Gogâltan and Ignat 2011, 7) while others still place it around 3500 BC (Vulpe 2010, 218, 222–223, fig. 30). According to these last opinions, some of the cultures regarded as belonging to the final stages of the Copper Age actually belong to the Bronze Age.

[10] In order to place the finds chronologically the author used the ^{14}C data (including the calibration diagrams) from the *IPCTE Radiocarbon Database* (http://arheologie.ulbsibiu.ro/radiocarbon/download.htm), against the relative dates and cultural synchronizations published by various specialists.

The analysis of textile imprints

Even though there is a great resemblance between the textile products and the half-rigid or rigid basket-like or mat-like structures (basketry or wickerwork), the author supports the definition of Elisabeth Barber in separating these two categories of artefacts. According to Barber, textiles represent all types of woven and non-woven materials that look like "thin sheets of material made from fibres, which are soft and floppy enough to be used as coverings for people and things".[11]
As previously stated, the only evidence of archaeological textiles for the Neolithic and Copper Age uncovered in Transylvania were textile imprints found on the base and on the sides of pottery. During this research 27 imprints have been analysed. They were discovered within 11 archaeological sites belonging to the Starčevo-Criş, Vinča, Turdaş, Tiszapolgár cultures as well as to the Foeni and Iclod cultural groups (Table 1.1).

Woven textile structures *(Fig. 1.23.1–3)*
The research of archaeological textiles, especially the woven structures, has lately seen a considerable progress from the application of new and advanced methods of interdisciplinary scientific research.[12] Applying these methods also depends on the conservation status and of the preservation form of the archaeological textiles. From this perspective, the textile imprints have restricted possibilities of investigation. Moreover, some factors like the properties and the quality of the textile product, the clay shrinkage factor, the deformation caused by the burning process (for ceramics) and so on can alter the original aspect of the textile product. This is why, in the case of textiles imprints, only the most visually noticeable properties were registered, these included: the structure of the textile product (binding type, the technological procedure through which the textile product was made, the thickness of the thread systems, the type of edge), the characteristics of the threads (torsion direction, torsion angle, thickness), the decoration, some technological errors, the joining and some wear traces. Therefore, each imprint was registered within a database according to a thoroughly defined set of criteria.[13]

For classifying the woven textiles imprints the structural categorisation proposed by Lena Hammarlund, who defined 28 different categories of fabrics, has been adopted.[14] The primary differentiation of the fabrics was made according to: 1. *the binding type* (the characteristic of the Neolithic period is the plain weave) (Fig. 1.2); 2. *the fineness group* (defined according to the fibres' thickness) and 3. *the thickness group* (defined according to the value of the cover factor).[15]

As Table 1.1 illustrates, there was an opportunity to analyse only four such woven textile imprints, even though, at least for Tiszapolgár culture levels there are records of more imprints. With the exception of the narrow woven textile found in the Neolithic site of Limba (Fig. 1.23.1), all the others are dated into the Copper Age. All the structures were made using the tabby weave technique, but displayed different morphological and technological particularities, thus dividing them into:

I warp-faced narrow tabby band;
II balanced tabby weave. According to the ratio between thickness and density the woven textile imprints were distributed into four different classes (Fig. 1.3).

[11] Barber 1991, 5.
[12] Andersson Strand *et al.* 2010.
[13] For more details see Mazăre 2010.
[14] Hammarlund 2005, 117, Mazăre 2010, 22–23.
[15] Hammarlund 2005, 115.

Table 1.1: Cultural and site distribution of the analysed textile imprints.

		Starčevo-Criş	Vinča		Turdaş	Foeni	Iclod	Tiszapolgár	
Site code	**Site name**	*IIIB–IVA*	*A3–B1*	*B1–B2*	-	-	*I*	-	*B*
ALN	Alba Iulia-*Lumea Nouă*					2 1			
DAC	Dăbâca-*Cetate*								1
DOR	Doroltu-*Castel*							1	
HGC	Hunedoara-*Grădina Castelului*	1							
LBT	Limba-*Bordane*			1					
LVL	Limba-*Vărăria*			4					
MSP	Miercurea Sibiului-*Petriş*		3						
TAG	Ţaga						1		
TLL	Turdaş-*La Luncă*				1				
TRD	Turdaş				10				
VSG	Valea Sângeorgiului				1				
Total		1	8		12	3	1	1	1

Twined textile	21	String (Cord)	1
Woven textile	4	Uncertain textile structure	1

The woven textiles were created using simple or plied yarns. With the exception of the narrow cloth, made of z-twisted yarns, all the others were made using s-twisted yarns. The twist angle varies between 30° and 53°. The thickness of system A threads is almost identical to those threads from system B. The thinner threads (0.32mm) are found within the weaving imprint from Lumea Nouă belonging to the Foeni cultural group (Fig. 1.23.2), and the thicker ones (1.4mm) recovered at Doroltiu belong to the Tiszapolgár culture settlement (Fig. 1.23.3).

Twined textile structures *(Fig. 1.23.4–7)*

They represent the majority of textile structures identified as imprints on Transylvanian pottery fragments (Table 1.1). With the exception of the Foeni imprints, belonging to Copper Age, all the others are within the Neolithic period.

Twining is a recently defined textile technique in the Romanian archaeological literature.[16] Analysing the twined textiles' structure meant running through the same methodological

[16] Mazăre 2011c.

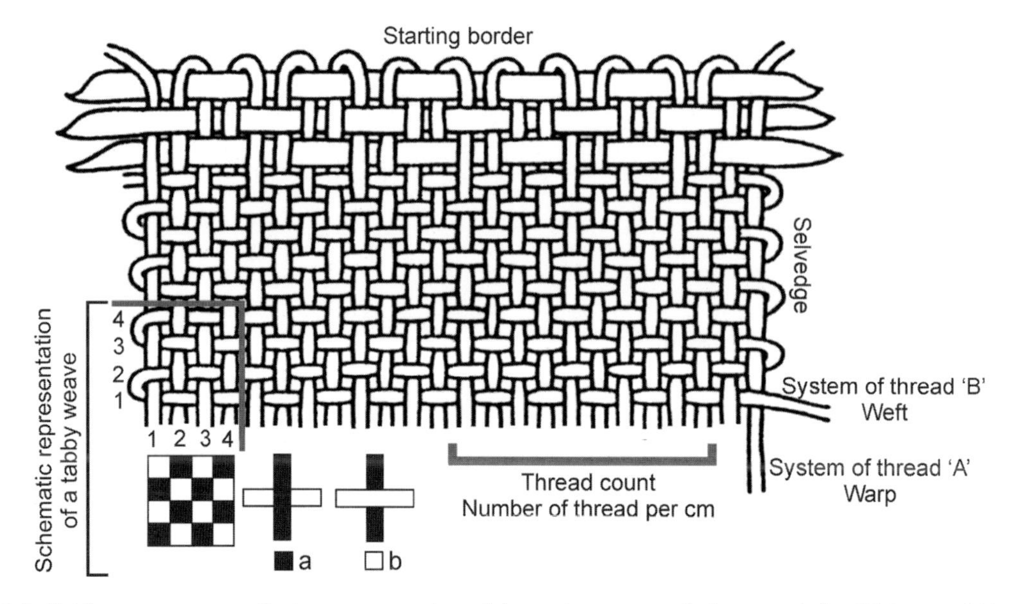

Fig. 1.2: Tabby weave: naturalistic representation of the main structural elements (after Walton and Eastwood 1983); schematic representation by squares (after Cioară 1998).

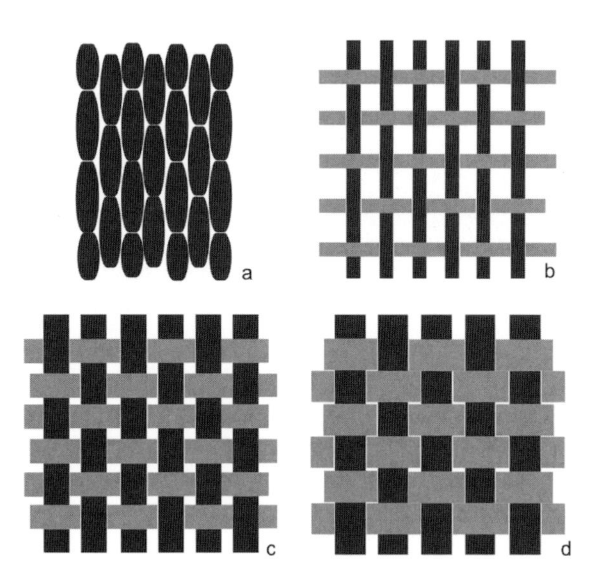

Fig. 1.3: Types of woven textiles identified as imprints on Neolithic and Copper Age pottery fragments: a. I-5c = medium-coarse and dense narrow band, warp-faced plain weave (LBT.1050); b. II-2a = thin and open plain weave (ALN.1001); c. II-6b = coarse and medium-dense plain weave, (DAC.58024); d. II-7c = very coarse and dense plain weave (DOR.61329).

stages as in the case of woven textiles. Due to the fact that both the manufacturing technique and the structural aspect are different from those of the woven materials, the twined textiles were treated separately. Therefore, they were characterised based on the following criteria: raw materials, the thread diameter, the thickness and density of textiles, the orientation of the rows of active elements, the edges (margins), technological details (and faults), use-wear traces.

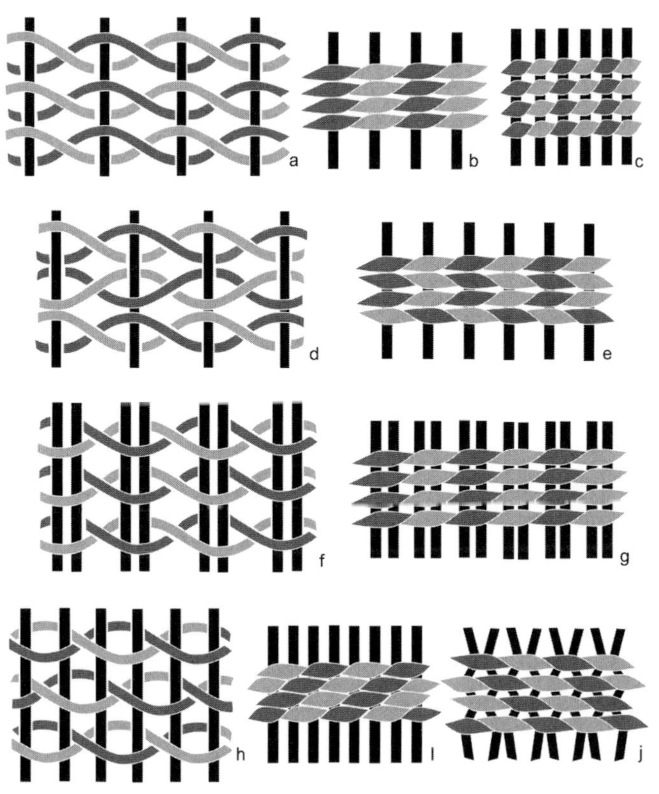

Fig. 1.4: Examples of twined structures belonging to class II2 (two-thread weft twining): a. Open simple Z-twist twining (II2-z-A3); b. Tight simple S-twist twining (II2-s-A1); c. Closed simple S-twist twining (II2-s-A2); d. Open simple ZS-twist twining (II2-zs-A3); e. Tight simple ZS-twist twining (II2-zs-A1); f. Open Z-twist twining over two passive elements (II2-z-B3); g. Closed S-twist twining over two passive elements (II2-z-B2); h. Open diagonal Z-twist twining (II2-z-C3); i. Tight diagonal S-twist twining, with parallel warp threads (II2-s-C1a); j. Closed diagonal Z-twist twining, with transposed warp (II2-z-C2b); (drawing: P. Mazăre from Seiler-Baldinger 1991; Médard 2010).

An older study by James M. Adovasio was ustilised for classifying the twined structures[17] along with the work of Irene Emery and Annemarie Seiler-Baldinger regarding the classification of the textile structures and techniques[18] and the methodology of investigating the twined structure discovered in the Neolithic lake dwellings of the Swiss Plateau.[19] Thus, the twined structures have been divided according to the following classification model, displayed by Table 1.2.

Applying this system assigns a code to each twined structure, as in the following examples:

I_2-z-A1 Simple Twined Structure, Two Z Twist Warps;
II_2-s-C3 Open Diagonal Twining, Two S Twist Wefts (see Fig. 1.4).

[17] Adovasio 1977, 15–52.
[18] Emery 2009, 195–212; Seiler-Baldinger 1994, 31–32, 50, 61–62.
[19] Médard 2010, 61–63, 74–86.

Table 1.2: Typological classification levels for defining the twined textile structures.

Classification level	**The defined typological category**	**Classification criteria**	**Numbering (Coding system)**
1.	Technological class	The active-passive relation between the thread systems	I, II, III... ($I_{2-3...}$; $II_{2-3..}$ ..)
2.	Technological type (binding type)	Twist direction	z, s, zs
3.		The passive strand layout	A, B, C...
4.		The distance between the rows	1, 2, 3....
5.	Subtypes/variants	Structural and technological features	a, b, c.../$_{1,\ 2,\ 3...}$

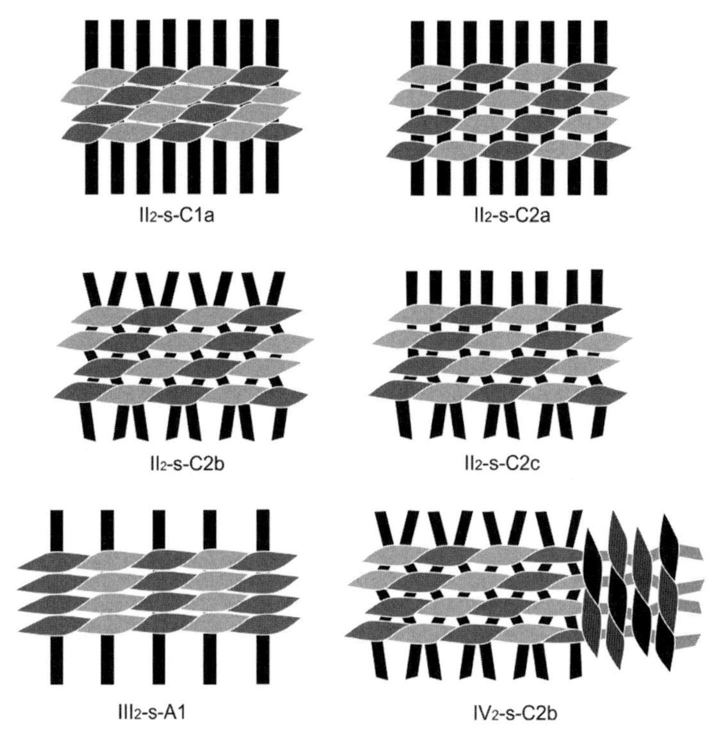

Fig. 1.5: Examples of the twined textile structures identified as imprints on Neolithic and Copper Age pottery from Transylvania.

Among the textile imprints from Transylvania three types of twined textiles and several subtypes were identified (Fig. 1.5). Of these the majority were created in diagonal twining with more or less closed rows. A single imprint revealed a simple twined structure (III2-s-A1; ALN-0018, Foeni culture). Also, a single imprint revealed a twined structure with an inversed active system. (IV_2-S-C2b; TRD-5271, Turdaş culture). In all of the structures the active strands (weft threads) were twist in S direction.

All of the twined textiles were made using stripes or bundles of vegetal fibres, some looking similar to decorticated stems/fibres, used in a raw form. The strands' diameters are between 0.7 and 3.6mm with an average between 1 and 2.67mm. All these textiles are thicker than all other woven textiles analysed, although there are variations in thickness that allow a separation into four classes. The thickest are more similar to mats than textiles structures. Some display rows of curvilinear active elements, a clue that they were manufactured freely, without any tension frame or device (Fig. 1.23.4–7).

Uncertain textile structures

A textile imprint from the Starcevo-Criş IIIB-IVA settlement at Hunedoara-*Grădina Castelului* was analysed.[20] It is an unidentified structure, and represents the oldest textile imprint from Transylvania so far. Even if the structure and its functionality are uncertain the fragment reveals a rugged character, most likely produced using unspun fibres, with a diameter between 1 and 3.9mm.

String type elements

Although it is not actually a textile structure, a segment of a string imprinted on a pottery fragment belonging to the Iclod cultural group was included in this study. It has a diameter of 3.5mm and was made using two elements secondary twisted in the Z direction, with a torsion angle of 24°.

The analysis of textile tools

The textile tools are all artefacts which had a functional role in the technological chain of manufacturing textiles and identifying them archaeologically is not always an easy task. The most certain functional interpretation is that of the spinning and weaving tools: spindle-whorls, loom-weights and spools. The author analysed these categories of textile tools during this research project. Bone, antler and stone tools that might have been used as textile production tools were also considered in this study. Their role is rather uncertain, as they lack use-wear analyses or similar specialised studies.

There are several methodological models of analysing textile tools. One of the most recent and well structured systems, organised in the form of a database, is that of the Centre of the Textile Research in Copenhagen (*CTR* Textile Tools Database). A Microsoft Access database was created using a fairly similar analysis and registration protocol. The intention was to record exhaustively all data related to the tools (spindle-whorls, loom-weights, spools). In the database, each artefact is characterised by: piece code, location, settlement type, the context of discovery; cultural and chronological frame; preservation status, typological assignment; raw material; morphology; surface treatment; decoration and signs, firing; dimensions; details of the perforation; wear traces; functional interpretations and observations; holding institution, collection, inventory number and bibliography.

The database contains over 690 records of textile tools, but parts of these were excluded from the analysis due to their uncertain cultural and chronological coordinates. Therefore, the final number of analysed artefacts was reduced to 652. Of these, 458 artefacts are of certain cultural affiliation, with a total of 12 cultures and/or cultural groups. The remaining 194 are recorded as uncertain from the point of view of their cultural affiliation (Table 1.3).

[20] Bărbat 2012, 59, Pl. 5:4–5.

Table 1.3: Numerical distribution for categories of textile tools in relation to their uncertain cultural affiliation (certain, uncertain).

Culture/Cultural group		Number of textile tools								Total
		Spindle-whorls		**Loom-weights**		**S-W/L-W**		**Spools**		
S-C	Starčevo-Criş	6		108	2	8	1			125
VIN	Vinča	4	1	91	28	10	2			136
LN	CCTLNI	2								2
CCL	Linear Pottery Culture (*Notenkopf*)	5		8						13
TRD	Turdaş	5	2	22	26	1				56
ICL	Iclod	1	1							2
SUP	Suplac	3	5	3	1					12
OSH	Oradea-Salca-Herpály			4						4
FOE	Foeni	1		7	8					16
PET	Petreşti	1	3	67	25		1			97
ARI	Ariuşd-Cucuteni	8	2	65					1	76
TSZ	Tiszapolgár		2							2
BDK	Bodrogkeresztúr			27						27
VIN/TRD	Vinča/Turdaş		1		53		4			58
CCL/PCC	Linear Pottery Culture/ Precucuteni?		1							1
TRD/FOE	Turdaş/Foeni?		2		3					5
TRD/PET	Turdaş/Petreşti				12					12
PET/FOE	Petreşti (Foeni?)				2					2
ICL/PET	Iclod/Petreşti		1				1			2
PET/COT	Petreşti/Coţofeni?								2	2
CPA	Copper Age (?)		1							1
	Total	36	22	403	160	19	9		3	**652**
		= uncertain cultural affiliation (u.c.a)								

In total, from the 51 archaeological sites investigated, 563 loom-weights, 3 spools and 58 spindle-whorls and potential spindle-whorls (perforated ceramic fragments, representing 34% of spindle-whorls) were analysed. Although recorded as loom-weights, a number of 28 artefacts have an uncertain functionality (either classified as loom-weights or spindle-whorls because they were either too big to be considered spindle-whorls, too small to be loom-weights or heavy enough to be considered as loom-weights but with a shape more easily related to spindle-whorls).

Archaeological context

From 235 textile tools, 36% were recovered from 81 features and structures of various types; most of them from surface houses. In contrast to spindle-whorls, that usually appear alone within a feature, the majority of loom-weights are in groups of at least two. Although a feature/structure

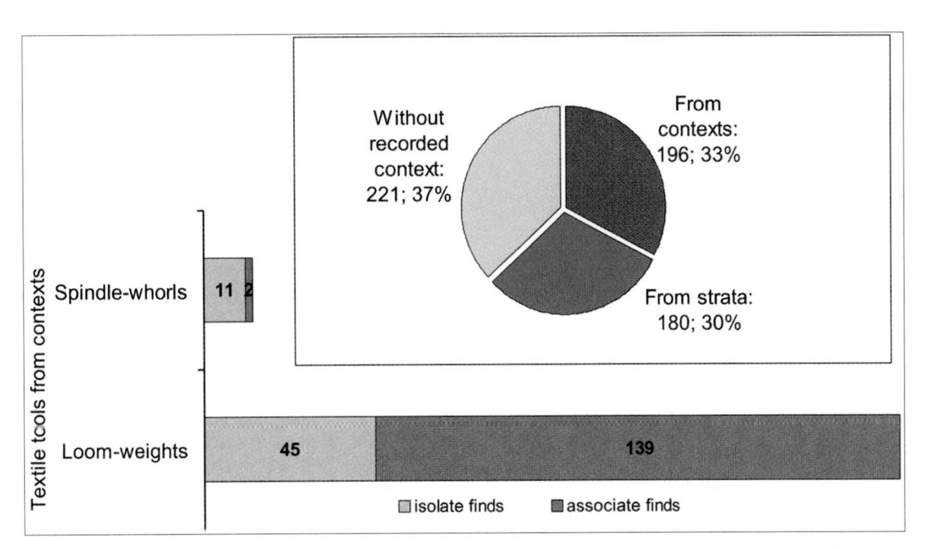

Fig. 1.6: Distribution of textile tools in regards to the archaeological contexts and the ratio between the number of individual and multiple artefacts found within features/structures.

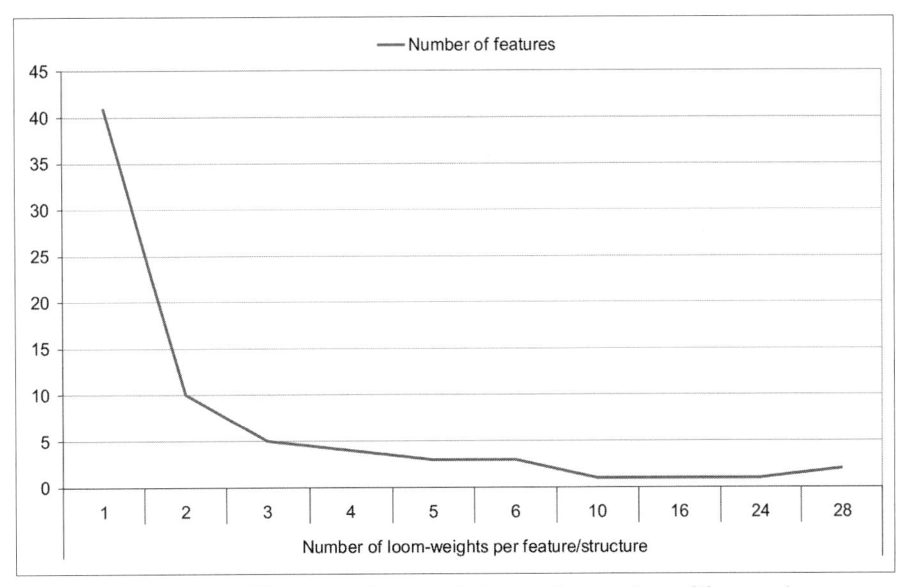

Fig. 1.7: Frequency of loom-weights in relation to the number of features/structures.

can contain more than one loom-weight, they are found functionally associated in only a few exceptional cases (Figs 1.6–7). For example, two Copper Age houses (of Ariuşd and Petreşti cultures) provided groups of 28 loom-weights. Other unusual contexts that provided weights and fragments of weights are a ritual pit from Limba (Vinča culture), a pole pit from Petreşti and several ovens from Ariuşd.

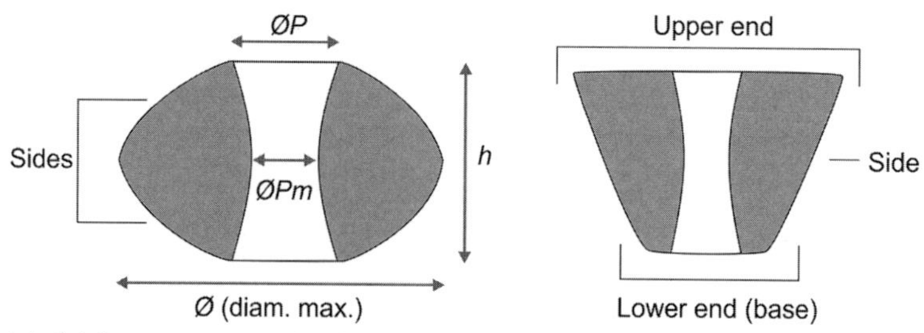

Fig. 1.8: Model for measuring the dimensions of spindle-whorls and their associated names.

Table 1.4: Hierarchic typological system for classifying the spindle whorls.

I	1	A	1	a
Category *(Raw material)*: I – baked clay; II – pottery fragment; III – stone IV – bone etc.	***Class*** *(Size = weight)*: 1. Very small: $w < 10g$ 2. Small: $10 \leq w < 25g$ 3. Medium: $25 \leq w < 50g$ 4. Large: $50 \leq w < 75g$ 5. Very large: $w \geq 75g$	***Group*** *(The flattening degree = h/diam.)*: (Fig. 1.9)	***Morphological type*** *(Morphology)*: (Fig. 1.10)	***Subtype*** *(The profile's aspect)*: (Fig. 1.11)

Spindle-whorls

Spindle-whorls are a category of artefacts poorly represented in the Neolithic and Copper Age settlements from Transylvania. In total 58 artefacts, of which 38 are fired clay spindle-whorls and 20 pierced rounded shards, have been collected and analysed. Although the numerical repertory is not representative for such a small number of artefacts, one can observe that most spindle-whorls were recovered in Copper Age habitation layers or features and most pierced rounded shards come from Neolithic settlements.

The analysis of the spindle-whorls regarded mainly their functional attributes, which were registered following all measurement rules illustrated in Fig. 1.8. When the artefacts are fragmentary, an estimation of the overall loom-weight and the maximum diameter was taken. The following abbreviations were used: ***w*** – weight (g); ***Ø*** – the maximum diameter of the spindle-whorl (mm); ***h*** – height = thickness (mm); ***ØP*** – the (exterior) maximum diameter of the perforation (mm); ***ØPm*** – the minimum diameter of the perforation (mm).

A spindle-whorl classification was formed from the model that had been proposed by F. Médard.[21] This model was adapted and modified to create a hierarchic typological system that has several levels of classification (Table 1.4). According to this system, each artefact is defined by a typological code.

Examples:

I1-A-3b very small, flat fired clay spindle-whorl of convex shape with a concave upper end
II4-B-3b perforated ceramic fragment, big, medium-flattened, with an irregular form, curved profile.[22]

[21] Médard 2006, 50–54.
[22] For a detailed presentation of the spindle-whorls' classification system see Mazăre 2012.

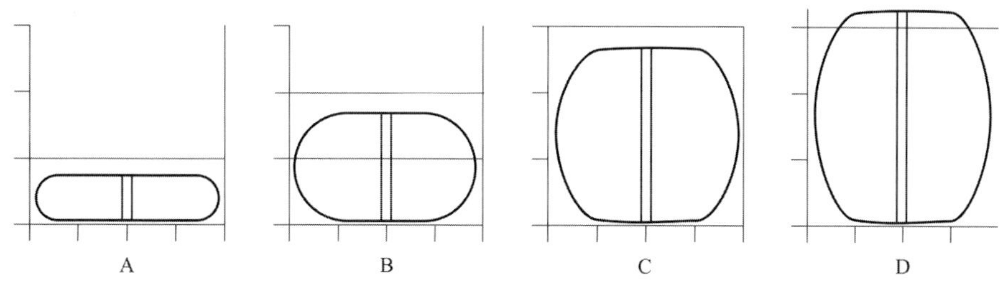

Fig. 1.9: Defining typological groups of spindle-whorls in accordance with the ratio between height and diameter (h/diam.) (drawing: P. Mazăre apud Médard 2006).

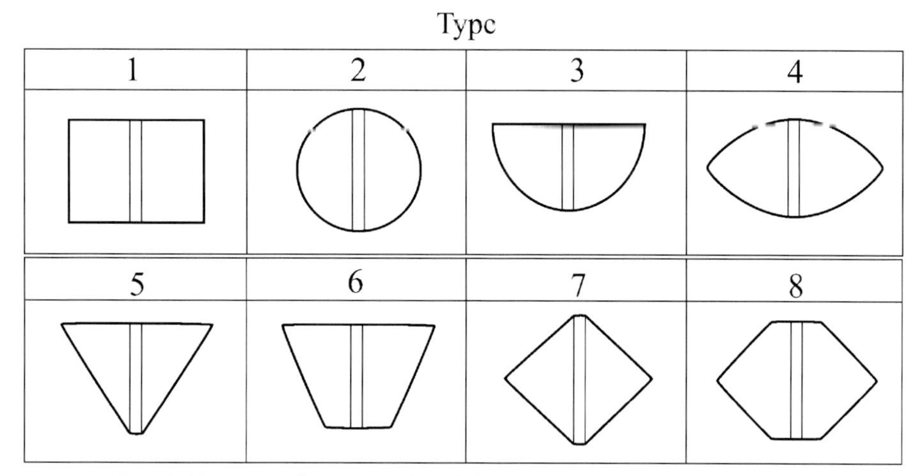

Fig. 1.10: Basic shapes used in defining the types of spindle-whorls.

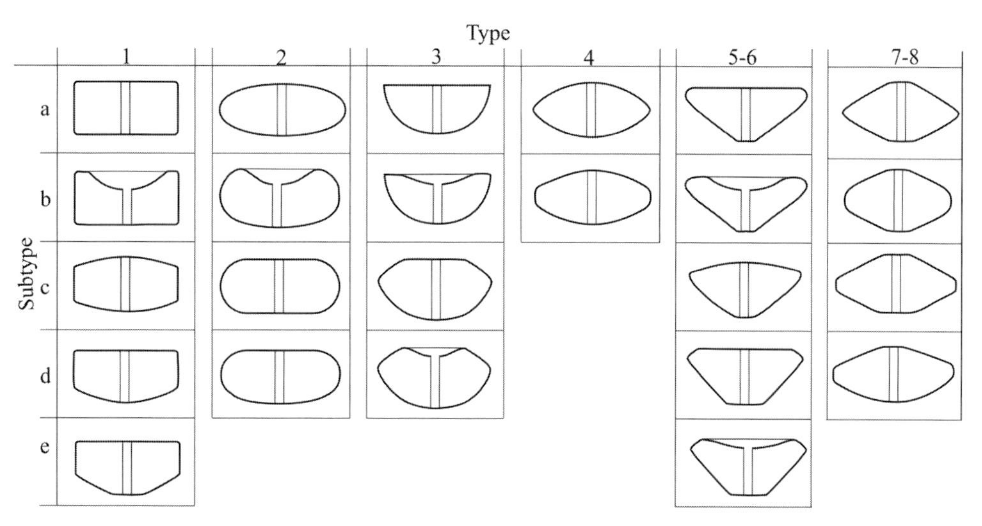

Fig. 1.11: Examples of subtypes defined for spindle-whorls belonging to group B (flattened spindle-whorls, h/diam. < 0.65).

For fired clay spindle-whorls eight base types were identified, some with sub-types and variants (Fig. 1.13). Most of them can be classified as small sized (class 2) = under 25g and medium sized (class 3) = 25–50g. On average the heaviest are those of biconical shape from the Linear Pottery culture (groups B–C), and the lightest are those of discoid shape (group A) belonging to the Ariuşd Culture. Even so, the heaviest spindle-whorl was recorded for Ariuşd Culture, estimated around 174g, much heavier than the values recorded for the entire lot of spindle-whorls (Fig. 1.12).

In the case of pierced rounded shards another system of classification was developed, in accordance with their morphological and functional attributes: shape, width and finishing degree. Thus most of the pierced rounded shards are of circular shape (type 1), only a few displaying an ellipsoidal morphology (type 2) and one irregular (type 3). With the exception of two artefacts of large size (Starčevo-Criş culture), the majority have weight values under 20g, lighter than most of the fired clay spindle-whorls.

Loom-weights

The loom-weights represent the majority of textile tools investigated (563 items). With the exception of a fragment of (un-fired) clay loom-weight found at Turdaş (and most likely belonging to the Turdaş culture), all the others are made of fired clay. As in the case of spindle-whorls, the loom-weights were analysed based on functional attributes which the technological weaving optimum depends upon, as the tensioning and the equal distribution of warp fibres. The weight and the thickness are seen as main functional attributes for the loom-weights. Other important features are the width and/or diameter and height and the diameter of the hole (Fig. 1.14).

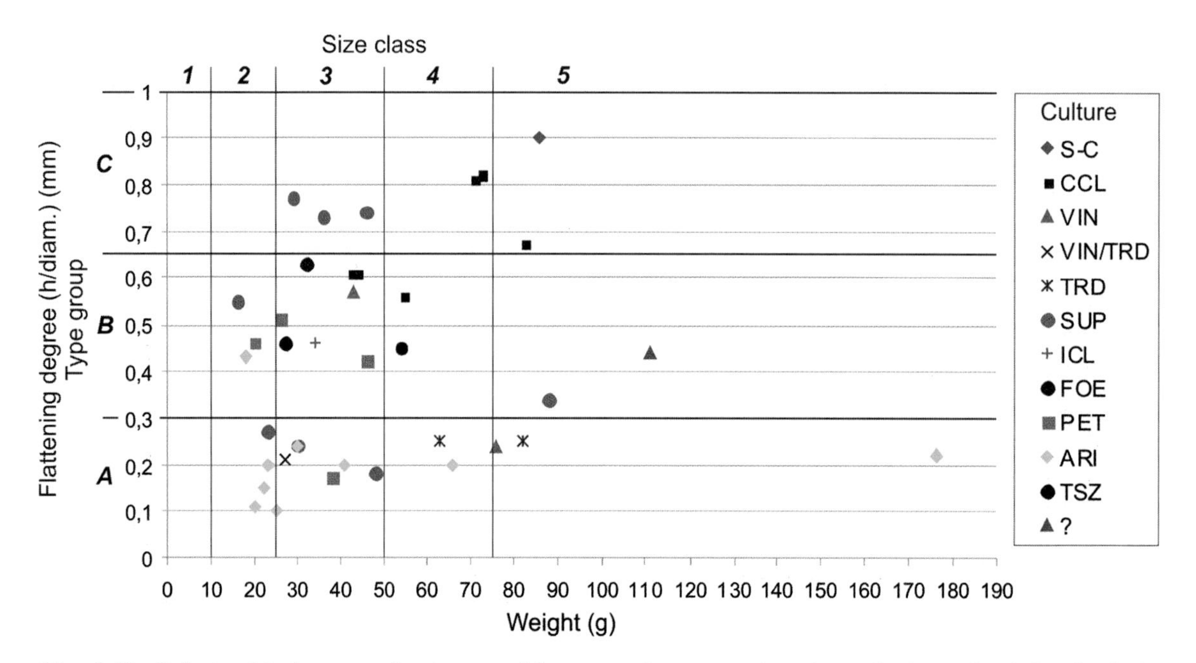

Fig. 1.12: Relationship between the degree of flattening (type group) and weight (size class) for fired clay spindle-whorls against their cultural affiliation.

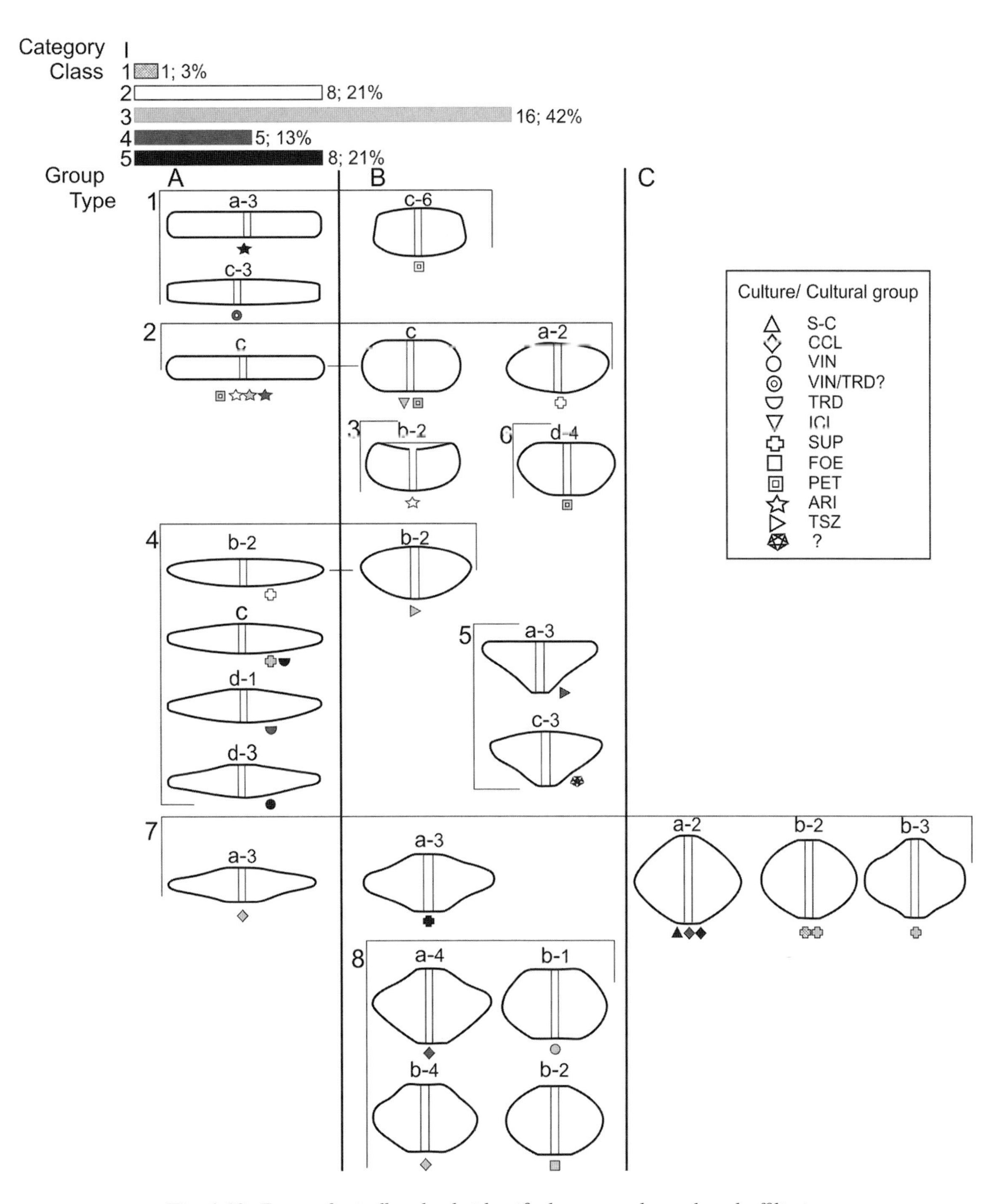

Fig. 1.13: Types of spindle-whorls identified compared to cultural affiliation.

For classifying loom-weights, similarly to the spindle-whorls, a hierarchic typological system with several classification levels were adopted. In the end, a typological code is assigned to each artefact according to the structure depicted by Table 1.5.

Examples:

I1-A-1.a	upper perforated (with a single perforation), very small flat, of irregular form and elongated loom-weight
I4-C-6.c	upper perforated, large-sized, thick flattened, conical, short and wide loom-weight
III3-B-3	medium-sized, centrally perforated, medium flattened, of circular form loom-weight

Table 1.5: Hierarchic typological system for classifying the loom-weights.

I	1	A	1	a
Category (*Presence/lack and the position of the attaching hole*): I – Single upper hole II – Two upper holes III – Central hole IV – Without hole	***Class*** (*Size = weight*): 1. Very small: w < 50g 2. Small: 50 ≤ w < 250g 3. Medium: 250 ≤ w < 750g 4. Large: 750 ≤ w ≤ 1250g 5. Very large: w > 1250g	***Group*** (*The flattening degree = thick/wide*): (Fig. 1.15)	***Morphological type*** (*Morphology*): (Fig. 1.17)	***Subtype*** (*The elongation degree = width/height*): (Fig. 1.16)

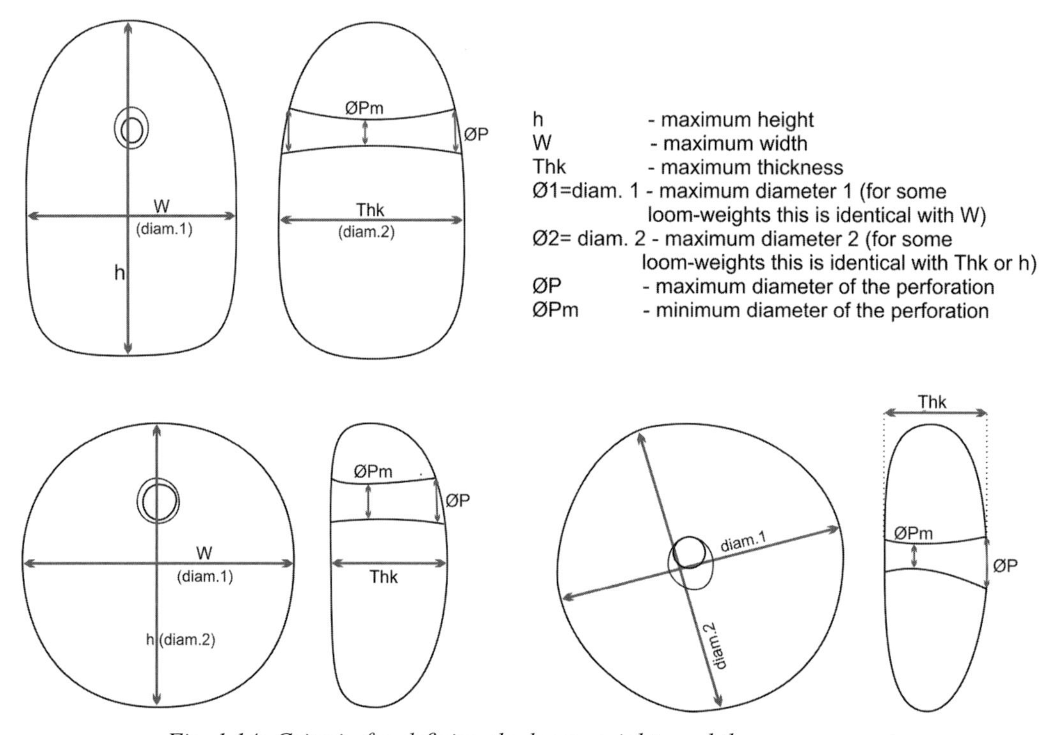

Fig. 1.14: Criteria for defining the loom-weights and the measurement.

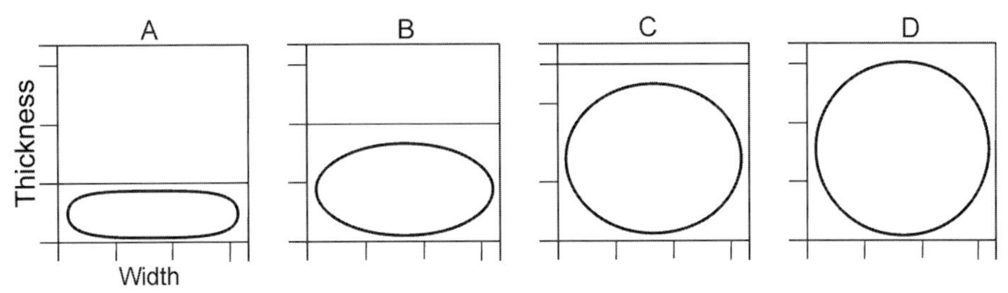

Fig. 1.15: Defining typological groups according to the ratio between the thickness and width of the loom-weights.

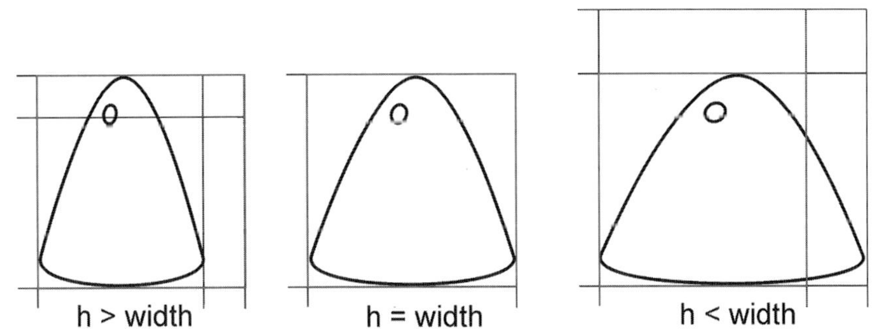

Fig. 1.16: Defining subtypes according to the elongation (slimness) degree of the loom-weights, the ratio between height and width respectively.

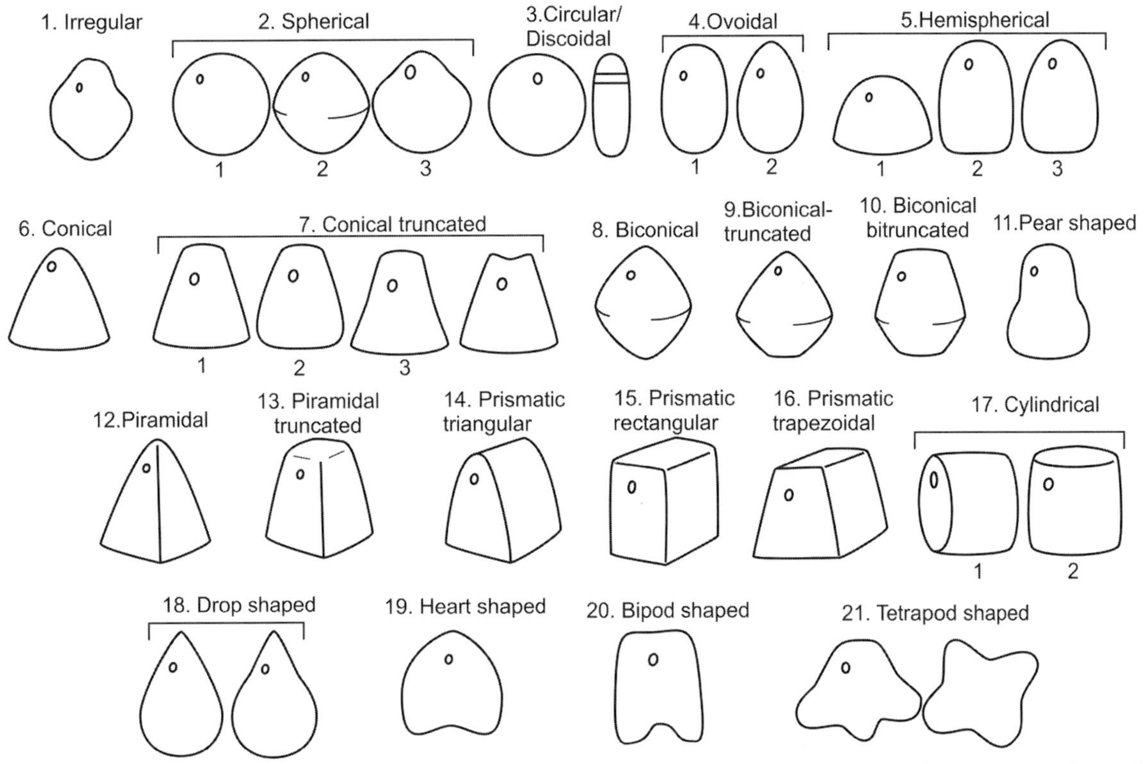

Fig. 1.17: Loom-weight types defined according to the primary morphology (examples of the upper-perforated loom-weights).

Given the large number of artefacts and their diverse typological variations, the analysis was conducted according to their cultural affiliation. For each culture several types of loom-weights were identified, some of them being rather similar in terms of artefact morphology. The centrally perforated loom-weights, belonging to the Vinča, Turdaş, Foeni and Petreşti cultures presents the highest similarity in terms of morphology, weight and thickness. The most diverse types were recorded for upper perforated loom-weights of the Ariuşd (Fig. 1.19) and Petreşti cultures, also presenting the highest variety of subtypes and variants.

The weight of the loom-weights are similar, most of them found in between 150 and 700g. The majority of loom-weights are classified as medium sized (class 3), between 250 and 600g. There are also exceptions, for example the loom-weights belonging to the Linear Pottery culture, all under 60g. Also for Starčevo-Criş culture, the upper perforated loom-weights are of small size and weigh between 80 and 250g thus being generally smaller even compared to the majority of centrally perforated ones from the same culture. Of small size (under 250g) are the upper perforated loom-weights from the Vinča and Foeni cultures and some of those belonging to the Ariuşd culture. All the centrally perforated weights of Ariuşd culture and most of the Bodrogkeresztúr weigh under 250g. The heaviest loom-weight was found in Ariuşd culture, 937g.

The thickness of all loom-weights is between 20 and 80mm. For Petreşti and Ariuşd upper perforated loom-weights elongation and flattening was observed, thus entering group B (according to the ratio between thickness and width). Also in group B there is a majority of centrally perforated loom-weights. In general these have a larger perforation than the upper perforated ones (Fig. 1.18), and in the case of the Vinča and Turdaş cultures they are mostly decorated.

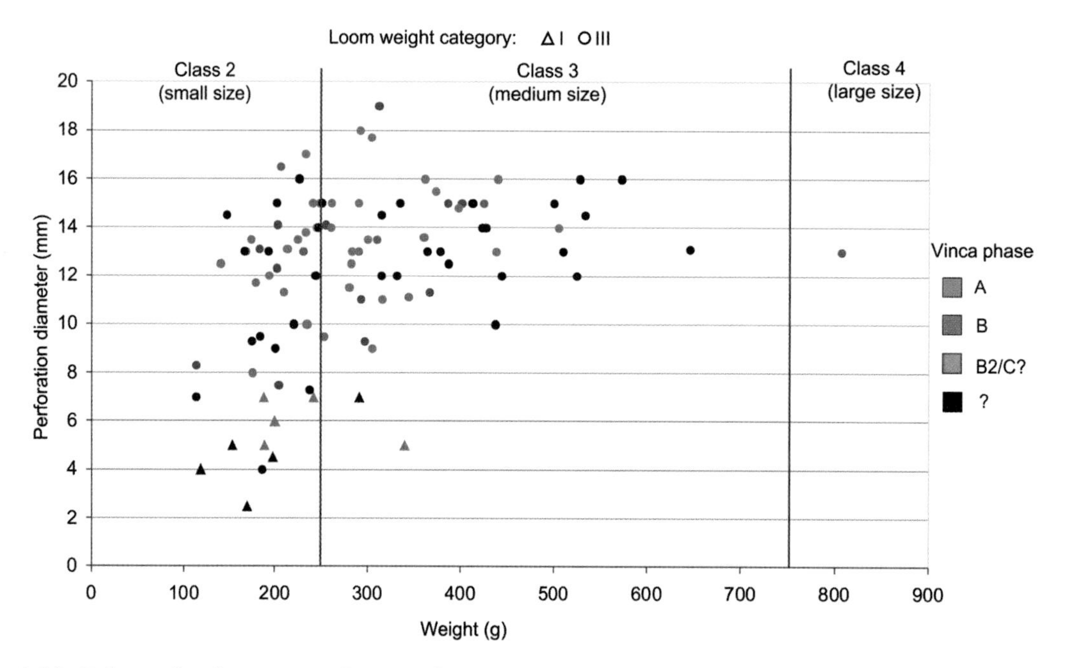

Fig. 1.18: Relationship between perforation diameter, weight (size class) and typological category (I, III) for Vinča culture loom-weights.

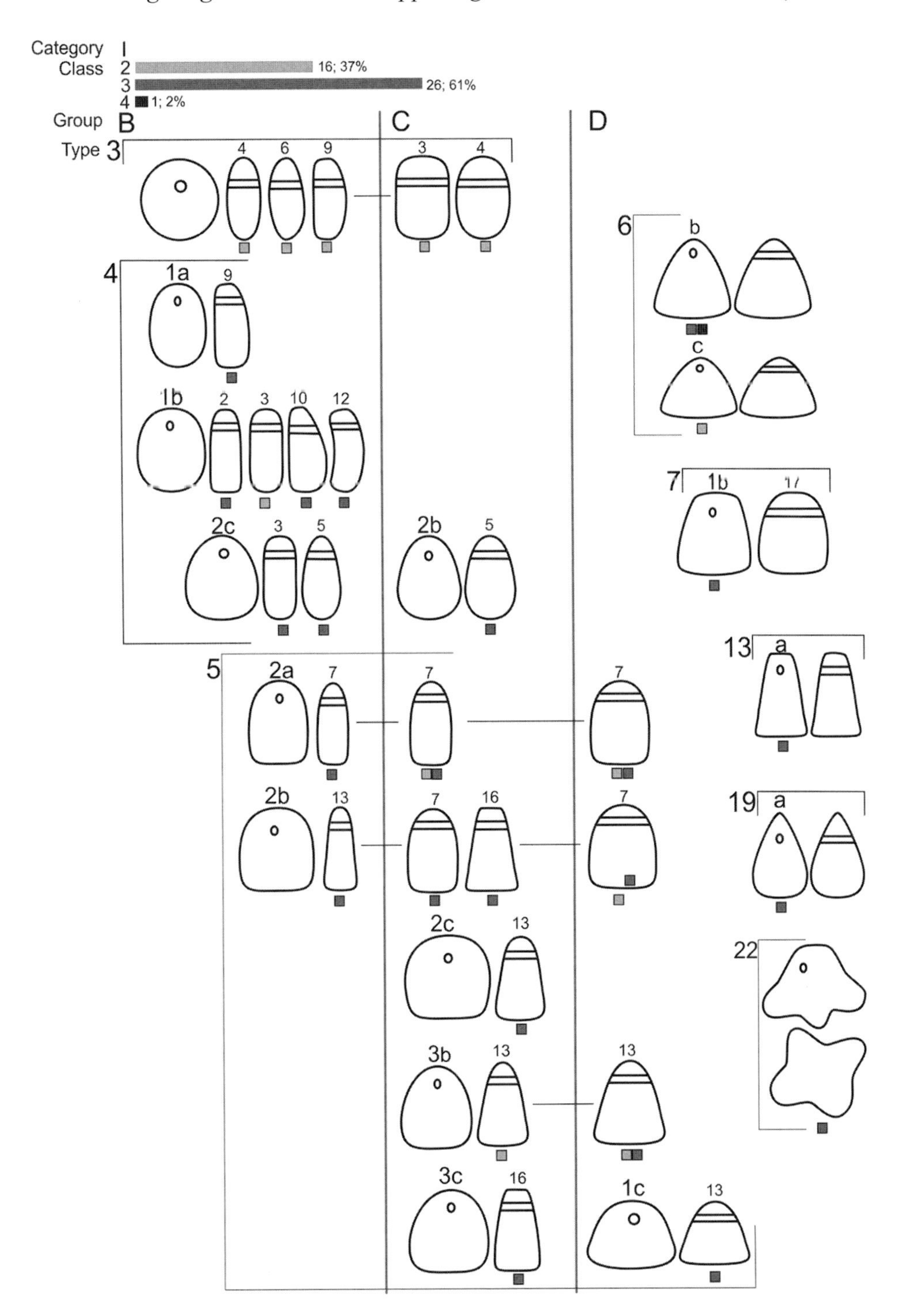

Fig. 1.19: Types of upper perforated loom-weights belonging to Ariuşd culture.

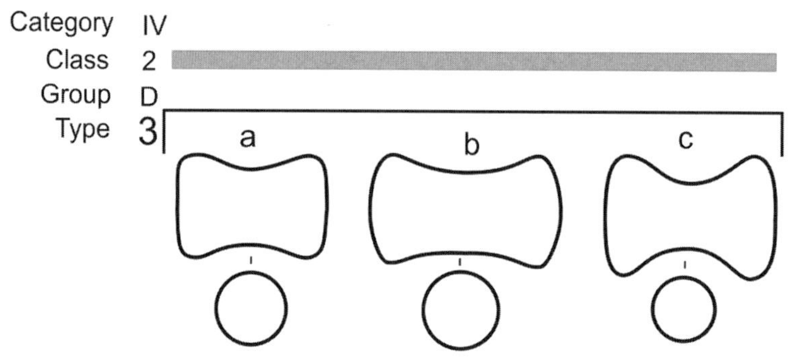

Fig. 1.20: Types of spools.

Spools

All small fired clay artefacts designated as *spools* have in general a maximum length of 10cm and weighted between 8 and 245g. They mostly present with cylindrical shapes, often with prominent ends, resembling the spools or reels currently used for coiling threads.

Only three artefacts that have the characteristics of spools were analysed (Fig. 1.20). One of these artefacts originates from the Ariuşd culture settlement at Şoimeni-*Dâmbul Cetăţii* (SDC-8765), and the other two from Tărtăria (TAR-13991) and Pianul de Jos (PJP-10385), with an uncertain cultural affiliation (Petreşti or Coţofeni cultures). All of these artefacts are of small sizes, with weights between 55g and 75g. They display similar sizes: the maximum diameter varies between 32/30 and 40/41mm and height between 46 and 56mm.

Other tools potentially used in the textile manufacturing technology

Besides spindle-whorls and weights for looms, identifying other tools among the artefacts recovered archaeologically is rather difficult due to many circumstances, the most crucial being the lack of wear trace analysis to clearly discern the artefact's functionality. This is the reason why these artefacts were not included in this research strategy and I have not approached them with the same analytical eye as in the case of spindle-whorls and loom-weights. Only additional observations were made considering these tools, mainly based on bibliographical sources and in a small percentage on direct analysis. They are structured from general to particular, from defining the main artefacts involved in textile production to a case-study of artefacts from bone tools found within the Neolithic settlements of Limba. These sites have provided a number of 174 bone tools, extensively studied and published.[23] They originate from the Starčevo-Criş III B and Vinča (phases A2–A3 and B1–B2) habitation layers. Of these a number of 89 artefacts may have been used in textile production practices: pin beaters, weaving needles, shed or patterning sticks used in small weaving implements, warp spacers, tips of combs used for fibre separation, shuttles, weaving knives, instruments for detaching the fibres from stalks/ bark, needles used in *nålbinding* or looped-needle netting.

[23] Mazăre 2005.

The functional interpretation of textile tools

Spindle-whorls

Fired clay spindle-whorls

The literature offers plenty of discussions for the usage of spindle-whorls, from simple notions to complex experimental interdisciplinary studies.[24] Among these are the recent studies of the researchers from the Centre for Textile Research (*CTR*).[25] The studies of Médard,[26] T. Chmilelewski and L. Gardyński[27] or A. Verhecken[28] with physical descriptions of artefacts and analyses of the moment of inertia and rotational speed, based on their mechanical properties are also important. The limitations of these studies are that they mainly deal with a single type of spinning (with suspended spindle), thus excluding the functional evaluation of spindle-whorls in relation with other types of spinning that might have been used in prehistory.

These studies provide an argument for the current interpretation of Neolithic and Copper Age spindle-whorls from Transylvania. These spindle-whorls are divided into two main categories, corresponding to typology groups and to different mechanical properties. In one category there are the flattened discoid spindle-whorls of group A, and in the other, the medium and tall ones from groups B and C. Items from group C are usually heavier than the rest, with an average weight of 1.6 to 1.7 times that of groups A and B. Taking into account the relationship between the radius of spindle-whorls and the moment of inertia on one side and the relation between the radius and rotation speed on the other side it can be calculated that, on average, the rotation of group B spindle-whorls is about 1.3 times faster and 1.8 shorter than the flat discoids of group A. In exchange the added weight from group C (with an increased height) indicates a higher moment of inertia and thus a longer time of rotation compared to group B. These observations suggest that, if the technique of spinning would have been that of suspended spindle, the different spinning whorls would have been used to produce threads of various qualities.

Observations were also made on the relationship between weight, diameter and height of spindle-whorls and the diameter of the perforation. Other observations were made on the perforation's degree of alignment in relation to the centre of the spindle-whorl. All usage traces and/or external notches on discoid spindle-whorls were also analysed. The characteristics of Neolithic and Copper Age spindle-whorls from Transylvania might actually indicate two ways of spinning, with the resulted threads being of varied quality, and probably originating from different fibres:

1. spinning with suspended short and thick spindles, with the spindle-whorl either on the upper or lower part; these would have produced finer threads (possibly from flax?);
2. spinning with suspended or supported longer and thinner spindle, with the spindle-whorl located on the upper side. These would be used to spin/twist long fibres or filaments of fibres (possible tree bast?) or plying the yarn.

---

[24] Liu 1978; Raymond 1984; Barber 1991; Bier 1995; Crewe 1998; Keith 1998; Grömer 2005; Martial and Médard 2007; Breniquet 2008; Chmielewski 2009.

[25] Mårtensson *et al.* 2005–2006; Mårtensson *et al.* 2006a, b.

[26] Médard 2006, 105–118.

[27] Chmielewski and Gardyński 2010.

[28] Verhecken 2010.

Pierced rounded shards

In this case the balance between diameter and height, that allows modelled clay artefacts to be used as spindle-whorls, are exceptionally found. Perhaps most of these pieces were used for other purposes and only a few can be related to actual spinning practices. An interpretation for the items lighter than 20g is that they might have been used as pairs of discs fixed on the spindle and acting as supplemental weight next to a spindle-whorl. Other uses are also possible besides this one.[29]

Loom-weights

Currently, most of the 'weights' (made from fired clay) found in archaeological sites are named and defined functionally by Romanian archaeologists mainly as loom-weights (the upper perforated/bored ones) and fishing net sinkers (the centrally perforated ones). Besides these, there are other functional possibilities, mentioned by the literature: "firedogs" ("andiron") or other functions related to fire, "link-stones" ("loop-stones") used for fixing the thatched roofs, counter-weights, door-stoppers, weapons or prestige items; tools for twisting fibres/yarns.[30] The main criteria for differentiating loom-weights from the other types are both the context of discovery (the most obvious contexts are those that provide weights in rows or groups) and the wear traces, although all of these can be interpreted differently.[31] As opposed to upper perforated weights, the centrally perforated ones rarely provide use-wear marks that would sustain a suspended usage. This also provides a clue that they were actually used for something quite different.

The function of weights within the warp weighted looms

Ethnographical data as well as the experimental studies by Médard[32] or those performed within *CTR*[33] have shown that the weight (mass) and maximum width are the fundamental functional parameters of loom-weights. The density and uniform, balanced distribution of threads depends on these properties, and a relation can be established with the ease of weaving and the width of the resulted textile. Choosing the loom-weight according to width and weight is done in relation with the type of weaving that is aimed at and the type of fibres used (Table 1.6; Fig. 1.21).[34]

Evaluating the functional parameters of the loom-weights and estimating the aspect and properties of textiles based on these parameters. Case studies

It is the merit of Mårtensson *et al.* (2009) of setting the bases of a method for reconstructing the production of a tabby-weave using different loom setups, starting from the functional attributes (weight and width) of a given loom-weight. The calculation proposed allows also the evaluation of the efficiency of weights usage in the production of textiles. This method was used on representative samples from each studied culture. As a novel element the method was also applied for sets of loom-weights (Table 1.7).

According to this evaluation, apart from a single exception, all the weights analysed could have been used to tension the yarn threads in a vertical loom. According to the calculations, the most

[29] Raymond 1984, 19–20, fig. 5; Crewe 1998, 12.
[30] Médard 2000, 38–39; Mazăre 2013, 27–36.
[31] Mazăre 2013, 37–46.
[32] Médard 2000, 88–97.
[33] Mårtensson *et al.* 2007a; Mårtensson *et al.* 2009.
[34] Mårtensson *et al.* 2009, 390.

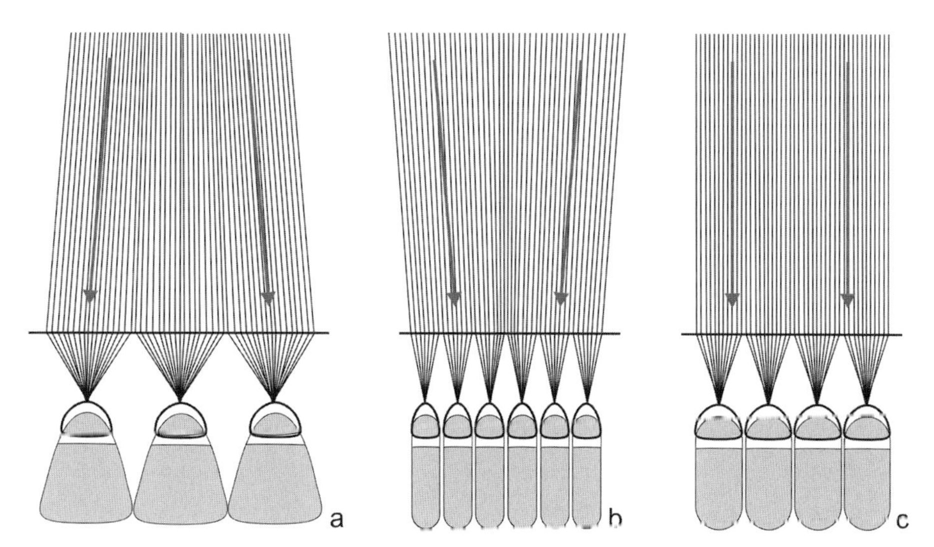

Fig. 1.21: Relationship between the width of loom-weights, the orientation of yarn threads and the width of the textile at the upper (starting) and lower (ending) border (drawing by P. Mazăre after Médard 2000; Mårtensoon et al. *2007a; Mårtensoon* et al. *2009).*

Table 1.6: Relationship between the type of fabric (type of fibres) and the loom-weight type (defined by weight and width) used in woven textile production (after Mårtensson et al. *2009).*

Type of fabric	**Type of yarns**	**Type of loom-weights**
Coarse, open fabric	Thick yarns	Heavy, thick loom-weights
Coarse, dense fabric	Thick yarns	Heavy, thin loom-weights
Thin, open (weft-faced) fabric	Thin yarns	Light, thick loom-weights
Thin, dense fabric	Thin yarns	Light, thin loom-weights

efficient weights, able to properly tension threads of variable thicknesses, are the elongated and flattened weights such as those from the Petreşti culture as well as the round upper or centrally perforated weights from the Vinča, Turdaş and Bodrogkeresztúr cultures. The quantity of threads necessary for producing one square metre of textile is also dependant of the density and thickness of threads used.

Spools

The main functions of spools as interpreted by J. Carrington Smith[35] and more recently by M. Gleba[36] were considered, and several hypotheses can be concluded. If one accepts the idea that they were indeed connected to the production of textiles the most plausible interpretation for the

[35] Carington Smith 2000, 228.
[36] Gleba 2008, 140–143.

Table 1.7: Calculation of various loom setups with representative items of the loom-weights group found at Păuca-Homm, Petreşti culture (after Mårtensoon et al. *2009).*

Loom-weights group – House L1/1965, Păuca-*Homm,* Petreşti culture												
No. of loom-weights: 28												
Weight:	238–493g		Medium Weight: 388g									
Thickness:	2.4–5.3cm		Medium Thickness: 3.7cm									
Fabric width:	No. of loom-weights/2 layers of warp threads × 3.7cm = 51.8cm ≈ 50cm											
Artefact code	PHO-9883			PHO-9879			PHO-9873					
Type code	I3-B-5.2a-7			I2-A-4.2a-2			I3-B-5.2a-8					
Weight:	493g			242g			373g					
Thickness:	4.45cm			2.4cm			4.05cm					
Warp thread tension	*10g TFU*			*20g TFU*			*30g TFU*			*40g TFU*		
No. of warp threads/loom-weight	50	25	37	25	12	19	17	8	13	13	6	10
No. of warp threads ×2 loom-weights	100	50	74	50	24	38	34	16	26	26	12	20
Warp threads/cm	23	21	18	11	10	9	8	7	6	6	5	5
No. of warp yarn	1150	1050	900	550	500	450	400	350	300	300	250	250
Technical evaluation	Unlikely – too many threads/cm			Optimal			Optimal			Optimal		
Amount of warp yarn = amount of weft yarn				1000m			750m			666m		
Yarn consumption for 1m² cloth =				2040m			1428m			1332m		
Time consumption for spinning the yarn =				51h			28h			25h		

use of spools is as small weights to tension the threads in textiles created by weaving or by using other techniques.[37] According to this functional role they should be found in archaeological context as groups or ensembles. The issue of their functionality is left open by the fact that in the Neolithic and Copper Age habitation layers from Transylvania have been recovered only as isolated finds so far.

The functional role of Neolithic and Copper Age textile products

The archaeological discoveries from Europe compared to the ethnographical sources and historical writings show that the textile products were used as domestic and practical items as well as personal articles of clothing. Their function could also pass over the daily life and become symbolic and spiritual artefacts. In general, it can be assumed that there is a correspondence between the quality of a textile product and its value and function.

[37] Carington Smith 2000, 228; Ræder Knudsen 2002, 228–229; Mårtensson *et al.* 2007b; Gleba 2008, 141.

The role of textiles in pottery manufacturing

The textile imprints analysed, as well as the numerous imprints of mats on Neolithic and Copper Age vessels are proof of frequent usage of perishable fibres products in the technology of pottery manufacturing. From the various interpretations given by archaeologists on the basis of experiments and ethnographical analogies several ways of using textiles can be distinguished:

1. As support for setting the vessel to dry after shaping;
2. As support on which the vessels were raised (a primitive variant of a rotational device);
3. As implements used to create an imprint for better adhesion between separately created vessel components;
4. As actual items within the structure of vessels, for consolidation of walls and bottoms (in this case being fired along with the vessels);
5. They also served for decorating the vessels.[38]

   Even if we do not know the degree of usage, it is obvious that textiles were a common, usual presence. It is certain that these textiles were either of an inferior quality, at the end of their intended usage or representing pieces from items created for a different purpose. Even so they are proof that textiles, especially woven ones, were quite a common presence among these communities.

The Neolithic and Copper Age anthropomorphic representations and their importance in reconstructing the functions of textiles

The anthropomorphic representations are the main source of interpretation on the usage and functionality of textiles and their actual role as clothing, and this is the case for the South-Eastern Europe. The archaeological literature is abundant in interpretations on clothing representations on anthropomorphic figurines.[39] Based on this literature and the actual analysis of the figurines, several types of garments and clothing accessories specific to these representations have been identified. A repertory for the representative cultures of the Neolithic and Copper Age cultures for Romania was also created. The difference between textile clothing and that created using other materials is quite difficult. Several criteria for establishing these differences were adopted and the following questions were adressed:

- Which of the clothing pieces depicted on figurines or other representations were made from textiles and what was the technique used in their production?
- Are the realistic representations of full garments (dresses) from the Copper Age female figurines a consequence of a wider phenomenon taking place at the end of the 5th millennium and the beginning of the 4th millennium BC? Could this phenomenon be linked also to the emergence of weaving imprints on Cucuteni-Trypillian and Tiszapolgár vessels or the frequency of weight ensembles from Kodjadermen-Gumelniţa-Karanovo VI (KGK VI) culture settlements, some of them engraved with female silhouettes?
- Is the clothing depicted on the figurines the actual clothing worn by the members of the community on a daily basis? Is there a correspondence between the clothing depicted and

[38] For more details see Mazăre 2011b; Mazăre *et al.* 2012.

[39] With strict reference to Romanian literature there are several authors that have interpreted the ornaments on these figurines as actual depictions of clothing: Mateescu 1959; Cucoş 1970; Dumitrescu 1974, 1979; Comşa 1974, 1988, 1995, 1995–1996; Luca and Dragomir 1987; Luca 1990; Mantu 1993; Monah 1997; Andreescu 1997, 2002; Sorochin 2001, Sorochin and Borziac 2001; Frânculeasa 2004; Sztáncsuj 2009.

the status and social identity of the one wearing it (in terms of sex, role and social status)? In this respect, are these figurines an expression of societal stratifications within prehistoric communities and if so in what manner did the textile contribute to the expression of these differences?

Discussion on the Neolithic and Copper Age textile production in Transylvania

The data presented in this paper, although reduced to only a few categories of artefacts, provides sufficient arguments to support the existence of a textile production in the Neolithic and Copper Age communities in Transylvania.

Types of textile structures and techniques of production

Based on the analysis of textile imprints from the Neolithic and Copper Age, two types of textile structures that were made using two different fabrication techniques could be identified: twining and weaving. They complement the data already known from Romania with regard to fabrication techniques and textile structures used in the Neolithic and Copper Age (Fig. 1.22).[40]

Imprints of fabric reveal two types of structures that indicate the use of two different methods of weaving, involving different tools: woven fabric bands using small implements, and loom weaving for larger textiles. Fired clay weights found in most Neolithic and Copper Age sites suggest the use of a vertical warp weighted loom as the main technique for producing larger woven textiles.

Much like the twined archaeological textiles discovered in the Swiss Plateau[41] or those found in the form of imprints in the Vinča cultural area south of the Danube,[42] the ones identified in the form of imprints in Transylvania were made without the use of a tension frame.

Raw materials – Selection and differentiated exploitation in textile production

The lack of textile remains in the analysed geographical area makes it impossible to identify precisely what types of raw materials were employed. However, textile imprints show two different patterns in the usage of fibres: raw fibres (for twined textiles), and processed fibres (spun threads/yarns) (for woven textiles). In both cases plant fibres were utilised, but it is possible that the raw material was of a different sort, an indication of this aspect being the textile artefacts from the Neolithic of the circum-Alpine area (4th–3rd millennium BC). In that case, twined textiles were largely made from tree bast fibres, while woven fabrics were made almost exclusively from flax (*Linum usitatissimum* L.) yarns.[43]

Therefore, it is possible that the textiles produced in Neolithic and Copper Age Transylvania followed the same strategy in the usage of the fibres. However, other cultivated textile plants from the spontaneous flora might also have been used, as shown by the prehistoric archaeological textiles found for different periods.[44] For instance, the fibre or decorticated stem characteristics seen with twined textile imprints from Transylvania corresponds with the assumptions made by J. M. Adovasio

[40] Mazăre 2011a.
[41] Médard 2010, 145.
[42] Adovasio and Maslowski 1988.
[43] Médard 2010, 145–146.
[44] Alfaro Giner 1980, 1984; Barber 1991; Körber-Grohne 1991; Roche-Bernard and Ferdiere 1993; Mannering 1995; Shishlina *et al.* 2002; Bazzanella *et al.* 2003; Rast-Eicher 2005; Gleba 2008.

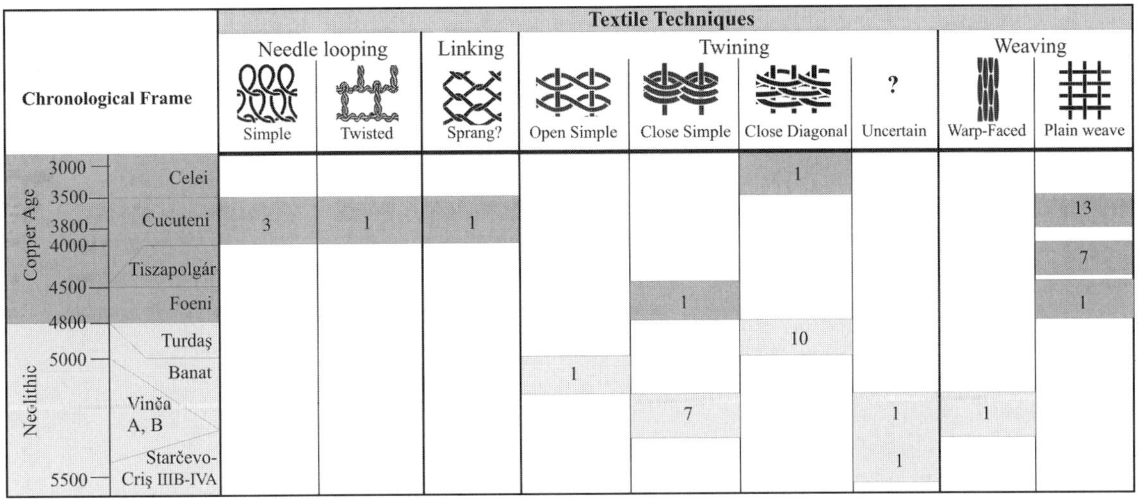

Chronological Frame		Needle looping: Simple	Needle looping: Twisted	Linking: Sprang?	Twining: Open Simple	Twining: Close Simple	Twining: Close Diagonal	? Uncertain	Weaving: Warp-Faced	Weaving: Plain weave
Copper Age (3000–3500)	Celei						1			
Copper Age (3500–4000)	Cucuteni	3	1	1						13
Copper Age (4000–4500)	Tiszapolgár									7
Copper Age (4500–4800)	Foeni					1				1
Neolithic (4800–5000)	Turdaş						10			
Neolithic (5000)	Banat				1					
Neolithic	Vinča A, B					7		1	1	
Neolithic (5500)	Starčevo-Criş IIIB-IVA							1		

Fig. 1.22: The frequency of techniques and textile structures as identified for Neolithic and Copper Age settlements in Romania (after Mazăre 2011a).

and R. F. Maslowski that twined textiles might also have been made using decorticated stems of *Artemisia* sp.[45] On the other hand, the recent find in the site of Hódmezővásárhely-Gorzsa (Tisza culture, 5th millennium BC) of velvetleaf seeds (*Abutilon theophrasti* Medic.)[46] could support the early use of the *Malvaceae* as cultivated textile plants. Moreover, researchers believe that the importance of nettle as textile plant in prehistoric times was greater than that currently estimated, and its resemblance with other vegetal fibres making its identification almost impossible up to very recently.[47] Given these circumstances it is imposible to determine how often flax was used as a textile plant by Neolithic and Copper Age communities of Transylvania since the archaeobothanical data from Romania is hardly sufficient to support an earlier cultivation of flax.[48] As flax is considered to be part of the so-called 'Neolithic crop package'[49] it must have been brought over to Transylvania with the arrival of the earliest Neolithic communities. The reduced amount of fibre provided by the oleaginous flax variety cultivated during the Neolithic[50] leads us to believe that it was used only for certain textiles, probably thin and open woven fabrics, as seen in the case of the imprint found on Foeni pottery at Alba Iulia (Figs 1.3.b, 1.23.2).

---

[45] Adovasio and Maslowski 1988, 353.
[46] Medović and Horváth 2012.
[47] Médard 2006, 27; Bergfjord and Holst 2010; Bergfjord *et al.* 2012.
[48] To my knowledge, up to this day, in the area of Transylvania there is a single flax seed published, identified for the middle Bordogkeresztúr culture habitation at Cheile Turzii-*Peştera Ungurească* (Nisbet 2009, 172–173, table 5).
[49] Zohary 1996, 143; Zohary and Hopf 2000, 241.
[50] As the recent archaeobotanical studies of Herbig and Maier (2011) for the Late Neolithic wetland settlements in southwest Germany show, the transition from cultivating linseed to that of the fibre flax type began in the Horgen culture (3400–2800 BC). The data would indicate the fact that flax was cultivated during Neolithic mainly for its seeds; its fibres were also used, but not in large quantities.

Preparation and transformation of raw materials. Yarn production

With the exception of the transformation of raw fibre into yarn, proven by the existence of spindle-whorls and woven textile imprints, there are no other direct evidence of the methods used in fibre processing for the Transylvanian area. A method of processing/spinning the fibres, similar to that practiced in ancient Egypt, and also highlighted by the analysis of U. Leuzinger and A. Rast-Eicher[51] in the case of Neolithic flax vestiges in the northern Alps, is most likely corresponding to that practiced by the Neolithic communities in Transylvania. This idea is supported by the existence of the S plied yarn, observed in textile imprints, and the methods of spinning suggested by the study of spindle-whorls.

The use of spindle-whorls of different sizes and shapes within Neolithic and Copper Age communities of Transylvania could be related to several possible scenarios:

1. use of different kinds of fibres;
2. production of different quality yarns;
3. use of different techniques;
4. gender differentiated handling of textile tools within the same community.
   However, the small number of spindle-whorls found raises questions about the importance of spinning and indirectly about the importance of weaving in the Neolithic and Copper Age communities in Transylvania, although the number of loom-weights found is considerably higher.

Textile production. Weaving and the differentiated use of the weights in the loom

Production of various quality fabrics using fibres of different properties and probably of a different nature is demonstrated by the morphological and ponderous variety of the weights (if they were indeed used as parts of a loom). The fact that this variety is registered at a cultural level (in the same cultural area or even within the same site) could be an indication that fabrics of different qualities were being produced and used within the same communities. The diversity of weavings corresponding to the varied typology of weights seems to have been higher for the Copper Age in comparison with the Neolithic period. At the end of Neolithic (*c.* 5000–4500 BC) several technological changes and improvements, were later picked up and developed during the Copper Age, providing a superior textile production. These changes are suggested by the use of larger clay weights, with top perforations and a much more flattened appearance than those of the Neolithic period.

Even if an attempt to explain the dilemma of a parallel existence, within the same settlement, of two types of weights (with top and central hole), the question of their functionality remains open. Although they could have been used as weights in the loom, I suspect that centrally perforated weights, mostly from the sites of the Vinča and Turdaş cultures, had other functional purposes than those perforated at the top.

Although loom-weight rows were not discovered in the Neolithic and Copper Age sites of Transylvania, the two sets of 28 loom-weights found in two Copper Age dwellings (in Păuca-*Homm* and Ariuşd) could be an indication of two such looms. In Romania several sites (mainly of Copper Age) were also found with dwellings containing between 20–32 loom-weights: Caransebeş-Balta Sărată, house L18, B1/B2 Vinča culture (26–28 loom-weights),[52] Padea-*Dealul Viei*, house

[51] Leuzinger and Rast-Eicher 2011.
[52] Lazarovici and Lazarovici 2007, 172, fig. IIIa. 56.

L2, Dudeşti-Vinča culture (20 loom-weights),[53] Turdaş-*La Luncă*, house L_2/1994–1995, Petreşti culture (27 loom-weights),[54] Radovanu-*La Muscalu*, house C, Boian-Gumelniţa culture (36 loom-weights in first level; 20 loom-weights in second level),[55] Pietrele-*Gorgana*, house B/2006, KGK VI culture (23–24 loom-weights),[56] Sultana-*Malul Roşu*, house L2, KGK VI culture (30 loom-weights),[57] Poduri-*Dealu Ghindaru*, house L2/2006, B1 Cucuteni culture (32 loom-weights),[58] Sălcuţa-*Piscul Cornişorului*, house L2, Sălcuţa culture (28 loom-weights),[59] Almăjel, a house of Sălcuţa culture (31 loom-weights).[60] In conclusion, the group of weights analysed here integrates within a broader technological area, defined by similar preferences or rather subject to the same technological standards.

The time and the amount of raw materials necessary to produce fabrics

According to ethnographic analogies, the whole process of textile production was long, hard and ran sequentially throughout the entire year. For the prehistoric period it is difficult to approximate the time allocated for textile production. According to experimental data[61] and the calculations regarding the loom-weights, the time needed to produce enough yarn to weave a square metre of fabric can be estimated between 2–7 days, depending on the thickness of yarn and fabric density. The act of weaving required, in turn, its specially allocated time.[62] The time taken to complete the fabric was determined by the quality of the yarn being woven, the fabric density, and, of course, its physical dimensions. For instance, for a woven cloth 50cm wide (like the one suggested by the 28 loom-weights group found at Păuca-*Homm*, Petreşti culture) with a medium density of 12 threads per cm, one weaver could possibly weave about 120–130cm per day.

The comparative studies on the fibre development of different fibre flax and linseed types of modern *Linum usitatissimum* L. show a great degree of variability into the total yield of fibre, variability influenced by the cultivation and harvest methods, by the retting and the extraction processes.[63] According to the experimental results of D. L. Eason and R. M. Molloy, the total fibre dray matter yields by an average seed rate of 1000 flax seeds per square metre[64] could vary between 600 and 2200 kg/ha for fibre flax and between 500 and 860 kg/ha for linseed.[65] Developing this information one may approximate that 1000 plants/m^2 could produce an average of 140 grams and about 70 grams of total fibre for fibre flax and linseed respectively. In other terms, a single fibre flax plant/stem could yield in average about 0.14g of fibres and the linseed only about 0.07g. Considering this and applying several formulas, an attempt to estimate the quantity of fibres and the number of plants required for the production of a square metre of woven cloth, resembling those found as imprints in Transylvania was made (Table 1.8).

[53] Nica and Niţă 1979, 41, fig. 6.
[54] Luca 2001, figs. 9–11.
[55] Comşa 1990, 50.
[56] Hansen *et al.* 2007, 49–52; Toderaş *et al.* 2009, 46, 55, 60.
[57] Andreescu *et al.* 2010, 10.
[58] Monah *et al.* 2007, 275; Dumitroaia *et al.* 2009, 41.
[59] Berciu 1961, 181–182, 237–249, figs. 37, 40–41, 73–81.
[60] Galbenu 1983.
[61] Mårtensson *et al.* 2009, 393.
[62] Andersson 2003, 46.
[63] Sankari 2000; Eason and Molloy 2000.
[64] The average of the three seed rates (of 500, 1000 and 1500 seeds/m^2) tested by Eason and Molloy 2000.
[65] Eason and Molloy 2000, 366–367, fig. 4.

Table 1.8: Estimative calculation of the flax quantity needed for weaving a square metre of textile similar to the one found from pottery imprints at Alba Iulia-Lumea Nouă (ALN.1001) Foeni culture group; Dăbâca-Cetate (DAB.58024) and Dorolțiu-Castel (DOR.61329), Tiszapolgár culture.

Woven textile imprint	**Length of thread/m^2 of fabric (m)**	**Spinning time/m^2 of fabric (h)**	**Weights of fibres/m^2 of fabric (g)**	**Number of plants; Cultivated area (1000 plants/m^2)**	
				Linseed 0.07g	**Fibre Flax 0.14g**
ALN.1001: thread count: 11/8.5; average thread diam.: 0.32mm	1950m	55h	150g	2142 plants; 2.14m^2	1071 plants 1.07m^2
DAB.58024: thread count: 5.5/6.5; average thread diam.: 1.17mm	1200m	22h	266g	3800 plants; 3.8m^2	1900 plants 1.9m^2
DOR.61329: thread count 5/6.5; average thread diam.: 1.4mm	1150m	20h	383g	5471 plants; 5.47m^2	2735 plants 2.73m^2

This data pertains to modern flax and it is well-known that back in the Neolithic and Copper Age plants were less developed than today and thus produced less fibre. Possibly less than 0.07 grams yield fibre per plant/stem was likely for Neolithic and Copper Age flax. Thus the area needed to be cultivated in order to produce 1m^2 of fabric could reach or exceed 5m^2, given a density of about 1000 plants/stems per square metre. To this a number of variable factors that are impossible to quantify, like the density of seeded plants, climate and weather and soil quality can be added. These make predictions of fibre production harder to establish. It is possible to imagine that, if an entire settlement were to use flax as textile raw material, the resulting cultivated land would have to be of large size, and the labour and time involved in cultivating, maintaining and preparation of fibres would have been considerable. The important amount of processed flax plant remains, found in several late Neolithic (*c.* 3500–2500 BC) settlements of the Swiss Plateau and south-east Germany,[66] correspond to this scenario. Unfortunately it does not fit with the absence of vegetal macro-remains of flax in the Neolithic and Copper Age of Transylvania (and Romania in general),[67] both being periods that predate the habitations of the circum-Alpine area. On the other side other species of textile plants could have been used, some even with a higher potential for fibre quantity. For example, a single nettle stem (*Urtica dioica* L.) is capable to provide between 0.45–1.30g spinnable fibres (an average of 0.744g),[68] while the velvetleaf (*Abutilon theophrasti* Medic.) is also known to provide a quantity of 1.8–2.4 t/ha of retted, dried fibres from 12 t/ha of green plant.[69] This implies that their cultivation and gathering were involving less time and effort than flax and also a smaller cultivated area. Even more, their usage together with the tree bast would diminish the time and labour allocated for obtaining flax fibres considerably.

[66] Herbig and Maier 2011; Maier and Schlichtherle 2011.
[67] This absence is surely a consequence of the early stage of archaeobotanical research in Transylvania.
[68] Hurcombe 2010, 135.
[69] Medović and Horváth 2012, 219.

A different observation that rises from these calculations is that, given the quality of a textile, there is an inverse ratio between the quantity of raw material needed and the time spent in spinning and weaving. More precisely, the coarser the weaving is (implying larger diameter fibres), the lesser time is involved but also a larger quantity of processed fibres is required. For fine weavings less material is used but more time and effort is needed. The comparison is valid only with the use of the same raw material.

Textile production: a common, prestigious, or ritualistic activity?

Ethnographic sources indicate that the activities dedicated to textile production generally occurred outdoors, within settlements.[70] The locations of discovered loom-weights, especially the concentrations of weights, show that weaving on warp-weighted looms was an activity mostly performed indoors. Therefore, the question arises of whether the weaving was performed in family homes or inside special buildings.

Loom-weights are not found in every dwelling of Neolithic and Copper Age settlements and this has led some researchers to believe that the weaving was a craft held by only a small group of individuals, being practiced only in buildings designated for the textile activities.[71] Thus, holding the monopoly over the knowledge related to the production of specific categories of textiles, with special purpose and function, and perceived within the society as prestige goods, might have been a premise for the emergence of an elite of textile craftsmen. At the same time, the discovery of clusters of loom-weights in some areas of worship, such as the shrines from Parţa[72] and Uivar[73] (Banat province, Romania) might equally suggest a specialization of a particular social class (sacerdotal elite?), and a symbolic, ritual function of weaving.[74]

If refering to workshops within the houses or the sanctuaries, it is clear that textile production in these spaces was limited and aimed at textile categories with a specific destination and for the benefit of selected individuals within that community. Given the fact that even a small piece of woven cloth implied plenty of raw materials, time and labour, it is hard to believe that there was only one specialised building for creating textiles, with only one loom, and that this was also able to supply all the needs of the entire community. One suggestion is that probably each household had an individual way to produce textiles and it did not necessarily imply weaving. The twining technique, identified as imprints in the pottery of the Vinča and Turdaş cultures (*c.* 5400–4900 BC) was intensely used also in later times, as shown by the discoveries from the Swiss Plateau, those of the Bronze Age from Ukraine (Yamnaya culture, *c.* 2500 BC)[75] and Russia[76] and also by ethnographical similarities with the North-Western Canada coast Amerindians.[77] Therefore it is possible that twining represented the current way of producing textile for the Neolithic and Copper Age of Transylvania. The usage of unspun fibre strands for the production of daily coarse textiles could also generally explain the scarcity of spindle whorls finds within the Neolithic settlements.

[70] Endrei 1968; Cordry and Cordry 1973; Dunsmore 1985; Hecht 1989; Broudy 1993.

[71] Todorova 1978, 71; Comşa 1990, 51.

[72] Lazarovici *et al.* 2001, 209–214.

[73] I am grateful for the information provided by Prof. Florin Draşovean, Muzeul Banatului, Timişoara.

[74] It is possible to make deeper interpretations on the subject of loom-weights found in cult areas by taking into account the analogies of Antiquity, even if they are of a much later time (Gleba 2008, 178–181; 184–185).

[75] Gleba and Nikolova 2009.

[76] Orfinskaya *et al.* 1999, 76; Shishlina *et al.* 2002, 2003.

[77] Gulbrandsen 2010, 151–153, 158–159; Médard 2010, 90–91, 93, figs 69–70, 74.

The discovery context for the loom-weights studied here is not sufficiently clear to allow a comprehensive argumentation of the hypotheses stated above. A big question mark is raised by the weights or weight fragments found isolated, both inside and outside homes. When a craft implies care for the tools involved and the valuing of those tools, the opposite must also be true – displaying negligence toward or abandoning them suggests that they were ordinary, even worthless. Another question is raised by those weights found within kilns or nearby hearths and also by those with several other context and associations, either as complete pieces or fragmentary, isolated or in groups. These might suggest at least three possible situations:

1. a post-functional character (the breaking and consequent discarding of loom-weights);
2. a functional re-conversion (their re-use for a different purpose);
3. an extra-functional manipulation that has a profound symbolism and ritual connotation, suggested by the spiritual character of this craft in general. For the latter explanation, one could find examples of loom-weights recovered from ritual pits and foundation pits of houses from Transylvania. Even though they appear much later, the deposition of loom-weights as part of the foundation ritual of houses, temples and city walls from the 7th to 3rd century BC Italy[78] can be considered as analogies.

Concluding speculations

Evidence of textile production is hard to read and interpret, and can even provide contradictory information. They are far from offering a clear view of the textile production characteristics of Neolithic and Copper Age communities, like the specific production process, place and time reserved for textile activities, as well as their extent and degree of specialisation. Even so, it is obvious that textiles were produced and used in the Neolithic and Copper Age communities in Transylvania, this area being part of a larger unit[79] in which textiles are documented way back to the Mesolithic (unwoven) and Early Neolithic (woven and unwoven fabrics).

Although difficult to capture, there are several pieces of evidence that could indicate an evolutionary shift in the craft of textile production, and an increased production of woven fabrics in the Neolithic communities, compared to the Copper Age ones, not just for Transylvania but also for the neighbouring regions:

1. the presence of woven textile imprints on Copper Age pottery (in the areas of Tiszaplogár and Cucuteni-Trypillian Culture);
2. morphological and ponderous differences between loom-weights, the Copper Age ones being adapted for production of more robust fabrics;
3. the groups of loom-weights reported in several settlements (most of them in the KGK VI area), anthropomorphic representations of women clothed in dresses (mainly in the areas of KGK VI and Cucuteni-Trypillian civilisations), which could be an indication of woven garments usage and also that of social differentiation, etc.

---

[78] Gleba 2008, 182–183.

[79] The author is considering here both the European area and also the Near East, as the origin of the Neolithic phenomenon and also of other innovative flows that have propagated periodically towards the Balkans and South-Eastern Europe.

If we are to believe the statement by J. Winiger, according to whom, throughout the Neolithic, woven fabrics remained secondary to those made by twining,[80] and rely on evidence from textile imprints, it can be stated that the spontaneous vegetation was the main source of raw textile material during the Neolithic. Flax would have been but a plant with limited textile potential, used only to produce certain rare and valuable types of textiles, a statement sustained also by Médard in connection to the flax weavings of the Swiss Neolithic period.[81] This would also justify their absence as textile imprints on Neolithic pottery from Transylvania. Changes observed during the Copper Age could be related to standardisation, at least for some settlements, of the cultivated textile plants (either flax or other textile plants). A movement towards the cultivation of textile plants would have been a natural consequence of the depletion of resources provided by the spontaneous vegetation due to the form of economy specific to Copper Age settlements (especially those of *tell* type). The difficulties involved in growing textile plants (flax in particular), as well as the entire process of extracting the fibres further magnified their value. Sets of loom-weights discovered in dwellings at the periphery and the outside of settlements (examples are found either in Copper Age of Southern Romania[82] and Bulgaria[83] or in Swiss Neolithic)[84] may indicate a monopoly on the textile raw material as well as on the weaving technology.[85] Perhaps it is premature to promote such theories, but it is possible that the development of this 'invisible' craft of textile production, that can hardly be documented archaeologically, contributed to the emergence of a social hierarchy and elite among Copper Age communities, which, in turn, are well represented from an archaeological point of view. This inequality projected onto a cultural-symbolic level and illustrated mainly by the rich clothing of anthropomorphic figurines could suggest that women were the ones knowing the craft.

Although highly speculative due to a lack of sufficient archaeological material available for analysis, the theories presented reflect a current textile research in Romania. Further continuation of the research involving interdisciplinary studies, by attracting specialists in archaeobotany, zoology, anthropology, microwear traces etc. will lead to the enrichment of our knowledge of the evolution of prehistoric textile production and to the confirmation and/or rebuttal of the theories that exist today.

[80] Winniger 1995, 178.
[81] Médard 2010, 146.
[82] The site of Radovanu I and II (Comşa 1990, 50–51).
[83] Golyamo Delchevo IV and Ovcharovo X (Todorova 1978, 71).
[84] Chalain 19 (Médard 2010, 150).
[85] This supposition might be true if these dwellings were in fact workshops or weaver houses located close to the lands where textile crops were cultivated or the textile raw material could be found.

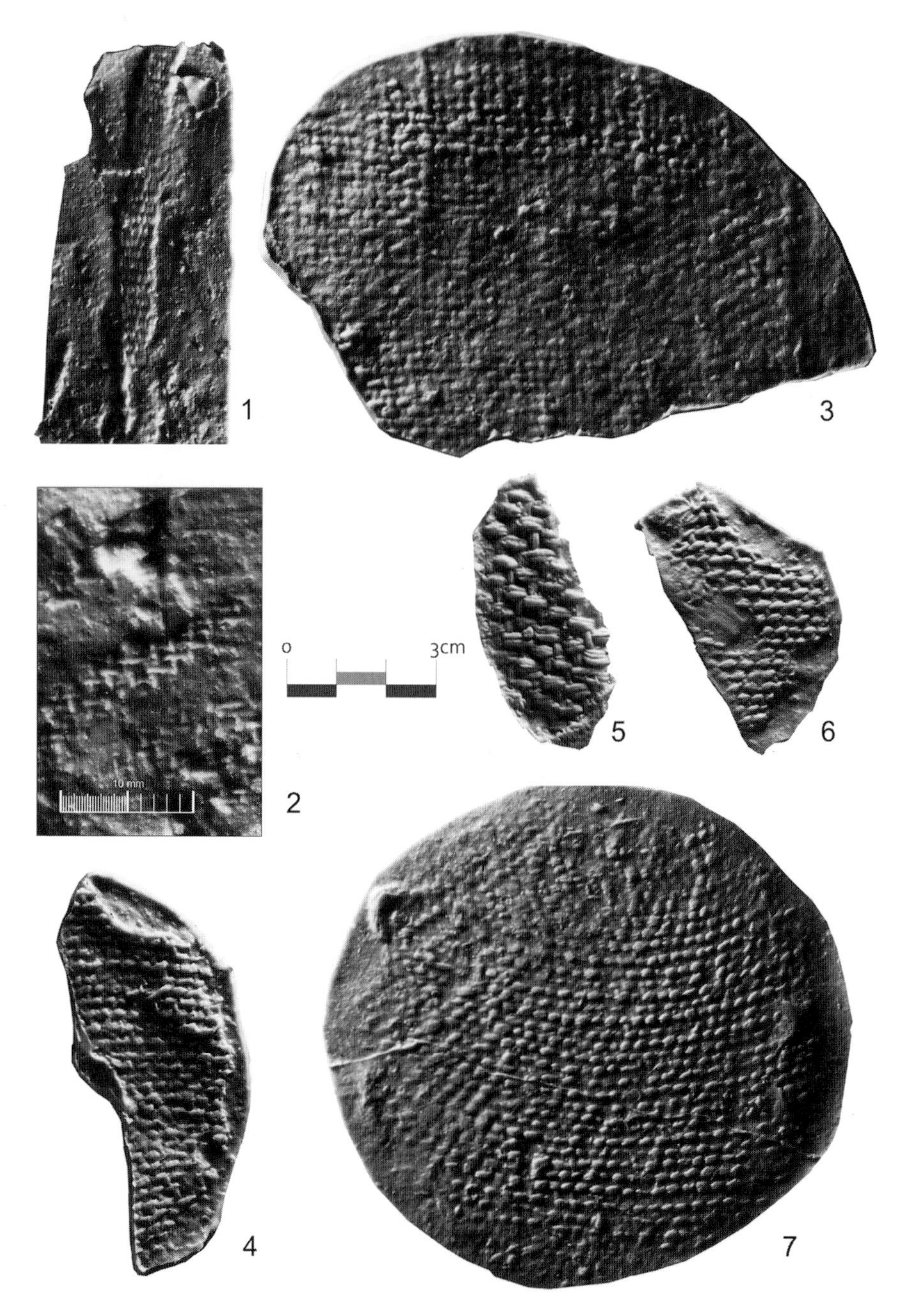

Fig. 1.23: Examples of textile imprints found in Neolithic and Copper Age sites (positive casts): Woven textiles – 1. LBT.1050, Limba-Bordane, B1–B2 Vinča culture; 2. ALN.1001, Alba Iulia-Lumea Nouă, Foeni group; 3. DOR.61329, Dorolţu-Castel, Tiszapolgár culture; Twined textiles; 4. LVL.3385, Limba-Vărăria, B1–B2 Vinča culture; 5. TRD.5257, Turdaş, Turdaş culture (?); 6. TLL.16309, Turdaş-La Luncă, Turdaş culture; TRD.5278, Turdaş, Turdaş culture (?). Repository: University "1 Decembrie 1918" of Alba Iulia (1–2, 4); The National Museum of Transylvanian History, Cluj-Napoca (3, 5–7).

Fig. 1.24: Examples of loom-weights found in Neolithic and Copper Age sites: 1. BRB.4265, Brateş-Bende, Starčevo-Criş culture; 2. TUR.18255/5, Turia-Grădina Palatului Apor, House L1/1986; Starčevo-Criş culture; 3. ZDC.1080/80, Zăuan-Dâmbul Cimitirului, House L1/1980, Starčevo-Criş culture; 4. OCN.14990, Olteni-Cariera de Nisip (Site B), Pit G34/2008, Linear pottery culture; 5. LVL.2846, Limba-Vărăria, House L1/2001, B1–B2 Vinča culture; 6. DTA.2849, Deva-Tăualaş, Turdaş culture (?); 7. ALN.7413/2, Alba Iulia-Lumea Nouă, Foeni group (?); 8. PHO.9879, Păuca-Homm, House L1/1965, Petreşti culture; 9. ODC.1308, Olteni-Dealul Cetăţii (Vármegye), Ariuşd culture; 10. ARI.7417, Ariuşd-Dealul Tyiszk, House L4, Ariuşd culture; 11. CTU.143109, Cheile Turzii-Peştera Ungurească, Hearth V8/1995, Bodrogkeresztúr culture. Repository: The Székely National Museum, Sfântu Gheorghe (1–2, 10); The History and Art Museum of Zalău (3); The National Museum of Eastern Carpathians, Sfântu Gheorghe (4, 9); University "1 Decembrie 1918" of Alba Iulia (5–6); The National Museum of the Union, Alba Iulia (7); The National Museum of Transylvanian History, Cluj-Napoca (11).

Abreviations

ATN	*Archaeological Textiles Newsletter.*
BCŞS	*Buletinul Cercurilor Ştiinţifice Studenţeşti.* University '1 Decembrie 1918' of Alba Iulia.
JAS	*Journal of Archaeological Science*, Academic Press.
OJA	*Oxford Journal of Archaeology*, Oxford.
SCICP	*Studii şi Comunicări de Istorie a Civilizaţiei Populare din România.* Brukenthal Museum, Sibiu.
SCIV(A)	*Studii şi Cercetări de Istoria Veche (şi Arheologie).* The Romanian Academy's Institute of Archaeology 'Vasile Pârvan', Bucharest.
StComCaransebeş	*Studii şi Comunicări de Istorie şi Etnografie*, Caransebeş.
Veget Hist Archaeobot	*Vegetation History and Archaeobotany*, SpringerLink.

Bibliography

Adovasio, J. M. 1977 *Basketry Technology: A Guide to Identification and Analysis.* Chicago.

Adovasio, J. M. and Maslowski, R. F. 1988 Textile Impresions on Ceramic Vessels. In A. McPherron and D. Srejovic (eds), *Divostin and the Neolithic of Central Serbia*, 345–357.

Aghiţioaie, V. and Draşovean, F. 2004 Date despre impresiunea unei ţesături în aşezarea neolitică târzie de la Foeni–'Cimitirul Ortodox' (jud. Timiş). *Patrimonium Banaticum* 3, 47–49.

Alfaro Giner, C. 1980 Estudio de los materiales de cestería procedentes de la Cueva de los Murciélagos (Albuñol, Granada). *Trabajos de Prehistoria* 37, 109–161.

Alfaro Giner, C. 1984 *Tejidos y cestería en la Península Ibérica. Historia y de su técnica e industrias desde la prehistoria hasta la romanización*, Bibliotheca Praehistorica Hispana 21. Madrid.

Andersson, E. 2003 *Tools for Textile Production from Birka and Hedeby*, Birka Studies 8. Stockholm.

Andersson Strand, E., Gleba, M., Mannering, U., Munkholt, C. and Ringgaard, M. (eds) 2010 *North European Symposium for Archaeological Textiles X*, Ancient Textiles Series 5. Oxford.

Andersson Strand, E., Frei, K. M., Gleba, M., Mannering, U., Nosch, M.-L. and Skals, I. 2010 Old Textiles – new Possibilities. *European Journal of Archaeology* 13 (2), 149–173.

Andersson Strand, E. and Nosch, M.-L. (eds) 2013 *Tools, Textiles and Context: Textile Production in the Aegean and Eastern Mediterranean Bronze Age*, Ancient Textiles Series 13. Oxford.

Andreescu, R. R. 1997 Plastica gumelniţeană din colecţiile Muzeului Naţional de Istorie a României. *Cercetări Arheologice* 11, 309–323.

Andreescu, R.-R. 2002 *Plastica antropomorfă gumelniţeană: analiză primară.* Bucureşti.

Andreescu, R.-R., Lazăr, C., Oană, V. and Moldoveanu K. 2010 Câteve consideraţii asupra unor descoperiri din locuinţa nr. 2 din aşezarea eneolitică de la Sultana-Malu Roşu. *Cercetări Arheologice* 17, 9–20.

Baldia, C. M. and Jakes, K. A. 2007 Photographic Methods to Detect Colorants in Archaeological Textiles, *JAS* 34 (4), 519–525.

Bărbat, I. Al. 2012 Descoperiri aparţinând neoliticului timpuriu din colecţiile arheologice ale muzeului devean. *Terra Sebus* 4, 23–63.

Barber, E. J. W. 1991 *Prehistoric Textiles. The Development of Cloth in the Neolithic and Bronze Age with Special Reference to the Aegean.* Princeton.

Barber, E. J. W. 1994 *Women's Work: The First 20,000 Years. Women, Cloth, and Society in Early Times.* New York.

Barber, E. J. W. 2007 Weaving the Social Fabric. In C. Gillis and M.-L. Nosch (eds), *First Aid for the Excavation of Archaeological Textiles*, 220–228.

Bazzanella, M., Mayr, A., Moser, L. and Rast-Eicher, A. (eds) 2003 *Textiles. Intrecci e tessuti dalla preistoria europea.* Riva del Garda.

Berciu, D. 1961 *Contribuţii la problemele neoliticului în Romînia în lumina noilor cercetări.* Bucureşti.

Bergerbrant, S. 2007 *Bronze Age Identities: Costume, Conflict and Contact in Northern Europe 1600–1300 BC*, Stockholm Studies in Archaeology 43. Lindome.

Bergfjord, C. and Holst, B. 2010 A New Method for Identifying Textile Bast Fibres Using Microscopy. *Ultramicroscopy* 110 (9), 1192–1197.

Bergfjord, C., Mannering, U., Frei, K.-M., Gleba, M., Scharff, A., Skals, I., Heinemeier, J., Nosch, M.-L. and Holst, B. 2012, Nettle as a Distinct Bronze Age Textile Plant, *Scientific Reports* 2 (664), 1–4 (http://www.nature.com/srep/2012/120928/srep00664/full/srep00664.html).

Bichler, P., Grömer, K., Hoffmann-De Keijer, R., Kern, A. and Reschreiter, H. (eds) 2005 *Hallstatt Textiles. Technical Analysis Investigation and Experiment on Iron Age Textiles*, BAR International Series 1351.

Bier, C. 1995 Textile Arts in Ancient Western Asia. In J. M. Sasson (ed.), *Civilisations of the Ancient Near East*, Vol. III, 1567–1588.

Brandt, L. O., Tranekjer, L. D., Mannering, U., Ringgaard, M., Frei, K. M., Willerslev, E., Gleba, M. and Gilbert, M. T. P. 2011. Characterising the Potential of Sheep Wool for Ancient DNA Analyses. *Archaeological and Anthropoplogical Sciences* 3, 209–221.

Breniquet, C. 2008 *Essai sur le tissage en Mésopotamie de premières communautés sédentaires au milieu du IIIe millénaire avant J.-C.*, Travaux de la Maison René-Ginouvès 5. Paris.

Broudy, E. 1993 *The Book of Looms. A History of the Handloom from Ancient Times to the Present*. Hanover

Cardon, D. 2007 *Natural Dyes. Sources, Tradition, Technology and Science*. London.

Carington Smith, J. 2000 The Small Finds: The Spinning and Weaving Implements. In C. Ridley, K. A. Wardle and C. A. Mould (eds), *Servia I, Anglo-Hellenic Rescue Excavations 1971–73 Directed by Katerina Rhomiopoulou and Cressida Ridley*, British School at Athens Supp. 32, 207–263.

Chmielewski, T. J. 2009 *Po nitce do kłębka... O przędzalnictwie i tkactwie młodszej epoki kamienia w Europie Środkowej*. Warszawa.

Chmielewski, T. and Gardyński, L. 2010 New Frames of Archaeometrical Description of Spindle Whorls: A Case Study of the Late Eneolithic Spindle Whorls from the 1C Site in Gródek, District of Hrubieszów, Poland. *Archaeometry* 52 (5), 869–881.

Cioară, L. 1998 *Structura ţesăturilor*. Iaşi.

Comşa, E. 1974 *Istoria comunităţilor culturii Boian*. Bucureşti.

Comşa, E. 1988 Unele date despre îmbrăcămintea din epoca neolitică de pe teritoriul Moldovei. *Hierasus* 7–9, 39–55.

Comşa, E. 1990 *Complexul neolitic de la Radovanu*, Cultură şi Civilizaţie la Dunărea de Jos 8. Călăraşi.

Comşa, E. 1995 *Figurinele antropomorfe din epoca neolitică pe teritoriul României*. Bucureşti.

Comşa, E. 1995–1996, Les figurines anthropomorphes des cultures de Turdaş et Vinča (Ressemblances et différences). *Sargetia* 26 (1), 91–104.

Cordry, D. and Cordry, D. 1973 *Mexican Indian Costumes*. Austin.

Cordwell, J. M. and Schwarz, A. (eds) 1979 *The Fabrics of Culture. The Anthropology of Clothing and Adornment*. Hague.

Crewe, L. 1998 *Spindle Whorls: A Study of Form, Function and Decoration in Prehistoric Bronze Age Cyprus*. Jonsered.

Cucoş, Ş. 1970 Reprezentări antropomorfe în decorul pictat cucutenian de la Ghelăieşti (jud. Neamţ). *Memoria Antiquitatis* 2, 101–113.

Cybulska, M. 2010 Reconstruction of Archaeological Textiles. *Fibres and Textiles in Eastern Europe* 18, 3 (80), 100–105.

Cybulska, M., Jedraszek-Bomba, A., Kuberski, S. and Wrzosek, H. 2008, Methods of Chemical and Physicochemical Analysis in the Identificationof Archaeological and Historical Textiles. *Fibres and Textiles in Eastern Europe* 16, 5 (70), 67–73.

Diaconescu, D. 2009 *Cultura Tiszapolgár în România*, Bibliotheca Brukenthal 41. Alba Iulia.

Dumitrescu, V. 1974 *Arta preistorică în România*. Bucureşti.

Dumitrescu, V. 1979 *Arta culturii Cucuteni*. Bucureşti.

Dumitroaia, G., Munteanu, R., Preoteasa, C. and Garvăn, D. 2009 *Poduri-Dealu Ghindaru. Cercetări arheologice in Caseta C 2005–2009*. Piatra-Neamţ.

Dunsmore, S. 1985 *The Nettle in Nepal. A Cottage Industry*. Surbiton.

Eason, D. L. and Molloy, R. M. 2000 A Study of the Plant, Fibre and Seed Development in Flax and Linseed (*Linum usitatissimum*) Grown at a Range of Seed Rates, *Journal of Agricultural Science, Cambridge* 135, 361–369.

Emery, I. 2009 *The Primary Structures of Fabrics. An Illustrated Classification*. Washington D.C.

Endrei, W. 1968 *L'évolution des techniques du filage et du tissage du Moyen-Age à la révolution industrielle*. Paris.

Frânculeasa, A. 2004 Plastica antropomorfă şi zoomorfă din epoca neo-eneolitică din patrimoniul Muzeului Judeţean de Istorie şi Arheologie Prahova. *Cumidava* 27, 26–46.

Frei, K. M., Frei, R., Mannering, U., Gleba, M., Nosch, M.-L. and Lyngstrøm, H. 2009a Provenance of Ancient Textiles – A Pilot Study Evaluating the Strontium Isotope System in Wool. *Archaeometry* 51 (2), 252–276.

Frei, K. M., Skals, I., Gleba, M. and Lyngstrøm, H. 2009b The Huldremose Iron Age textiles, Denmark: an Attempt to Define their Provenance Applying the Strontium Isotope System. *JAS* 30, 1–7.

Galbenu, D. 1983 Aşezarea de tip Sălcuţa de la Almăjel, jud. Mehedinţi. *Cercetări Arheologice* 6, 141–156.

Gillis, C. and Nosch, M.-L. (eds) 2007a *Ancient Textiles. Production. Crafts and Society, Proceeding of the First International Conference on Ancient Textiles, held at Lund, Sweden and Copenhagen, Denmark, on March 19–23, 2003.*

Gillis, C. and Nosch M.-L. (eds) 2007b, *First Aid for the Excavation of Archaeological Textiles*. Oxford.

Gleba, M. 2008 *Textile production in Pre-Roman Italy*, Ancient Textiles Series 4. Oxford.

Gleba, M. 2012 From Textiles to Sheep: Investigating Wool Fibre Development in Pre-Roman Italy Using Scanning Electron Microscopy (SEM). *JAS* 39 (12), 3643–3661.

Gleba, M. and Mannering, U. (eds) 2012 *Textiles and Textile Production in Europe: From Prehistory to AD 400*, Ancient Textiles Series 11. Oxford.

Gleba, M., Munkholt, C. and Nosch, M.-L. (eds) 2008 *Dressing the Past*, Ancient Textiles Series 3. Oxford

Gleba, M. and Nikolova, A. 2009 *Early Twined Textiles from Sugokleya (Ukraine). ATN* 48, 7–9.

Gogâltan, F. 2008 Fortificaţiile tell-urilor epocii bronzului în bazinul carpatic. O privire generală. *Analele Banatului S.N.* 16, 81–100.

Gogâltan, F. and Ignat, A. 2011 Transilvania şi spaţiul nord-pontic. Primele contacte (cca. 4500–3500 a. Chr.). *Tyragetia* S.N. 5 (1), 7–38.

Good, I. 2001 Archaeological Textiles: A Review of Current Research. *Annual Review of Anthropology* 30, 209–226.

Grömer K. 2005 *Efficiency and Technique – Experiments with Original Spindle Whorls*. In P. Bichler, K. Grömer, R. Hoffmann-De Keijer, A. Kern, and H. Reschreiter (eds) *Hallstatt Textiles. Technical Analysis Investigation and Experiment on Iron Age Textiles*, BAR International Series 1351, 81–90.

Grömer, K. 2010 *Prähistorische Textilkunst in Mitteleuropa. Geschichte des Handwerkes und Kleidung vor der Römern*. Wien.

Gulbrandsen, D. 2010 *Edward Sheriff Curtis. Visions of the First Americans*. London.

Gumă, N. 1977 Evoluţie şi permanenţă în meşteşugul ţesutului şi arta decorării ţesăturilor pe teritoriul judeţului Caraş-Severin. *StComCaransebeş* 2, 146–148.

Hammarlund, L. 2005 Handicraft Knowledge Applied to Archaeological Textiles. *Nordic Textile Journal* 8, 87–119.

Hansen, S., Toderaş, M., Reingruber, A., Gatsov, I., Georgescu, C., Görsdorf, J., Hoppe, T., Nedelcheva, P., Prange, M., Wahl, J., Wunderlich, J. and Zidarov, P. 2007 Pietrele, Măgura Gorgana. Ergebnisse des Ausgrabunden im Sommer 2006. *Eurasia Antiqua* 13, 43–112.

Hecht, A. 1989 *The Art of the Loom Weaving, Spinning and Dyeing across the World*. London.

Herbig, C. and Maier, U. 2011 Flax for Oil or Fibre? Morphometric Analysis of Flax Seeds and New Aspects of Flax Cultivation in Late Neolithic Wetland Settlements in Southwest Germany, *Veget Hist Archaeobot* 20 (6), 527–533.

Hopkins, H. (eds) 2013 *Ancient Textiles, Modern Science*. Oxford.

Hurcombe, L. 2010 Nettle and Bast Fibre Textiles from Stone Tool Wear Traces? The Implications of Wear Trace on Archaeological Late Mesolithic and Neolithic Micro-Denticulate Tools. In E. Andersson Strand, M. Gleba, U. Mannering, C. Munkholt and M. Ringgaard (eds) *North European Symposium for Archaeological Textiles X*, Ancient Textiles Series 5, 129–139.

Keith, K. 1998 Spindle Whorls, Gender, and Ethnicity at Late Chalcolithic Hacinebe Tepe. *Journal of Field Archaeology* 25, 497–515.

Körber-Grohne, U. 1991 The Determination of Fibre Plants in Textiles, Cordage and Wickerwork. In J. M. Renfrew (ed.) *New Light on Early Farming. Recent Developments in Palaeoethnobotany*, 93–104.

Larsson Lovén, L. 2002 *The Imagery of Textile Making: Gender and Status in the Funerary Iconography of Textile Manufacture in Roman Italy and Gaul*. Göteborg.
Lazarovici, C.-M. and Lazarovici G. 2006 *Arhitectura neoliticului și epocii cuprului din România, I. Neoliticul*, Bibliotheca Archaeologica Moldaviae 4. Iași.
Lazarovici, C.-M. and Lazarovici G. 2007 *Arhitectura neoliticului și epocii cuprului din România, II. Epoca cuprului*, Bibliotheca Archaeologica Moldaviae 8. Iași.
Lazarovici, G., Drașovean F. and Maxim, Z. 2001 *Parța. Monografie arheologică*, Vol. 1.1–2, Bibliotheca Historica et Archaeologica Banatica 12.
Leuzinger, U. and Rast-Eicher, A. 2011 Flax processing in the Neolithic and Bronze Age pile-dwelling settlements of eastern Switzerland. *Veget Hist Archaeobot* 20 (6), 535–542.
Liu, R. K. 1978 Spindle Whorls: Part I. Some Comments and Speculations. *The Bead Journal* 3, 87–103.
Luca, S. A. and Dragomir, I. 1987 Date cu privire la o statuetă inedită de la Liubcova-Ornița (jud. Caraș-Severin). *Banatica* 9, 31–42.
Luca, S. A. 1990 Contribuții la istoria artei neolitice. Plastica din așezarea de la Liubcova- "Ornița". *Banatica* 10, 6–44.
Luca, S. A. 2001 *Așezări neolitice pe valea Mureșului (II). Noi cercetări arheologice la Turdaș-Luncă (I) Campaniile anilor 1992–1995*, Bibliotheca Musei Apulensis 17. Alba Iulia
Maier, U. and Schlichtherle, H. 2011 Flax Cultivation and Textile Production in Neolithic Wetland Settlements on Lake Costance and in Upper Swabia (South-West Germany). *Veget Hist Archaeobot*, 20 (6), 567–578.
Mannering, U. 1995 Oldtidens brændenældeklæde. Forsøg med fremstilling af brændenældegarn. *Naturens Verden*, 161–168.
Mannering, U., Possnert, G., Heinemeier, J. and Gleba, M. 2010 Dating Danish Textiles and Skins from Bog Finds by Means of ^{14}C AMS. *JAS* 37, 261–268.
Mantu, C.-M. 1993 Plastica antropomorfă a așezării Cucuteni A_3 de la Scânteia (jud. Iași). *Arheologia Moldovei* 16, 51–67.
Marian, C. 2006 Figurinele "Venus" – mărturii iconografice ale existenței tehnologiilor textile în paleolitic, *Buletinul Centrului de Restaurare Conservare* 2, 32–39.
Marian, C. 2008 Archaeological Arguments Concerning the Textile Technologies of Cucuteni Civilisation. Nalbinding Technique. In V. Chirica and M.-C. Văleanu (eds), *Établissements et habitations préhistoriques. Structure, organisation, symbole, Actes du colloque de Iași, 12–14 Decembre 2007*, Bibliotheca Archaeologica Moldaviae 9, 327–334.
Marian, C. 2009 *Meșteșuguri textile în cultura Cucuteni*, Iași.
Marian, C. 2010 Meșteșuguri textile – preistorie și actualitate. *Monumentul* 10 (2), 535–540.
Marian, C. 2012 Cercetări privind ornamentarea cu șnurul a ceramicii de tip Cucuteni C. In *Cucuteni 5000 Redivivus: științe exacte și mai puțin exacte.* Culegere de lucrări prezentate la Simpozionul 'Cucuteni 5000 Redivivus: științe exacte și mai puțin exacte', ediția a 7-a, 13–15 Sept. 2012, Chișinău, 41–45.
Marian, C. and Anăstăsoaei, D. 2007 Artă și creativitate în tehnologia textilă a culturii Cucuteni. *Conservarea și Restaurarea Patrimoniului Cultural* 7, 220–226.
Marian, C. and Anăstăsoaei, D. 2008 Cercetări privind tehnologiile textile ale civilizației Cucuteni. In *Comunicări prezentate la cel de-al doilea simpozion 'Cucuteni 5000 Redivivus: științe exacte și mai puțin exacte', ediția II, Chișinău, 2–3 Octombrie 2007*, 69–76.
Marian, C. and Bigbaev, V. 2008 Cercetarea modalităților de împletire a unor materiale ale căror impresiuni s-au păstrat pe ceramica culturii Cucuteni-Tripolie. In *Cucuteni – 5000 Redivivus. Științele exacte și mai puțin exacte. Comunicări prezentate la Simpozionul Internațional 'Cucuteni – 5000 Redivisus', ediția a III, Chișinău, 11–12 septembrie 2008*, 42–49.
Marian, C. and Ciocoiu, M. 2004a Impresiuni de materiale textile pe ceramica arheologică descoperită la Cucuteni, Partea I. *Revista română de textile – pielărie* 2, 95–108.
Marian, C. and Ciocoiu, M. 2004b Impresiuni de materiale textile pe ceramica arheologică descoperită la Cucuteni, Partea II. *Revista română de textile – pielărie* 3, 19–32.
Marian, C. and Ciocoiu, M. 2005 Arta textilă – parte integrantă a istoriei civilizațiilor străvechi, *Revista română de textile – pielărie* 1, 105–118.

Marian, C., Anăstăsoaei, D. and Gugeanu, M. 2004 Cercetarea structurii unor materiale textile ale căror impresiuni s-au păstrat pe ceramica arheologică de Cucuteni. *Buletinul Centrului de Restaurare Conservare* 2 (1–2), 58–67.

Marian, C., Anăstăsoaei, D., Gugeanu, M. and Bigbaev, V. 2005 Cercetarea modalităţilor de realizare a unor materiale împletite ale căror impresiuni s-au păstrat pe ceramica culturii Cucuteni-Tripolie, *Monumentul* 6, 433–441.

Mårtensson, L., Andersson, E., Nosch, M.-L. and Batzer, A. 2005–2006 *Technical report. Experimental archaeology. Part 1*. Tools and Textiles – Texts and Contexts research program. The Danish National Research Foundation's Centre for Textile Research, University of Copenhagen, http://ctr.hum.ku.dk/tools/Technical_report_1_experimental_archaeology.pdf/.

Mårtensson, L., Andersson, E., Nosch, M.-L. and Batzer, A. 2006a *Technical report. Experimental archaeology. Part 2:1 Flax*. Tools and Textiles – Texts and Contexts research program. The Danish National Research Foundation's Centre for Textile Research, University of Copenhagen, http://ctr.hum.ku.dk/tools/Technical_report_2-1_experimental_archaeology.pdf/.

Mårtensson, L., Andersson, E., Nosch, M.-L. and Batzer, A. 2006b *Technical report. Experimental archaeology. Part 2:2 Whorl or bead?*. Tools and Textiles – Texts and Contexts research program. The Danish National Research Foundation's Centre for Textile Research, University of Copenhagen, http://ctr.hum.ku.dk/tools/Technical_report_2-2__experimental_arcaheology.PDF/.

Mårtensson, L., Andersson, E., Nosch, M.-L. and Batzer, A. 2007a *Technical Report. Experimental Archaeology. Part 3 Loom weights*. Tools and Textiles – Texts and Contexts Research Program. The Danish National Research Foundation's Centre for Textile Research, University of Copenhagen, http://ctr.hum.ku.dk/tools/Technical_report_3__experimental_archaeology.PDF/.

Mårtensson, L., Andersson, E., Nosch, M.-L. and Batzer, A. 2007b *Technical Report. Experimental Archaeology. Part 4 Spools*. Tools and Textiles – Texts and Contexts research program. The Danish National Research Foundation's Centre for Textile Research, University of Copenhagen, http://ctr.hum.ku.dk/tools/Technical_report_4__experimental_arcaheology.PDF/.

Mårtensson, L., Nosch, M.-L. and Andersson Strand, E. 2009 Shape of Things: Understanding a Loom Weight. *OJA* 28 (4), 373–398.

Martial, E. and Médard, F. 2007 Acquisition et traitement des matières textiles d'origine végétale en Préhistoire: l'exemple du lin. In V. Beugnier and P. Crombé (eds), *Plant Processing from a Prehistoric and Ethnographic Perspective. Prodeedings of a whorkshop at Ghent University (Belgium) November 28, 2006*, BAR International Series 1718, 67–82.

Mateescu, C. N. 1959 Săpături arheologice la Vădastra. *Materiale şi Cercetări Arheologice* 5, 61–73.

Maxim, Z. 1999 *Neo-eneoliticul din Transilvania. Date arheologice şi matematico-statistice*, Bibliotheca Musei Napocensis 19. Cluj-Napoca.

Mazăre, P. 2005 Artefacte din os descoperite în aşezarea neolitică de la Limba (jud. Alba). In C. I. Popa, G. Rustoiu (eds), *Omagiu profesorului Ioan Andriţoiu cu prilejul împlinirii a 65 de ani. Studii şi cercetări arheologice*, 237–312.

Mazăre, P. 2008 Impresiuni de ţesături pe fragmente ceramice descoperite în situl preistoric de la Limba (jud. Alba). *Apulum* 45, 315–330.

Mazăre, P. 2010 Metodologia de investigare a textilelor arheologice preistorice. *Terra Sebus* 2, 9–45.

Mazăre, P. 2011a Textile Structures and Techniques Identified in Neolithic and Copper Age Sites from Romania. *Marisia* 31, 27–48.

Mazăre, P. 2011b Textiles and Pottery: Insights into Neolithic and Copper Age Pottery Manufacturing Techniques from Romania. *ATN* 53, 28–34.

Mazăre, P. 2011c, O tehnică preistorică de confecţionare a textilelelor: tehnica şnurată. *Terra Sebus* 3, 63–89.

Mazăre, P. 2012 Definirea şi clasificarea artefactelor preistorice destinate torsului: fusaiolele. *Terra Sebus* 4, 103–131.

Mazăre, P. 2013 *Interpretări funcţionale ale 'greutăţilor' din lut ars. Annales Universitatis Apulensis.* Series Historica, 17 (2), 27–67.

Mazăre, P., Lipot, Ş. and Cădan, A. 2012 Experimental Study on the Use of Perishable Fibre Structures in Neolithic and Eneolithic Pottery. In V. Cotiugă and Ş. Caliniuc (eds), *Interdisciplinary Research in Archaeology, Proceedings of the First Arheoinvest Congress, "Alexandru Ioan Cuza" University of Iaşi, Romania, 10–11 June 2011, BAR International Series 2433, 159–168.*

Médard, F. 2000 *L'artisanat textile au Néolithique: L'exemple de Portalban II (Suisse), 3272–2462 avant J.-C.*, Préhistoires 4. Montagnac.

Médard, F. 2006 *Les activités du filage au Néolithique sur le Plateau suisse: Analyse technique, économique et sociale*, Monographies du CRA 28. Paris.

Médard, F. 2010 *L'art du tissage au Néolithique IVe-IIIe millénaires avant J.-C. en Suisse*. Paris.

Medović, A. and Horváth, F. 2012 Content of a Storage Jar from the Late Neolithic Site of Hódmezővásárhely-Gorzsa, South Hungary: A Thousand Carbonized Seeds of Abutilon theophrasti Medic. *Veget Hist Archaeobot* 21 (3), 215–220.

Michel, C. and Nosch, M.-L. (eds) 2010 *Textile Terminologies in the Ancient Near East and Mediterranean from the Third to the First Millennnia BC*, Ancient Textiles Series 8. Oxford.

Monah, D.1997 *Plastica antropomorfă a culturii Cucuteni-Tripolie*, Bibliotheca Memoriae Antiquitatis 3. Piatra-Neamţ.

Monah, D., Dumitroaia, G., Munteanu, R., Preoteasa, C., Garvăn, D., Uţă, L. and Nicola, D. 2007 Poduri, com. Poduri, jud. Bacău. Punct: Dealu Ghindaru. In M. V. Angelescu and F. Vasilescu (eds) *Cronica Cercetărilor Arheologice din România. Campania 2006, A XLI-a Sesiune Naţională de Rapoarte Arheologice, Tulcea, 29 Mai–1 Iunie 2006*, 214–215.

Müller, M., Murphy, B., Burghammer, M., Riekel, C., Gunneweg, J. and Pantos, E. 2006 Identification of Single Archaeological Textile Fibres from the Cave of Letters Using Synchrotron Radiation Microbeam Diffraction and Microfluorescence", *Applied Physics A* 83, 183–188.

Murphy, T. M., Ben-Yehuda, N., Taylor, R. E. and Southon, J. R. 2011 Hemp in Ancient Rope and Fabric from the Christmas Cave in Israel: Talmudic Background and DNA Sequence Identification. *JAS 38 (10)*, 2579–2588.

Nica, M. and Niţă, T. 1979, Les établissements Néolitiques de Leu et Padea de la zone d'interférence des cultures Dudeşti et Vinča. Un nouvel aspect du Néolitique moyen d'Olténie. *Dacia NS* 23, 31–64.

Nica, M. 1981 Date despre descoperirea celei mai vechi ţesături de pe teritoriul României, efectuată la Sucidava-Celei, din perioada de trecere de la neolitic la epoca bronzului (2750–2150 î.e.n). *SCICP* 1, 121–125.

Nisbet, R. 2009 New Evidence of Neolithic and Copper Age Agriculture and Wood Use in Transylvania and the Banat (Romania). In F. Draşovean, D. L. Ciobotaru and M. Maddison (eds), *Ten Years After: The Neolithic of the Balkans, as Uncovered by the Last Decade of Research. Proceedings of the Conference held at the Museum of Banat on November 9th–10th, 2007*, Bibliotheca Historica et Archaeologica Banatica 49.

Orfinskaya, O. V., Golikov, V. P. and Shishlina, N. I. 1999 *Комплексное исследование текстильных изделийэпохи бронзы евразийских степей* [Comprehensive Research of Textiles from the Bronze Age Eurasian Steppe]. In N. I. Shishlina (ed.), *Текстиль эпохи бронзыевразийских степей* [*Textiles of the Bronze Age Eurasian Steppe*], Papers of the State Historical Museum 109, 58–184

Petrescu-Dîmboviţa, M. 1950 Date noi asupra înmormântărilor cu ocru roşu în Moldova. *SCIV* 1 (2), 110–125.

Petrescu-Dîmboviţa, M. 2010 Neo-eneoliticul. Periodizarea şi cronologia relativă şi absolută. In M. Petrescu-Dîmboviţa and Al. Vulpe (eds) 2010, *Istoria românilor*. Vol. I. *Moştenirea timpurilor îndepărtate*, 103–206.

Prisecaru, D. 2009a Consideraţii metodologice cu privire la analiza unor obiecte de uz casnic din epoca bronzului în spaţiul românesc. *BCŞS* 15, 17–25.

Prisecaru, D. 2009b Meşteşuguri casnice în epoca bronzului pe teritoriul României. Prelucrarea materiilor textile. *Corviniana* 13, 93–102.

Ræder Knudsen, L. 2002 La tessitura con le tavolette nella tomba 89. In P. von Eles (ed.), *Guerriero e sacerdote. Autorità e comunità nell'età del ferro a Verucchio. La Tomba del Trono*, 220–234.

Rast-Eicher, A. 2005 Bast before Wool: the First Textiles. In P. Bichler, K. Grömer, R. Hoffmann-De Keijer, A. Kern and H. Reschreiter (eds) *Hallstatt Textiles. Technical Analysis Investigation and Experiment on Iron Age Textiles*, BAR International Series 1351, 117–131.

Rast-Eicher, A. 2008 *Textilien, Wolle, Schafe der Eisenzeit in der Schweiz*, Antiqua 44. Basel

Rast-Eicher, A. and Bender Jørgensen, L. 2013, Sheep Wool in Bronze Age and Iron Age Europe. *JAS* 40 (2), 1224–1241.

Raymond, L. C. 1984 *Spindle Whorls in Archaeology*. Greeley.

Roche-Bernard, G. and Ferdiere, A. 1993 *Costumes et textiles en Gaule Romaine*. Paris.

Sankari, H. S. 2000 Bast Fibre Content, Fibre Yield and Fibre Quality of Different Linseed Genotypes, *Agricultural and Food Science in Finland* 9, 79–87.

Săvescu, I. 2004 Războiul de ţesut cu greutăţi. *Carpica* 33, 65–77.

Schier, W. and Draşovean, F. 2004 Vorbericht über die rumänisch-deutschen Prospektionen und Ausgrabungen in der befestigten Tellsiedlung von Uivar, jud. Timiş, Rumänien (1998–2002). *Praehistorische Zeitschrift* 79 (2), 145–230.

Schneider, J. 1987 The Anthropology of Cloth. *Annual Review of Anthropology* 16, 409–448.

Seiler-Baldinger, A. 1994 *Textiles. A Classification of Techniques*. Bathurst.

Shishlina, N. I., Orfinskaya, O. V. and Golikov, V. P. 2002 Bronze Age Textiles from North Caucasus: Problems of Origin. In *Steppe of Eurasia in Ancient Times and Middle Ages, Proceedings of International Conference Saint Petersburg*, 253–259.

Shishlina, N. I., Orfinskaya, O. V. and Golikov, V. P. 2003 Bronze Age Textiles from North Caucasus: New Evidence of Fourth Millennium BC Fibers and Fabrics. *OJA* 22 (4), 331–344.

Smith, J. S. 2002 Changes in Workplace: Women and Textile Production on Late Bronze Age Cyprus. In D. Bolger and N. Serwint (eds), *Engendering Aphrodite: Women and Society in Ancient Cyprus, CAARI International Conference, 19–23 March* 1998, 281–312.

Sorochin, V. 2001 Plastica antropomorfă din aşezarea cucuteniană de la Brânzeni VIII, jud. Edineţ. *Memoria Antiquitatis*, 22, 137–153.

Sorochin, V. and Borziac, I. 2001 Plastica antropomorfă din aşezarea cucuteniană de la Iablona I, jud. Bălţi. *Memoria Antiquitatis* 22, 115–136.

Sztáncsuj, J. S. 2009 Human Reprezentations in the Ariuşd Culture in Transylvania. In G. Bodi (ed.), *In medias res praehistoriae. Miscellanea in honorem annos LXV peragentis Professoris Dan Monah oblata*, 409–434.

Tiedemann, E. J. and Jakes, K. A. 2006 An Exploration of Prehistoric Spinning Technology: Sinning Efficiency and Technology Transition. *Archaeometry* 48 (2), 293–307.

Toderaş, M., Hansen, S., Reingruber, A. and Wunderlich, J. 2009 Pietrele–Măgura Gorgana: o aşezare eneolitică la Dunărea de Jos între 4500 şi 4250 î.e.n. *Materiale şi Cercetări Arheologice S. N.* 5, 39–90.

Todorova, H. 1978 *The Eneolithic Period in Bulgaria in the Fifth Millennium B.C.*, BAR. International Series 49. Oxford.

Ursulescu, N. 1986–1987 Contribuţia cercetărilor arheologice din judeţul Suceava la cunoaşterea evoluţiei neo-eneoliticului din Moldova. *Suceava* 13–14, 69–74.

Ursulescu, N. 1993 Continuitate şi restructurări cultural-etnice în neoliticul şi eneoliticul României. *Suceava* 20, 14–21.

Văleanu, M.-C. and Marian, C. 2004 Amprente umane, vegetale şi de textile pe ceramica eneolitică de la Cucuteni-Cetăţuie. In M. Petrescu-Dâmboviţa and M.-C. Văleanu (eds) *Cucuteni-Cetăţuie. Săpăturile din anii 1961–1966. Monografie arheologică*, Bibliotheca Memoriae Antiquitatis XIV, 318–327.

Vanden Berghe, A., Gleba, M. and Mannering, U. 2009 Towards the Identification of Dyestuffs in Early Iron Age Scandinavian Peat Bog Textiles. *JAS* 36, 1910–1921.

Verhecken, A. 2009, The Moment of Inertia: a Parameter for Functional Classification of Worldwide Spindle Whorls from all Periods. In E. Andersson Strand, M. Gleba, U. Mannering, C. Munkholt and M. Ringgaard (eds) *North European Symposium for Archaeological Textiles X*, Ancient Textiles Series 5, 257–270.

Vulpe, Al. 2010 Epoca bronzului. Consideraţii generale. Bronzul timpuriu. In M. Petrescu-Dîmboviţa and Al. Vulpe (eds) 2010, *Istoria românilor*. Vol. I. *Moştenirea timpurilor îndepărtate*, 209–232.

Walton, P. and Eastwood, G. 1983 *A Brief Guide to the Cataloguing of Archaeological Textiles*.Oxford.

Winiger, J. 1995 Die Bekleidung des Eismannes und die Anfänge der Weberei nördlich der Alpen. In K. Spindler, E. Rastbichler-Zissernig, H. Wilfing, D. zur Nedden and H. Nothdurfter (eds) *Der Mann im Eis. 2: Neue Funde und Ergebnisse Reihe*, 119–187.

Zaharia, F. and Cădariu, S. 1979 Urme de textile pe ceramica neolitică descoperită în judeţul Caraş-Severin. *Banatica* 5, 27–34.

Zohary, D. 1996 The Mode of Domestication of the Founder Crops of Southwest Asian Agriculture. In D. R. Harris (ed.), *The Origin and Spread of Agriculture and Pastoralism in Eurasia*, 142–158.

Zohary, D. and Hopf, M. 2000 *Domestication of Plants in the Old World: The Origin and Spread of Cultivated Plants in West Asia, Europe and the Nile Valley*.

2. Spindle Whorls From Two Prehistoric Settlements on Thassos, North Aegean

Sophia Vakirtzi, Chaido Koukouli-Chryssanthaki and Stratis Papadopoulos

The economic importance of fibre crafts in prehistoric communities is manifested by the spindle whorls which usually appear in abundance during the excavation of settlements. Ropes, cords and yarn for weaving can be manufactured from plant or animal fibres, with various techniques,[1] one of which requires the implementation of the spindle.[2] The spindle does not survive in the archaeological record since it is usually made of wood, a perishable material.[3] Spindle whorls, however, are traditionally and cross-culturally made of clay, stone or bone,[4] and have much better chances of good archaeological preservation. For this reason they constitute one of the most critical categories of archaeological data for the study of prehistoric yarn production. They are technological, economic and cultural markers of a craft which is otherwise archaeologically invisible.

The significance of spindle whorls as technological markers of yarn production lies in the function of the spindle. Experimental spinning performed with replicas of prehistoric spindle whorls has provided important information on the relationship among raw materials, whorls and products.[5] Fibres are attached to the spindle and yarn is formed while they are twisted along with the rotation of the tool. The rotation is enhanced if a whorl is added on the spindle.[6] The spinner must pick a whorl large and heavy enough to allow for the fibres to twist better and faster than before, yet not cause them to break. Small and light spindle whorls are suitable for softer and shorter raw materials and finer products, while larger and heavier ones are optimal for harder and longer fibres and coarser yarns.[7] Evidently, the size of the spindle whorl is crucial for the success of the endeavor. It must be chosen with regards both to the raw materials and the end product. It thus becomes apparent that spindle whorls of significantly different sizes are suitable for spinning different kinds of fibres into different kinds of yarns. Thus the degree of whorl size differentiation reflects the degree of product differentiation in terms of raw material, quality of end product, or both.

[1] Barber 1991, 41.
[2] Barber 1991, 42.
[3] Barber 1991, 51; Tzachili 1997, 118.
[4] Barber 1991, 43.
[5] Andersson Strand 2010, 14–15. Also, Andersson *et al.* 2010, 165.
[6] Barber 1991, 43.
[7] Barber 1991, 52.

A preliminary evaluation of yarn production at two prehistoric settlements on the island of Thassos is attempted on this basis. The two settlements are Aghios Ioannis on the south-east coast and Skala Sotiros on the west coast of the island (Fig. 2.1). At Aghios Ioannis a concentration of architectural features (stone substructures of probably perished superstructures), hearths, pottery, stone and clay tools of various functions (Fig. 2.2) was attested in an area of 50 square metres in the so-called "North Sector" of the site.[8] The settlement is dated between the Final Neolithic and the earliest phase of the Early Bronze Age in terms of relative chronology according to pottery styles, or in the last third of the 4th millennium BC in terms of absolute dating.[9] Skala Sotiros on the west coast is a fortified settlement of the Early Bronze Age. A circuit wall, architectural remains of large stone buildings, pebbled and clay floors, narrow alleys are some of its characteristics (Fig. 2.3). Two distinct but subsequent occupational phases are attested architecturally, Skala Sotiros II and Skala Sotiros III, while there is evidence for an even earlier one. In terms of absolute dating phase II is dated between 2580 and 2030 BC, and phase III is dated between the two last centuries of the 3rd millennium and the two first centuries of the 2nd millennium.[10] The two settlements partly overlap chronologically, but it is certain that Aghios Ioannis preceded Skala Sotiros by a few hundreds of years.

Aghios Ioannis yielded 39 spindle whorls from a surface of about 50 square metre+s and Skala Sotiros 169 items from a much broader area.[11] The two assemblages are examined within the frame of a PhD research.[12] The typology and the size range of the whorls were determined for each assemblage and their subsequent comparison revealed interesting similarities and differences, which are discussed in this paper.

Methodology

The typological classification of the spindle whorls is based on the shape of the section of each object. The main typological categories are thus discerned and designated. Further sub-categories and type variations are created with reference to morphological details. There are a few cases when typological classification is difficult either due to poor preservation, or because of poor manufacture quality, which resulted in shape asymmetry. For such whorls an "undiagnosed" category is created. The size can be determined on more accurate criteria such as the metrological data which are assimilated by recording the maximum and minimum diameters, the height, the weight and the maximum and minimum diameters of the central perforation. In case of fragmentary objects, any available dimensions are also measured. In this paper, two categories of metrological data are used, the diameter and the weight, because these are the most crucial functional parameters of spindle whorls. Only intact items are taken into account in this approach, since the original dimensions and mass of the objects are not available in the case of fragmentary pieces.

Practically each object has a unique size, in the sense that the combination of specific diameter, height and weight values is hardly ever repeated in a second case. But given the fact that value

[8] Lespez and Papadopoulos 2008, 675.
[9] Maniatis and Papadopoulos 2011, 21–37.
[10] Koukouli-Chryssanthaki *et al.* (forthcoming)
[11] A surface of a little more than 500 square metres has been excavated at Skala Sotiros from 1986 to 2006. The overall volume of the deposits excavated there is estimated to be larger than the one at Aghios Ioannis as well.
[12] Vakirtzi S., University of Crete

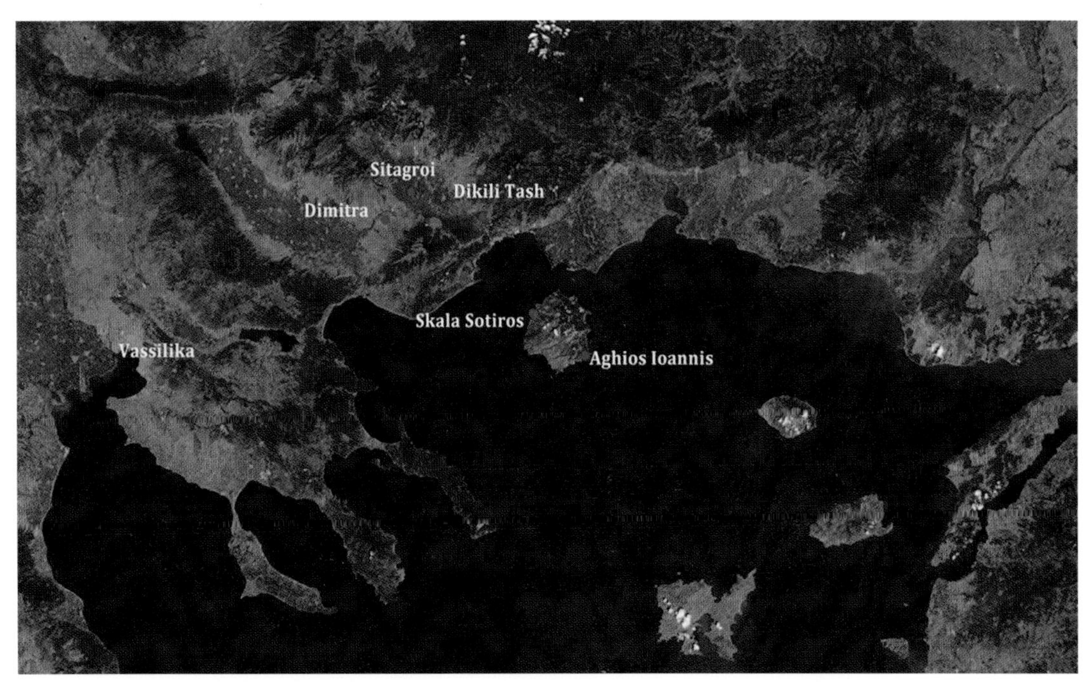

Fig. 2.1: Map of the region with Thassos showing Skala Sotiros on the west and Aghios Ioannis on the east coast.

Fig. 2.2: Aghios Ioannis, aspect of the habitation level.

Fig. 2.3: Skala Sotiros, aspect of the circuit wall.

Table 2.1: Spindle whorl Diameter and Weight scales.

Diameter categories (in cm)	Weight categories (in grs)
0–0.5	0–10
0.5–1	10–20
1–1.5	20–30
1.5–2	30–40
2–2.5	40–50
2.5–3	50–60
3–3.5	60–70
3.5–4	70–80
4–4.5	80–90
4.5–6	90–100
6–6.5	100–110
6.5–7	110–120
7–7.5	120–130
7.5–8	130–140
8–8.5	140–150
8.5–9	150–160

deviations can be minor from one case to the other, spindle whorls of similar sizes can be grouped together in one size class. In order to standardize the size classes of these objects, scales of diameter and weight values were formulated using division units of 0.5cm in the first case and 10g in the second case (Table 2.1).[13] In this way diameter and weight categories were created and the recorded values were grouped accordingly. On this basis it is possible to observe the dominant size classes within each assemblage. It is also possible to compare the represented size classes of each site on a percentage scale. Given the fact that spindle whorl size is relevant to its functional potential, one can ultimately estimate the dominant trends of spinning production in terms of comparatively different raw materials and/or manufactured yarns, on an intra-site and inter-site level.

[13] Experimentation suggests that a difference of 10g maximum in the overall spindle weight can affect the result of spinning in terms of yarn fineness (Andersson and Nosch 2003, 198).

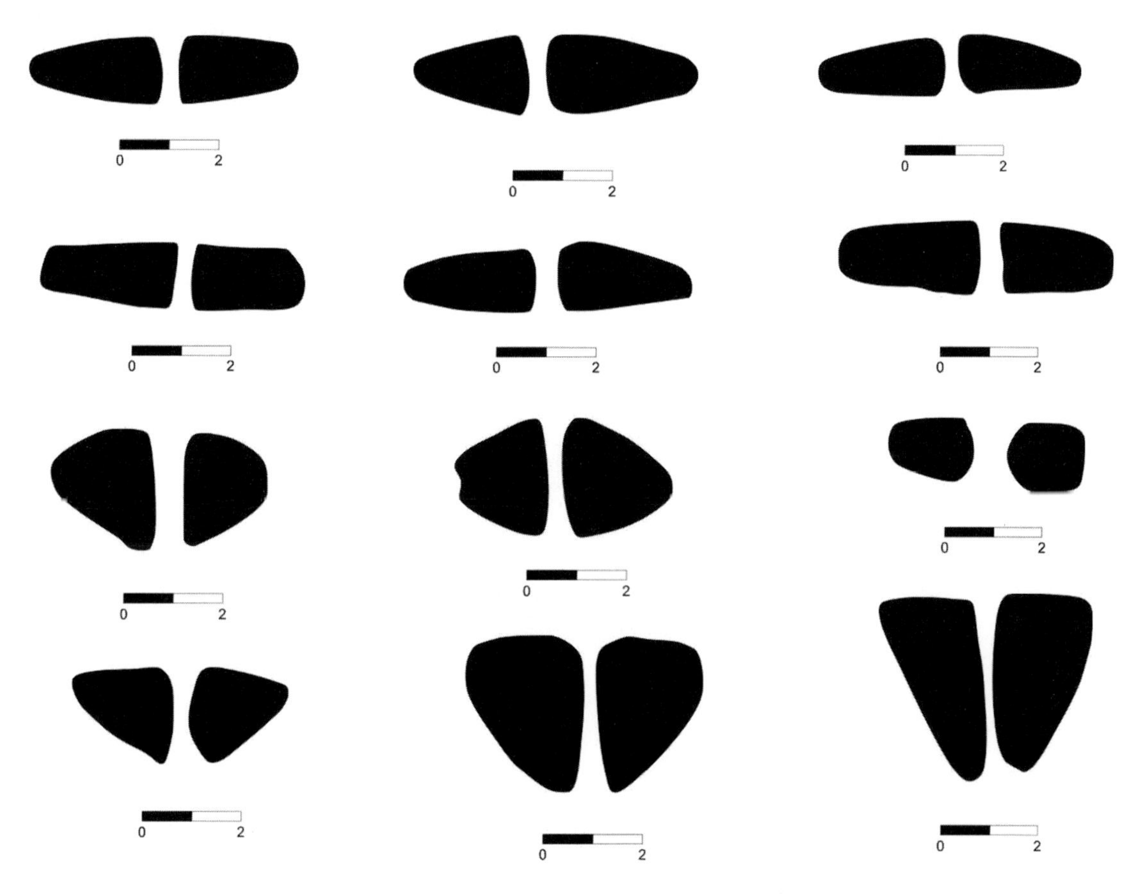

Fig. 2.4: Aghios Ioannis spindle whorl typology.

The archaeological data

a) Aghios Ioannis

The 39 spindle whorls of Aghios Ioannis are made of fired reddish-orange or brownish clay with small inclusions.[14] They are classified into three main typological categories: the discoid, the conical and the biconical. Three pierced sherds and one spool[15] are also included in the assemblage. The discoid type can be further distinguished into varieties according to the shape of the object's section, which can be lentoid, plano-convex or plano-concave. The biconical type includes symmetrical and asymmetrical varieties of bicones, with regard to the point of the carination. The conical type includes low, middle and high cone varieties, with regard to the diameter/height ratio (Fig. 2.4). The discoid is the predominant type. The conical and the biconical types come next in frequency, and they were found in almost equally small numbers.

[14] Clay characterizations which appear in this paper are based on macroscopic observations of S.Vakirtzi.

[15] Spools have various interpretations in the archaeological literature. They are interpreted either as spindle whorls, loomweights or reels. This preliminary paper aims at a first evaluation of the material and for this reason they are included in the presentation, regardless of issues of interpretation which will be addressed elsewhere.

Only 18 of the total 39 spindle whorls were found intact. The most common type, the discoid, is the worst preserved. The metrological data of the intact items was recorded and the main points derived from this analysis are the following:

- The diameters range from 3.5 to 7.5cm.
- The most frequent diameter values occur between 4 and 5cm and between 5.5 and 6cm. In the first cluster pierced sherds, conical and biconical whorls are included, while in the second only discoid whorls are represented.
- The discoid whorls have the largest diameters which do not fall below the limit of 5cm. On the contrary, all the other types are smaller than 5cm in diameter.
- Weight values range between 20 and 80g. Only two items fall below the limit of 20g and they are two pierced sherds which weigh 19 and 19.2g respectively.
- The most frequent weight values are observed between 30 and 60g, and more particularly between 40 and 50g. All typological categories are represented in this cluster.

On the basis of thc abovc data, a distinction between the discoid and the rest of the types emerges: the discoid whorls are more numerous and larger in terms of diameter. The rest of the spindle whorls are fewer and have smaller diameters. This distinction could point to a differentiation in use. However, in terms of weight there is no such distinction among types. Moreover, weight values cluster rather within a narrow value range.

Although the sample of intact spindle whorls from this site is small, it could be argued that yarn production at Aghios Ioannis was relatively homogeneous. It can be deduced that the fibres spun necessitated relatively heavy whorls. A very small degree of deviation from this dominant trend can be observed, by the presence of a few spindle whorls which fall out of the dominant weight categories. However the distinction between the preponderant, large discoid whorls and the rest of the types which are less numerous, must not be underestimated, as it certainly reflects a choice mechanism, triggered either by functional or cultural factors.

b) Skala Sotiros

The 169 spindle whorls of Skala Sotiros are made of fired red or brown clay with inclusions. They are classified into four main typological categories, the biconical, the conical, the spheroid and the cylindrical, and in the less numerous truncated conical, discoid, spool type and pierced sherd categories. Four objects from this assemblage are typologically undiagnosed. Each one of the four main types is distinguished into low, middle or high varieties according to the diameter/height ratio. The biconical type can present particular morphological details such as a hollow cavity around the hole on one end of the whorl, or a slightly elevated rim around the central perforation, resulting in a "collar-like" profile. The biconical type can also include the symmetrical/asymmetrical varieties depending on the point of the carination. The degree of curvature of both the carination and the sides can also result into further variations. Characteristic examples of the spindle whorl typology at Skala Sotiros are shown in Fig. 2.5. The predominant type in this assemblage is the biconical. Conical and spheroid spindle whorls come next in frequency and the cylindrical is represented by only a few items.

The intact spindle whorls of the assemblage are 77. All types except for the conical are preserved at a 50% of their original mass or more. From the metrological recording of their diameters and weights emerge the following points:

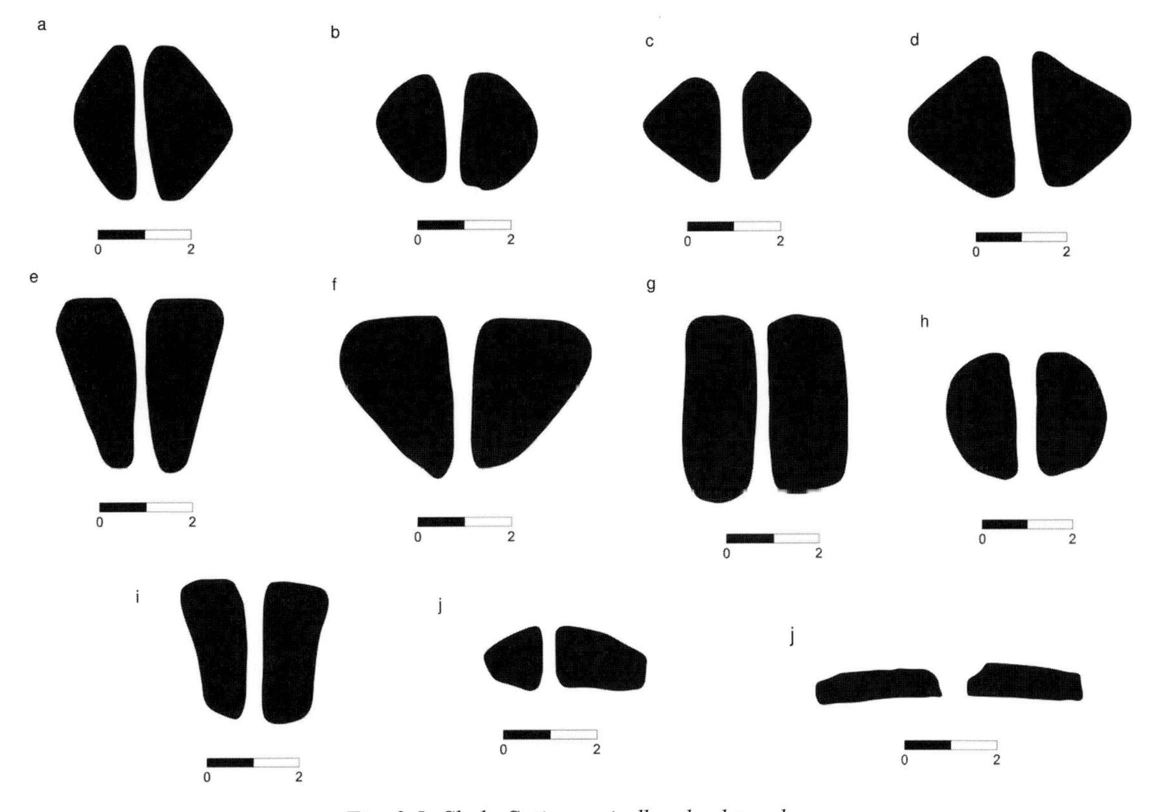

Fig. 2.5: Skala Sotiros spindle whorl typology.

- The diameters range from 1.5 to 6.5cm. Within this range the most frequent values are observed between 2.5 and 4cm, and particularly between 3 and 3.5cm. The types found within this diameter category are the biconical, the truncated cone and the spheroid.
- The preponderant biconical type is represented in almost all diameter categories. The next most popular types, the spheroid and the conical are contradictory in this sense, because the spheroid has small diameters in general, while the conical has larger ones. The cylindrical type is present in various diameter categories in the middle of the scale.
- The weight range starts from values below 10g and reaches values over 150g. The biconical type is present throughout this range.
- The most frequent weight values are observed between 20 and 30g. This weight category includes mostly biconical spindle whorls. The numbers of whorls weighing more than 60g decrease significantly.
- As in the case of diameter values, the spheroid and the conical types manifest contradictory weight trends, with the spheroid type clustering around smaller weight values (under 40g) and the conical around larger weight values (over 60g). All the remaining types weigh under 80g. Therefore, on the basis of the intact spindle whorls, it can be deduced that only the biconical and the conical types were being manufactured in the biggest possible sizes.

In conclusion, the above analysis points to a high degree of diversification of whorl types and sizes. The picture that emerges implies that yarn production focused mainly on products achievable with spindle whorls belonging to the dominant category, i.e. biconicals weighing between 20 and 30g and with diameters measuring between 3 and 3.5cm. A much smaller production of finer or coarser qualities than the basic one can be deduced from diameter and weight categories with a minor percentage of representation. The factor of differentiation could be the raw materials, the desired thread quality, or both. However at this point of research it is not possible to go any further into interpretation. Skala Sotiros appears to be a complex settlement both in a synchronic and in a diachronic aspect, and the attribution of all the spindle whorls into their proper, detailed spatial and stratigraphical contexts, an endeavor still in progress, should help us reach a more refined estimation of the yarn production at the site. Despite this, the first results provide an important frame both for the evaluation of spinning activities at the settlement, and for a broad comparison with the Aghios Ioannis material.

c) Comparison of the spindle whorl assemblages of Aghios Ioannis and Skala Sotiros

The comparison of the two assemblages in terms of typological composition and size diversity shows that both similarities and differences occur between them. The similarities can account for some sort of "common denominator" in their spinning equipment. The differences, on the other hand, show that for some reason spinning at these two settlements differed in certain aspects. The comparison of types, diameter and weight values, is pictured in Charts 2.1, 2.2 and 2.3 and is commented below.

The comparison of the typology shows that the biconical, the conical, the discoid, the pierced sherd and the spool were common at both sites. The cylindrical, the spheroid and the truncated cone are present only at Skala Sotiros. Even within the common types, however, significant differences occur, in terms of percentage representation. The popular discoid whorl of Aghios Ioannis is almost non existent at Skala Sotiros. The biconical, being a minority in the Aghios Ioannis typological

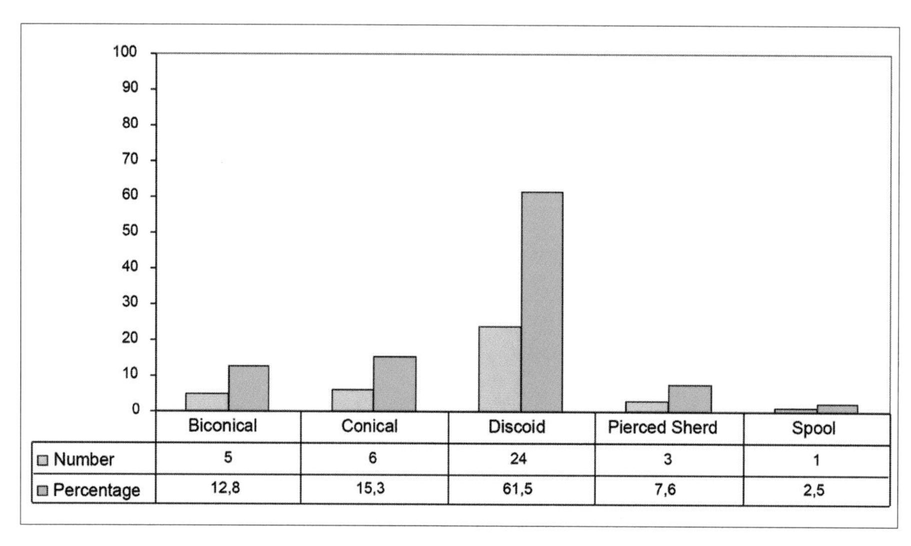

	Biconical	Conical	Discoid	Pierced Sherd	Spool
Number	5	6	24	3	1
Percentage	12,8	15,3	61,5	7,6	2,5

Chart 2.1: Aghios Ioannis – Typological categories.

N
Spool
Pierced Sherds
Biconical
Conical
Discoid
3,5 to 4
4 to 4,5
4,5 to 5
5 to 5,5
5,5 to 6
6 to 6,5
6,5 to 7
7 to 7,5
Diameter values (in cm)

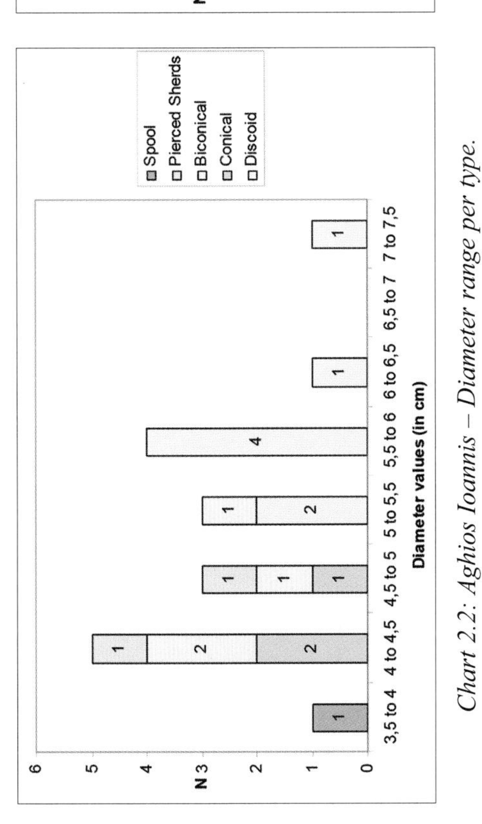

Chart 2.2: Aghios Ioannis – Diameter range per type.

N
Spool
Pierced Sherds
Biconical
Conical
Discoid
10 to 20
20 to 30
30 to 40
40 to 50
50 to 60
60 to 70
70 to 80
Weight values (in grs)

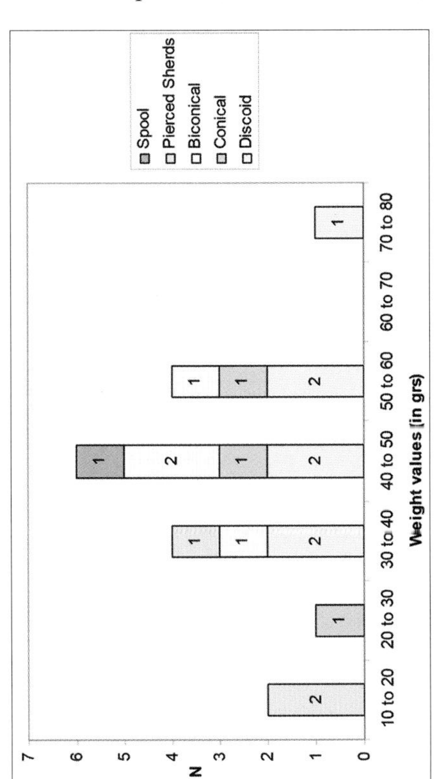

Chart 2.3: Aghios Ioannis – Weight range per type.

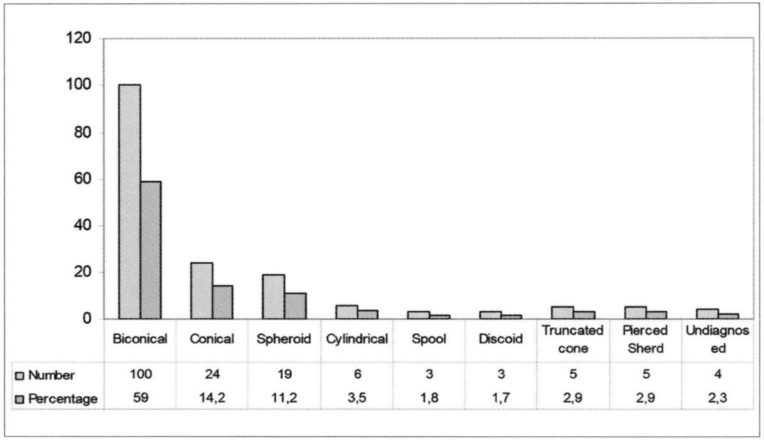

Biconical
Conical
Spheroid
Cylindrical
Spool
Discoid
Truncated cone
Pierced Sherd
Undiagnosed
Number 100 24 19 6 3 3 5 5 4
Percentage 59 14,2 11,2 3,5 1,8 1,7 2,9 2,9 2,3

Chart 2.4: Skala Sotiros – Typological categories and undiagnosed items.

N
0,5 to 1
1 to 1,5
1,5 to 2
2 to 2,5
2,5 to 3
3 to 3,5
3,5 to 4
4 to 4,5
4,5 to 5
5 to 5,5
5,5 to 6
6 to 6,5
Biconical 1 1 8 16 13 4 7 4 1
Conical 1 2
Cylindrical 2 1 1 1
Spheroid 2 2 3
Spool 3
Truncated cone 3 1

Chart 2.5: Skcla Sotiros – Diameter range per type.

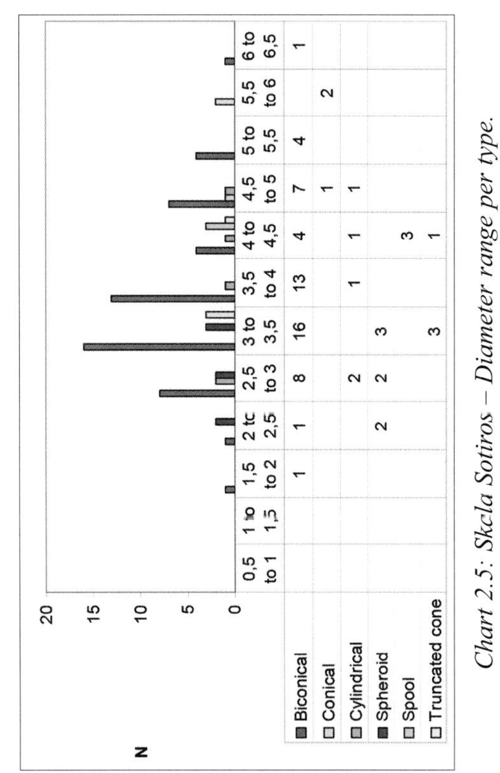

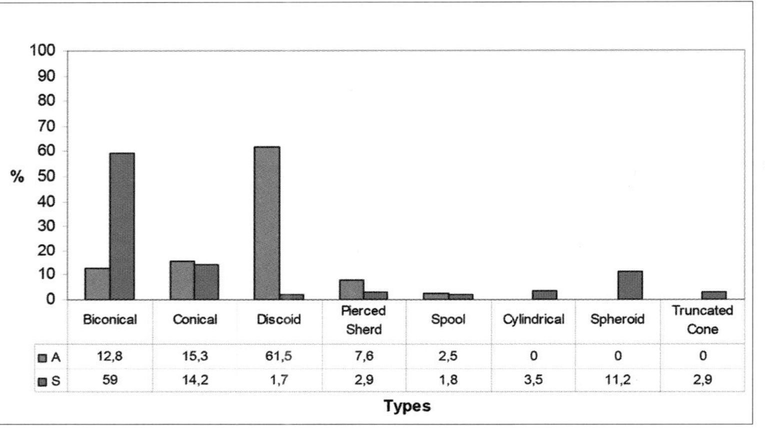

Chart 2.6: Skala Sotiros – Weight range per type.

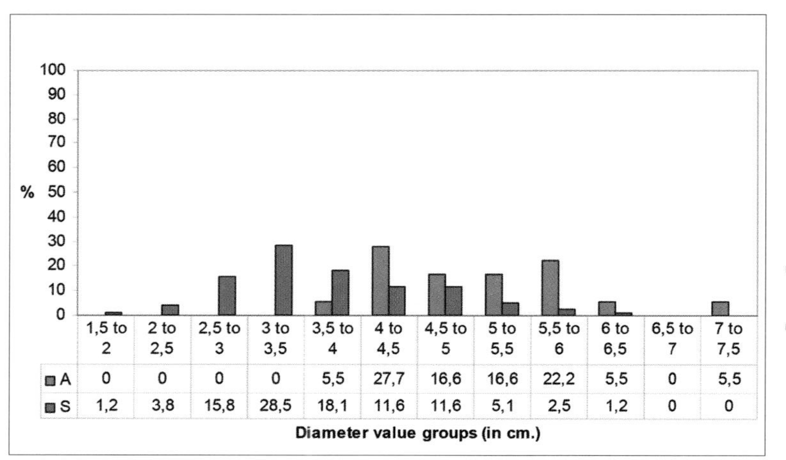

Chart 2.7: Typological categories percentages at Aghios Ioannis and Skala Sotiros.

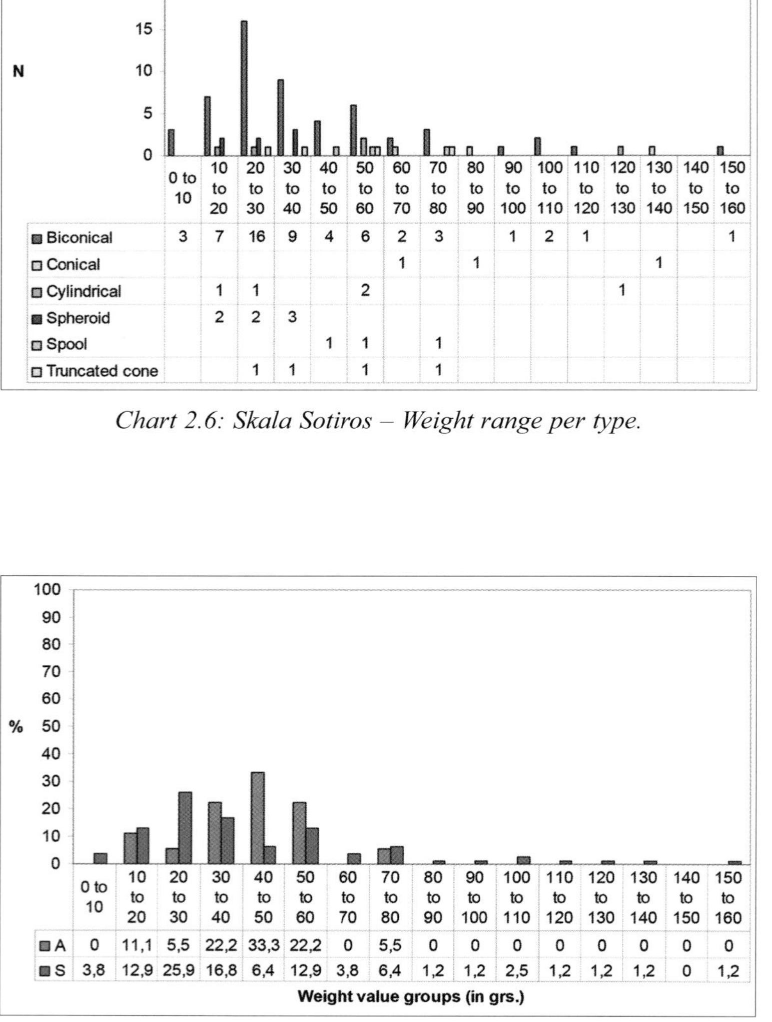

Chart 2.8: Weight group percentages at Aghios Ioannis and Skala Sotiros.

Chart 2.9: Diameter group percentages at Aghios Ioannis and Skala Sotiros.

repertoire, becomes the dominant type at Skala Sotiros. Interestingly enough, the conical type has similar percentages of representation in the two assemblages. The pierced sherd appears to have a larger representation at Aghios Ioannis (Chart 2.1).

In terms of diameter dimensions, two observations are striking: a) the much wider range of diameter size at Skala Sotiros and b) the comparatively bigger diameters of the Aghios Ioannis assemblage. The "popular" large sizes of Aghios Ioannis are decreased on average at Skala Sotiros. On the other hand, no need for spindle whorls with diameters as small as the minimum Skala Sotiros category where needed at Aghios Ioannis (Chart 2.2).

The wider size range of Skala Sotiros is confirmed in terms of weight, too. Although spindle whorls with larger diameters are observed at Aghios Ioannis, the heaviest examples, although in small numbers, come from Skala Sotiros. Whorls weighing more than 80g are not recorded from Aghios Ioannis. But within the common area of their weight range (10–80g) this impression is changed. The Aghios Ioannis spindle whorls have higher percentages of representation in the heavier categories, and lower in the lighter categories. It is interesting, however, that both assemblages manifest almost equal percentages in the weight category of 70–80g. The most striking contrast between the two assemblages occurs in the weight categories where each assemblage has its highest representation, i.e. the weight categories 20–30g in Skala Sotiros, and 40–50g in Aghios Ioannis (Chart 2.3).

The diversified representation of the various weight categories shows that despite the fact that these communities possessed whorls of similar size ranges, the focus of the production at each settlement was different. What seems to be the "main line of production" at one site is a minor operation at the other site. The basic production at Skala Sotiros is clearly achieved with lighter whorls than at Aghios Ioannis. But for some reason, very heavy objects were also necessary at the former site, although in very small quantities.

Discussion

On the basis of the analysis of intact spindle whorls, it appears that spinning at Aghios Ioannis was concentrated on a production which either relied on harder fibres or aimed to coarser yarns than at Skala Sotiros, the latter being characterized moreover by a less homogeneous production: a larger variety of either fibre or yarn qualities and thus fabric qualities must be assumed on the basis of the wider whorl size range at Skala Sotiros. The few, very heavy spindle whorls attested there could be attributed to some special production, related to either plying, or to spinning very hard fibres or ropes. Although both plant and animal fibres can be rendered more or less fine, depending on their processing before spinning, it is generally accepted that plant fibres such as flax are harder to spin than wool, and that plant fibres necessitate heavier spindle whorls.[16] If the difference of spinning equipment between the two settlements is to be interpreted on the basis of raw material, it could be argued that plant fibres were the dominant material at Aghios Ioannis while wool was more popular at Skala Sotiros. Alternatively, it could be argued that plant and animal fibres were being exploited at both sites, but finer varieties of these materials were available at Skala Sotiros. On the basis of spinning equipment alone, there is no evidence to support one argument more than the other. But an examination of the wider chronological context of the two assemblages could provide more criteria of interpretation.

[16] Barber 1991, 52.

According to radiocarbon dating, Aghios Ioannis was founded towards the second half of the 4th millennium BC.[17] The ceramic assemblage of the site is characterized by elements of the transitional Final Neolithic/Early Bronze Age period.[18] Aghios Ioannis therefore predates Skala Sotiros. The settlement of Aghios Ioannis therefore seems to belong to the transitional period which is characterized by both remnants of the Neolithic tradition and advances towards the Early Bronze Age. It should not be considered impossible that the spinning equipment of the site belongs rather to the Final Neolithic tradition. Comparison with the assemblage from Skala Sotiros, well fixed into the Early Bronze Age, certainly provides arguments of differentiation. But why such a chronological and cultural distance should matter as far as spinning is concerned?

The transition from Neolithic to Early Bronze Age economies was affected, among other factors, by the outcome of the long and slow process of animal domestication. It has been argued that in this process, the exploitation of the domesticated species gradually expanded from single-target to multiple benefits. The hypothesis of such a transition includes a set of innovations which altogether consist of "a critical phase of change" in human economic practices, also designated as "the Secondary Products Revolution",[19] the exploitation of wool for yarn and textiles being among these innovations. According to this model, a popular shift from plant fibres to wool fibres for textile production must have occurred two or three millennia after the beginning of the Neolithic cultural phase.[20] Albeit geographically distant, the near eastern region provides evidence in support of this model, as recently accumulated research on textiles and fibre exploitation shows.[21] Moreover, archaeozoological studies support the hypothesis that the development of woolly fleeces must have been achieved around the end of the Neolithic and the beginning of the Early Bronze Age.[22] The settlement at Aghios Ioannis falls within this economically transitional period.

The metrological data presented from Thassos Island seem to match this interpretative scheme. If indeed a shift from plant fibres to wool fibres occurred in the beginning of the Early Bronze Age, it would have left its "mark" on the spinning equipment. Such a "mark" would be the decrease of the average spindle whorl size from one period to the other, and this was indeed demonstrated in the case of the two settlements of Thassos, presented in the above analysis: the introduction and the popularization of smaller whorl sizes in the spinners' toolkit at Skala Sotiros may be due to the availability of softer fibres.

Fibre craft studies and archaeozoological research can have a mutually beneficiary interaction, as has been suggested elsewhere.[23] Therefore, in the case of Thassos, too, the archaeozoological analyses from both sites could further contribute in the interpretation of the spinning tools. At Skala Sotiros animal husbandry was proven to have an orientation towards mixed economy, therefore the exploitation of wool for yarn is a reasonable suggestion.[24] The presence of ovicaprids has

---

[17] Papadopoulos and Maniatis (forthcoming)
[18] Papadopoulos 2007, 323.
[19] Sherratt 1983, 90–104.
[20] Sherratt 1983, 93.
[21] Frangipane *et al.* 2009, 16–17. Sudo 2010, 190.
[22] Bokonyi 1986, 79–80. For a recent criticism of the Secondary Products Revolution model see Halstead and Isaakidou 2011, who nevertheless admit that "wool mortality in sheep [...] is perhaps compatible with the SPR model" (Halstead and Isaakidou 2011, 67–68).
[23] Frangipane *et al.*, 2009, 27. See also Andersson-Strand, Frei, Gleba, Mannering, Nosch and Skals 2010, 154–155.
[24] Yannouli 1994, 332.

been testified at Aghios Ioannis as well, but the archaeozoological analysis is still in progress.[25] Therefore, aspects of this analysis which are crucial for the investigation of fibre crafts, such as the age composition of sheep and goats, are not published yet and the archaeozoological material of Aghios Ioannis cannot contribute in this discussion yet.

The interpretation suggested in this paper remains an open working hypothesis, eligible for future testing against more artifactual and archaeozoological data. But for the moment, the analysis of the spindle whorl assemblages from these two settlements seems to support, on a local scale, a shift from plant to animal fibres in the advent of the Early Bronze Age. More importantly, however, it stresses the potential of interdisciplinary collaboration between archaeozoological research and textile tool analysis in an effort to address the question of fibre domestication and exploitation in precise scales of time and space, and in particular geographical and temporal frames.

Acknowledgement

This research has been co-financed by the European Union (European Social Fund – ESF) and Greek national funds through the Operational Program "Education and Lifelong Learning" of the National Strategic Reference Framework (NSRF) – Research Funding Program: Heracleitus II. Investing in knowledge society through the European Social Fund.

Bibliography

Andersson Strand, E. 2010 The Basics of Textile Tools and Textile Terminology: From fibre to fabric. In C. Michel and M.-L. Nosch (eds) *Textile Terminologies in the Ancient Near East and Mediterranean from the Third to the First Millennia BC*, Ancient Textile Series 8. Oxford, 10–23.

Andersson, E. and Nosch, M.-L. 2003 With a Little Help from my Friends: Investigating Mycenaean Textiles with Help from Scandinavian Experimental Archaeology. In K. P. Foster and R. Laffineur (eds), Aegaeum 24, *Metron: Measuring the Aegean Bronze Age. Proceedings of the 9th International Aegean Conference / 9e Rencontre égéenne internationale, New Haven, Yale University, 18–21 April 2002*. Liège, 197–205.

Andersson Strand, E., Frei, K. M., Mannering, U., Nosch, M.-L. and Skals, I. 2010 Old Textiles – New Possibilities. *Journal of European Archaeology*, 13, 149–173.

Andersson, E., Felluca, E., Nosch, M.-L. and Peyronel, L. 2010 (b) New Perspectives on Bronze Age Textile Production in the Eastern Mediterranean. The first results with Ebla as a Pilot Study. In P. Matthiae, F. Pinnock, L. Nigro and N. Marchetti (eds) *Proceedings of the 6th International Congress on the Archaeology of the Ancient Near East, May 5th–10th 2008*, Vol. 1 Near Eastern Archaeology in the Past, Present and Future. Wiesbaden, 159–176.

Barber, E. 1991 *Prehistoric Textiles*. Princeton.

Bokonyi, S. 1986 Faunal Remains. In C. Renfrew, M. Gimbutas and E. Ernestine (eds), *Excavations at Sitagroi. A Prehistoric Village in Northeast Greece*. Los Angeles, 63–96.

Frangipane, M., Andersson-Strand, E., Laurito, R., Möller-Wiering, S., Nosch, M.-L., Rast-Eicher, A. and Wisti Lassen, A. 2009 Arslantepe, Malatya (Turkey): Textiles, Tools and Imprints of Fabrics from the 4th to the 2nd Millennium BCE. *Paléorient* 35.1, 5–29.

Halstead, P. and Isaakidou, V. 2011 Revolutionary Secondary Products: the Development and Significance of Milking, Animal-Traction and Wool Gathering in Later Prehistoric Europe and Near East. In T. Wilkinson, S. Sherratt and J. Bennet (eds), *Interweaving Worlds. Systemic Interactions in Eurasia, 7th to the 1st Millenia BC*. Oxford, 61–76.

[25] Personal communication with Eleni Psathi who has undertaken the study of the faunal remains from Aghios Ioannis.

Koukouli-Chryssanthaki, C., Malamidou, D. Papadopoulos, S. and Maniatis, Y. (forthcoming), Οι νεότερες φάσεις της Πρώιμης Εποχής του Χαλκού στη Θάσο. Νέα δεδομένα. In C. Doumas, A. Giannikouri and O. Kouka (eds), *The Aegean Early Bronze Age: New Evidence. Proceedings of the International Conference, Athens, April 11th–14th*. Athens.

Lespez, L. and Papadopoulos, S. 2008 Étude géoarchéologique du site d'Aghios Ioannis, à Thassos. *Bulletin de Correspondance Hellénique* 132, 667–692.

Maniatis, Y. and Papadopoulos, S. 2011 ^{14}C Dating of a Final Neolithic – Early Bronze Age Transition Period Settlement at Aghios Ioannis on Thassos (North Aegean). *Radiocarbon*, Vol. 53, 21–37.

Papadopoulos, S. 2007 Decline of the Painted Pottery in Eastern Macedonia and North Aegean. In H. Todorova, M. Stefanovich and G. Ivanov (eds), *The Struma / Strymon River Valley in Prehistory*. Sofia, 317–328.

Papadopoulos, S. and Maniatis, Y. (forthcoming) Η πρώιμη Εποχή του Χαλκού στη Θάσο. Οι αρχαιότερες φάσεις. In C. Doumas, A. Giannikouri and O. Kouka (eds), *The Aegean Early Bronze Age: New Evidence. Proceedings of the International Conference, Athens, April 11th–14th*. Athens.

Sherratt, A. 1983 The Secondary Exploitation of Animals in the Old World. *World Archaeology*, Vol. 15, No.1, *Transhumance and Pastoralism*, 90–104.

Sudo, H. 2010 The Development of Wool Exploitation in Ubaid-Period Settlements of North Mesopotamia. In R. Carter and G. Philip (eds), *Beyond the Ubaid. Transformation and Integration in the Late Prehistoric Societies of the Middle East*, Studies in Ancient Oriental Civilization, 63. Chicago, 169–179.

Tzachili, I. 1997 *Υφαντική και Υφάντρες στο προϊστορικό Αιγαίο 2000 – 1000 π.Χ.* Ηράκλειο.

Yannouli, E. 1994 *Aspects of Animal Use in Prehistoric Macedonia, Northern Greece*, Unpublished doctoral dissertation, University of Cambridge.

3. Textile Texts of the Lagaš II Period

Richard Firth

The Lagaš II period largely precedes Ur III but, in terms of modern interest, it has been rather overshadowed by this later more prolific period. However, there are a substantial number of cuneiform clay tablets from Girsu, dating to Lagaš II (*c.* 2200–2100 BC), that are directly concerned with textiles. These show that many of the textiles and textile processes of Ur III were already known during Lagaš II.

The aim of this paper is three-fold. Firstly, consideration will be given to the administration of the textile industry. This will be done by focussing on a group of 19 textile tablets from the same year and probably from the same two month period, during the 16th year of Gudea, the ruler (ensí) of Lagaš.[1] Secondly, the paper considers the terms used to describe textile quality during the Lagaš II period, as these are rather more complex than the Ur III terms after they were standardised during the reign of Šulgi. Thirdly, the paper considers some of the textiles that were used during the Lagaš II period.

Administration of the Textile Industry

There are over a hundred tablets, dating from the Lagaš II period, that are directly concerned with textiles.[2] By closer analysis, it is also possible to associate some of the other tablets from this period with the administration of the textile industry.

Many of the Ur III tablets were excavated illegally and sold on the antiquities market. It is fortunate that most of the Lagaš II tablets from Girsu were excavated, although, because of the early date of the excavation, details of find-places are extremely limited. Nevertheless, it is possible to gain some insight by an analysis of museum inventory numbers. Amongst the Lagaš II tablets there are examples of clusters of tablets from the same year. One particularly clear example shows 19 tablets bearing the year name, mu mi-ì-tum sag-ninnu ba-dím-ma (or abbreviated forms of this), which corresponds to the 16th year of Gudea's reign. Further investigation shows that 8 of these 19

[1] It is convenient here to use the attribution of year names given by Sigrist and Damerow (http://cdli.ucla.edu/tools/yearnames/yn_index.html), although it is recognised that these are subject to some uncertainty for the Lagaš II period.
[2] The abbreviations used for identifying texts here are based on those used by CDLI, cdli.ox.ac.uk/wiki/abbreviations_for_assyriology (downloaded October 2014).

Table 3.1

Museum no.	Publication	Year	Month	Contents
L7521	MVN 6 492		5	Rations
L7522	MVN 6 493	Gudea 16	5	Textiles by weight
L7526	MVN 6 497			Wool tablet
L7527	MVN 6 498	Gudea 16	4	Textiles by weight
L7529	MVN 6 500			List of slaves by name
L7531	MVN 6 502	Gudea 16	4	Textiles
L7533	MVN 6 504	Gudea 16	4	Textiles
L7534	MVN 6 505			List of men
L7535	MVN 6 506			List of workers
L7537	MVN 6 508			Offerings
L7540	MVN 6 511	Gudea 16	5	Textiles by weight
L7542	MVN 6 513			List of workers
L7545	MVN 6 516	Gudea 16		Textiles
L7549	MVN 6 520	Gudea 16 (†)		Textiles for fulling
L7551	MVN 6 522	Gudea 16	4	Textiles by weight
L7557	MVN 6 528			Offerings

(†) The reading of the year name for MVN 6, 520 is [ ] sag-ninnu [ba-d]ím-ma. In view of the above discussion, it is almost certain that the reading should be modified to [mu mi-ì-tum] sag-ninnu [ba-d]ím-ma, i.e. Gudea year 16, in order to fit in with the remainder of the tablets in this group.

tablets appear within a restricted range of the Istanbul museum inventory numbering, L7521–L7560 (ITT 4, 7521–7560). The items in this part of the inventory list show a mixture of approximately equal numbers of tablets from Girsu during the Lagaš II and Ur III periods.

In practical terms, the sheer volume of tablets being excavated at Girsu would have forced an orderliness on the way the tablets were processed and stored on the excavation site and then transported to the museum. Nevertheless, it is important to ask whether this arrangement of the tablets reflects the way the tablets were found or whether it was imposed by archaeologists at some later stage. In this case, the fact that the sequence of Lagaš II tablets was thoroughly interspersed with Ur III tablets can be taken as evidence that this grouping of Lagaš II tablets was not imposed by archaeologists re-arranging the tablets after they had been excavated.[3] Thus, the most likely explanation for this mixture is that two (or more) parts of the Girsu tablet archive were being excavated around the same time and found their way together, first into the excavation storage trays and later into the museum inventory lists.

Thus, there are 19 Lagaš II tablets within the range L7521–L7560 and, most importantly for this paper, these show a clear emphasis towards textile records. Furthermore, all the dates noted in the tablets in the range L7521–L7557 are not only from the 16th year of Gudea's reign but, more specifically, they are from the two months (ezem) šu-numun and ezem $munu_4$-gu_7 which correspond to the 4th and 5th months (see Table 3.1).

[3] Firth (1998, 2002) describes the successive re-arrangements of the Linear B tablets from Knossos by Sir Arthur Evans and an analysis of the Iraklion museum numbers of these tablets.

Table 3.2

Museum no.	Publication	Year	Month	Contents
L7985	MVN 7 382	Gudea 16	5	Textiles by weight
L8042	MVN 7 435	Gudea 16	5	Textiles by weight
L8047	MVN 7 440	Gudea 16	4	Textiles by weight
AO 3319	RTC 198	Gudea 16	4	Textiles for fulling

Table 3.2 lists four additional textile tablets from the same two month period as those listed above that probably formed part of the same archive.

The latter tablet in Table 3.2 is in the Louvre Museum because the excavated tablets from Girsu were divided between the museums in Istanbul and Paris. The author of RTC, Thurcau-Dangin, was particularly interested in year-names and it seems highly likely that he 'cherry-picked' tablets for the Louvre collection that gave the widest range of year names.[4]

The tablets listed in Tables 3.1 and 3.2 will form the basis for the discussion that follows. However, it will be necessary to supplement these by introducing other Lagaš II tablets to represent features of textile administration which would otherwise not be sufficiently represented. One particular example of this is in the collection, storage and distribution of wool and flax to the weavers, which is considered next.

Collection, storage and distribution of wool and flax

AOAT 25, p. 80, 6 records *c.* 33 tonnes of wool in store. MVN 6 497 and RTC 185 are not well preserved but appear to be recording transactions involving wool.

Tablets such as RTC 182 and 183 each record the distribution of about 1 tonne of wool, including wool to lú-^{d}dumu-zi, an overseer of the weavers. MVN 7 393 also records wool that is being issued to lú-^{d}dumu-zi from the é-mí.[5] There are 16 Lagaš II tablets that name lú-^{d}dumu-zi and he features in the discussion that follows.[6]

Table 3.3 lists records of wool being issued by weight for the manufacture of specific textiles. It is not clear whether this is raw wool or wool that has been washed.

There is much less evidence for flax than for wool. However, MVN 8 85 records the distribution of 2254 bundles of flax to šu-na most of which was from the palace adminstrative centre. It is interesting to note that both wool and flax have been shown to come from palace stores, demonstrating that the tablets are recording a textile industry that was a royal enterprise.

Listing of manufactured textiles

There are three types of tablets listing textiles within the group 7521–7577. These probably correspond to:

- Textiles listed according to the leaders of the weaving teams and their overseer
- Textiles listed by weight according to the overseer of the workshop
- Tablets recording the distribution of textiles, noting the overseer of the workshop.

[4] There is further discussion of this in Firth 2013b.

[5] The é-mí in Lagaš is an administrative centre. Strictly, the term means the queen's household but Maekawa (1973–4) argues that it came to mean the "temple" of Ba-ú administered by the queen of the earthly ruler of Lagaš.

[6] MVN 6 358, 363, 377, 493, 504, 520; MVN 7 311, 378, 384, 393, 435; RTC 182, 183, 190, 209, 264.

Table 3.3

Tablet (date)	Textile	Wool issued	Issued to
MVN 6 531 (Gudea 16.01)	1 ^{túg}aktum guz-za lu[gal] 1 ^{túg}šà-gi-da$_5$ ⸢ús⸣ 1 ^{túg}šà-ga-dù ús 10 ^{túg}aktum guz-za-àm [x ^{túg}aktu]m guz-za tur	$26^{2}/_{3}$ ma-na $1^{1}/_{2}$ ma-na 96 ma-na $21^{2}/_{3}$? ma-na	ugula ba-zi-ge […]-HAL?
MVN 7 437	1 ^{túg}níg-lám sag$_{10}$ 7 ^{túg}mu-du$_8$-um 8 ^{túg}aktum guz-za	2 ma-na 2? gín 26 ma-na lá 1 gín 80 ma-na	^{d}nanše-á-dah
MVN 7 568	2? ^{túg}níg-lám [...] 1 ^{túg}níg-lám guz-za 2 ^{túg}aktum ki-ná 24 ^{túg}aktum guz-za 5 ^{túg}mu-du$_8$-um ús	6 ma-na 7 ma-na 63 ma-na 227 ma-na 20 ma-na	[šu]-na

Table 3.4

Tablet (date)	Textile	Weaving team leader	Overseer
MVN 6 502 (Gudea 16.04)	6 ^{túg}mu-du$_8$-um ú[s] 2 ^{túg}gú-la ús 3 ^{túg}gú-è ús 5 ^{túg}aktum guz-za	ú-da	ugula lugal-ezem
	6 ^{túg}mu-du$_8$-um ús 2 ^{túg}gú-lá ús 3 ^{túg}gú-è ús 5 ^{túg}aktum guz-za 1 túg munus	ur-^{d}ba-ú	
	10 gada du	lugal-ì-bí-l[a]	
	[…] ⸢2⸣ ⸢^{gada}⸣sag-gá šà-gu 2 ^{gada}sag-gá gada úr-bala	nigir-di-dè	ugula$^{!}$(MAŠ) ur$_5$-bi-šè
	1 ^{túg}mu-du$_8$-um ús 1 ^{túg}gú-lá ús	ú-da	
	9 gada du 1 gada x dam GIŠ.LUM	lú-[^{d}]nin-gír-su	ug[ula] gù-dé-a
	3 gada [s]ag-gá šà-gu 2 gada [s]ag-gá gada úr-bala	nigir-[d]i-dè	
	1 tú[g m]u-du$_8$-um ús 1 ⸢túg⸣gú-lá ús	ur-^{d}ba-ú	

MVN 6 502 and 516 fall into the first category. Table 3.4 details the contents of MVN 6 502[7]; it will be shown below that some of these textiles also appear on MVN 6 498 and 522, where textiles are listed by weight.

The textiles produced by the weavers are also listed by weight. This could be to check that the weight of textiles produced is comparable to the weight of materials provided and/or a check on the

[7] The overseers, ur$_5$-bi-šè and lugal-ezem also appear on MVN 6 516 together with their weaving team leaders.

Table 3.5

Tablet (date)	Textile	Weight	Overseer
MVN 6 498 (Gudea 16.04)	[^{t}]^{úg}mu-du$_8$-um 4 [^{túg}]gú-la 6 [^{túg}g]ú-è 10 ^{túg}aktum guz-za 1 túg munus 10 gada du	[l]á 2 ma-na 1 ma-na $2^1/_3$? ma-na 1 gú 31 ma-na 1 ma-na 15 gín 16 ma-na 13 gín	[ ]⌜lugal⌝-ezem
MVN 6 522 (Gudea 16.04)	9 gada du 1 ^{gada}dam-šè-lum 6 ^{gada}sag-gá šà-gu 2 ^{gada}sag-gá gada ⌜umbin⌝ bala 1 ^{túg}mu-du$_8$-um 1 ^{túg}gú-lá	14 ma-na $1^2/_3$ ma-na 8 gín $^1/_2$ ma-na 2 gín $^1/_2$ ma-na 5 gín 2 ma-na lá 5 gín 1 ma-na lá 9 gín	ugula gù-dé-a

productivity of the workshop. Within the L7521–7557 series, there are four such tablets listing the weights of textiles and each one records the month and year when the tablet was written. Table 3.2 above lists a further three tablets with the same characteristics.[8] It is interesting to note that the only other known Lagaš II tablet of this form is MVN 7 390 also from Gudea 16, but written in the 7th month (iti UR). The consistent inclusion of the month and year on all of these tablets implies that these tablets record a particularly important step within the administrative process. It is reasonable to assume that the information on these tablets was used to monitor the productivity of the textile overseers and probably formed a basis for payment.

It is evident from these tablets that the workshops of ugula gù-dé-a and lú-^{d}ba-ú specialised in linen textiles. However, there are some examples of woollen textiles produced in the workshop of ugula gù-dé-a and also linen textiles being produced in workshops that otherwise specialise in woollen textiles. Thus, it is clear that there was not a rigid delineation between workshops weaving woollen and linen textiles.[9]

It is interesting to list the contents of MVN 6 498 and 522 in some detail (Table 3.5).[10] It is clear that the identical textiles, which are found on MVN 6 502 for overseers lugal-ezem and gù-dé-a also appear on MVN 6 498 and 522. These tablets all have the same date (Gudea 16, 4th month) and the numbers of each type of textile match almost exactly.[11] From an administrative point of view, it is interesting to note that MVN 6 502 includes the details of the quality of each of the woollen textiles but this information is completely absent from MVN 6 498 and 522. Similarly, the tablets listing the weights of textiles do not identify the weaving teams that produced the textiles. Furthermore, since the textiles of the same type are bundled together for weighing (irrespective of the team that produced them), it is generally not possible to calculate the exact weights of textiles produced by each team.[12]

[8] MVN 6 493, 498, 511, 522; MVN 7 382, 435, 440.

[9] However, it is possible that some of the workers specialised in either working with wool or with linen because there are differences in the techniques involved.

[10] 1 gú = 30kg, 1 ma-na = 500g, 1 gín = 8.33g.

[11] Note that 1 gada x dam GIŠ.LUM on MVN 6 502 can be identified with 1 ^{gada}dam-šè-lum on MVN 6 522. Similarly, 2 ^{gada}sag-gá gada úr-bala on MVN 6 502 can be identified with 2 ^{gada}sag-gá gada ⌜umbin⌝ bala on MVN 6 522. [Note that capitals letters are used in transliterations of a sign if it is not clear how it should be rendered.]

[12] Strictly, it remains possible that such tablets records did exist but have not been preserved.

Table 3.6

Tablet (date)	Textile	Recipient	Workshop
MVN 6 504 (Gudea 16.04)	1 ^{túg}mu-du$_8$-um sag$_{10}$ 1 ^{túg}búr ensí	mu-du$_{10}$-ga a-[KU]-gu-ni	ugula lú-^{d}dumu-zi
	1 ^{túg}|NÍG.SAG.LÁ.SAL| nin 4 ^{túg}aktum guz-za-àm	ur-gá eren-da	ugula al-la
	1 ^{túg}mu-du$_8$-um sag$_{10}$ 2 ^{túg}aktum guz-za 2 ^{túg}gú-l[á sa]g$_{10}$ 2 ^{túg}gú-[...] 2 ^{túg}aktum guz-[za-àm] 3 ^{túg}aktum guz-za 3 ^{túg}aktum guz-za 2 ^{túg}gú-da$_5$ dingir-ra 2 ^{gada}sag-gá šà-gu	ur-gá eren-da mu-du$_{10}$-ga ur-níg ha-al-ka úr-sa$_6$-sa$_6$ lugal-é-gíd-e	ugula šàr-rum
	1 ^{túg}mu-du$_8$-um ús 1 ^{túg}gú-è sag$_{10}$	mu-du$_{10}$-ga ^{geš}gigir-re	ugula gù-dé-a

It seems reasonable to assume that tablets such as MVN 6 502 precede the tablets with details of textile weights because the former include details of the people directly responsible for the teams weaving the textiles, whereas these details are omitted from the latter, where the sole person named as being responsible for the manufacture of the textiles is the overseer.

MVN 6 504 sets down the distribution of the textiles that have been manufactured (see Table 3.6). In this case, the people who have received (i$_3$-dab$_5$) the textiles are named as well as the overseers of the workshops where the textiles were made.

On MVN 6 504, the name of the recipient, mu-du$_{10}$-ga appears separately three times and the names of ur-gá and eren-da both appear twice. Thus, the primary listing of the textiles is under the names of the overseers, rather than the recipients. This is interesting because, from an administrative point of view, it would appear to indicate that the emphasis is being given to ensuring that each of the textiles that has been produced is distributed, rather than simplifying the actual process of distribution.

The above tablet can be compared with the contents of MVN 6 377 (see Table 3.7), where the date has not been preserved, although it is evidently from a similar date because the same group of people are named.

There is also a receipt (MVN 7 63, Gudea 16) for 2 ^{gada}sag-gá-šà-gu received by lugal-é-gíd-e cf. Table 3.7 (see also MVN 7 390).

In Tables 3.6 and 3.7 there are not only textiles from the same workshops but the list of recipients is very similar, i.e. mu-du$_{10}$-ga, ur-gá, ur-níg, eren-da, a-KU-gu-ni, ^{geš}gigir-re, ha-al-ka. This strongly suggests that these are not the 'end-users' of the textiles but part of a distribution chain.[13]

[13] The term gìri is very often used on tablets of Lagaš II and Ur III to denote the man who is a simple intermediary between the man disbursing and the man receiving. However, the term gìri is not used in MVN 6 377 and 504, where the named men are listed as recipients, although it is suggested that they are probably not the 'end-users'. This tends to

Table 3.7

Tablet (date)	Textile	Recipient	Workshop
MVN 6 377	[...] 1 ^{túg}⸢búr⸣ ensí	[...]-sa$_{6}$ a-KU-gu-ni	ugula lú-^{d}dumu-zi
	1 ^{túg}mu-du$_{8}$-um ù-lá sag$_{10}$ 1 ^{túg}mu-du$_{8}$-um sag$_{10}$ 4 ^{túg}aktum guz-z[a]	mu-du$_{10}$-ga eren-da	ugula al-la
	2 ^{túg}gú-è sag$_{10}$ 3 ^{túg}aktum guz-[za] 2 3 2 1 ^{gada}sag-gá šà-gu 1 [1] ^{túg}mu-du$_{8}$-um sag$_{10}$ 1 ^{túg}gú-è sag$_{10}$ 1 ^{túg}mu-du$_{8}$-um ús	eren-da mu-du$_{10}$-ga ur-gá ha-al-ka ur-níg lugal-é-gíd-e lugal AŠ ur-gá ^{geš}gigir-re mu-du$_{10}$-ga	ugula gù-dé-a
	1 ^{túg}aktum guz-za sag$_{10}$ lugal 1 ^{túg}aktum guz-za ús lugal	ha-al-ka	[ugu]la ⸢elam⸣

There is some evidence for the next stage in this distribution chain in the tablet, RTC 197 (Gudea year 16, 2nd month), which describes textiles being sent to officials from ur-níg and ^{geš}gigir-re, amongst others. In addition, MVN 6 388 sets out the receipt by [l]ú-sa$_{6}$-sa$_{6}$ in Nippur of a number of textiles from ur-gá.

Finally, in this section, it is interesting to note MVN 7 378 which records lú-^{d}dumu-zi delivering of a small number of textiles to named individuals.

The fulling/finishing of textiles

There is one tablet within the series L7521-7557 that relates to textiles being sent to a fuller/finisher of textiles (lú azlag$_{2}$),[14] MVN 6 520. Its text is not very well preserved, but it lists textiles being sent to the fuller/finisher who works for the overseer, lú-^{d}dumu-zi. This latter point is interesting because lú-^{d}dumu-zi has already been noted as an overseer of weavers, so this tablet shows that his responsibilities included both the weaving and the fulling/finishing operations.

RTC 198 (Gudea 16, 4th month) also lists textiles being sent to the fuller/finisher. It appears to include the name of the weaving team leader for each textile and also the overseer, although they are not specifically identified as such on this tablet. The overseers include ba-zi-ge who is identified as such (ugula ba-zi-ge) on MVN 6 531.

MVN 14 220 describes the sending of two textiles, one from a fuller and the other from ^{d}šará-ì-sa$_{6}$, to gù-dé-a. Since this concerns textiles, then there is a possibility that this is ugula gù-dé-a, though, in principle, it could be any man named Gudea including the ruler of Lagaš.

MVN 6 282 is a list of payments that includes ^{túg}níg-lám being given to both fullers in the list. MVN 7 43 records the giving of 5 Akkadian baskets of barley to the fuller/finisher, šeš-šeš.

imply that the chain of distribution here is more complex than found on tablets using the term, gìri.

[14] For a discussion on fulling/finishing at Girsu see Firth 2013a.

Waetzoldt (1972, 173–174) notes that fullers could have used barley to brew a beer that was used in the cleansing process, although, in this case, the use of the barley could simply have been for consumption. It is interesting also to note two tablets recording the fulling agents, im-babbar$_2$ (gypsm) and naga (alkali): RTC 221 and 222, where, in the former, the alkali may be associated with the washing of wool.

Rations and offerings

One of the clearest examples of a ration tablet in the Lagaš II textile industry is MVN 6 105. Lines 1–13 set out the rations of barley for the workforce of ugula šu-na (see, for example, MVN 7 440 above). In the 2nd month, the five most senior women each had a ration of 80 litres of barley, the 25 other women each had a ration of 40 litres of barley and the four children each got 20 litres. Thus, the work force included a total of 32 women and 4 children. Three of the women are described as door-keepers or porters (ì-du$_8$, *ePSD*). Three of the women are described as elam.[15] Seven of the women are described as elam kas$_4$, which can be interpeted as elam couriers and these women had two childen. The weavers are not identified as such but, by elimination, this leaves 17 women who were weavers. The size of the monthly barley ration is typical of that found in more general surveys for ancient Mesopotamia.[16]

MVN 6 492 has similar details for the workshop of [ugula] šu-na for the 5th month, except that here the size of the workforce is much larger, so that there are 5 senior women weavers, 112 women weavers, 25 children with half rations, 10 children with quarter rations, 2 porters, 20 women from the highlands with 12 children and 7 elam couriers with 2 children.[17] Although the overall size of the workforce is very different between these two tablets, there are clearly some signs of continuity. For example, there are 5 senior women and 7 elam couriers with 2 children on both tablets. This large change in workforce could be interpreted as showing an expansion (reduction) in the size of a single weaving workshop. However, it is possible that all of these weavers did not work in a single textile workshop but a series of small workshops. Following this line of reasoning, the change in the size of the workforce could have resulted from a difference in the number of workshops being considered.

MVN 6 358 records flour being given to lú-^{d}dumu-zi and MVN 7 311 records wheat being given to his wife, a-ba-ba (dam lú-^{d}dumu-zi). MVN 6 508 and 528 are offerings of barley and wheat flour, butter and dried dates to a number of deities. It is not clear that these are concerned with the textile industry, even though they are Lagaš II tablets within the range L7521–7557.

Lists of workers

MVN 6 529 (Gudea 20) records the payment of silver to a list of men including 83g to ur$_5$-bi-šè, overseer of weavers, whose name was noted above on MVN 6 502 and 516. RTC 211 also records the payment of silver to a list of men and, whilst there is no direct confirmation that they were

[15] Michalowski (2008) suggests that a translation of elam as Elamite is too specific and that it is preferable to interpret it as women from the highlands.

[16] See, for example, Gelb 1965, also Maekawa 1980, 94–96.

[17] See also MVN 6 335, for month 11, which is similar to the above, though not so well preserved. Nevertheless, it appears to be showing a workforce very similar to that found on MVN 6 492. It is interesting to note that the number of small children has changed from 10 to 15. This could be interpreted literally as an increase of 5 small children, however, it is possible that these are 'rounded' numbers and the increase is approximately rather than exactly five.

involved in the textile industry, there are a number of names on the list that have already been seen in the tables above: eren-da, ur-^{d}ba-ú and šu-na.

MVN 6 500 sets out the issuing of slave women (sag-munus) and their children to named individuals. The connection with the textile industry is not made explicit, although the fact that these are female slaves together with this tablet appearing amongst a group of textile tablets must increase that likelihood. On the other hand, it is not clear that the men listed on MVN 6 505 and 513 have any connection with the textile industry.

MVN 6 506 is a list of officials headed by the administrator of the temple of ^{d}nanše. The only one with a clear link to the textile industry in lugal-ti, who is specifically identified as an overseer of weavers (ugula uš-bar).

Textile Quality

Waetzoldt (1972) gives a discussion of textile quality, however some aspects of this were questioned by Carroué (1994). Therefore, the aim of this section is to re-consider textile quality for the Lagaš II period in the light of Carroué's contribution.

The main factor determining textile quality is the quality of the wool or linen used. Quality differentials could then be enhanced by the amount of fulling woven fabrics received.[18]

The terms used to describe textile quality were standardised around the 32nd year of Šulgi's reign.[19] Following this, there were 5 levels of quality:

lugal (or šàr)	royal quality
ús (lugal)	the quality following royal quality
3-kam ús	3rd quality
4-kam ús	4th quality
du (or gin)	normal quality

Occasionally the top quality (lugal) textiles are denoted by the name of the king Šu-Suen or Ibbi-Suen (Pomponio 2010). Alternatively, if the textiles are garments specifically for females, then nin (lady) can be used in place of lugal.[20]

Prior to this standardisation, a different set of qualities were used:

sag_{10} (or sig_5 or saga)	good quality
ús sag_{10}	the quality following good quality
ús	the following quality
du (or gin)	normal quality

In addition, there were the qualities, ensí, lugal and nin to denote the highest quality, together with ús lugal for the following quality.

Table 3.8 lists details of textiles designated as having quality 'ensí'.[21]

[18] Firth 2013a.

[19] The earliest examples of textiles with quality designated as 3-kam ús or 4-kam ús are on SNAT 259 (Š32), BCT 1, 134 (Š33). The latest known use of ensí to denote quality is Š11.

[20] Since Šu-Suen, Ibbi-Suen and nin were used in the same context as lugal (or šàr) then it seems preferable to use lugal rather than šàr.

[21] This table excludes SAT 3, 1402 (ŠS4), 1 gada du ensí, where the quality is designated as "du" and "ensí" is the

Table 3.8

^{túg}gú-da$_5$ ensí	MVN 6 520 (GU16)
^{túg}šà-ga-dù ensí	MVN 6 493 (GU16)
^{túg}níg-lám ensí	MVN 6 327; ITT 5 6713 (Š8), 6810 (Š11); RTC 276 (Š4-Š7)
^{túg}búr ensí	MVN 6 377, 493, 504, 520 (all GU16 except 377)
^{túg}bar-dab ensí	RA 65, 20 8 (Š8)
^{gada}šà-ga-dù ensí	DCS 49 (Š9); ITT 5 6738 (Š10), 6805 (Š9), 6813 (Š10), 6820 (Š7), 6826; MVN 7 29 (Š7); RTC 276 (Š4-Š7)

Table 3.9 (∗)

^{túg}aktum lugal	RTC 276 (Š4-Š7)
$^{\text{[túg]}}$aktum guz-za lugal	RTC 198 (GU16) (†)
^{túg}aktum guz-za sag$_{10}$ lugal	MVN 6 377, 493 (GU16), 531 (GU16), AO 3379
^{túg}aktum guz-za ús lugal	MVN 6 377
^{túg}gú-da$_5$ lugal	MVN 6 520 (GU16), AO 3323 (GU 15)
^{túg}lugal	MVN 7 413
^{gada}ù-lá lugal	RTC 232 (Ur-Namma.I)
^{gada}šà-ga-dù lugal	MVN 7 157
^{túg}|NÍG.SAG.LAL.SAL| nin	MVN 6 504 (GU16), MVN 7 152; RTC 198 (GU16)
[^{túg}a]ktum guz-za [sig$_{10}^{?}$] dumu lugal	AO 3379

(∗) For the AO (Louvre) tablets in this table see Carroué (1994, 59).
(†) Using Carroué's reading (1994, 59).

Table 3.9 demonstrates the use of lugal (or šàr) and nin in association with textiles during the same period that is represented in Table 3.8.

Waetzoldt (1972, 46–49) suggests that in these tablets, the terms lugal and nin designate the quality of the textiles. Carroué (1994, 57–60) rejects this suggestion and argues instead that these are examples of textiles specifically for a king and queen. In doing this he notes specifically the unpublished tablet, AO 3379, which lists textiles for the king's son (dumu lugal).[22] Carroué's discussion is part of a larger study demonstrating that some of the Lagaš II tablets describe gifts for visiting royalty. In the present context, Carroué's interpretation of these tablets is attractive because it removes some of the confusion apparent in Waetzoldt's proposals, by explaining why some Lagaš II textiles are described as lugal despite the ruler being an ensí.[23]

recipient. It also excludes ITT 4 7007 (as listed by Waetzoldt 1972, 49) following the new reading given as MVN 6, 7.

[22] For completeness, in AO 3379, it is noted that it would be possible to read dumu as tur (small). In this case Carroué's reading becomes [^{túg}a]ktum guz-za [sig$_{10}^{?}$] tur lugal, and the strength of his argument rests on the reliability of the reading, [sig$_{10}^{?}$]. Further consideration of this must await the publication of AO 3379. Carroué (1994, 57) also suggests that lugal is not used to denote the quality of textile documents during the early years of Šulgi. However, this omits to note ^{túg}aktum lugal on RTC 276 (Š4-Š7).

[23] However, note that Carroué (1994, 57, 59) lists ^{túg}|NÍG.SAG.LAL.SAL| nin on MVN 6 504 as an example of a textile

Table 3.10

Earlier textile qualities (Lagaš II and early Ur III)	Standardised textile qualities (later Ur III)
ensí, lugal (or šàr), nin	lugal (or šàr), nin
sag_{10} (or sig_5, saga)	ús (lugal or šàr, nin)
ús sag_{10}	3-kam ús
ús	4-kam ús
du (or gin)	du (or gin)

It seems reasonable to presume that textiles designated as lugal were of a high quality. Thus, even though lugal may have been used to denote (visiting) royalty, there would have been an implication of quality. For example, MVN 6 520.8–9 lists "1 ᵗᵘ́ᵍgú-da_5 lugal, 1 ᵗᵘ́ᵍgú-da_5 ensí" and it would seem reasonable to presume that these two ᵗᵘ́ᵍgú-da_5 were of a similarly high quality, although one was for a king and the other for the ensí.

The question arises whether the quality of textiles designated ensí is equivalent to sag_{10} or superior to it. There would seem to be two obvious possibilities. One is to assume that the textile quality implied by ensí is equivalent to sag_{10}. On this basis, during the Lagaš II period, there would have been four levels of textile quality and then, at some later stage, in the early Ur III period, the textile quality grading would have been thoroughly re-configured to give the later, standardised 5 levels of textile quality. However, there is the problem that ensí and sag_{10} can both appear on the same tablet.[24] Therefore, the most satisfactory way to resolve this question is to assume that there were 5 basic quality levels for textiles throughout the Lagaš II and Ur III period which were determined by wool quality. The underlying assumption here is that the change of denoting quality did not arise because of a radical re-appraisal of textile quality but rather because it was recognised that the old system was somewhat confusing. On this basis, it is possible to draw up the equivalences given in Table 3.10.[25]

The textiles of the Lagaš II tablets

The aim of this section is to consider some of the textiles that are included in tablets from the Lagaš II period. For reasons of space, the paper excludes discussion of textiles such as ᵗᵘ́ᵍguz-za, ᵗᵘ́ᵍníg-lam and ᵗᵘ́ᵍbar-dul_5 which are widely found both in the Lagaš II and Ur III periods and considered elsewhere.[26]

It is worthwhile beginning by considering examples of names of textiles that include two textile terms, i.e.

ᵗᵘ́ᵍníg-lám guz-za
ᵗᵘ́ᵍaktum guz-za

for a queen, although this tablet also lists ᵗᵘ́ᵍbúr ensí, so it is possible that nin is from the family of the ensí of Lagaš, rather than visiting royalty. Similarly on MVN 7 152, ᵗᵘ́ᵍ|NÍG.SAG.LAL.SAL| nin appears in a list of textiles, with no evidence of a lugal or ensí, so again nin is not necessarily associated with visiting royalty. For a discussion on the rendering the term as ᵗᵘ́ᵍ|NÍG.SAG.LAL.SAL| see Firth 2012.

[24] MVN 6 327, 377, 493, 504, 520; RTC 276.

[25] It is open to question where Lagaš II textiles labelled as sag_{10} lugal or ús lugal fit into this scheme. However, intuitively, one might presume that a textile designated as sag_{10} lugal was of the same quality as one designated lugal.

[26] See for example Waetzoldt 1972, 1980–1983a, 1980–1983b, Firth and Nosch 2012.

^{túg}bar-dul$_5$ guz-za
^{túg}níg-lám uš-bar
^{túg}šà-ga-dù uš-bar

In a discussion on Ur III textiles from Garšana, Waetzoldt (2010) hypothesised that, in such cases, the first term denotes the piece of clothing and the second term refers to the weave of the fabric. Applying this hypothesis to the Lagaš textiles would imply that guz-za and uš-bar are types of fabric and níg-lám, aktum, bar-dul$_5$ and šà-ga-dù are garments.[27]

^{túg}aktum guz-za

There are 20 examples of texts listing ^{túg}aktum guz-za. Of these, 17 are from a known location and all of these are from Girsu. In addition, all but one of these examples are from the Lagaš II period.

There are a number of examples of weights of ^{túg}aktum guz-za. These show that the weights of higher quality ^{túg}aktum guz-za (i.e. ús, sag$_{10}$, sag$_{10}$ lugal) are *c.* 7.5kg (MVN 6 493), whereas the weights of ordinary (unqualified) ^{túg}aktum guz-za are *c.* 4.5kg (MVN 6 498, MVN 7 440).

There are also details of the weight of wool provided for weaving a ^{túg}aktum guz-za. For ordinary ^{túg}aktum guz-za the weight is slightly larger than the finished product (*c.* 4.8kg).[28] It follows from this that the wool has been washed before it was allocated, because raw wool looses half its weight when it has been washed and combed.[29]

The weight of wool allocated for the lugal quality textile (MVN 6 531) is 13.3kg, which is considerably larger than the typical weight of *c.* 7.5kg. This seems to imply that a substantial proportion of the wool was discarded prior to spinning the thread for the higher quality textile.

Durand (2009, p. 139) interprets ^{túg}aktum as a garment which covers completely, possibly a sort of a cloak.

^{túg}bala

^{túg}bala appears in texts from ED IIIb onwards. In the Lagaš II period, it appears on ITT 5 6674 and MVN 6 108.[30]

^{túg}bar-dul$_5$ guz-za

There are 17 tablets including ^{túg}bar-dul$_5$ guz-za. Two tablets are from Ur, and the remaining 13 tablets with known provenience are from Girsu. Four of the tablets are from the Lagaš II period, and (with the exception of LB 2505 and UET 3, 1671) all those with known dates, are from the 10th year of Šulgi or earlier. For the Lagaš II tablets most of the ^{túg}bar-dul$_5$ guz-za listed are of sag$_{10}$ quality and on one tablet this textile is described as elam. Two of the Ur III tablets include information on weights and in both cases the average weight of a ^{túg}bar-dul$_5$ guz-za is 1kg.

[27] However, note ^{túg}bar-dul$_5$ |NÍG.SAG.LAL.SAL| on Ur III tablet HSS 4, 6, which does not conform with this hypothesis.
[28] MVN 6 531; MVN 7 437, 568.
[29] Firth and Nosch 2012.
[30] This textile is noted by Gelb *et al.* 1991, 294.

^{túg}bar-si (Akk. *paršīgu*, “a sash often used as a headdess”, CAD P 203).

Amongst the Lagaš II tablets, ^{túg}bar-si appears on: MVN 6 12, 59, 388, 493, 520; MVN 7 152, 378.[31] Weights for a ^{túg}bar-si can be calculated from Ur III tablets: ITT 5, 6713 (7 gín on lines 5f. and 2.5 gín on lines 9f.); HLC 68 pl. 43 (5 gín); MVN 5, 292 (9 gín); CUSAS 3, 747 (4 gín); and TMH NF 1–2, 227 (3.5 gín for a ^{túg}bar-si bu-ra). This gives a range of 21–75g with an average weight of 43g.[32]

^{túg}búr

^{túg}búr appears on a total of 16 inscriptions (excluding two lexical tablets[33]). Eight of the inscriptions are from the Lagaš II period, including an inscription on a ‘statue’ (RIME 3/1.1.7, St.L iii.4’). There are two qualities listed: ensí and sag$_{10}$.

It is interesting to note that the only three tablets containing ^{túg}búr from the Ur III period all date to the first ruler of the Ur III period, Ur-Nammu (year i; MVN 7, 459; RTC 232, 270). There are two Old Akkadian tablets (OIP 14 153, 181) that give details of the weight of the textiles, giving an average weight of 146g on OIP 14, 153 and 108g on OIP 14, 181.[34]

^{túg}gú-da$_5$

^{túg}gú-da$_5$ is found on 10 tablets. There are three Lagaš II tablets (MVN 6 504, 520; RTC 197) and two tablets from the early years of Šulgi (ITT 5 6713, Š8; RTC 276, Š7). In addition, there are three Old Akkadian tablets and two tablets from later in the Ur III period (SANTAG 6, 48; UNT 39 rev.I.5). With the exception of SANTAG 6, 48 (Umma) all these tablets are from Girsu (but see below).

There are two indications of weight. On ITT 5 6713 there are four ^{túg}gú-da$_5$ with an average weight 1.5 kg. On ITT 5 9297, 1 ^{túg}na-áš-ba-ru-um plus 4 ^{túg}gú-da$_5$ weigh 24+ ma-na. On the basis of estimates given below a ^{túg}na-áš-ba-ru-um is relatively light, weighing *c.* 330g. This implies that on this tablet, the ^{túg}gú-da$_5$ weighs *c.* 3kg.

Taken literally, ^{túg}gú-da$_5$ is a garment that goes around the neck. If account is taken of the weight, then this might imply that ^{túg}gú-da$_5$ is a cloak. On SANTAG 6, 48, ^{túg}gú-da$_5$-anše is a textile that goes around the neck of a donkey.

^{túg}gú-lá (Akk. *ḫullānu*, a blanket or wrap of (linen or) wool, CAD H, 229)[35]

^{túg}gú-lá (^{túg}gú-la) appears on about 50 tablets and 11 of these are from Girsu in the Lagaš II period. There are also a number of examples from EDIII and Ur III. Of the 48 tablets 37 are from Girsu from a wide range of periods. The general impression is that ^{túg}gú-lá were widely used in Girsu over a long period but, on the available evidence, they only gained some limited usage in other locations during Ur III.

---

[31] According to Waetzoldt (1972 128, note 419) the readings on MVN 6 493.4 and 520.12 should be ^{túg}bar-si-šà-^{kuš}suhúb, i.e. footcloths (cf. the published readings ^{túg}bar-si ⌜x x⌝ [ ] and ^{túg}bar-si šà-KUŠ.DINGIR [ ], respectively).

[32] However, MVN 6 493 appears to suggest that 1 ^{túg}bur2 ensí plus 70 ^{túg}bar-si have a combined weight of *c.* 35 gín. Even if it was suggested that the 70 ^{túg}bar-si alone weighed ~35 gín, then each ^{túg}bar-si would still only weigh 0.5 gín (4.2g). [Note Waetzoldt’s interpretation considered in the previous footnote.]

[33] SF 64, ED IIIa, from Fara; SLT 11, Early Old Babylonian from Nippur.

[34] Note also UET 3, 1682, which includes 20 ^{túg}búr-zi, with an average weight of 192g.

[35] See Waetzoldt 1980–1983a 22.

It is possible to derive an average weight based on MVN 6 522, where a ^{túg}gú-lá weighs 51 gín, and MVN 7 382, where 4 ^{túg}gú-la weigh 3 ma-na 7 gín. This gives an average weight for these 5 textiles of 400g.[36]

^{túg}mu-du$_8$-um (Old Akk., *mudû*, CAD M2, 168).

There are over 30 tablets listing ^{túg}mu-du$_8$-um and the overwhelming majority of these were excavated from Girsu. A large proportion of these tablets are from the Lagaš II period. In addition, there are three Ur III tablets recording this textile and these are all from Girsu; RTC 270 (Ur-Nammu i), ITT 5 6812 (Š11) and MCS 8, 89, BM 100462. The clear implication is that, on the available evidence, ^{túg}mu-du$_8$-um were essentially restricted to Girsu and were primarily found during the Lagaš II period and the early years of Ur III.

A number of the Lagaš II tablets include the weight of ^{túg}mu-du$_8$-um and these vary from 0.9 to 2.6 kg, with an average weight of 1.16 kg (based on 14 textiles). There are also two tablets which include the weight of wool required to make a ^{túg}mu-du$_8$-um and this varies from 1.86 to 2 kg per item.[37] In addition, the Ur III tablet, MCS 8, 89, BM 100462 includes the weight of 7 ^{túg}mu-du$_8$-um, which have an average weight of 0.88kg.

Amongst the Lagaš II tablets, the qualities of the ^{túg}mu-du$_8$-um listed, when specified, are either sag$_{10}$ or ús, with 19 examples of the former and 131 of the latter.

Gelb (1957, MAD 3, 169) makes the tentative suggestion that ^{túg}mu-du$_8$-um, *mudû* could be a head covering (cf. *muttatu*, *muttu*, headband?, CAD M2, 310, 313). However, in view of the weights given above, this seems unlikely.

^{túg}na-áš-ba-ru-um (Old Akk., *našparu*, CAD N2, 77).

In the Lagaš II period, this textile is found on Girsu tablets, MVN 6 108, 343, MVN 7 407. These appear to be amongst the latest known uses of this textile.

There are no weights given for the ^{túg}na-áš-ba-ru-um on these Lagaš II tablets. On Nik 2, 86 (Old Akkadian) 3 linen na-áš-ba-ru-um plus 3 linen šà-ga-dù together weigh 2.3kg. Assuming that an Old Akkadian woollen šà-ga-dù weighs *c.* 450g (see below) and that a woollen and linen šà-ga-dù are approximately the same weight, then it can be estimated that the weight of a linen na-áš-ba-ru-um was approximately 330g.

The textile, *našparu* is clearly related to the word, *našparu*, 'messenger, envoy'. CDA 245 suggests that *našparu* is a garment for a messenger or envoy. An alternative suggestion from Foster (2010) is that *našparu* is a 'sending container' or within a textile context, a 'garment bag'.[38]

---

[36] There is also a weight associated with this textile on MVN 6 498, which appears to suggest that 4 [túg] gú-lá weigh as little as 1 ma-na (500g), however, the text is damaged.

[37] The tablets listing weights of wool are MVN 7 437, 568.

[38] As an example, Foster (2010, 140–141) discusses the Old Akkadian tablet, NBC 1141, which gives a listing, headed by 2 *našparu*, of 120 ^{túg}nig-lám, 120 ^{túg}šà-ga-dù, 120 ^{túg}šà-gi-da$_5$ and 7 ^{túg}uš bar. However, it is evident that 2 'garment bags' weighing roughly 330g would be inadequate for carrying the large numbers of textiles listed.

^{túg}níg-lám guz-za

There are only two examples of tablets listing ^{túg}níg-lám guz-za. One is from Girsu in the Lagaš II period (MVN 7 568) and the other is from Umma during Ur III (SAT 3 2070). The Girsu tablet includes the information that one ^{túg}níg-lám guz-za requires 3.5kg of wool.

^{túg}níg-lám uš-bar

There are 15 examples of tablets including ^{túg}níg-lám uš-bar. Of these, 13 are from Girsu. Ten of these inscriptions are from the Lagaš II period, four from the Old Akkadian period and one from Ur III.[39]

^{túg}|níg.sag.lal.sal|

Headband (see discussion by Firth 2012).

$^{túg/gada}$šà-ga-dù (Akk. *šakattû*, CAD Š1, 158; also ^{gada}šà-ga-dù *nêbahum*, belt or sash[40]).

This textile is found frequently in the texts. In the Lagaš II period, the woollen ^{túg}šà-ga-dù appears on: ITT 5 6853; MVN 6 57, 440, 493, 531; RTC 197, 198. Similarly, the linen ^{gada}šà-ga-dù appears on ITT 5 6826, 6851; MVN 6 314, 327; MVN 7 157.

On MVN 6 493, a single ^{túg}šà-ga-dù of ensí quality weighs 83g. On the Ur III tablet, MVN 5 292, a ^{túg}šà-ga-dù weighs 141.6g. However, three Old Akkadian tablets (OIP 14 145, 146 and 181) imply that ^{túg}šà-ga-dù have an average weight of *c.* 450g.

Durand (2009, 162) interprets ^{gada}šà-ga-dù as *nêbahum* implying that it is a belt or sash. Foster (2010) suggests instead that it is a short-sleeved undershirt or mid-body wrapping.

Concluding remarks

It is worthwhile concluding with a few general remarks about the nature of the administration of the textile industry compared to the following Ur III period.

Waetzoldt (1972) notes that the nature of the administration of textile workshops varies with location in the Ur III period and so it is necessary to use the Lagaš II tablets, rather than try to apply conclusions drawn from other locations and periods. On the basis of the above discussion, at Girsu in the Lagaš II period, it appears that at least one overseer (lú-^{d}dumu-zi) had responsibility for both the weaving and fulling workshops. In addition, in the discussion of MVN 6, 105 and 492 above, there was also an indication that ugula šu-na might have overseen female textile workers in more than one workshop. It follows that if lú-^{d}dumu-zi and šu-na were responsible for more than one workshop then they would have employed people to oversee each of the individual workshops in their absence. These may in part correspond to the people designated above as weaving team

[39] In principle, a ^{túg}níg-lám uš-bar could be a ^{túg}níg-lám for a weaver (uš-bar). However, ^{túg}níg-lám is a common type of fabric and there are many weavers so this interpretation would not explain why there are so few ^{túg}níg-lám uš-bar and why they were largely concentrated on Girsu in the Lagaš II period. Note also ^{túg}šà-ga-dù uš-bar; there are only three tablets including these textiles. All three are from Girsu, with two from the Old Akkadian period and one from Lagaš II.

[40] Durand (2009, 162) equates ^{gada}šà-ga-dù with *nêbahum*, belt or sash (see also CAD Š1 158, N2 143).

leaders (see for example, Table 3.4). In addition, the ration tablets list women who receive double the rations of the other women and these presumably had a role as supervisors. Thus, it is possible to perceive something recognisable as a management structure.

In terms of gender in the textile industry, the situation in the Lagaš II persisted through the Ur III period. The textile industry was usually managed by men and men were also responsible for fulling. The task of spinning and weaving fabrics was undertaken by women, géme, and this word has an implication that the women were slaves, recompensed by monthly rations. This implication is re-enforced by the listing of even small children, who only required quarter rations. Such children would be unlikely to have made a large contribution to the work but nevertheless there was an obligation to supply them with rations, and it seems more likely that this arose from ownership than benevolence.[41]

The aim of this paper has been three-fold. Firstly, it has considered the administration of the textile industry by focussing on a group of 19 textile tablets from the same year and probably from the same two month period, during the 16th year of Gudea's reign. Secondly, the paper has considered the terms used to describe textile quality during the Lagaš II period and, thirdly, the paper has considered some of the textiles that were used during the Lagaš II period.

Acknowledgements

I wish to thank Marie-Louise Nosch for giving the support of the Danish National Research Foundation's Centre for Textile Research to this work and also Cécile Michel for her helpful comments.

Abbreviations

CAD	Chicago Assyrian Dictionary, Oriental Institute of Chicago
CDA	Concise Dictionary of Akkadian, Wiesbaden
ePSD	http://psd.museum.upenn.edu/epsd/index.html, Pennsylvania Sumerian Dictionary Project (downloaded August 2012)
RIME 3/1	Edzard, D. O. 1997 *Gudea and his Dynasty*, *RIME* Vol. 3/1, University of Toronto Press

[41] It is interesting to note, by contrast, that rope and cord making was a male task (túg-du$_8$). At Ur, during Ur III, these workers were organised along with other craftsmen, metalworkers, goldsmiths, stone-cutters, carpenters, blacksmiths, leatherworkers, and reedworkers (Van de Mieroop 1987, xiii).

Bibliography

Carroué, F. 1994 La situation chronologique de Lagaš II: un élément du dossier, *Acta Sumerologica* 16, 47–75.

Durand, J.-M. 2009 *La Nomenclature des Habits et des Textiles dans les textes de Mari: Matériaux pour le Dictionnaire de Babylonien de Paris Tome 1*. Paris.

Firth, R. J. 1998 The Find-Places of the Tablets from the Palace of Knossos, *Minos* 31–32, 7–122.

Firth, R. J. 2002 A Review of the Find-Places of the Linear B tablets from the Palace of Knossos, *Minos* 35–36, 63–290.

Firth, R. J. 2012 ^{túg}|NÍG.SAG.LAL.SAL|: A Headband written using a logogram, *Archaeological Textiles Review* 2012, 57–61.

Firth, R. J. 2013a Considering the Finishing of Textiles based on Neo-Sumerian Inscriptions from Girsu. In M.-L. Nosch, H. Koefoed and E. Andersson Strand (eds), *Textile Production and Consumption in the Ancient Near East: Archaeology, Epigraphy, Iconography*, Ancient Textiles Series 12. Oxford, 140–160.

Firth, R. J. 2013b Notes on the Year Names for Lagaš II and the Early Years of Ur III, forthcoming.

Firth, R. J. and Nosch, M.-L. 2012 Spinning and Weaving Wool in the Ur III Administrative Texts, *Journal of Cuneiform Studies* 64, 65–82.

Foster, B. R. Clothing in Sargonic Mesopotamia: Visual and Written Evidence. In C. Michel and M.-L. Nosch (eds) *Textile Terminologies in the Ancient Near East and Mediterranean from the Third to the First Millennia BC*, Ancient Textiles Series 8. Oxford, 110–145.

Frayne, D. R. 1997 Ur III Period (2112–2004), *RIME* vol. 3/2, Toronto.

Gelb, I. J. 1957 *Glossary of Old Akkadian*, Material for the Assyrian Dictionary, No. 3, Chicago (=MAD 3).

Gelb, I. J. 1965 The Ancient Mesopotamian Ration System, *JNES* 24, 230–243.

Gelb, I. J., Steinkeller, P. and Whiting, R. M. Jr. 1991 *Earliest Land Tenure Systems in the Near East: Ancient Kudurrus*, Oriental Institute Publications 104, Chicago.

Maekawa, K. 1973–4 The Development of the é-mí in Lagash during Early Dynastic III, *Mesopotamia* 8–9, 77–144.

Maekawa, K. 1980 Female Weavers and their Children, *Acta Sumerologica* 2, 81–125.

Michalowski, P. 2008 Observations on 'Elamites' and 'Elam' in Ur III Times. In P. Michalowski (ed.), *On the Third Dynasty of Ur: Studies in honor of Marcel Sigrist*, JCS Supplemental Series Vol. I. Boston, 109–123.

Pomponio, F. 2010 New Texts Regarding the Neo-Sumerian Textiles. In C. Michel and M.-L. Nosch (eds), *Textile Terminologies in the Ancient Near East and Mediterranean from the Third to the First Millennia BC,* Ancient Textiles Series 8. Oxford, 186–200.

Thureau-Dangin, F. 1903 *Recueil de Tablettes Chaldéennes*, Paris.

Van de Mieroop, M. 1987 *Crafts in the Early Isin Period: A study of the Isin craft archive from the reigns of Išbi-Erra and Šū-ilišu*, Orientalia Lovaniensia Analecta 24. Leuven.

Waetzoldt, H. 1972 *Untersuchungen zur Neusumerischen Textilindustrie*, Studi Economici e Tecnologici I. Rome (= UNT).

Waetzoldt, H. 1980–1983a Kleidung A. Philologisch, *RlA* 6, 18–31. Berlin.

Waetzoldt, H. 1980–1983b Kopfbedeckung A. Philologisch, *RlA* 6, 197–203. Berlin.

Waetzoldt, H. 2010 The Colours and Variety of Fabrics from Mesopotamia during the Ur III Period (2050 BC). In C. Michel and M.-L. Nosch (eds) *Textile Terminologies in the Ancient Near East and Mediterranean from the Third to the First Millennia BC*, Ancient Textiles Series 8. Oxford, 201–209.

4. In Search of Lost Costumes. On royal attire in Ancient Mesopotamia, with special reference to the Amorite kingdom of Mari.

Ariane Thomas

On sentait que Madame Swann ne s'habillait pas seulement pour la commodité ou la parure de son corps ; elle était entourée de sa toilette comme de l'appareil délicat et spiritualisé d'une civilisation.
Marcel Proust, *A l'ombre des jeunes filles en fleurs*

One felt that Mme Swann did not dress simply for the comfort or the adornment of her body; she was surrounded by her garments as by the delicate and spiritualised machinery of a whole form of civilisation.
Marcel Proust, *Within a Budding Grove*

Introduction

In the late nineteenth century, Léon Heuzey, curator at the Louvre Museum, took a great interest in Ancient Near Eastern costume. He subsequently worked on numerous publications, gave various lectures, and even recreated costumes on living models at the Ecole des Beaux-Arts (Fig. 4.1). Thus, he inaugurated a scientific approach to ancient costume, which until then had been explored only by artists.[1] Yet, despite several ensuing studies, Ancient Near Eastern costume remains generally unfamiliar. Following Heuzey, the first scholars dealt with the subject in rather broad terms and without combining all available sources.[2] Then, with the exception of some general articles in encyclopedias,[3] more recent publications focused on specific chronological and geographical fields or themes.[4] Given this background of several general preliminary studies along with a few specialized ones on limited themes, the subject of Ancient Near Eastern costume remains relatively unexplored. Drawing on research on the deliberately broad subject of Mesopotamian royal costume between the third and first millennium BCE,[5] this paper will focus on examples from the Amorite kingdom of

[1] For instance, Gustave Courbet and Joséphin Péladan were inspired by images of Assyrian costumes (Gustave Courbet, *La Rencontre ou Bonjour M .Courbet*, 1854, oil on canvas, 132 × 150.5cm., Montpellier, Musée Fabre, inv. 868.1.23; *Portrait du Sâr Peladan*, vers 1895, photograph, Paris, Musée d'Orsay, inv. PHO1992-18). These were also imitated for the opera *Semiramide* by G. Rossini as early as 1860 (André-Salvini 2008 (2), 481).

[2] See Reimpell 1916; Houston and Hornblower 1920; Speleers 1923; Lutz 1923 or Van Buren 1930.

[3] See Bier, Collon 1995; Ogden, Allgrove Mc Dowell 1996; Sarkhosh-Curtis 1996; Collon 1996; Barber 1997; Sass 1997; Irvin 1997; Waetzoldt and Strommenger 1998; Waetzoldt and Boehmer 1998; Green 2000; Bienkowski 2000.

[4] Canby 1971; Strommenger 1971; Maxwell-Hyslop 1971; Mazzoni 1979.

[5] Ariane Thomas, PhD thesis, *Research on the Royal Costume in Mesopotamia from Akkadian Times to the end of the Neo-Babylonian Empire*, University Paris-Sorbonne, Paris IV, 2012.

Mari. The aim of this contribution is to stress the need for an interdisciplinary approach to research costume as it is understood here. Such a study requires a specific methodology to collect basic data spanning as many approaches as possible for a more in-depth analysis. Thus, we will initially discuss three general points: the definition of costume itself, the interest in studying it, and the main sources available for the study of Ancient Near Eastern costume. We will then focus on the specific case of royal costume in the Mari kingdom in Amorite times.

Fig. 4.1: Léon Heuzey wrapping a model. © Heuzey 1922, quatrième de couverture.

What does the term "costume" imply?

Originally, the word "costume" denoted a whole way of being and now qualifies the manner of dressing of a country, a period, or condition.[6] As such, it can designate any items worn contributing to a person's artificial appearance. According to this broad understanding of the term, a costume may include clothes, hats, hair arrangements, belts, shoes, jewels, weapons, and anything else that may be worn, including cosmetics and perfumes.

This approach to costume as global attire probably coincides with the Mesopotamian conception based on clothing ensembles identified in written sources.[7] Additionally, depending on its function, a particular costume piece could potentially be more important than others while garments, although systematically worn and visually dominant, could form a minor part of the costume. For example, the wearing of some protective gem may have been as least as important as the clothes themselves since the gem's purported magical protection or ostentatious power would take precedence over the practical protection from cold or the sun and the "decency" afforded by clothes. In the same way, the symbolic importance of certain body parts such as the head – and consequently hats and hair arrangements – could take precedence over garments.

Sources

Understanding a costume requires not only an accurate analysis of a specific costume but also an exploration of any other sources that may explain its significance. Paradoxically, Ancient

[6] Rey 2001, 648.
[7] See Durand 2009, 12.

Mesopotamian costume is omnipresent in iconographic and epigraphic documents, but its material aspect has almost completely disappeared as only very few remains of costume pieces have been discovered. Nevertheless, it is still possible to employ several types of documents, written and figurative, for the identification of existing costumes and their linkage to a broader historical context.

Archaeological survivals

Although Mesopotamian costume might include other materials, especially metals and/or minerals, it was primarily made of organic materials, such as textile and leather. Unfortunately, due to geo-climatic conditions unfavourable to their conservation in Mesopotamia and the Ancient Near East, remains of such clothing are very rare or highly deteriorated.[8] Furthermore, costume pieces composed of metal or stone, such as jewels and weapons, have also largely disappeared because these materials were reused.

The few exceptional surviving vestiges are priceless direct sources of knowledge on costumes in Mesopotamia. However, due to their rarity, these remains may be neither representative of their period or region nor of royal costume. For instance, very few examples of the jewels or weapons that have been found in Mari and date back to the Amorite period could have been royal.[9]

Iconography

Iconography thoroughly illustrates costume, especially royal attire. Different types of images give us practical information on how it was worn, in which circumstances, and by whom. Nevertheless, while being very informative, these documents are more questionable than material remains because they are indirect sources on costume.

First, it is necessary to question the realism of representations that were mainly intended to deliver a religious or political message. In this respect, monuments might provide a fair image of a reality now lost, especially as regards royal costume because religious and/or formal requirements had to be faithfully reproduced. The correlation sometimes possible between representations made on different mediums using various techniques, such as paintings and sculptures,[10] seems to confirm the value of information provided by iconography.[11] This value is further substantiated by a correspondence with several archaeological finds and more hypothetically with written records. However, theintentions underlying Mesopotamian images may also have been motivated by a system of iconographic conventions which produced a relatively distorted vision of reality. One should always keep in mind the fundamental specificity of Mesopotamian images, especially compared to our modern western way of thinking. Unlike the Platonic concept of *mimesis*, according to which images imitate reality, Mesopotamian images would have had an ontological rather than aesthetic value. Hence, these images would not have presented a copy inferior to the original subject, but rather a repetition with enough equivalence to serve as a substitute for the original reality.[12] Thus, even if artists might have been inspired by reality, their images were stylized so as to achieve a

[8] Breniquet 2008, 55–58.

[9] See for instance at the Louvre Museum, AO 18318 (medallion; silver; Palace of Mari), AO 18440 (pin; bronze; Palace of Mari, throne room) or Alep Museum, M 790 (pin; electrum; Palace of Mari, throne room).

[10] Although they are not exactly identical, see, for instance, representations of a short piece of cloth on a cylinder-seal (Amiet 1960, 230, fig. 13), a stele (André-Salvini 2008 (2), 68, no. 23) or a painting (André-Salvini 2008 (2), 72, no. 27).

[11] For instance, see below on circular medallions and the *kubšum* hat.

[12] Bahrani 2003, 203, 205, 210.

more or less successful and faithful transposition of reality. For instance, the folds of wrapped clothing often appear more rigid than what would occur in reality.[13] Also, many figures are very schematic and consequently quite imprecise, but the presence of some details on relatively inaccurate representations may indicate their importance.[14] More generally, only three-dimensional sculpture can give a comprehensive insight into the shape and arrangement of costumes. Therefore, numerous examples in various forms obviously contribute to a better understanding of each type of clothing and allow for identifying its corresponding period and the way in which it was worn.

Using iconography is also difficult because representations are often fragmentary and incomplete. They also may have been modified from their original state by alterations. For instance, paintings from the palace of Mari are among the very few pieces of evidence for the colours of royal costume, although the rapid degradation of these fragile colours after their discovery and further contact with air has sometimes altered original hues.[15]

In addition, as contexts depicted on monuments are relatively limited and largely depict the king before gods engaged in ritual or in battle scenes, they may reveal only a few aspects of the royal wardrobe. Some images were perhaps even intended to portray a timeless archetype of kingship,[16] reflecting archaic clothes rather than contemporary ones. Royal ceremonial dress and robes worn by gods are cases in point.[17]

Finally, except for a minority of works inscribed with the names of royal figures,[18] most images lack any inscription that would allow an identification of the figures.

Epigraphy

As words survive better than clothing,[19] epigraphic sources supply a wealth of information on Mesopotamian costume.

An abundant nomenclature of fabrics/garments[20] and to a lesser extent other clothing items is present in texts, but many of these terms are still not well understood. Archives may provide information on sizes, weights, colours, qualities, prices of costume pieces, or time required to produce them.[21] A few written sources also relate to the way in which costume pieces were worn, their users, their provenance, and even technical specificities or decoration.[22] But many texts are "dry enumerations"[23] of terms without any other indication of meaning. As a result, it is difficult to glean any descriptive element since most of these texts were written by and for insiders who knew

[13] As noted by Heuzey 1925, 166. For instance in Mari, the numerous folds that probably existed are not depicted even on paintings, arguably a technique more easily suited to reproducing such details.
[14] For instance jewels on female figures (Louvre Museum, inv. AO 18393 or AO 18999: Barrelet 1968, pl. LXIV nos. 695 and 690).
[15] See for instance Pierre-Muller 1990, 499.
[16] Barrelet 1987, 58–59.
[17] Collon 1996, 883.
[18] Such as for the Mari kingdom, the statues of *shakkanakku*-rulers (see below note 44).
[19] Barber 1991, 260.
[20] It is difficult to distinguish whether the terminology refers to fabrics or to clothes with a specific shape (Durand 2009, 10; Michel and Veenhof 2010, 261–264).
[21] For instance a few clothes are said to be light (Rouault 1977 (1), 25, l. 9), heavy (Kupper 1983, 139, l. 1), large (Durand 1983, 318, l. 1) or small (Rouault 1977 (1), 49, l. 9) in the texts of Mari. On qualities and colours, see Durand 2009, 14–16, 172–174; for prices and time of manufacture of the clothes, see Durand 1983, 194.
[22] For instance Durand 1997, no. 138, 276–277 (=Dossin *et al.*, 1964, 12 (= Durand 1997, no. 138)); see below.
[23] As regrets for instance Jean-Marie Durand concerning the Amorite archives of Mari (Durand 1983, 394).

what to expect. Such a lack of detail precludes a full understanding of these documents, which are moreover frequently incomplete or deteriorated.

Due to this fragmentary information, translations of a single term might evolve over time. Indeed, any interpretation is likely to be contradicted or revised by still unknown texts, which may also multiply the number of terms.

Although it has been suggested,[24] it is not certain whether obsolete or new words reflected the existence of fashions, even though the nomenclature of costume would probably have changed depending on time and region. Depending on the location, the same item might have been referred to differently[25] or, conversely, different types of clothing could have been designated in the same way. For instance, within the same area, the name of a given item may evolve over a certain period of time while still referring to the same object.[26] The movement of people, trade or diplomatic exchanges in Ancient Mesopotamia may have been factors in this.

Moreover, as in the case of iconography, not all texts may necessarily have depicted the costumes of their time. For example, even though they might have been inspired by costumes of the period in which they were recorded (especially in the second millennium BCE), epic and mythological texts evoke ancient times and spheres removed from human reality. Consequently, they do not necessarily reflect the actual costumes of their age. On the contrary, Mesopotamian rulers, notably Zimrî-Lîm of Mari, exchanged many letters in which costumes were discussed. Tablets have also been found that discuss the involvement of some king's servants in related issues. The terms found in these letters and in economic documents, which constitute the major part of the records utilized for this paper, were certainly linked to contemporary reality.

Methodology

Combining the three complementary sources – figurative monuments, ancient texts and material remains of costumes – provides the only way of rediscovering Mesopotamian royal costume despite its almost complete material disappearance.[27] Unfortunately, each individual source has its own limitations, and it is difficult to connect these partial and largely indirect sources as their correlation is sparse. One must also remain objective in order to avoid anachronistic prejudices in analyzing these documents, including the unconscious use of contemporary patterns to understand Ancient Mesopotamian costumes or the use of technical, cultural, or historical assumptions that are sometimes deeply rooted in the archaeological literature.

Due to the large number of fields of study related to costume, it is necessary to initially focus on some specific aspects. As a means for exploring costumes in Ancient Mesopotamia, the present investigation is based on iconographic data, systematically linked to material remains and written sources when relevant. This typological catalogue of Mesopotamian royal costume pieces forms the basis for exploring other aspects of the subject.

[24] Durand 2009, 9.
[25] Michel and Nosch 2010, xi.
[26] Perhaps due to variations in one given type of cloth (for instance see below on decorated dresses).
[27] Foster 2010 combined textual data and images to study the Sargonic royal costume and its evolution.

The case of the Mari Kingdom in Amorite times

This article focuses on the specific case of the Mari kingdom in Amorite times.[28] This specific case, delimited in time and space, enjoys a very rich and well-researched corpus of texts. By connecting this written information with iconography and material remains, a typology of costume pieces was established. Some of the points raised by the study are presented here.

The king's clothes: garments cut in a specific shape and custom-made fabrics with rich ornaments

As Amorite texts of Mari contain a very rich nomenclature of clothing and/or fabrics, it is not surprising that the iconography should depict various types of clothes, among which several were probably very sophisticated. As such, representations seem to depict high quality cloth fragments of various sizes that were designed to be wrapped around the body while other images show fabrics that were cut in a specific shape and did not need to be wrapped.

Some monuments of Mari depict headgear and clothes made of textiles that are marked with two vertical and parallel lines of alternate colours.[29] Such fabrics were probably woven[30] or made of different strips cut apart and then sewn together. The fabric called *kaunakes* might also have been woven[31] in a particularly complex way or made of a specific material, not necessarily textile, as its various aspects on images may refer to different types of manufacture.[32]

Some clothes had to be cut into a special shape and then sewn. One example of this would be the dress of *shakkanakku* (ruler) of Mari Iddin Ilum.[33] Its length and complexity of its arrangement, if reflecting reality, suggest a particular pattern of design.[34] Apart from this unique example, several monuments seem to show sleeved dresses with round or V-cut collars.[35] The sleeves may have been cut separately before being sewn onto the dress, but if so, could this correspond to the term *ahâtum,* which might have designated removable sleeves?[36] This type of garment may also have been cut from a piece of textile directly in the shape of a sleeved dress. In this case, the sleeves would have been part of the complete garment. In any event, the dress would have been first cut into a predefined shape, including holes for the neck and arms, with sleeves (if these were not made separately). This shape would then have been sewn, probably on the sides, to close it. Such

---

[28] Ancient Mari (modern Tell Hariri) was a very important city, located in modern Syria, on the western bank of the Euphrates, a fundamental transit hub for people and exchange. Founded at the beginning of the third millenium BCE, it was sacked by Hammurabi of Babylon in 1759 BCE. After this destruction, it was inhabited only sporadically. During the so-called Amorite period until its destruction by Hammurabi, the city was led by several Amorite kings who had conquered it: under Iahdun-Lim (*c.* 1815–1799), the kingdom of Mari seems to have been prosperous and powerful. But his son Sumu Yamam was defeated after only two years by Samsî-Addu of Ekallatum who put his youngest son Yasmah-Addu on the throne. After this Assyrian domination, Zimrî-Lîm (*c.* 1775–1761), who is said to have belonged to the dynasty of Iahdun-Lim, became the last king of Mari for fourteen years until he was defeated by his former ally Hammurabi of Babylon.

[29] Amiet 1977, pl. 64; Pierre-Muller 1990, 481–483, pl. VIII et IX.

[30] Concerning this matter, a piece of textile found in Susa showed stripes of different densities woven in the same textile (Breniquet 2008, 61).

[31] Breniquet 2008, 61.

[32] Ariane Thomas, forthcoming.

[33] Amiet 1977, no. 413.

[34] Parrot 1959, 20: hypothesis for a pattern of *shakkanakku* (ruler) Iddin-Ilum's dress by Jean Lauffray.

[35] For instance: André-Salvini 2008 (2), 68, no. 23 (it should be noted that this stele, related by its inscription to Samsî-Addu of Ekallatum who conquered Mari, does not come from the kingdom of Mari); Aruz 2008, 31–32, no. 7; Spycket 1948, 92, fig. 5. These sleeves appear to be short: they stop before the elbow to cover only the upper arm.

[36] Durand 2009, 29.

long dresses were meant to be slipped on over the head and the arms, and worn with nothing over them, except perhaps a shawl. Such dresses may therefore correspond to the Sumerian term (túg) gú.è.a or its Akkadian equivalent *nahlaptum*. Indeed, these terms would signify "the dress by which the neck goes out", that is to say "where one slips on" according to Amorite texts from Mari.[37] It could thus designate a garment cut to a specific shape.[38]

However, these cut and sewn garments were certainly not the only high quality clothes. Despite their apparently simple shape, wrapped garments could also be the result of a very sophisticated tailoring according to iconographic and written evidence, especially for the so-called royal dress.[39] The latter resembles a long robe wrapped around the body while leaving one shoulder uncovered. It was probably made of one large and apparently plain piece of textile. This style using a single large piece of cloth appeared in the time of Akkad,[40] but the dress itself would be a legacy from Neo-Sumerian Kingdoms.[41] Iconography depicts the rulers of Mari wearing this dress from the end of the third millennium BCE[42] into the Amorite period[43] as well as in other contemporary kingdoms of Mesopotamia.[44] Monuments of Mari also indicate the use of the dress in a slightly different shape. On several representations (Fig. 4.2)[45] this long and asymmetrically wrapped dress leaves the legs of the wearer uncovered. It thus reveals a short garment worn under the dress, in accordance with written sources, such as a letter from Queen Shibtu according to which she sent her royal husband two garments – one was to be worn on top of the other.[46] Another type of wrapped dress appears on images of Mari. It is shorter, falling approximately to the knee, and held in place by a belt.[47]

Despite their differences, these wrapped dresses, self-supporting or held in place by a belt or pins, must have been made from the arrangement of a single quadrangular fabric, given its visible angles.[48] Such garments support the idea that the túg or *ṣubâtum*, generally accepted as a generic name for textiles or garments,[49] could indicate both the fabric and the dress since the two could not be dissociated. But it does not exclude the possibility that these fabrics could have been made-to-measure for the intended wearer in a style dependent upon the type of clothing desired. Despite their basic shape, fabrics used as clothing had to be of sufficiently accurate dimensions in order to achieve the desired appearance of proper length. This presumption is suggested by its exceedingly precise indications conveyed by the king of Mari.[50]

---

[37] Durand 1983, 402. It might be an overcoat (Durand 1983, nos. 318, 322, 397; Durand 2009, 11, 67–72).
[38] Durand 2009, 69.
[39] For instance: Aruz 2003, 427; André-Salvini 2008 (1), 16.
[40] See for instance Amiet 1977, no. 365; Breniquet 2008: 66; Foster 2010: 127.
[41] See for instance Aruz 2003, nos. 304, 306.
[42] See the statues of *shakkanakku* (rulers) Ishtup-Ilum and Puzur-Ishtar (Amiet 1979, figs 55–56).
[43] See for instance Parrot 1959, 148, pl. XXXIX/788.
[44] See representations of Hammurabi of Babylon or other Mesopotamian rulers in Amorite period (for instance André-Salvini 2008 (1), figs 13, 15, 23, 28).
[45] See also for instance Amiet 1960, 230, fig. 13.
[46] Dossin 1978, 17 (Durand 2000, 306 (1129)).
[47] André-Salvini 2008 (2), 72, nos. 27 and 68, no. 23.
[48] Some scholars thus suggested rectangular patterns for these pieces of cloth (Houston and Hornblower 1920, 49, fig. 26a: suggestion of pattern for the garment of Gudea of Lagash).
[49] http://psd.museum.upenn.edu/epsd/nepsd-frame.html (túg (5078x: ED IIIa, ED IIIb, Old Akkadian, Lagash II, Ur III, Early Old Babylonian, Old Babylonian) wr. túg "textile, garment" Akk. ṣubâtu).
[50] Thus "the detailed requirements of the King [of Mari in the Amorite period] on his ceremonial dress lead to conceive of this garment as a made-for-measure one" (Bry 2005, 74).

Fig. 4.2: Wall painting of Investiture Scene; wall painting on white plaster; H. 175, L. 250; Mari, royal palace, court 106, southern wall; Amorrite period; Musée du Louvre, département des Antiquités orientales, inv. AO 19826. © Musée du Louvre/dessin C. Florimont.

Not only are these wrapped dresses intriguing because of the way they were made, but also because of their identification with some written mentions of royal garments. It is indeed quite tempting to compare the clothing represented on royal figures in ceremonial contexts (investiture, sacrificial procession) to the precious royal dresses worn for ceremonies according to the royal letters of Mari. However, these letters mention that the dresses were decorated with added or embroidered patterns that are not visible on representations of royal dress, although these images depict other ornaments, such as large fringes or decorative trimmings (Fig. 4.2). A painting fragment[51] might even illustrate a gather detail according to the hypothetical understanding of the term *himṣum* in a letter of Zimrî-Lîm.[52] Even if these decorative details are depicted, other ornaments, as valuable as they were, might have been omitted in iconography. Such may have been the case of added decorations, including precious metal or stones, although sources also sometimes refer to textiles.[53] Yet, other

[51] Parrot 1958, 103 fig. 80.

[52] Durand 2009, 45.

[53] Such as the flower *arzallum* made of wool while *zîmum* ornaments were very probably metallic according to Durand 2009, 140 and 142.

Fig. 4.3: Male figure; ceramic; H. 6,9; l. 3,2; Mari; Amorrite period; Musée du Louvre, département des Antiquités orientales, inv. AO 18406. © Musée du Louvre/Philippe Fuzeau.

representations do show this kind of decoration (Figs 4.3–4). They especially depict circular designs comparable to golden discs – the so-called "buttons" –discovered in large quantities in the Amorite royal tombs of the kingdom of Ebla[54] and also in smaller quantities in Mari.[55] Such images of dresses with circular ornaments could correspond to the *taddêtum* or *tandûm (taddi'u)* for which the king of Mari specifically requested a solid workmanship because of the weight of its ornaments.[56] Nevertheless, it is interesting that this dress is also described with hems in the style of Yamhad, which might have been the festoons visible on monuments of Mari. This garment is also thought to have been made for the coronation.[57] Consequently, this kind of dress could correspond to the long royal wrapped dress, such as the one worn by the king on the "Investiture painting" of Mari (Fig. 4.2), even if no ornaments other than festoons are visible on the dress.[58] Although it does not depict an exact coronation in front of the population as described in the letter about *taddêtum* dress, this image shows an investiture scene of the king amidst the gods, and was visible in the courtyard leading directly to the throne room in the palace.

Other terms to be considered here include *hatûm,*[59] which would have designated clothing decorated with trimmings, or *uṭba*, which was a luxurious dress omnipresent in texts that might have been the usual garment for the king and dignitaries.[60] One of these terms could potentially refer to images of other wrapped dresses that carried separately produced festoons as, for example, on the royal figure and dignitaries of a sacrificial cortege.[61]

Nevertheless, other texts of Mari describe various figurative patterns[62] that could relate to Mesopotamian Amorite representations although they were discovered outside of the kingdom of Mari.[63] Among them, a description of a *nahlaptum* dress with *sagikkum* ornaments made of thirty

[54] Catalogue Rome, 1995, 483, nos. 403 and 404.

[55] Such as a so-called gold button with repoussé decoration (Maxwell-Hyslop 1971, 87, fig. 65 c). Other examples were probably taken away from Mari in antiquity because of their preciousness.

[56] Durand 2009, 112. The author considers this cloth as being the *tuttubum.*

[57] Durand 2009, 112.

[58] Burnt in antiquity, this painting is unfortunately highly damaged.

[59] Durand 2009, 45.

[60] Durand 2009, 130.

[61] André-Salvini 2008 (2), 72, no. 27.

[62] Durand 1983, no. 342, l. 1 (referring to a 1st quality *mardatum* shawl representing a *lamassatum*); Beaugeard 2010, 285 and 288 (about the words *zîmum* and *nahzabum*).

[63] For instance on a figurine found in Larsa: Barrelet 1968, 315, no. 578, pl. IV (Louvre Museum, inv. AO 20193).

pieces of blue stones and coral and weighing 11 shekels and 5/6 of silver,[64] suggests that different types of clothing could be richly decorated. This demonstrates the difficulty in connecting terminology and iconography since the correlation may only be partial but also because of the variation in shapes or decoration for the same type of clothing.

Fig. 4.4: Fragment of a figurine; ceramic; H. 5,2; l. 3,4; Mari; Amorrite period; Musée du Louvre, département des Antiquités orientales, inv. AO 18421. © Musée du Louvre/Philippe Fuzeau.

Royal hats and the question of gender distinction

As mentioned above, it is not easy to connect images of clothing worn for special occasions with the ceremonial garments described in written sources. The same issue arises with royal hats in iconography and in the texts of Mari. Iconography shows the king wearing a cap surrounded by a brim, inherited from Neo-Sumerian rulers (in the same way as the long wrapped dress). Except for one that displays stripes,[65] the cap appears to be made of a plain, seemingly rigid, fabric. Since the brim of the cap consistently appears high[66] and thick,[67] it was probably rather a part of a hat made in a specific shape than a simple turned-up brim. It also could have been separately fabricated before being added to the round shape. Alternatively, the overall shape of the whole hat might have been made of a hard surface then covered with textile. This type of hat had to be quite rigid in order to stay in shape, and it apparently consisted of one large piece of material rather than bands of textile.[68] Thus, it could also have been made of some rigid material, such as felt, leather or even thick wool without any support.

In the texts of Mari, a *kubšum* hat appears to have been worn by the king for major ceremonies. One letter of King Zimrî-Lîm seems especially to prove the importance of this hat and describes him as worrying about the delay in its production as he had an upcoming encounter with other kings in Sagarâtum. The king also demanded that stones (certainly precious since another letter mentions onyx and carnelian)[69] and gold be sent quickly to adorn the hat.[70] Should this royal *kubšum* be considered as the royal cap surrounded by a brim as has been suggested?[71] Despite the fact

[64] Dossin et *al*., 1964, 12 (=Durand 1997, no. 138))

[65] Amiet 1960, 230, fig. 13. Another monument of Mari shows a likely striped hat but its shape is seemingly different (Amiet, 1977, pl. 64). Elsewhere in Mesopotamia, another monument shows a cap with a large brim much closer to the royal hat model which presents comparable strips (Louvre Museum, inv. 9061).

[66] Covering almost the entire forehead and part of the ears, it must have measured just under half the height of the whole hat, that is to say about 18–25cm high according to the average dimensions of the forehead.

[67] It might have been 2–5cm large.

[68] Durand 2009, 53 e.

[69] Durand 1983, 223.

[70] Rouault 1977 (1), 8 (= Durand 1997, 111) ; Bardet *et al*. 1984, 202.

[71] Durand 2009, 52.

Fig. 4.5: Male head; ceramic; H. 4,5; l. 2,5; Mari; Amorrite period; Musée du Louvre, département des Antiquités orientales, inv. AO 19004. © Musée du Louvre/Philippe Fuzeau.

that this hat is depicted without any ornament, it characterizes the king in Mesopotamian iconography, and it is possible that the *kubšum* existed with different degrees of ornamentation. However, the king of Mari on the "Investiture painting" is shown wearing a hat that is slightly taller and oval-shaped rather than round (Fig. 4.2). Furthermore, other monuments from Mari (Fig. 4.5) depict similar hats that are decorated with circular patterns, which are likely to match the *kubšum*'s ornaments. This taller and oval-shaped hat could be the *kubšum* of Mari. As already mentioned, it is perhaps a local and richer variant of the royal Mesopotamian round cap with a brim that was also used in Mari. In this case, there would be, on the one hand, the cap with a brim and, on the other hand, the very valuable *kubšum*. However, the term *kubšum* could also more generally refer to the fabric of these hats, which was probably both quite rigid and made in the same way except for their shape and decoration. The difficulty in distinguishing between them underlines the importance of understanding whether a term designated the item of clothing or the fabric.

Unlike the king, the other members of the royal sphere, men or women, are shown wearing hats without a brim. Some of these hats[72] seem to be quite rigid while others appear to have been made of different pieces of fabric wrapped around the head like turbans.[73] Should the latter be compared to the *hazîqatum* turban[74] or to the *agûm*[75] since texts indicate that it was composed of strips held in place by a golden *kamkammatum* or *namarum* ornament?

Iconography also depicts men[76] and women[77] wearing a headband. For men at least, it might have indicated high status, since it was worn by the princes of Qatna or Alalakh.[78]

While both women could wear turbans or headbands apparently similar in shape to those of men (even the royal *kubšum* was worn by a great priestess),[79] written sources indicate that only women wore veils. Thus, the wife of king Zimrî-Lîm was veiled,[80] as well as the king's concubines.[81] Considering that King Samsî-Addu of Ekallatum conquered Mari, it seems that the veil was used

[72] André-Salvini 2008 (2), 72, no. 27.
[73] Parayre 1982, 77, no. 65; André-Salvini 2008 (2), 72, no. 27 (first man behind the king).
[74] Durand 2009, 44.
[75] Durand 2009, 28.
[76] Parrot 1958, 20, fig. 18.
[77] Parrot 1959, 22, pl. XIII (Louvre Museum, inv. AO 19521).
[78] Aruz 2008, fig. 72; Amiet 1979, fig. 63.
[79] Durand 2009, 53.
[80] Durand 1988, 103–104.
[81] Vogelsang-Eastwood 2008, 23.

during the marriage ceremony,[82] as was the case in Assyria according to Assyrian sources.[83] Other texts also mention a veil as something worn by the bride.[84] Unfortunately, without any existing images, iconography is of no help; it is consequently rather difficult to know the aspect of these veils and what they covered. Texts attest only that they could have been decorated as a luxurious textile and with figurative patterns called *zêmum*. The embroidered fabric called *mardatum* could also have been used as a veil.[85] It is hard to determine whether the veil was worn only during the wedding ceremony or if it was a constant obligation for married women.[86]

Following these remarks, one may wonder whether there was a gender distinction in costume at the court of Mari. More generally in Mesopotamia, several texts describe a form of transvestism in a religious context. One Amorite period text recounts an unusual worship ritual dedicated to the goddess Ishtar, during which, the participants wore costumes said to be specific to the opposite sex in a procession of symbols of the goddess.[87] These practices may have been particularly linked to the cult of the goddess Ishtar who had a dual aspect as both warrior and seductress. Yet these texts indicate a clear recognition of the distinction between male or female clothing or, at least, of particular costume elements that were gender specific. Nevertheless, it has been stressed that the Amorite texts of Mari do not contain any indication about gender distinction in costume pieces between men and women, except for a few references that are considered to allude more to difference of size than to real specificities.[88]

However, both iconography and texts do suggest that specific costume elements were likely exclusively reserved for women while others were prohibited. Among them, veils and perhaps double-breasted clothes in the front and back[89] (possibly a legacy of coats crossed over in the back and worn by Neo-Sumerian princesses)[90] seem to be have been worn only by women. Conversely, the elements reserved for men would reflect the position of women in the Mesopotamian royal sphere. As women were neither armed and therefore warriors, nor governing as suggested by the lack of insignia, women of the court enjoyed a privileged status, which is suggested by the luxuriousness of some female clothing and jewellery. They are represented with jewels apparently heavier than those of men, such as multi-strand necklaces that covered more and were certainly less convenient.

Jewels and ceremonial dress in Mari: a specific taste for adornment?

While some statues of *shakkanakku* (rulers) from the late third millennium BCE[91] wear robes decorated with rather simple fringes and without any jewels, the Amorite iconography of Mari represents the king and other members of the royal sphere as quite richly dressed. Indeed, various written sources indicate that both clothing and hats appear to have been abundantly adorned with

[82] Durand 2009, 55–56 suggests the existence of a *kutummum* veil for the bride.
[83] Michel 2006, 161.
[84] Démare-Lafont 2008, 239. The author notably quotes the description of Enkidu's face veiled as a bride in the Epic of Gilgamesh.
[85] Beaugeard 2010, 285 and 288 but see also Durand 2009, 56, 64 and 140.
[86] As in the Medio-Assyrian period (Démare-Lafont 2008, 246–248).
[87] Groneberg 1997, 291–303. According to the author, it could be a purification ritual in which the King would participate.
[88] Durand 2009, 12–13.
[89] Spycket 1948, 92, fig. 5; Aruz 2008, 31–32, no. 7. However, the very small number of examples is certainly not sufficiently representative.
[90] See Parrot 1948, 194, fig. 39 A, 186, fig. 39 B, 190–191, fig. 41a (Louvre Museum, inv. AO 43, AO 226, AO 295, AO 297).
[91] Amiet 1977, nos. 80 and 415.

various types of trimming elements,[92] including trimmings in the manner of Yamhad.[93] Precious ornaments could also decorate clothes, hats or shoes.[94] Written sources indicate that the colours and the high quality fabrics of these numerous embellishments give the impression that ceremonial dress in the Amorite court of Mari was gaudy compared to that of previous eras or of contemporary Mesopotamian courts that favoured, for instance, a sparser use of trimmings.[95] Thus, although it is difficult to determine any local or specific "fashion" at that time, it has been noted that the "taste for abundant ornaments belong[ed] to the Syrian [including Mari] rather than the Mesopotamian iconographic area [even if] this type of dress was worn by the king of Babylon".[96] Thus, one can compare clothes from the wardrobe of the kingdoms of Mari and Eshnunna to garments worn by the rulers of Babylon, or earlier in Lagash and Ur.[97] The former appear to have been richly decorated with (sometimes double) rows of festoons and possibly lined with braid and tassels while the others, although they were apparently the same type of clothes and might have been made of noble fabric, appear less adorned. Furthermore, iconography of Mari testifies that this ceremonial manner of dressing was complemented by many jewels, which can be listed in a brief typological inventory.[98]

Images show that men in a ceremonial context could wear simple bracelets of beads, as documented in written descriptions. Thus, the white hue of one of the two bracelets worn by the leader of a sacrificial procession on a wall painting of Mari's royal palace[99] suggests it could have been silver, gold or even iron, such as examples described in texts,[100] since the metallic brilliance may have been depicted by the colour white. His second bracelet seems to have been made with three groups of beads arranged in different colours. The first four are white and could therefore be metallic; an orange bead likely symbolizes carnelian; and the last one resembles agate. Other monuments show women and goddesses wearing multiple bracelets closely together. Their various colours, as depicted on paintings of Mari,[101] suggest they might have been made of different materials.

Similarly, necklaces with multiple strands of often decreasingly sized beads,[102] which covered the entire neck like "chokers",[103] were certainly made of various beads since they are depicted with different colours on paintings. Images show this type of collar was frequently worn in association

[92] If some credence is given to iconography, it shows not only a majority of more or less identical festoons with rounded edges, but also festoons which seem to "fly in the wind" because of a greater than the average length (Parrot 1958, 100, figs 77–78), while others are represented as rectangular (Parrot 1958, 95, fig. 72).
[93] See for instance Rouault 1977 (= Durand 1997, no. 136).
[94] See above and Durand 1985, 164, note 64. Copper is said to have decorated the boots of the Great Priestess.
[95] See for instance Aruz 2003, figs 103, 106, 107, nos. 304–308 (rulers of second dynasty of Lagash with no jewellery and a "simple" fringed dress); Aruz 2008, fig. 10 (Hammurabi of Babylon with jewels and a fringed dress without trimmings).
[96] André-Salvini 2008 (2), 72, no. 27. Southern Mesopotamia may have followed patterns from inner Syria (or the Levant, highly influenced by Egypt). For example, long-sleeved dresses appeared very early in Mari compared to southern kingdoms; likewise, the development of festoons in Mari could prefigure the borders so well-represented in the Neo-Assyrian period.
[97] André-Salvini 2008 (2), 33–36, 69, 71–73.
[98] This article does not claim to be exhaustive, but only to summarize some data.
[99] André-Salvini 2008 (2), 72, no. 27.
[100] Dossin 1952, 5; Birot 1960, 20, l. 7 and 10; Dossin 1978, 61.
[101] See for instance Amiet 1977, no. 65.
[102] See for instance Amiet 1977, no. 65.
[103] Maxwell-Hyslop 1971, 85.

with necklaces apparently made of rectangular beads; the whole collar appears striated. Both of these collars only grace women in the court of Mari. On the contrary, necklaces with pendants and beads appear specifically on men. Thus, single-strand necklaces with a large circular pendant at the centre are characteristic of male dignitaries in religious scenes.[104] They may have been made of very large beads[105] – perhaps metallic beads or chain(s) as they are also depicted in white. Another figure wears an ochre-dyed collar, which may have symbolized leather.[106] Other necklaces were perhaps more specifically royal and might correspond to the necklaces of magical stones described in texts.[107] This type of necklace could be represented by an exceptionally preserved jewel said to come from Dilbat.[108] This elaborate granulated example was made of a triple-row of fluted melon-shaped beads and seven crescent-shaped pendants with a fork lightning symbol, a pair of presumed Lama goddesses,[109] two circular rosettes, and a circular pendant with rays.[110] Some pendants found in Mari[111] are very comparable to the crescent-shaped and circular ones. These circular pendants adorned with astral designs correspond to artefacts in gold or silver, which were widespread in the Middle East in the second millennium BCE. It has been assumed that these pendants could correspond to the GUR_7-ME term, *šamšum* in Akkadian.[112] Indeed, it designated golden medallions, more or less precious, that are supposed to be amulets with religious significance. Based on this assumption, an example found in Larsa resulting from a very fine granulation work[113] might reflect these valuable models that may symbolize a whole country.[114]

Many collars are represented with long counterweights. These large elements, which were very valuable according to textual sources,[115] likely contributed to the ceremonial dress of the Mari court. Their practical function probably prevailed as they appear on certainly heavy necklaces with multiple rows worn by high-ranking women,[116] as well as on lighter necklaces with pendants[117] and beads such as those worn by male figures.[118] Counterweights are depicted as long and smooth-looking strands,[119] of a reddish brown colour reminiscent of a leather cord, or similar to a lock of hair[120] that could have been made of braided metallic threads.[121]

---

104 Maxwell-Hyslop 1971, 85; André-Salvini 2008 (2), 72, no. 27.
105 Parrot 1958, pl. D, 1.
106 André-Salvini 2008 (2), 72, no. 27 (dignitaries from left to right).
107 Durand 1983, nos. 247, 236. Concerning this question, see Schuster-Brandis, 2008.
108 Aruz 2008, 24, no. 4.
109 Compare with André-Salvini 2008 (2), 82, nos. 37–38 (Louvre Museum, inv. 4636).
110 Maxwell-Hyslop 1971, 89.
111 Maxwell-Hyslop 1971, 87, figs 65 a, b.
112 Durand 1990; Charpin 1990.
113 Durand 1990, 158.
114 Durand 1990, 149, 157.
115 Durand 1983, 233. Counterweights could be very large, with ten to thirty beads of different shapes, made of gold and precious stones. In fact, the counterweight was probably slightly heavier than the collar, as suggested by texts from Mari indicating that the composition of a counterweight or *pitu* included a higher number of beads than those of the collar itself (Durand 1983, 219 and 247).
116 For instance: Spycket 1948, 92, fig.5; Parayre 1982, pl. 77, no. 65; Aruz 2008, 31–32, no. 7; Parrot 1959, 23–25, pl. XV.
117 For instance: Barrelet 1968, 358, no. 694, pl. LXIV; Parrot 1958, pl. D, 1.; Amiet 1960, 230, fig. 12 (?).
118 Parrot 1958, pl. D, 1 and 92, fig. 69, no.14 (too fragmentary to identify any figure or collar); Maxwell-Hyslop 1971, 87, fig. 66.
119 Aruz 2008, 31–32, no. 7.
120 Spycket 1948, 92, fig. 5
121 Spycket 1948, 92.

Paintings from Mari show male figures wearing circular single bead earrings whose colour recalls precious stones such as carnelian or lapis-lazuli.[122] The king of Mari, Zimrî-Lîm, himself wore earrings.[123] On the other hand, women are represented with much larger earrings, including triply fluted ones. These earrings also appear with complicated arrangements composed of several pieces formed as a crescent and fixed into a circular ring. Another example is a very large earring with multiple fluted bodies.[124] For women at least, this last example (as well as several schematic figurines, which contain the multiple holes designed for earrings)[125] testifies that multiple earrings could be worn on each ear. Iconography and texts[126] also demonstrate that earrings were always worn in pairs.

Texts of Mari mention several poorly represented jewels, such as an ankle bracelet sent to a king of Karana.[127] If the king had personally worn this item, this type of jewel may have been part of the royal wardrobe. Although well documented in texts,[128] rings are not visible in Mari iconography; nor are beads in the shape of fruits such as dates,[129] which might have resembled some Egyptian remains.[130] Texts also mention pendants in the shape of animals, such as flies,[131] that could be illustrated by fly-shaped material examples dating to the end of the third millennium BCE.[132] It is also possible that jewels were worn in the hair, as suggested by one female head showing a chignon held in place by a kind of large hair ring.[133]

Although jewellery contributed to the royal pomp of appearance in Mari, the luxurious royal settings may have been offset by a kind of simplicity or even austerity. A letter of a Yaminite nomad criticizes the faults of a sedentary lifestyle affording a taste of expenditure and idleness as opposed to the simple and authentic values of nomadic life.[134] Knowing that Amorite kings were descendants of nomads or semi-nomads, could the Amorite kings have possessed these same thoughts, regardless of their actual relations with nomads? In such a case, the appearance of the royal court might have shown humility in certain situations, although this is not well documented and requires further investigation. Nonetheless, it may be necessary to moderate the statement that ceremonial costumes worn at the court of Mari were considered as luxurious. Indeed, the analysis of the king's wardrobe, which texts have described as the richest of the court, is estimated as being suitable but not sumptuous[135] in that it consisted of approximately one garment per week (in practice less than this due to the combination of several pieces). In fact, the fortune of King Zimrî-Lîm, which is the best documented among the Amorite kings of Mari, appears to be rather modest as compared to the kings of Yamhad or Ekallatum even if he likely became richer over

[122] André-Salvini 2008 (2), 72, no. 27; Parrot 1958, 104, fig. 83.
[123] Arkhipov 2012, 73.
[124] Maxwell-Hyslop 1971, 85; Aruz 2008, 31–32, no. 7.
[125] Barrelet 1968, 359, pl. LXIV, nos. 695 and 690 (Louvre Museum, inv. AO 18393 and AO 18999).
[126] Spycket 1948, 92, fig. 5 for instance ; Arkhipov 2012, 79.
[127] Talon 1985, 280.
[128] See for instance Villard 1984, 535. Arkhipov 2012, 73, 102.
[129] Bottéro 1957, 247, l. 12; Durand 1983, 235.
[130] For instance at the Louvre Museum, inv. E 24591.
[131] Durand 1983, nos. 223, l. 42–43. To be compared to Qatna: Bottéro 1949, 15. See Lion and Michel 1997, 722–723.
[132] See for instance at the Louvre Museum, inv. AO 18309; 18273. It is tempting to relate to those pendants with jewels in the shape of insects found in the Middle Bronze Age Crete (see for instance Aruz 2008, 102, fig. 32).
[133] Parrot 1959, 22, pl. XIII (Louvre Museum, inv. AO 19521).
[134] Marello 1991, 115–125.
[135] Durand 2009, 23.

time.[136] Despite some exceptionally luxurious costume pieces with many precious ornaments and colours, which certainly required expensive materials and specialists,[137] the royal wardrobe in the kingdom of Mari may not have been overly precious despite the suggestions of iconography. This humbler aspect of the royal costume of Mari may well have been linked to a will for humility.

Outfits and accessories: belts, baldrics, shoes, weapons, seals, insignia and other items
The basic royal costume seems to have included clothes, hats, and often jewels. Apart from certain circumstances such as rituals, it also extended to shoes, belts and weapons, at least for men. Underwear must also have composed the wardrobe for men and women.[138] Five groups of costume pieces are discussed here: shoes, belts, weapons (with baldrics), insignia and other items.

Shoes
Shoes were perhaps not worn indoors as suggested by the possible equivalence between the words *šênum* and kuš-e-sír, which literally means "the leather object [worn] when going out in the street".[139] If shoes were worn neither in the palace nor for rituals, the lack of their representation in Mari might be significant. Sandals are described on several Mesopotamian monuments, including the Hammurabi stele.[140] It is feasible that *mešênum* referred to sandals or more generally low shoes,[141] rather than boots. Boots are possibly described in texts as kuš-šuhub$_2$ or *šuhuppatum*[142] and also represented in iconography as worn by a goddess figure[143] since some women could wear them.[144] Unless referred to only as *kaballum,*[145] they could also be viewed as gaiters, which were made of wool, possibly dyed in blue.[146] Additionally, the term *kaballum* often coexists in texts with the word *šuhuppatum,* which is explicitly said to be made of leather, as boots should have been.[147] Sometimes called Cretan,[148] the latter might have come from Crete or have been inspired by some Cretan models. Contemporary Cretan images apparently depict boots and/or gaiters worn on shoes.[149]

Belts
Belts do not seem to have been systematically worn. This may have depended on the nature of the costume (whether or not it needed to be held in place), and probably on the nature of the belt itself. According to images and texts of Mari, at least three types of belt might have existed. First, a kind

[136] Lerouxel 2002, 459.
[137] For instance for the *mardatum* fabric (Durand 2009, 64).
[138] Durand 2009, 12, 33, 72, 76 and 118–119 (notably about *didûm* for women and *nahramum* for men which may have served as slips).
[139] Durand 2009, 170.
[140] André-Salvini 2008 (2), 36.
[141] Durand 2009, 166.
[142] Durand 1983, 422–423, nos. 330, 331, l. 8 and 333, l. 8', 19', 39'; Durand, 2009, 168, 171 (he underlines that the translation of this term as boots or gaiters is convenient but not certain).
[143] Amiet 1960, 230, fig. 13.
[144] Rouault 1977 (1), 22.
[145] Rouault 1977 (1), 66, 27, l. 15 (= Durand 1997, no. 184); Durand 1983, 423; Durand 2009, 48–49. The term *karikkum* might also designate some gaiter (Durand 2009, 50).
[146] Durand 1983, no. 365.
[147] Rouault 1977 (1), 35 (= Durand 1997, no. 222).
[148] Durand 1983, no. 342, 1. 5–6.
[149] See for instance Aruz 2008, 132, fig. 42.

of cord wrapped around the waist that could have been made of either leather or a dyed and hard textile, according to its brown and flexible appearance.[150] Although it is a mere hypothesis, a second variety of belt may have been fabricated from textile since it has fringes,[151] which might evoke the *patinnum.*[152] A third type of belt would be the metallic ones mentioned in written sources.[153] Nevertheless, the terminology suggests that other types of belts existed in Mari, including the *naṣmadum.*[154] This term, which was documented at the time of Yasmah-Addu (who was the son of Samsî-Addu of Ekallatum) and is illustrated on a stele that is said to be of Samsî-Addu,[155] would have designated a kind of shawl that served as a belt to attach weapons.

Baldrics and weapons

Baldrics and weapons, which were logically used for fighting, were worn with short or opened garments particularly suitable in this context. Yet the precious ceremonial weapons may have been held in other contexts as potentially illustrated in the painting of a sacrifice leader.[156] Indeed, he seems to wear a sword probably hanging from his belt or from an invisible baldric. Its white colour evokes a metallic brilliance, such as that of gold or silver as quoted in written records.[157] Among the ceremonial weapons belonging to the king,[158] the mace-head could correspond to the *katâpum* since it is described in texts as having a skull and a body as its head and handle respectively.[159] Texts also mention a curved weapon brought for the coronation of King Asqur-Addu,[160] a *hubûsum,* which could be a sacrificial dagger worn by the king at his waist,[161] as well as other daggers and gold or silver-plated spears.[162]

Insignia

Although not systematically, insignia of the royal status, such as the controversial rod and ring, and signs of prestige might also be part of the costume. The rod and the ring appear on investiture scenes as if they were given to the king by the gods, but it is possible that he never actually held them as he is not seen touching them in the scenes.[163] However, these items may have materially existed if they correspond to the *haṭṭum* and *kipattum*-ring described as being made of bronze and gold.[164] According to a rare mention, the rod, which appears similar to a sceptre, might have actually

[150] André-Salvini 2008 (2), 72, no. 27 (leader of the procession).
[151] Parrot 1958, 9, fig. 7.
[152] Durand 2009, 169.
[153] Bottéro 1957, 238; Durand 1983, 342–343.
[154] Durand 2009, 75–76.
[155] André-Salvini 2008 (2), 68, no. 23 (on the side of the victorious king's dress, an element looks like an arrow and may be a part of this type of belt. It seems comparable to the one visible on King Naram-Sin of Akkad on his victory stele against the Lullubis (Louvre Museum, inv. Sb 4)).
[156] André-Salvini 2008 (2), 72, no. 27.
[157] Durand 1983, nos. 345, 222.
[158] See for instance Arkhipov 2012, 105 (belonging to Samsî-Addu of Ekallatum); Lerouxel 2002, 439 (sent for the death of King Yarim-Lim of Yamhad).
[159] Bottéro 1957, 238; Durand 1983, 342–343.
[160] Durand 2009, 59 (A. 203:33).
[161] Arkhipov 2012, 108.
[162] Arkhipov 2012, 110, 115, 121 (about some *imittum, marhašum* and *qaštum*).
[163] Van Buren 1949, 450, Wiggermann 2006, 414.
[164] Arkhipov 2012, 107.

been a component of the king's costume.[165] Among the costume pieces serving particularly royal prestige, a fan or flyswat is described in texts as a *nêZaBBum*.[166] Written sources also mention a parasol if this is really what designated the an-dùl object, said to be composed of *sappum,* probably the structure to stretch the fabric which could be woven in a *mardatum* fabric[167] or as a *zîrum ša andulli*.[168]

Seals and other costume items

The royal costume in Mari also included seals and possibly other items, such as gloves, which might be designated by the rare mention of *rittum* and described as fabricated from textile or leather.[169]

Clothing ensembles

As discussed above, the royal wardrobe in Mari may not have been overly sumptuous with a relatively small number of pieces for a year, except for large quantities of shawls.[170] Concerning the number and the quality of costume pieces worn together, the most easily detectable in texts and iconography is the way of dressing with a piece of cloth meant to be worn over another one, namely an undergarment.[171] This secondary dress, which was worn against the body, could have been a long robe or a short tunic or skirt of varying length (Fig. 4.2 – such a short piece of costume is visible under the king's dress). It may also have been a type of underwear.[172] As such, it would have been totally or partially covered by the overcoat. This pattern seems more common for men, but long garments worn by women could simply conceal such an undergarment.

Analysis of the associations of costume pieces also reveals different costumes, some of which might have been specific to ceremonies, travels, fighting, etc.[173]

Finally, the issue of the clothing outfit particularly calls into question the documentary sources. Texts of Mari mention rather precisely what neither material evidence nor iconography reveal – for instance, shoes which were indisputably part of the royal wardrobe. Similarly, a flyswat or fan, as well as some jewels, are not visible on Amorite images but are described in texts. On the contrary, iconography alone shows the possible associations of different costume pieces as they were worn, while the existing material remains, although scarce, provide direct evidence and can correct the deficiencies of written and figurative sources as in the case of medallions/ circular pendants.

[165] Arkhipov 2012, 123.
[166] Durand 2009, 76–77.
[167] Durand 1990 (2), 161, note 13.
[168] Durand 2009, 141.
[169] Durand 2009, 90, 169.
[170] Durand 2009, 23.
[171] This distinction was already used by Heuzey 1935, 102 before Durand 1997, 268. It appears clearly in a letter of Queen Shibtu to the king of Mari (Dossin 1978, no. 17, l. 12 (= Durand 2000, 1129)) and it even seems to be depicted in the legend of Gilgamesh who is said to put on new clothes followed by a coat (*Epopée de Gilgamesh*, VI, col. I, l. 1–5).
[172] Durand 2009, 72 d) on the term *nahramum*.
[173] A. Thomas, PhD, *op. cit.*

Hair arrangements, cosmetics and perfumes: natural or artificial?

In Mesopotamia, great importance was given to hair and consequently to its care and arrangement, for reasons of hygiene, notably against headlice and parasites,[174] as well as for self-adornment or symbolical background. Indeed, hair appears to have been considered as a sign of virility for men and was particularly important for the king. This is suggested by the reproaches of King Samsî-Addu of Ekallatum toward his son Yasmah-Addu whom he made king of Mari. Samsî-Addu wrote in a most likely metaphorical fashion that his son was still a child with no beard on his chin instead of being a man and a king.[175] According to the Amorite iconography of Mari, men appear to have had short hair, but whether the king also had short hair or a chignon under his hat remains largely unknown. On the contrary, women seem to have worn their hair long in a chignon. This chignon was occasionally detailed as being plaited, which was perhaps accomplished by one of the hair-braiding specialists mentioned in some texts.[176] In fact, though images depict apparently simple hair arrangements, it is not clear whether these arrangements were natural or artificial as several texts mention the use of wigs.[177]

Finally, perfumes and cosmetics contributed, even if in an immaterial way, to building one's appearance, particularly in the royal sphere, as witnessed through the words of a king's courtier concerning the manner in which he is covered by his master's perfume.[178] Nonetheless, it is difficult to find any positive illustration of this.[179] Texts of Mari testify only that the king used perfume quite regularly, even when away from Mari, as did the queen and the women of the Harem. Perfume is also said to have been used in religious and banqueting ceremonies.[180]

The question of local fashions

After focusing on several of the pieces that could have formed part of the royal wardrobe in Mari, this section briefly addresses the issue of local fashions. This may call into question the specificity of the royal costume of Mari as compared to other contemporary kingdoms, as well as ancestral costumes.[181] As mentioned above, the ceremonial royal costume of Mari appears to have been more richly adorned than in Southern Mesopotamian kingdoms. But these added ornaments also appear on monuments associated with the courts of Ekallatum or Eshnunna as well as Yamhad or even Byblos according to written sources about *taddêtum*. In fact, it has been suggested that the royal costume of Mari could have illustrated the importance of Eshnunna's traditions in Mari.[182] Though the richly-festooned royal costume may have come from the wealthy Diyala region, braided ornaments would probably have been local, being attested in Mari, Yamhad and up to the coast in Ougarit.[183] Some fashions probably developed in a particular area due to political connections between kingdoms. One example of this is the ceremonial and highly adorned dress of the king of Mari or the *mardatum* fabric apparently characteristic of North and West Mesopotamia.[184]

---

[174] Contenau 1950, 71.
[175] Dossin 1950, no. 61, l. 10; no. 73, l. 44–45; no. 108, l. 6, no. 113, l. 7–8.
[176] Durand 2009, 39, note 36.
[177] Durand 1991, 35; 2009, 45–46.
[178] Ziegler 1996.
[179] Some details such as the black outlines of paintings of Mari may reflect the use of cosmetics.
[180] Joannès 1993, 263–264.
[181] Durand 2009, 51, 54, 109.
[182] Durand 2009, 112.
[183] Durand 2009, 95.
[184] Durand 2009, 63.

In addition, ancient texts mention many pieces of clothing under foreign names. Even if they were local productions influenced only by foreign models evoked by their name, the variety of provenances recalled by such names[185] suggests that fashions or at least pieces of clothing circulated widely. They were notably exchanged between kingdoms as diplomatic gifts.[186] Specific types of clothing and fashion patterns, such as sleeved dresses which appeared very early in Mari compared to southern kingdoms,[187] may have circulated through these exchanges.

Conclusion

This article summarises various data on the royal wardrobe and focuses on specific points within the case study of the Amorite kingdom of Mari. It was largely intended to demonstrate that although facing an almost complete disappearance of its original components, the study of costume could lead to identifications supported by evidence but also a number of interesting conjectures based upon sometimes tenuous foundations. It would certainly be beneficial to expand the study of this vast subject, combining as many methods and therefore sources of information as possible.

Bibliography

Allgrove Mc Dowell, J. 1996 Ancient Near East Textiles. In J. Turner (ed.) *The Dictionary of Art* 1. New York, 879–883.
Andre-Salvini, B. 2008 (1) *Le Code de Hammurabi*. Paris (1st ed. 2003).
Andre-Salvini, B. (ed.) 2008 (2) *Babylone*. Paris.
Amiet, P. 1960 Notes sur le répertoire iconographique de Mari à l'époque du palais. *Syria* 37, 215–232.
Amiet, P. 1977 *L'art antique du Proche-Orient*. Paris.
Amiet, P. 1979 *Introduction à l'histoire de l'art de l'antiquité orientale*. Paris.
Arkhipov, I. 2012 *Le Vocabulaire de la métallurgie et la nomenclature des objets en métal dans les textes de Mari. Matériaux pour le Dictionnaire de Babylonien de Paris III, ARM 32*. Paris.
Aruz, J. (ed.) 2003 *Art of the First Cities. The Third Millennium BC from the Mediterranean to the Indus*. New York.
Aruz, J. (ed.) 2008 *Beyond Babylon, Art, Trade, and Diplomacy in the Second Millennium BC* New York.
Bahrani, Z. 2003 *The Graven Image. Representation in Assyria and Babylonia*. Philadelphia.
Barber, E. J. 1991 *Prehistoric Textiles. The Development of Cloth in the Neolithic and Bronze Ages with Special Reference to the Aegean*. Princeton.
Barber, E. J. 1997 Textiles: textiles of the Neolithic through Iron Ages. In E. M. Meyers (ed.) *The Oxford Encyclopedia Archaeology in the Near East* 5. New York/Oxford, 191–195.
Bardet, G., Joannes, F., Lafont, B., Soubeyran, D. and Villard P. 1984 *Archives royales de Mari XXIII*. Paris.
Barrelet, M. T. 1968 *Figurines et reliefs en terre cuite de la Mésopotamie antique*. Paris.
Barrelet, M. T. 1987 En marge de l'étude de quelques empreintes de cylindres-sceaux trouvés dans le palais de Mari. In *Mari annales de recherches interdisciplinaires* 5. Paris, 53–63.
Barthes, R. 1967–1983 *Système de la Mode*. Paris.

[185] Clothes in the fashion of Elam, Haššum, Byblos, Huršânû, Lullum, Kiš, Subartum, Iamhad or mountains (Durand 2009, 69, 100, 101, 106, 71, 111, 112–113 and 70) Clothes in fashion of Elam (Durand 2009, 69 and 100). *Kaballu* shoes would also be a type of fashion adopted from the mountains (Durand 2009, 49) while some elements of shoes said to be Cretan were already mentioned (see above note 147) such as a lance (Arkhipov 2012, 110).
[186] Lerouxel 2002.
[187] Southern Mesopotamia may have followed some patterns from inner Syria (or the Levant, strongly influenced by Egypt) as sleeves are depicted from the third millennium in Egypt to the early second millennium in Mari.

Beaugeard, A.-C. 2010:Les textiles du Moyen-Euphrate. In C. Michel and M.-L. Nosch (eds) *Textile Terminologies in the Ancient Near East and Mediterranean from the Third to the First Millennia BC.* Ancient Textiles Series 8. Oxford, 283–289.

Bienkowski, P. 2000 Jewelry. In P. Bienkowski and A. Millard (eds) *Dictionary of the Ancient Near East.* London, 162.

Bier, C. 1995 Textile arts in Ancient Western Asia. In J. M. Sasson (ed.) *Civilizations of the Ancient Near East* III. New York, 1567–1588.

Birot, M. 1960 *Textes administratifs de la salle 5 du palais, Archives royales de Mari* IX. Paris.

Bottero, J. 1949 Les inventaires de Qatna. *Revue d'assyriologie et d'archéologie orientale* 43, 1–2, 1–40 and 137–215.

Bottero, J. 1957 *Archives royales de Mari VII, Textes économiques et administratifs.* Paris.

Breniquet, C. 2008 *Essai sur le tissage en Mésopotamie des premières communautés sédentaires au milieu du IIIe millénaire avant J.-C.* Paris.

Bry, P. 2005 *Des règles administratives et techniques à Mari.* Sabadell.

Catalogue Rome, 1995 *Ebla all'origine della civiltà urbana, Trent'anni di scavi in Siria dell'Università di Roma «La Sapienza».* Rome.

Canby, J. V. 1971 Decorative Garments in Assurnasirpal's Sculpture. *Iraq* 33, 31–49.

Charpin, D. 1990 Recherches philologiques et archéologie: le cas du médaillon "GUR$_7$ME». *Mari Annales de Recherches Interdisciplinaires* 6, 159–160.

Collon, D. 1995 Clothing and grooming in Ancient Western Asia. In J. M. Sasson (ed.) *Civilizations of the Ancient Near East* I. New York, 503–515.

Collon, D. 1996 Dress. In Jane Turner (ed.) *The Dictionary of Art* 1. London, 883–886.

Contenau, G. 1950 *La Vie* quotidienne *à Babylone et en Assyri*e. Paris.

Demare-lafont, S. 2008 A cause des anges: le voile dans la culture juridique du Proche-Orient ancien. In O. Vernier, M. Bottin and M. Ortolani (eds) *Etudes d'histoire du droit privé en souvenir de Maryse Carlin.* Paris, 235–253.

Dossin, G. 1939 Les archives économiques du palais de Mari. *Syria* 20, 97–113.

Dossin, G. 1950 *Archives royales de Mari I, Correspondance de Samsi-Addu.* Paris.

Dossin, G. 1952 *Archives royales de Mari V, Correspondance de Iasmah-Addu.* Paris.

Dossin, G. 1978 *Archives royales de Mari X, Correspondance féminine.* Paris.

Dossin, G., Bottero, J., Birot, M., Burke, L., Kupper, J.-R. and Finet, A. 1964 *Archives royales de Mari XIII, Textes divers.* Paris.

Durand, J.-M. 1983 *Archives royales de Mari XXI, Textes administratifs des salles 134 et 160 du palais de Mari.* Paris.

Durand, J.-M. 1985 La situation historique des Šakkanakku. *Mari Annales de Recherches Interdisciplinaires* 4, 147–172.

Durand, J.-M. 1988 Archives épistolaires de Mari 1/I. Archives royales de Mari 26/1. Paris.

Durand, J.-M. 1990 La culture matérielle à Mari (I): le bijou *HÚB.TIL.LÁ/"GUR$_7$-ME. *Mari Annales de Recherches Interdisciplinaires* 6, 125–158.

Durand, J.-M. 1990 (2) ARM III, ARM VI, ARMT XIII, ARMT XXII. In O. Tunca *De la Babylonie à la Syrie en passant par Mari, Mélanges offerts à Monsieur J.-R. Kupper à l'occasion de son 70e anniversaire.* Liège, 149–177.

Durand, J.-M. 1991 Perruques. *Nouvelles Assyriologiques Brèves et Utilitaires* 2, 35–36.

Durand, J.-M. 1997 *Les documents épistolaires du palais de Mari* I. Littératures anciennes du Proche-Orient (LAPO 16). Paris.

Durand, J.-M. 2000 *Les documents épistolaires du palais de Mari* III. Littératures anciennes du Proche-Orient (LAPO 18). Paris.

Durand, J.-M. 2009 *Matériaux pour le Dictionnaire de Babylonien de Paris, tome 1, la nomenclature des habits et des textiles dans les textes de Mari.* Paris.

Foster, B. R. 2010 Clothing in Sargonic Mesopotamia: Visual and written evidence. In C. Michel and M.-L. Nosch (eds) *Textile Terminologies in the Ancient Near East and Mediterranean from the Third to the First Millennia BC.* Ancient Textiles Series 8. Oxford, 110–145.

Green, A. 2000 Clothing. In P. Bienkowski, A. Millard (eds) *Dictionary of the Ancient Near East.* London, 75–76.

Groneberg, B. 1997 Ein Ritual an Istar. *Mari annales de recherches interdisciplinaires* 8, 291–303.

Heuzey, L. 1922 *Histoire du costume antique, d'après des études sur le modèle vivant.* Paris.

Heuzey, L. 1925 Costume chaldéen et costume assyrien. *Revue d'assyriologie et d'archéologie orientale* XXII 4, 163–168.
Heuzey, L. and J. Heuzey 1935 *Histoire du costume dans l'antiquité classique, l'Orient*. Paris.
Houston, M. G. and Hornblower, F. S. 1920 *Ancient Egyptian, Assyrian and Persian costumes and decorations*. London.
Irvin, D. 1997 Clothing. In E. M. Meyers (ed.) *The Oxford Encyclopedia of Archaeology in the Near East* 2. New York/Oxford, 38–40.
Joannès, F. 1993La culture matérielle à Mari (V): les parfums. *Mari Annales de Recherches Interdisciplinaire,* 7, 251–270.
Kupper, J.-R. 1983 *Archives royales de Mari, volume XXII/I, Documents administratifs de la salle 135 du palais de Mari.* Paris.
Lerouxel, F. 2002 Les échanges de présents entre souverains amorrites au XVIIIe siècle av. N. E. d'après les archives royales de Mari. *Florigelum Marianum* 6. SEPOA, 413–463.
Lion, B. and Michel, C. 1997 Criquets et autres insectes à Mari. *Mari annales de Recherches Interdisciplinaires* 8, 707–724.
Lutz, H. F. 1923 *Textiles and costumes among the Peoples of the Ancient Near East*. Leipzig–New York.
Marello, P. 1991 Vie nomade. In J.-M. Durand (ed.) *Recueil d'études en l'honneur de Michel Fleury, Florilegium Marianum* I, *Mémoires de NABU* 1. Paris, 115–125.
Maxwell-Hyslop, K. R. 1971 *Western Asiatic Jewellery c. 3000–612 BC*. London.
Mazzoni, S. 1979 Nota sull'evoluzione del costume paleosiriano. *Egitto e Vicino Oriente* II, 111–138.
Michel, C. 2006 Bigamie chez les Assyriens. *Revue historique de droit français et étranger* 84, 155–176.
Michel, C. and Nosch, M.-L. (eds) 2010 *Textile Terminologies in the Ancient Near East and Mediterranean from the Third to the First Millennia BC.* Ancient Textiles Series 8. Oxford.
Michel, C. and Veenhof, K. R. 2010 The Textiles Traded by the Assyrians in Anatolia (19th–18th centuries BC). In C. Michel and M.-L. Nosch (eds) *Textile Terminologies in the Ancient Near East and Mediterranean from the Third to the First Millennia BC.* Ancient Textiles Series 8. Oxford, 210–271.
Ogden, J. 1996 Ancient Near East $ II, 4, Jewellery. In J. Turner (ed.) *The Dictionary of Art* 1. London, 872–877.
Parayre, D. 1982 Les peintures non en place de la cour 106 du palais de Mari, nouveau regard. *Mari Annales de Recherches Interdisciplinaires* 1, 31–78.
Parrot, A. 1948 *Tello, vingt campagnes de fouilles (1877–1933)*. Paris.
Parrot, A. 1958 *Mission archéologique de Mari II, Le palais 2, Peintures murales.* Paris.
Parrot, A. 1959 *Mission archéologique de Mari II, Le palais 3, Documents et monuments*. Paris.
Pierre-Muller, B. 1990 Une grande peinture des appartements royaux du palais de Mari (salles 219–220). *Mari annales de recherches interdisciplinaires* 6, 463– 558.
Reimpell, W. 1916 *Geschichte der Babylonischen und Assyrischen Kleidung*. Berlin.
Rey, A. (ed.) 2001 *Le Grand Robert de la langue française* 2 (2e édition). Paris.
Rouault, O. 1977 (1) *Archives royales de Mari XVIII, Mukannisum, l'administration et l'économie palatiales à Mari.* Paris.
Rouault, O. 1977 (2) L'approvisionnement et la circulation de la laine à Mari d'après une nouvelle lettre du roi à Mukannisum. *Iraq* 39, 147–153.
Sarkhosh, Curtis V. 1996 Ancient Near East Dress. In J. Turner (ed.) *The Dictionary of Art* 1. London, 883–885.
Sass, B. 1997 Jewelry. In E. M. Meyers (ed.) *The Oxford Encyclopedia Archaeology in the Near East* 3. New York/Oxford, 238–246.
Schuster-Brandis, A. 2008 *Steine als Schutz und Heilmittel, Untersuchung zu ihrer Verwendung in der Beschwörungskunst Mesopotamiens im 1. Jt. v. Chr.* AOAT 46. Münster.
Speleers, L. 1923 *Le costume oriental ancien*. Brussel.
Spycket, A. 1948 Un élément de la parure féminine. *Revue d'Assyriologie et d'Archéologie orientale* 42, 89–97.
Strommenger, E. 1971 Mesopotamische Gewandtypen von der frühsumerischen bis zur Larsa Zeit. *Acta praehistorica et archaeologica* 2, 411–413.
Talon, P. 1985 *Archives royales de Mari XXIV, Textes administratifs des salles «Y» et «Z» du palais de Mari.* Paris.
Van Buren, E. D. 1930 Some Archaic Statuettes and a Study of Early Sumerian Dress. *Annals of Archaeology and Anthropology* XVII 3–4, 39–56.

Van Buren, E. D. 1949 The Rod and Ring. *Archiv Orientalni* XVII 2, 434–450.
Villard, P. 1984: *Archives royales de Mari XXIII.* Paris.
Waetzoldt, H. and Boehmer, R. M. 1998 Kopfbedeckung. *Reallexikon der Assyriologie und Vorderasiatischen Archäologie* 6, 197–210.
Waetzoldt, H. and Strommenger, E. 1998 Kleidung. *Reallexikon der Assyriologie und Vorderasiatischen Archäologie* 6, 18–38.
Wiggermann, F. A. M. 2007 Ring und Stab. *Reallexikon der Assyriologie und Vorderasiatischen Archäologie* 11, 414–421.
Ziegler, N. 1996 Ein Bittbrief eines Händlers. *Wiener Zeitschrift für die Kunde des Morgenlandes* 86, 479–488.

5. Elements for a Comparative Study of Textile Production and Use in Hittite Anatolia and in Neighbouring Areas

Giulia Baccelli, Benedetta Bellucci and Matteo Vigo

1. Introductory Overview

"Words survive better than cloth".[1] This statement is certainly valid for the ancient Near Eastern study of textiles.

Although our general knowledge regarding trade and use of textiles in the ancient Near East seems to be secure, particularly due to studies in the economic and administrative texts of Mesopotamia in the 3rd and 2nd millennium BC,[2] we do not yet have significant archaeological remains to confirm information provided by philologists. Over the last 50 years, scholars have been specifically investigating technical terms referring to textiles within texts.

If we look at the study of textile terms of the 2nd millennium Anatolia before the rise of the Hittite kingdom (17th–13th centuries BC), we observe the same lack; the information provided by textual evidence cannot be confirmed by iconography nor by the very scant archaeological remains.[3]

Monographs which address Assyrian trade in Anatolia during the 19th–18th centuries BC,[4] provide information on textile production, costs and selling prices, workmanship, quality and shape of thc fabrics, trade routes, "textile topography" (that is, the provenance and the final destination of particular fabrics).[5] We are able to detect details regarding certain types of fabrics among the records of the Old Assyrian traders (personal letters written in Akkadian) found in the private archives of the commercial quarters of *kārum* Kaneš (modern Kültepe, near Kayseri, Turkey).[6] These texts (written in a foreign language) speak of a foreign trade market controlled by a structured business system between the indigenous (Anatolians) and the traders (Assyrians).

What then can be said about the supposed Hittite textile production and economy of the following centuries in Anatolia?

---

[1] The present motto of emerita textile scholar Elizabeth Barber (Barber 1991, 260) was successfully recalled in the introduction of the recent proceedings on Textile Terminologies in the Ancient Near East and Mediterranean from the 3rd to the 1st millennium BC (Michel and Nosch 2010a).

[2] Among others Waetzoldt 1972; Veenhof 1972; Zawadski 2006; Breniquet 2008; Pomponio 2008; Verderame 2008; Biga 2010.

[3] It is worth to note that Cécile Michel (CNRS-ArScAn-Nanterre) and Eva Andersson Strand (CTR-SAXO Institute-University of Copenhagen) have recently started a systematic study of textile (and basketry) imprints on sealings from Kültepe.

[4] Above all, Veenhof 1972.

[5] Michel and Veenhof 2010; Wisti Lassen 2010a.

[6] For such letters, see Veenhof 1972, 103–115; Michel 2001; Wisti Lassen 2010b.

Thanks to the information provided by other cuneiform texts found in Anatolia, such as administrative accounts of goods stored in the Hittite palaces, Hittitologists have produced indexes of *realia* (i.e. a presentation of the evidence of everyday objects used by the Hittites), in which luxurious textiles (or fabrics) and clothes often occur.[7] Moreover, Hittite official texts such as the accounts of royal victories, the descriptions of cult activities and the diplomatic correspondence between royal courts almost always contain lists of precious textiles and garments as gifts given to gods or allocated to the palatial storehouses. Such documentation has contributed to the development of several important studies on textile terminology of Hittite Anatolia. Albrecht Goetze was one of the first Hittitologists who interpreted a number of Hittite words closely related to types of garments, most of which were worn by Hittite kings on different occasions (official ceremonies or worship).[8] Although he rigorously analysed these Hittite textile terms from a linguistic perspective, Goetze made misleading comparisons with modern textile categories, in the quasi-absence – at that time – of evident archaeological data and technological experimentations. Nevertheless, because of his study scholars were able to obtain more information about types of garments, their colours, and, most notably, their supposed place of origin.

During the 1980s two fundamental editions of the Hittite palace inventories were published.[9] This *corpus*[10] consists of few, often fragmentary, cuneiform tablets in the form of lists and memoranda of terms indicating items, supplies and materials, containers and places of storage, most of which are still awaiting a strict semantic interpretation.[11] We may, therefore, define Košak's first edition of the Hittite inventory texts as a preliminary research on the "Hittite economic history" characterized by a strong lexical slant.

A few years later the Czech scholar Jana Siegelová, in order to study some aspects of the metallurgy of Bronze Age Anatolia,[12] also investigated these inventories and increased the *corpus* (thanks to the discovery of new fragments and the study of many duplicates and joins). Her aim was to analyse each text to better understand the structure of the Hittite administration, the role of the various institutions and the officials involved in the process of storing of goods and their possible redistribution.[13] The final result is a useful survey of the Hittite administration during the 13th century BC; however, with regards to the study of textiles (that represent almost 80% of the items listed in these inventories) we have not yet made significant progress. Most of the terms that are thought to indicate the manufacturing, workmanship or shape of the fabrics remain difficult to determine, classify and translate.

Recent studies have updated and improved our knowledge of Hittite textile terminology.[14] This is due, in part, to progress in the field of Hittite language studies and the discovery of new economic or administrative clay tablets from excavations of the Hittite capital, Ḫattuša, and provincial centres. After recent archaeological investigations we are now able to better define the function of a number of urban structures (storehouses, treasuries, archives). However, we still have no idea where the textile workshops were located and, most notably, how they worked.

[7] See the principle editions of Hittite palace inventory texts: Košak 1982; Siegelová 1986.
[8] Goetze 1947a; Goetze 1955; Goetze 1956.
[9] See the bibliography provided in note 7.
[10] Such documents were first classified by Laroche in his CTH and thereafter updated by Košak (2002) according to the new archaeological discoveries.
[11] See recent observations by Mora 2007, 535–536.
[12] Siegelová 1984.
[13] Siegelová 1986, 547–568.
[14] See, for example, Klengel 2008; Vigo 2010.

1.1. Issues and Goals

The geographical and chronological range is limited, as far as possible, to Hittite Anatolia of the 2nd millennium BC (Middle and Late Bronze Age),[15] crossing these limits, when necessary. Likewise, as a comparative study we cannot exclude close examination of those neighbouring areas from which we are able to obtain much more archaeological information on textile production and use than from the core of Anatolia. These neighbours encompass places located in the periphery of the Hittite Empire, especially Northern Syria.

The selected framework allows us to compare information provided by epigraphy, archaeology and iconography.[16] However, an exhaustive research project on textile production and use in Anatolia during the 2nd millennium BC would require too extensive a study to be covered here. Hence, we present some elements for a comparative study, as a *vade mecum*, actually attempting to join the information provided by the rarely surviving archaeological finds in Anatolia and its neighbours with the written documentation. On the other hand, we also make an effort to fill gaps left by texts, in particular where "Hittite textile production" is concerned, matching archaeological data. For example, the study of the unearthed weaving tools could help to fill the almost absent information on crafting and weaving techniques in the written sources. The interdisciplinary approach is extended, where necessary, to an iconographical and iconological overview of the objects presenting processes of textile production.[17]

As with the majority of the interdisciplinary investigations, we would like all the issues to be solved or, at least, debated. This cannot be possible for many reasons, but we can provide glimpses on different matters. Since one of the most productive terminological Bronze Age categories for textiles seems to be "textile *topology*", we should establish whether items were simply channelled through the area or whether they are typical of that location because they were, for instance, crafted there.

Finally, pertaining to the use, as the majority of the written documentation deals only with luxurious textiles and does not give a complete overview of the many types of textile used in antiquity,[18] the exact definition of a "garment", "cloth" or "textile" is still a problematic issue, even if in recent times many studies have been devoted to clothing worn by rulers and elite.[19]
Any future interdisciplinary research should aim at understanding which are the untailored fabrics or the ready-to-wear costumes among those probably recorded in the Bronze Age archive documents of Anatolia and to better define the luxurious textiles carved on seal representations or on rock reliefs.

The same research methodology could be applied for colour indications, even if it is difficult to establish if a "coloured fabric" consists of dyed textile, a natural pigmentation, or both.[20]

---

[15] Period designations suffer many problems of synchronization between the different chronologies proposed by scholars of each single ancient Near Eastern culture. For an in-depth study on this topic see the international research project *Associated Regional Chronologies for the Ancient Near East* (ARCANE: http://www.arcane.uni-tuebingen.de/), even though limited for now to the 3rd millennium BC. For the archaeological period designations in Anatolia see, for example, Sharp Joukowsky 1996, 30–33.

[16] For the importance of these comparisons see Michel and Nosch 2010b, x–xi. Cf. the interdisciplinary research programme on the Bronze Age textile production at Ebla (Syria) realized through a collaboration agreement between the Italian Archaeological Expedition at Tell Mardikh-Ebla (MAIS) and the Danish National Research Foundation's Centre for Textile Research (CTR). See Andersson *et al.* 2010.

[17] Such studies have recently been proposed as regards the new reading of the proto-dynastic iconography (Breniquet 2008; Breniquet 2010) and the Sargonic iconography (Foster 2010).

[18] See Michel and Nosch 2010b, xiii, with the bibliography provided in notes 51, 52.

[19] Among others Biga 1992; Pasquali 2005; Sallaberger 2009; Michel and Veenhof 2010, 260–266; Vigo 2010.

[20] Cf. Michel and Nosch 2010b, xiii–xiv.

2. Sources for Textile Production from Archaeology, Epigraphy and Iconography

2.1. In Pursuit of Workshops

The context of archaeological findings is particularly significant when focusing on textile tools, because by inspecting the environment in which these remains are found, scholars can better understand the tools themselves.

Through the analysis of archaeological objects within their context, it is possible not only to delineate the importance of textile production in the 2nd millennium BC in Anatolia and Syria, but also to define the techniques and kind of products involved.

Besides the archaeological value of textile tools, this analysis will also look at real textile remains or impressions on other materials: they complete and enrich the whole picture of textile production and use in daily life and within funerary contexts.

It is not possible to identify leading sites for textile production of the second half of the 2nd millennium BC in Anatolia only through archaeological data. It is worth mentioning the early 2nd millennium "Old-Assyrian" sites of Kültepe/Kaneš and Kaman-Kale Höyük. The former yielded materials for epigraphic and archaeological documentation concerning textile production;[21] the latter yielded archaeological remains.[22] Scholars have identified some textile workshops within houses or housing units for both these sites.

Workshops are often determined by the presence of weaving tools (loom weights; holes for loom structures).[23] There are, in fact, two relevant situations in which the presence of loom weights can be found in an archaeological context. In the first one, the rows of the loom weights are intact, indicating that the loom was in use at the time of destruction/abandonment. The Gordion excavation provides an example dating to the 7th century BC; fourteen large loom weights were found in two 60cm long rows.[24] In the second case the loom was no longer in use and loom weights were found grouped on a floor, probably because they were stored in a basket or a ceramic container. Hence, they indicate a sort of storage room,[25] as in the so-called Gordion "Royal Storage House".[26]

The workshop of Gordion, now fully examined by Brendan Burke, is one of the most important for the study of the areas of textile production.[27] The great finds in Gordion indicate mass textile production and provide archaeological evidence for a large number of textile workshops.

Regrettably, we do not have similarly clear workshops unearthed in Late Bronze Age Anatolian sites. But other possible workshops are presented in section 2.2., analysing the contexts of some findings.

2.2. Archaeological Evidence: Tools and Contexts

From an archaeological perspective there are a number of tools that refer to spinning and weaving in Anatolia as well as the surrounding areas during the 2nd millennium BC. Therefore, the following section examines tools such as spindles, spindle whorls and loom weights beginning with their morphological features and analysing the archaeological evidence.

[21] In particular, Veenhof 1972; Michel and Veenhof 2010.
[22] Fairbairn 2004.
[23] Loom weights are surely the most abundant archaeological evidence regarding weaving because of the less perishable material of which they were made compared to the wooden structure of the loom.
[24] Bellinger 1962.
[25] Barber 1991, 101–102; Shamir 1996, 144.
[26] Bellinger 1962.
[27] Burke 2010, in particular pages 108–157.

Despite the small number of spindles found in archaeological sites, due to the perishable nature of the items, these objects are significant because of their symbolic value.[28] Evidence from Alaca Höyük, dating back to the second half of the 3rd millennium BC, is particularly important. In grave L archaeologists excavated a silver implement with an electro head and a disc at its centre, the features of which resemble a spindle with a spindle whorl.[29] A second spindle composed of precious metal was discovered in grave H in the same site.[30] Similar metal tools, dating back to the 3rd millennium BC, have been found in graves of Horoztepe,[31] Merzifon[32] and Karataş-Semayük.[33]

These remains are important because of their extraordinary features and funerary significance; they were found in the graves of high-ranking people. These tools did not meet functional needs because precious metals were not suitable for common use. They represent, instead, identity markers for social status.[34] There is constant evidence of spindles made of precious material in funerary contexts and grave goods throughout the ancient Near East and the possible votive and ritual meaning of these tools highlights the symbolic value often related to textile manufacturing.[35]

During the Late Bronze Age and as early as the end of the Middle Bronze Age, spindles of bone or ivory often decorated with engravings also appear in archaeological records. This particular kind of spindle, mainly known from the neighbour Jordan/Region in Megiddo's graves,[36] is attested during the Late Bronze III and the Iron I in Syria, Palestine and Cyprus,[37] but recent findings in Troy also indicate a larger spreading over Anatolia.[38]

The spindle whorl is a pierced tool used in the spinning process, located on a spindle to weight it down and ease the work. It allows the thread to spin in addition to accommodating the manufactured thread. In order to more precisely define this object it is helpful to delineate its main morph-dimensional features. Therefore, it is important to measure consistently the weight and diameter of the object and the diameter of the perforation. An additional consideration that has to be taken into account is the spindle whorl's inertia, hence the ability of the object to perform its function.

At Arslantepe/Malatya (Anatolia) a significant number of spindle whorls have been excavated in both private and public contexts, dating back from the 4th to the 2nd millennium BC.[39] The analysis conducted on the objects from Arslantepe is very important in the investigation of morphological parameters, like the diameter, the weight and the type of thread that can be obtained from their use. Through a careful treatment of the data, it was possible to underline a correspondence between the spindle whorl's shape and material. The majority of the spindle whorls from Arslantepe are made of bone and exhibit a convex or conical profile. However, others are made of clay, stone

[28] See *infra* 2.4; 3.2.
[29] Koşay 1951, 169, L. 8, Pl. 197, fig. 1.
[30] Koşay 1951, 159, H. 115, Pls. 124, 126. See in general Völling 2008, 84–89.
[31] For the deposition in Horoztepe, see Özgüç and Akok 1958, 43–44; Pl. V–VIII.
[32] Cf. Völling 2008, 256, with bibliography.
[33] These objects are usually about 15cm-long and they present a metal spindle whorl put in the middle of the object while a tip presents a more or less elaborated surface. Cf. Bordaz 1980, 256.
[34] Cf. Peyronel 2004, 53; Völling 2008, 87–88.
[35] Peyronel 2004, 198. Völling 2008, 97–100. See further considerations in sections 2.4.; 3.2.
[36] Guy 1938, 170–172, fig. 175:6, Pls. 84, 1–16, 95, 41–50.
[37] For Hama and Enkomi, cf. Peyronel 2004, 53; Völling 2008, 87–88 with previous bibliography.
[38] In this case the spindle whorl from Troy is probably made from hippopotamus-ivory. Cf. Balfanz 1995; Völling 2008, 257–258.
[39] Frangipane *et al.* 2009, 6 and table 1. A new study by R. Laurito on Late Bronze Age textile tools found at Arslantepe was issued after the submission of the present paper (March 2014).

and metal: the stone spindle whorls are mostly discoid or convex, while those made of clay often have a conical or bi-conical profile.[40]

Period VA, corresponding to the first centuries of the 2nd millennium BC, presents a variety of spinning tools.[41] Spindle whorls continue to be made of clay or stone, although, archaeologists have recorded a wider range of diameters and weights. "There may have been a change in the textile production with a larger variety of yarns being produced in later periods".[42]

A second example is the Anatolian site of Beycesultan, where we observe the presence of a large amount of fired clay spindle whorls, dating from the Middle to the Late Bronze Age. Almost all of them present a bi-conical profile and small dimension.[43] In most cases the spindle whorls are decorated on the surface with curvilinear incisions. This geometrical decoration involving lines, zigzags and dots is typical of the Anatolian region and was also found in Tarsus[44] as well as at Yanarlar.[45]

Turning to Syria, it is useful to recall a group of spindle whorls (55 objects) found at Ebla, dating to the first half of the 2nd millennium BC.[46] They are remarkably homogenous in shape and represent various typologies. The group appears to be rather standardized in terms of materials and shapes. It includes various types of stone such as agate, serpentine, basalt, soapstone and limestone, and commonly used materials such as clay and bone. The only spindle whorl dating back to the Late Bronze Age IA is made of serpentine and was found in a cistern-pit (P. 5213).[47]

The contexts of these findings are spread and it is interesting to note how their distribution can be non-homogeneous and without relevant concentration.[48] Few objects come from the votive cisterns in the holy area of the Ištar Temple (dating to the Middle Bronze Age).[49]
The spindle whorls were often linked to domestic and productive contexts or, as in the Western Palace of Ebla, to some craftsmanship quarters inside the palace.[50] Sometimes these tools are also connected to symbolic or ritual contexts as in the Royal Hypogeum located under the floor level of the southeast part of the palace. There are two spindle whorls that could be part of the funerary deposit of the "Tomba delle Cisterne"; one is made of limestone, the other of agate.[51] A third one, made of bone, was uncovered in the corridor, between the "Tomba della Princepessa" and the "Tomba del Signore dei Capridi".[52]

The Late Bronze Age *corpus* of spindle whorls from the site of Ugarit represents the best-preserved documentation for these instruments for the period.[53] They were found both in public

[40] Frangipane *et al.* 2009, 6–7, figs 2, 3, 8, 13.
[41] Such as spindle whorls, loom weights, brushes, beaters, spools, needles (Frangipane *et al.* 2009, 22).
[42] Frangipane *et al.* 2009, 26.
[43] Mellaart and Murray 1995, 118–120, 163, fig. O.13, 164, fig. O.14, 166, fig. O.16, 167 fig. O.17 195.
[44] They display profiles different from those of Beycesultan, despite the fact that they share the same geometrical decoration. Goldman 1956, 331–334, figs 447–450.
[45] Emre 1978, 113, Pl. 44. Spindle whorls with geometric decorations were found also at Gordion, in graves dating back to Early Bronze Age. Cf. Mellink 1956, 43, Pl. 24.
[46] Peyronel 2004, 161–168.
[47] Matthiae 1998, 570–572.
[48] On the contrary, in the Early Bronze Age levels a great concentration of spindle whorls was found in the same contexts. Cf. Peyronel 2004, 100–104; Andersson *et al.* 2010, 161–163.
[49] These cisterns were used until Late Bronze Age for votive and religious purposes after the destruction of the old Syrian city at the end of the 17th century BC. Cf. Peyronel 2004, 70.
[50] Peyronel 2004, 171.
[51] The use of rare materials is an indicator of a probable elite destination of these objects. Cf. Peyronel 2004, 172.
[52] Matthiae *et al.* 1995, 429.
[53] For Ugarit's spindle whorls, see in general Yon *et al.* 1987, figs 7, 22, 27, 49, 53, 57, 66, 68, 85. For stone spindle

buildings and private houses. Through the study of these textile tools it is possible to note the typological evolutions of this kind of instruments in the second half of the 2nd millennium BC.[54]

The majority of the spindle whorls found in Ugarit are made of stone and presents: a) a circular flat base with a dome-shape profile (almost conical in some cases),[55] b) a circular flat base, conical profile but concave sides.[56] Spindle whorls made of bone and ivory present similar shapes but are less tall in profile (sometimes, almost flat disks) and exhibit a polished surface.[57] Bone or ivory spindle whorls with a conical shape and concave sides are common in the Syro-Palestinian area in the last phase of the Late Bronze Age.[58] The production of tools made of precious materials (e.g. ivory) was particularly well known in Syria and Palestine during the Late Bronze Age and required highly specialized workshops. In Ugarit it was proven that there were specialized local manufacturers for these instruments. Similar typologies of spindle whorls found in Ugarit but also in Cyprus and in the south Palestinian area, suggest the possibility of contact among these regions.[59]

Taking into account the archaeological context, Ugarit provides two very interesting cases. The first example was the discovery of ten spindle whorls in Building F; the fact that at least six of the objects were from one room (No. 1222) suggests the existence of a specialized area which was devoted to spinning.[60] The second case regards a small group of spindle whorls found in the "Temple aux Rhytons". It suggests that spinning activities could have been practised in the room nearby the sanctuary, directly connected with cultic activities.[61]

The site of Alalaḫ is one of the most important centres for the production and diffusion of textile technology because of its strategic position for the trade routes between Syria and Anatolia.[62]

According to Woolley, spindle whorls made of different materials like stone, bone and clay were present in all the levels of the site.[63] A selection was found on the floor of some rooms in the palatial building of the king Niqmepa, dating to the Late Bronze Age I.[64] In Alalaḫ, spindle whorls exhibiting low dome-shape profiles are prominent.[65]

A loom weight is an object used in the weaving process to give tension to the warp in a warp-weighted loom. It must have a certain weight to keep the warp in traction. In the case of perforated loom weights a string, to which the warp is fastened, should pass through the hole. Loom weights were commonly found in Anatolian archaeological sites, with few examples from the Middle and Late Bronze Age Syria and Palestine.[66] It is likely that the warp-weighted loom was already in use

whorls cf. Elliot 1991, 41–45; for bone and ivory, Gachet-Bizollon 2007, 19, 116. See now Sauvage 2013, focusing on spindle whorls coming from Ugarit in French museum collections.

[54] Spindle whorls are not commonly recorded for other Syrian sites. In Hama there is no evidence of these objects for the Late Bronze Age II. Also in Qatna there are scanty traces of these instruments. Cf. Peyronel 2004, 175; Baccelli 2011.

[55] Elliot 1991, 43, fig. 13 (4–14).

[56] Elliot 1991, 44, fig. 13 (5–21).

[57] Gachet-Bizollon 2007, 19, 116, Tav. 75.

[58] Peyronel 2004, 148.

[59] Peyronel 2004, 178; Elliot 1991, 44–45.

[60] Yon *et al.* 1983, 213–214.

[61] Mallet 1987.

[62] Woolley 1955.

[63] Woolley 1955, 271, Pl. 68c. Unfortunately, only a few decorated spindle whorls are shown in this publication.

[64] In rooms 6, 7, 8, 16 and 17 were found spindle whorls usually made of bone and mostly decorated with incised geometric motifs. Woolley 1955, 119–122.

[65] Peyronel 2004, 177.

[66] Peyronel 2004, 200. For the evidences of warp-weighted loom in the 4th and 3rd millennium BC, see Breniquet 2008, 274–277, figs 71, 72; 294–295, figs 84, 85; 297–300, figs 87–89.

in Syria in the 2nd millennium BC, together with the ground horizontal loom. Loom weights in Anatolia and Cyprus date back to the Neolithic period.[67] Scholars suppose that the warp-weighted loom was brought to Syria through Anatolia. This kind of loom could have come from Europe (where it was in use since the Neolithic period) through the Aegean regions to Ancient Near East.[68]

The presence of looms *in situ* can be inferred from archaeological evidence not limited to loom weights. The case of Troy is clear: four rows of loom weights found on the floor of a room indicate the use of looms (*in situ)* and the specific designation of this room for weaving.[69]

More than one hundred loom weights coming from documented archaeological layers were found in the site of Arslantepe (Malatya).[70] They were made of different materials: stone, fired and unfired clay. The object shape was determined by its material; for example, unfired clay loom weights were usually hemispheric in profile, while fired clay loom weights were generally conical or discoid in shape. Loom weights dating back to the 2nd millennium BC show more diversification.[71] The majority of these tools come from the same domestic context. A large square room (A 58) contained 55 loom weights made of either stone or clay.[72]

In many Anatolian sites a significant amount of loom weights exhibit a typology characterized by a crescent profile with two perforations. This represents a variant common in central-western Anatolia during the 2nd millennium BC.[73] Remarkable is the case of 300 such items found in Karahöyuk, 70 of them in the same room.[74]

In the *Absidenhaus* in Demirci Hüyük were found 12 loom weights of the crescent typology together with a basin and a series of vessels. These objects were likely to be used in the preparation of thread and for weaving.[75]

A similar situation is found at Beycesultan, where a vessel and 31 unfired clay loom weights were excavated.[76]

Loom weights with conical and tronco-conical profiles were found in the 2nd millennium BC levels in Alişar Hüyük, Tarsus, Troy and Boğazköy.[77]

Almost 50 loom weights were found in Alaca Hüyük, dating back to the Hittite period. These objects were characterized by crescent or discoid shapes.[78]

Also belonging to this period are four weights found in Maşat Hüyük[79] (exhibiting a crescent profile) and those found in Korucutepe (showing spherical profile).[80]

---

[67] The very first evidence comes from the Neolithic levels of Çatal Hüyük with the presence of pierced loom weights. Burnham 1965, 173.

[68] Mellaart 1962, 56; von der Osten 1937, 42, 93, 214; Barber 1991, 300. From the site of Alişar Höyuk come some pyramidal loom weights with hole made of clay and dating back to the Neolithic period, founded direct on the floor of the domestic contexts.

[69] Blegen *et al.* 1950, fig. 461; Blegen 1963, 72.

[70] Frangipane *et al.* 2009, 8, 9, 13, 23, 25. fig. 9, fig. 25.

[71] The experimental analysis conducted on these tools from Arslantepe show that it was possible to weave the same kind of thread using loom weights with different weights and shapes. Cf. Frangipane *et al.* 2009, 22–25.

[72] Frangipane *et al.* 2009, 27.

[73] Völling 2008, 140. For the crescent shape loom weights in Anatolia, see now Wisti Lassen 2013.

[74] Trench C, Level I of the Room 25. Cf. Alp 1968, 73–76, Pl. 143, 439, Pl. 144–245.

[75] See Kull 1988, 10–11.

[76] Lloyd and Mellaart 1965, 51, fig. F2, 22; Völling 2008, 140. The same quantity of tools from Kusura were found in a context together with animal bones.

[77] For a general view of the sites with evidence of loom weights see Völling 2008, 137, tab. 3.

[78] Koşay and Akok 1966, 160–162, fig. 21.

[79] Özgüç 1982, 120, fig. 61.

[80] van Loon 1978, 90, fig. 130.

As suggested before, the discovery of such a large quantity of loom weights collected together could indicate either the employment or the storage and conservation of these tools in a specific room, probably designated for weaving activities.

Loom weights were also found in Syria, which allows scholars to draw geographical comparisons.

Only two loom weights from Ebla dating to the Middle Bronze Age are recorded.[81] Their shape is elliptic with the superior part of the tool quite rounded and the base flattened with an ovoid section. They were brought into light in the North Area P, in a layer linked to the later structures of the "Archaic Palace". Unfortunately, the complexity of the stratigraphic layers, does not allow a more precise collocation of these objects.[82]

In Ugarit, evidence of loom weights dates to Late Bronze I. A great number of clay tools with discoid shapes was analysed and found to be analogous to Cypriot loom weights.[83] It is reasonable to assume that these objects have a Cypriote provenance, rather than being an evolution of the conic loom weights employed usually in the Syrian region.

Loom weights from Alalaḫ were collected from a small area of the north corner of the southwest wall of the private building (Level XIIB Room 10). Fifty loom weights made of lightly fired clay were found together with some pottery fragments.[84] This evidence suggests that a specific corner of the house was designated to weaving with a warp-weighted loom. The second very interesting example is that of loom weights found in the palace of Niqmepa, (Level IV), in Room C8.[85] This evidence could suggest the use of a warp-weighted loom during the Late Bronze Age, perhaps due to Anatolian influence.[86]

In conclusion, we can confirm the existence of spinning activities in Anatolia. However, because of the limited quantity of spindles and spindle whorls, we cannot assume that the production was comparable to that of Syria, which enjoyed a large and complex spinning production.

Taking into account the Middle and Late Bronze Age, weaving technology can be summarized as follows: weaving techniques remain the same throughout the Middle and Late Bronze Age in Anatolia and scholars observe the continuous use of the warp-weighted loom in domestic contexts. This would have been the most common instruments employed in weaving production in this area, although the well-known horizontal loom and the vertical two-beam loom were also used.

2.3. Written Sources (Part One: The Production of Textiles)

As already pointed out in the "Introductory Overview", even though textile production was of prime importance in the ancient Near East,[87] not much evidence seems available both from archaeological[88]

[81] Peyronel 2004, 199.
[82] Peyronel 2004, 200.
[83] Elliot 1991, 40, 41, figs 12 (7–14); 13 (1–3).
[84] Woolley 1955, 23; Peyronel 2004, 201.
[85] Woolley 1955, 130, fig. 51B.
[86] Peyronel 2004, 201.
[87] See, in general, Bier 1995.
[88] We have many archaeological finds (textile tools) from different excavations. We know less about the real workshops. Cf. section 2.1. It is important here to remember the interesting database project of the Bronze Age Eastern Mediterranean textile tools. See Andersson *et al.* 2010, 160.

and epigraphic contexts,[89] excluding a few representational pieces of evidence.[90] The 2nd millennium Anatolia does not represent an exception.[91]

Agnete Wisti Lassen has recently rightly stressed: "The perishable nature of archaeological evidence means that certain aspects of some crafts are completely lost, and it is often not possible to reconstruct processes and social religious aspects of the ancient crafts on the basis of physical remains alone. Studies in terminology can therefore corroborate both the archaeological evidence we possess, and shed light on issues not illuminated by archaeology at all";[92] hence, in this section are analysed many passages belonging to different text categories, selected as samples among the Hittite written sources; always bearing in mind that: "Textile production belongs to the periphery of the literate world, as it is frequently associated with the private sphere and the female gender".[93]

Private letters found in the ancient site of Kaneš (modern Kültepe, Turkey), provide us with interesting information about what was on demand on the markets in Anatolia during the first half of the 2nd millennium BC. Probably, they reflect a first stage of economic administration, essentially structured on local textile production in Aššur and, at the same time, on large scale distribution in Anatolia. Sifting through these letters, one can find some references to weaving techniques of that time. But these documents just inform us about textile production in Aššur.

Textile production in Aššur during the Old Assyrian Period was based on the labour of women, who actually spun and wove in their own homes. If we read the texts that regulated such domestic commitments, we can infer specifications of what was in demand on the markets in Anatolia, for example, through the detailed description on how a woven textile had to be processed. This is the case of a letter from the merchant Puzur-Aššur to the craftswoman, lady Waqqurtum (TC 3, 17).[94]

In his letter, Puzur-Aššur instructs Waqqurtum in how she should make her textiles in order for him to sell them on market. Lines 11–13 seem to be concerned with the finishing treatments of one side of such a textile;[95] lines 14–18 with the warping;[96] lines 19–22 with the finishing treatments of the other side of textile[97] and lines 33–36 with the size of textile.[98] Hittite documentation lacks analogous and precise information.

---

[89] With the valuable exception of written sources coming from capital centres of Sumerian Mesopotamia (e.g. Isin), Northern Syria (Ebla), or Egypt (Amarna). See, therefore, Waetzoldt 1972; van de Mieroop 1987; Pomponio 2010 (amounts of wool supplied); Biga and Milano 1984; Archi 1985; Pomponio 2008 (entrusted textiles); Kemp and Vogelsang-Eastwood 2001.

[90] See section 2.4.

[91] Our knowledge of the Anatolian textile production during the Early Iron Age seems quite different, thanks to the excavations of the Gordion workshop. Cf. Burke 2005, Burke 2010, 108–157.

[92] Wisti Lassen 2010b, 270.

[93] This statement is purposefully paraphrased from Wisti Lassen 2010b, 271. In addition to this the author adds: "Also, as in many other ancient societies, Mesopotamia was home to a large textile production administered by palaces and temples and recorded by bureaucrats. Yet, the terminology of administrative records kept in such large organisations tends to be generalised and focus on raw materials and products rather than on actual work procedures and tool repertoire".

[94] Refer to Michel and Veenhof 2010, 249–250, for the latest treatment of this document.

[95] *ša ṣubātim pānam ištēnamma limšudū lā iqattupūšu*: "One must strike the one side of the textile, and not shear it".

[96] *šutûšu lu mādat iṣṣēr panîm ṣubātim ša tušēbilinni šaptam* 1 *mana*-ta *raddīma lu qatnū*: "Its warp should be close. Add per piece one pound of wool more than you used for the previous textile you sent me, but they must remain thin".

[97] *pānam šaniam i-li-la limšudū šumma šārtam itaš'û kīma kutānim liqtupūšu*: "Its second side one should strike only lightly. If it proves still to be hairy, let one shear it like a *kutānum*".

[98] *gamram ṣubātam ša tepišīni tiše inammitim lu urukšu šamānē ina ammitim lu rupuššu*: "A finished textile that you make must be nine cubits long and eight cubits wide".

LÚ/MUNUSUŠ.BAR (Male/Female Weaver)

We can assume from a passage of the "Hittite Laws" that weavers were considered professionals:

> "*If anyone gives* (*his*) *son for training either* (*as*) *a carpenter or a smith, a weaver or a leather worker or a fuller*(?), *he shall pay 6 shekels of silver as* (*the fee*) *for the training. If he* (the teacher) *makes him* (i.e. the son) *an expert* (and retains him in his own employ?), *he* (the teacher) *shall give to him* (i.e. to the parent) *one person.*"[99]

We know that weavers involved in the palace system were sometimes assigned to different duties as skilled labour. For example, this cult inventory reports:

> "[*In the ci*]*ty of Uwalma, His Majesty has assigned to the gods what follows: one estate, wherein ten deportees* [of?][100] *high ranking state dependents*(?);[101] *one estate, wherein 16 deportees of* (assigned as/belonging to?) *mountaineers; one estate, wherein ten deportees, servants of Mr Innara; one estate, wherein four deportees of the priest; one estate wherein ten deportees, weavers of the king. The total is: five estates, including 50 deportees and 50 previous sheep* (i.e. belonging to former estates or personal ownerships)."[102]

Similarly in the cult inventory of Pirwa it is stated:

> "*His Majesty has instituted the following things:* [...] *40 deportees* (as?) *weavers of the town of Ḫariyaša.*"[103]

Hence, we can infer that weavers were generally not free craftsmen:

> "*If anyone buys a trained artisan – either a potter, a smith, a carpenter, a leather-worker, a fuller*(?), *a weaver, or a maker of leggings*(?) – *he shall pay ten shekels of silver.*"[104]

A passage of the treaty between Muršili II and Targašnalli of Ḫapalla, included in the fugitives' clause,[105] states as follow:

> "*But* [if] *he is a cultivator, or a weaver, a carpenter, or a leather-worker –whatever sort of craftsman – and he does not* [deliver] *his assigned work,* [but] *runs off and comes to Ḫatti, I will arrest him and give him back to you.*"[106]

We have clear exemptions of this kind of provisions in case a man becomes a weaver in "holy cities", like Arinna:

> "*Formerly the house of a man who became a weaver in Arinna was exempt; also his heirs and relatives were exempt.*"[107]

Like many other professions among the Hittites, palace weavers had a hierarchy. A chief of the weavers is involved in a rite for the royal couple:

---

[99] KBo VI 26++, col. IV 27–31. Cf. Hoffner 1997, 158–159, with note 573.
[100] For this restoration see D'Alfonso 2010, 77–78.
[101] For the term LÚ.MEŠ GIŠTUKUL.GÍD.DA, see recently D'Alfonso 2010, 76–78.
[102] KUB XLVIII 105 + KBo XII 53, obv. 31–34.
[103] KUB LVII 108 (+) KUB LI 23, obv. II 7', 13'. Cf. Hazenbos 2003, 103–105.
[104] KBo VI 26, col. II 21–26 (§ 176b). Cf. Hoffner 1997, 140–141.
[105] KBo V 4, obv. 35–40 (§ 6).
[106] Beckman 1996, 66. The same statement recurs in a passage of the treaty between Muršili II and Kupanta Kurunta (KUB VI 44++, col. IV 34–45). Cf. Beckman 1996, 75.
[107] KBo VI 3+, col. III 3–4 (§ 51). Cf. Archi 1975, 331; Hoffner 1997, 63.

> "*Two palace officials are squatting before the queen. They are holding a* karza(n)- (*from*) *below.*[108] *The chief of the weavers gives plaited*[109] *white wool to the chief of the palace officials. The chief of the palace officials braids it once. The chief of the palace officials gives it to the king. The king* (*braids*) *it twice and winds* (*it*) *around*[110] *the* karza(n)-."[111]

In a parallel passage we find the chief of weavers in a similar context.[112] Weavers were sometimes in charge of cult offerings, as stated in a number of fragmentary passages of instructions for cultic celebrations.[113] Weavers enrolled in the palace system seem coordinated by the weavers' overseer (UGULA $^{\text{LÚ.MEŠ}}$UŠ.BAR), as testified at least once in a land grant tablet of the king Arnuwanda I and his wife Ašmunikal.[114] It is important to remark that the information provided by the *Landschenkungsurkunden* (land grant documents)[115] points to a corporate organization of skilled textile labour activities in Hittite Anatolia, even if we cannot exclude cases of housework commitments. Female weavers are surprisingly attested, together with cowherds ($^{\text{LÚ.MEŠ}}$SIPA.GU$_4^{\text{ḪI.A}}$) and shepherds ($^{\text{LÚ.MEŠ}}$SIPA.UDU$^{\text{ḪI.A}}$), only in a fragmentary passage of the Ritual of Zuwi:

> "*The female weaver cleanses the cowherds and the shepherds.*"[116]

Similarly, female and male[117] weavers, offspring[118] of the underworld goddesses, Ešduštaya and Papaya, cited in the "Ritual of Kingship" (CTH 414.1) should be considered as ritual functionaries:[119]

> "*Ḫalmašuit* (i.e. the royal throne) *says to the king: «Now bring their sons to the palace window: the skilled female and male*(?) *weavers». Before* (one group) *of them he* (the priest?) *places the* zapzaki[120] *and strews figs* (thereon?); *before the other he places* kinupi (crockery?) *and strews raisins and dried fruits* (thereon?) (*saying*): *«Soothe ye the king».*"[121]

---

[108] For the meaning of *karza*(*n*)- see below.
[109] As per Melchert 1998. We wonder if *taruppand*[*an*] could indicate in this context plucked wool.
[110] Here we follow CHP, "L-N", 360.
[111] IBoT II 96, col. V 5–13. Lastly Melchert 2001, 405.
[112] KUB XI 20, obv. I 2–16. See below the translation of the passage. Cf. Melchert 2001, 404.
[113] KUB LVIII 7, obv. II 15: [LÚ.ME]Š UŠ.BAR *da-an-zi*. Cf. Hazenbos 2003, 41–42; Groddek 2005, 21. An additional fragment citing male weavers in charge of cult offerings for the Storm-God of Aleppo (KBo XIV 142, col. IV 6) can be added.
[114] KBo V 7, rev. 27.
[115] For this group of texts (LSU) see the fundamental work by Riemschneider 1958.
[116] KUB XII 63+, obv. 14.
[117] KUB XXIX 1, col. II 13. According to Marazzi (1982, 164), the second Sumerogram in line 13, should be read $^{\text{LÚ.MEŠ}}$BAR.DUL$_8$. Therefore it could be a copyist's misunderstanding considering the Late Hittite copy in the same passage (KUB XXIX 2, col. II 5): $^{\text{LÚ.MEŠ}}$U[Š.BA]R$^?$. Cf. HZL, 100, No. 20. We wonder if $^{\text{LÚ.MEŠ}}$BAR.DUL$_8$/DAB might refer to another profession. Cf. MZL 275, No. 121. For the translation of $^{\text{LÚ.MEŠ}}$BAR.DUL$_8$, see the online edition by Görke (2012) in http://www.hethport.uni-wuerzburg.de/hetkonk/ *s.v.* CTH 414.1.A.
[118] KUB XXIX 1, obv. II 12: DUMU.DUMU^{MEŠ}-*ŠU* has to be referred to the weavers as per Marazzi 1982, 152–153, 164 in particular. Cf. Görke (2012) in this passage.
[119] Cf. Haas 2003, 24.
[120] A glassware? See already Goetze 1947b, 313–315.
[121] KUB XXIX 1, obv. II 13–17.

LÚÁZLAG [LÚTÚG][122] (Male Washer/Fuller?)

The LÚ.MEŠÁZLAG are usually thought to be fullers.[123] In past times some doubt has been cast on this.[124] We know, in fact, that in Ancient Mesopotamia (e.g. during the Ur III Period), woollen cloths were not heavily fulled.[125] Indeed, looking at the bulk of the Mesopotamian attestations of 2nd millennium BC, it is difficult to propose that this Sumerogram always refers to fulling activities. Accordingly, the Sumerogram in Hittite texts can hardly been interpreted as fuller(s):

> "*Even as the* LÚ.MEŠÁZLAG [*ma*]*ke linen sheer*[126] *and clean*[*se*][127] *it of fuzzes, and it becomes white, may the gods likewise cleanse away* [*this*] *person's bad disease.*"[128]

The situation reflected in this passage should be the cleaning of linen. Since this passage is the most comprehensive so far and we have no other Hittite sources to propose a translation LÚ.MEŠÁZLAG "fullers" or the more general "finishers", we would cautiously propose "washers".[129]

A LÚÁZLAG is mentioned in a land grant document of Arnuwanda I,[130] maybe belonging to the house of Šuppiluliuma (the "scribe on wooden tablet(s)"), among other people included in the estate given by Arnuwanda himself and the queen Ašmunikal to the queen's attendant, Kuwatalla. A "house of washer" (É LÚÁZLAG) is also attested, even if it appears in a very fragmentary ritual context.[131] In a tablet of the cult of Nerik, some "washers" seem involved in a ritual together with other palace attendants.[132]

The quasi-absence of attestations of female ÁZLAG in the Hittite written sources, cannot demonstrate that such activities were set aside for men because it was a hard job. Nevertheless, this lack of references should not be underestimated.[133]

In two different texts washers are mentioned along with a name of a town, namely Taštariša, which should have been laid in the territory of Nerik, somewhere around the modern towns of Zile and Tokat, in North-central Turkey.[134] Once more they are involved in cult activities.[135]

---

[122] For the current readings, see HZL, 198, No. 212.
[123] See, for instance, Gelb 1955, 234, with previous literature. Cf. Akk. *ašlāku* in CAD "A/2", 445–447, last page in particular.
[124] Cf. already Leemans 1960, 64, note 4.
[125] Cf. Waetzoldt 1972, 153.
[126] Christiansen (2006, 45) translates: "[säuber]n".
[127] This interpretation of [*arḫa*] *parkunu-* seems quite satisfactory, Cf. Christiansen 2006, 45: "entfernen".
[128] KUB XXVII 67, col. II 26–30.
[129] Two points are debatable. First, we cannot assure that GADA (with phonetic complementation! [*-an*]) refers to a linen cloth. It could be simply flax, even if it should usually come with the determinative (GIŠ). In the latter case we would suggest "washers" more than "fullers" (retting process?). Secondly, the fact that GADA after the LÚ.MEŠÁZLAG's treatment becomes white (*ḫarkīšzi*) could point to (a) fulling process(es) of a linen cloth instead.
[130] KBo V 7, obv. 19, rev. 13, 41.
[131] KBo IX 125, col. IV 3.
[132] KUB LVI 54, rev. 26, with duplicates.
[133] KUB XXV 11, col. III 5. Looking both at the hand-copy and at the photo of the tablet, we would not include even this unique attestation, because we cannot assure that the last sign in MUNUSMEŠÁZLAG? is indeed a variant of ÁZLAG or TÚG. The scanty attestations in Mesopotamian texts of 2nd millennium BC are noteworthy. Cf., for example, Waetzoldt 1972, 154.
[134] Bo 6002, obv. 4; KUB LX 131, r. col. x+1-2. Cf. Lebrun 1976, 187–188. For the suggested locations see RGTC VI, 412; VI/2, 164–165.
[135] In KBo XXXIV 242, rev.? 6 (duplicate of KUB LX 131, 1'-7') we read: [(*ma-a*)]-*an* 2 LÚ.MEŠÁZLAG 2 MUNUS.MEŠUŠ[BAR?

LÚGAD.TAR (Tailor?)

Assuming a correct reading of the signs, if we look at the attestations of the term LÚGAD.TAR, there is no certainty that it deals with any profession related to textiles production.[136] With regards to this, the lexical list KBo I 30 (9')[137] offers us a misleading lexical equation (LÚgad.tar =lu-ga-ad-tar =*nu-'-ú* =*dam-pu-pí-iš*). As rightly observed by Klinger[138] the obscure meaning of the logogram is confirmed by the fact that both the Akkadian and the Hittite terms (*nû'u*; *dampupi-*) are matched with two different logograms.[139] Based on the context of attestations,[140] we would rather suggest that LÚGAD.TAR originally may have had a professional connotation. Then it could be interesting to know the real meaning of the equivalent Hattian term LÚ*tušḫawa*a*dun tanišawe* listed in the "Instructions for the gatemen".[141] In any case, by the time the texts containing this term were written, LÚGAD.TAR probably transformed to indicate more generally a palace functionary.[142] Moreover, the curious form LÚ.MEŠ*kat-ta-ru-ut-ti-š*?*a-za* of KUB LV 5, col. IV 8'[143] speaks once more against the identification with the Hittite term *dampupi-*, despite the supposed misinterpretation because of the form LÚ.MEŠGAD.TAR=*ma*=*za* of the main text KUB XXV 27, col. III 14.[144]

We have no Hittite texts that allow a clear reconstruction of the whole textile manufacturing process. The Hittite documentation offers us sporadic references to textile tools and techniques. Once more, we know from ritual texts that the yarn (*kapina-*)[145] is separated (*mārk-*i*/mark-*, *partae-*zi);[146] the wool (SÍG/*ḫulana-*) can be drawn/drafted (*ḫuitt(iye/a)-*zi),[147] tied (*ḫamank-*i*/ḫame/ink-*),[148] cut off/removed ([*arḫa*] *tuḫs-*a(ri)),[149] spun (*mālk-*i*/malk-*),[150] and clean[s]ed ([*arḫa*] *parkunu-*).[151] Textile tools used during the manufacturing process encompass spindle ((GIŠ)*ḫue/iša-*),[152] distaff

---

[136] Differently Pecchioli Daddi (1982, 53) suggests, with reservations, to translate it "tailor". For a full discussion of this Hittite logogram see Weeden 2011, 227–229.

[137] MSL XII, 214–215.

[138] Klinger 1992, 191.

[139] KBo I 30, 8': LÚaš.ḫab lu-aš-ḫa-ab *nu-'-ú dam-pu-pí-iš*. The basic meaning of LÚAŠ.ḪAB (Akk. *išḫappu*<Old Babylonian *ašḫappu*) "rouge, villain" throws an interesting light on the possibility that the scribe simply has repeated the lexical equation in line 8 and 9. For Akk. *nû'u* meaning either "foreigner", "uneducated man", see Weeden 2011, 228, note 1017, with previous bibliography.

[140] Cf. Pecchioli Daddi 1982, 53.

[141] KBo V 11+KUB XXVI 23, col. I 17. Cf. Soysal 2004, 318, 838–839. Weeden (2011, 228–229) reports further suggestions.

[142] Cf. HZL, 174, No. 173: "ein Funktionär?". Otherwise one would think of a Hittite logogram from an unattested Akkadian professional designation †*qattārum* "incense-burner" or more specifically the one who offers meat and fumigates the statues of the gods and other stuff with animals' hairs. It would fit better with the context of the scanty occurrences of Akk. *qadurtu* (see below) and the Hittite LÚGAD.TAR than any other, but that is only speculation. Similarly, already Weeden 2011, 229.

[143] Nakamura (2002, 56) has transliterated LÚ.MEŠ*QÀT-TA-RU-UT-TI*, probably with reference to the Neo-Assyrian *hapax qadaruttu* (meaning unknown), cognate of *qadurtu*. Cf. CAD "Q", 45–46. We cannot even exclude †LÚ.MEŠGAD.TAR-*UT-TI-ša-za* with bilingual (Akkadian and Hittite) phonetic complementation as Weeden (2011, 228) has already suggested.

[144] Houwink ten Cate 1988, 176. Cf. Nakamura 2002, 56–57.

[145] Cf. HED "K", 65.

[146] Cf. CHD "L-N", 187; "P", 197–198.

[147] Clearly in KUB XXVII 67, col. II 15–24.

[148] See the examples in HED "H", 64–67.

[149] E.g. KBo IV 2, col. I 28–29, 31–32, 36–39; VBoT XXIV, col. I 22–24; KBo XXXIX 8, col. II 9–10; KBo XLVI 38, col. I 5–6.

[150] See the passages cited in CHD "L-N", 131–132.

[151] Cf. CHD "P", 174.

[152] HW² Band III/2, 632–633.

(^(GIŠ)*ḫulāli-*),[153] and spindle whorl (*panzakitti-*).[154] Then the spun wool (*malkeššar*?)[155] can be cleaned of impurities ((SÍG)*mariḫši-*),[156] looped forming knots ((SÍG)*pittula-*).[157] These bundles of wool can come in large quantities. In a palace inventory coloured wool is listed. It is not completely clear if the material is assigned by the queen to a palace attendant, namely Anni; or if Anni herself has already made wool yarns out of a roving (SÍG*MUKKU*?)[158] and she gives them to the palace.[159] Anyway, it is reasonable to suppose that Anni takes charge of some textile activities. A huge amount of wool seems to be looped, even if it contains impurities.[160] Unfortunately, apart from Anni, only few women among the 22 quoted in the Hittite palace inventories seem to be connected to textile activities.[161] The majority is mentioned in connection with finished products allocated in palace storehouses or given as gifts to the queen. About 12 women seem to be entrusted to textile production, despite the difficulty in interpreting the term SÍG *gaši(š)-* of KBo XVIII 199(+)KBo II 22 as untreated wool.[162] The best written source we have so far about textile techniques and tools is the aforementioned ritual (for the fertility?)[163] that involves the royal couple (CTH 669.9). In a relevant passage it is stated:

> "*The chief of the wooden tablet scribes and the chief of smiths bring* malkeššar (spun wool?). *They pass in front of the fireplace. The chief of the smiths gives it* (spun wool?) *to the chief of the wooden tablet scribes. The chief of the wooden tablet scribes in turn gives* (*it*) *to the chief of the waiters, and he hangs* (*it*) *from a table. The king and the queen take white and red wool from the* karza(n)-, *and join/tie* (taruppanzi) *them* (*together*) *and they m*[*a*]*ke them into loops/knots?* (pittuluš)."[164]

In a second passage of the same ritual something more interesting is reported:

> "*The chief of the palace officials takes a* (*wool*) kunzan *and ties it onto a* (piece of) *wood. The chief of the table-men hangs it* (i.e. the wood stick) *from a table. The chief of the weavers mix white and red wool. He gives the belt to the chief of palace officials and he puts it on/in his* antaka (loins or chamber?).[165] *One escorts out the chief of the weavers. The acrobat cries "aha!" The chief*(*s*) *of the palace officials escort*(*s*) *in a shepherd. He takes the* karza(n)- *and carries it out.*"[166]

Needless to say that we are dealing with a ritual. Thus, it must be underlined that the text itself has a strong magical value and a clear metaphorical connotation. According to Melchert the mixing of the white and red wool should symbolise the successful sexual union of male and female.[167] At any

---

[153] HW² Band III/2, 691. For the terms spindle and distaff see also Ofitsch 2001, with previous bibliography.

[154] CHD "P", 95–96.

[155] Cf. CHD "L-N", 132. We wonder if this deverbal abstract noun can be translated "wool ready to be spun". Cf. EDHIL, 550.

[156] Cf. CHD "L-N", 186–187.

[157] Cf. CHD "P", 365–366. The term means generally "loop". According to the attestations, the (SÍG)*pittula-* is something used to tightly fasten hands. However, it cannot be excluded that bunches of fibres could come in loops as well.

[158] It could simply indicate a bad quality of wool. Cf. CAD "M/2", 187–188.

[159] KUB XLII 66, rev. 7'-8': *kī=ma=kan* SÍG*MUKKU ku-*[xxx] *ANA* F*anni=kan arḫa d*[*āi*? See the parallel text KUB XLII 102, 6'': F*a*]*nni*? *peran arḫa dāi*. For the construct *arḫa* + *kan* (+ motion verb) refer to Hoffner and Melchert 2008, 369 (§28.65). Cf. Mora and Vigo 2012, 180–181.

[160] KUB XLII 102, 10'': 10 MA.NA SÍG*pittulaš QADU :mariḫ*[*ši*].

[161] Cf. Mora and Vigo 2012, 177, 180.

[162] See Mora and Vigo 2012, 177–180, for a close examination of the term.

[163] As per Melchert 2001.

[164] IBoT II 94, col. IV 4–15.

[165] For the meaning of *antaka/i-* in particular contexts see Melchert 2003.

[166] KUB XI 20, col. I 5–21=KUB XI 25, col. III 2–14.

[167] Melchert 2001, 407.

rate we believe that in the relevant passages a description of a real handwork is illustrated. Strange as it may seem, the chief of the weaver is really plaiting wool using a very simple technique. A stick made out of wood is hung on a table surface and then wools of different colours are mingled. Since red and white wool are taken from the *karza(n)*, we would agree with Melchert in considering this object a sort of niddy noddy.[168] We cannot be sure that the braided belt is the result of white and red wool only, or if the *kunza-* was plaited together. We would not even exclude that *kunza-* could be a particular device[169] hung to a surface (door, wall or table) to help the stick to maintain the tension. It is interesting to note that every person involved in the ritual plays a specific role, as usual in these kinds of ceremonies. In particular, the chief of the weavers has to mix together the red and white wool that we presume were passed to him by the royal couple and, at the very end, the shepherd carries out the basket of wool possibly containing just a bunch of fibres. Considering the difficulty of a correct interpretation of the passage, one could ask whether the process depicted in this ritual might instead point to a doubling technique (in fact cording), as opposed to a draft-spinning of two or more threads. So, the white and red wool taken from the basket by king and queen are joined together by simply plaiting the fibres.[170]

The Hurrian textile production is well attested in the Hittite epigraphic sources.[171] We cannot exclude that corporations of skilled Hurrian weavers in Ḫattuša and in other Hittite palatial centres did exist, producing items that were typical of their native lands.[172] In a land grant tablet of the king Arnuwanda I and his wife Ašmunikal in favour of Kuwattalla, the queen's attendant, among the estates of the scribe on wooden tablet(s), Šuppiluliuma, is listed the estate (literally "the house") of a certain Muliyaziti, the "Hurrian shirt maker" (LÚ*EPIŠ* TÚG.GÚ.È.A *ḪURRI*).[173]

Linen came primarily from Egypt; wool from Anatolia and Northern Syria. Many textiles made of linen or wool were probably dyed in the Eastern Mediterranean islands and coasts, such as Cyprus (Alašiya),[174] Ugarit or Lesbos (Lazpa).[175] In a passage of a prayer to the Sun-goddess of Arinna, the Hittite king Muršili II characterises the semi-nomadic population of the Pontic region, namely the Kaška, as "swineherds and linen weavers".[176] Because both occupations were generally, but not always, performed by women, this exceptional comment could be read as an insult. Remains of flax plants dating back to the Middle Bronze Age have been found on the Black Sea coasts.[177]

---

[168] See Malchert 2012, 177 with note 9. Indeed, we were not so convinced that the Hittite word *karza(n)* – (basically "(mass of) spun stuff") could have been related with the Luwian hieroglyph sign 314 (phonetic value /ka-/ or /ha-/) and its graphic representation (a wool basket rotated 90 degrees?). Cf. Melchert 1999, 128–130. The stands (or tables) frequently represented in the 1st millennium BC funerary *stelae* in the Syro-Anatolian area (see section 2.4.), are usually surmounted by horizontal bands topped by three loops. In fact, contrary to what Melchert claimed (Melchert 1999, 130), they cannot be interpreted as women's wool baskets nor as spinning bowls with internal or external fixed loops, just because in many cases it is so evident that loops actually represent breads and other food. Moreover these stands/tables are depicted even associated with men. Cf. Bonatz 2000a, 92. For this kind of baskets, see in general Barber 1991, 70–77.

[169] See, for instance, Haas 2003, 687: "Wollgegenstand". However, the presence of the determinative (SÍG) is not useful to support this suggestion. Perhaps it could simply indicate the leading thread to which white and red wool are plaited at.

[170] For the "doubling" *vs.* "draft-spinning" see in general Barber 1991, 47–48.

[171] See Klengel and Klengel 2009.

[172] On this matter, see Vigo 2010, 294, note 35, with previous bibliography.

[173] KBo V 7, obv. 3, 13, 41. Cf. Rüster and Wilhelm 2012, 241.

[174] See Vigo 2010, 291–293.

[175] Singer 2008.

[176] Cf. Singer 2002, 52.

[177] Compare in general Singer 2007, 169–170 with references.

A presumed Hittite textile production, inferred from the analysis of textile tools found in archaeological contexts of 2nd millennium Anatolia,[178] can hardly be confirmed by Hittite written sources. Although the following pattern is based solely on the evidence of the inventory texts and may not be representative, a region of textile production can be hypothesised in the Hittite Lower-Land (South Cappadocia) and in the Kizzuwatnean area (close to the Taurus mountain range, between Turkey and Syria).[179] Unfortunately, we do not know if textiles named after cities or countries were always crafted there or followed the fashion of those places.

In order to acquire more knowledge about wool production in Anatolia during the Hittite Empire we should try to carefully join together and compare many text categories (cult and palace inventories, festivals, etc.), but it would require a long-term research. However, sifting through the texts we can suggest that wool was probably conveyed in warehouses (É *tuppaš*)[180] by provincial administrators (LÚ^MEŠ AGRIG) together with livestock and dairy products.[181] Then the wool was sent to various palaces and institutions as "compulsory gifts", ready to be converted into finished products.[182] From another palace inventory we are informed that a considerable amount of wool was assigned to administrators, some identified by their place of residence or storehouses of the kingdom. In this case the type of colour is surprisingly never indicated, which could mean that this allotted wool was unprocessed, perhaps waiting for further processing.[183] We can say even less about any Hittite dyeing production, besides the aforementioned coloured products sent to Ḫatti from Cyprus, Ugarit or Lesbos.[184] It is also difficult to ascertain if the dyed textiles cited in many text categories are generally the result of colouring processes or made of natural pigmentation. This is of course a matter of old debate and there is no need to insist on it. What can really be inferred from our cursory browsing through the Hittite texts, it is that in many cases dyeing could have been applied to yarns before being woven ("dyed-in-the-wool").[185] Regrettably, we cannot even assume that the terms *ašara-* and *gaši(š)-* cited in the inventory texts KBo XVIII 199(+)KBo II 22 refer to the colours of unprocessed wool, ready to be treated by the women mentioned in these documents.[186]

Textiles are primarily quoted in the Hittite texts for their symbolic value.[187] The scanty textile manufacturing processes we are able to draw from rituals and other religious texts are only faded

[178] Cf. the preceding section.

[179] See Vigo 2010, 296, note 55. It is interesting to note that these areas actually reflect those of wool production during the Old Assyrian period. Cf. Wisti Lassen 2010a, 169, fig. 2.

[180] For É *tuppaš* as warehouse of bags/baskets (^GIŠ*tuppaš*), see already Otten 1988, 15; Mora 2006, 133; van den Hout 2010.

[181] See, for example, Singer 1984, 109–110.

[182] Cf. Siegelová 1986, 213–245. We would tentatively interpret the amount of wool listed in category 5.5. ("IGI.DU$_8$.A-*Einkommen*") of Siegelová (1986, 213–256) as unfinished products, but ready to be converted into finest garments, often ^TÚG E.ÍB(.KUN) (*MAŠLU*) SIG$_5$, and other accessories. Cf. Siegelová 1986, 213–214. *Contra* Košak 1982, 127. Particularly, we consider the formula XX MA.NA/XX GÍN SÍG ŠÀ.BA XX GÍN MUG of KBo IX 90 and KBo IX 89 as "XX minas/XX shekels of wool including XX minas/shekels of broken wool fibres (not suitable to be spun?)". Cf. Waetzoldt 1972, 56–57, also for MUG=*mukku*. Conversely, we did not find any convincing interpretation of the Sumerogram ^GIŠŠU.TAG.GA of KUB XLII 48. See therefore Siegelová 1986, 242; 245, note 5. For the proposed reading SÙḪ in the same text, we would not stay with Siegelová (1986, 244 and note 4) either.

[183] See Bo 6489 in Siegelová 1986, 324–327.

[184] See in particular Singer 2008, 29–31.

[185] The best example is provided by the palace inventories that list incoming unprocessed wool of different colours (red, blue, green and yellow). Cf. Siegelová 1986, 90–91; 213–214.

[186] Cf. Mora and Vigo 2012, 179–180.

[187] See the very useful list of attestations in Haas 2003, 638–690.

mirrors of textile activities carried out in the 2nd millennium Anatolia and surrounding areas and they may just reflect regional (i.e. specific) features, more connected to the ritual praxis than to any textile activities. Ešduštaya and Papaya (Hurrian Ḫudena-Ḫudellura), the Hattian goddesses of fate (*Gulšeš*) spin in the underworld with spindles and distaffs the life of the Hittite kings and queens.[188] Maybe they are also assisted by ritual weavers (*katra/i*-women)[189] during birth rituals.[190] Just like the Greek *Moĩrai* they controlled the thread of life of every mortal from birth to death. The textile tools they used have a symbolic value too. Spindles and distaffs are, in fact, used during incantations against impurity and diseases connected to sexuality.[191] Spindle and distaff as symbols of femininity *versus* bow and arrows (the symbols of masculinity) are key tools used during martial[192] and funerary rituals.[193]

Colours for cloths (or garments) also have a symbolic value. Despite the fact that the perception of colours differs greatly in cultures and it is therefore difficult to find exact equivalents,[194] we know from Hittite texts that the predominant colours of textiles and garments were red, blue, green and purple. Black, white, red, blue and yellow colours have, indeed, a strong symbolic connotation. Natural coloured and dyed textiles are often used during rituals for their "chromotherapic properties" against diseases, evil and impurity.[195]

2.4. Representational Evidence

In the art of the ancient Near East, there exist representations of textiles that provide indications as to how and when particular kinds of textile were used and by whom.

Representation of textile technologies too may give specific information to better understand how textiles were produced. In this section, we analyse representations of spinning and spinning tools in Hittite Anatolia, then of weaving and weaving tools in the neighbouring regions that sent textile products to the centre of the Hittite kingdom.

Spindles and spindle whorls made of metal are some of the most interesting finds in late 3rd millennium BC funerary deposits at Alaca Höyük and Horoztepe;[196] they are also quoted in texts dating back to the second half of 2nd millennium BC[197] and can be found in visual art of 2nd and 1st millennium BC.

A Middle Bronze Age cylinder seal impression from Kültepe depicts a woman holding a spindle (Fig. 5.1).[198] She has both hands raised, offering the spindle to the god seated in front of her behind an altar or a banquet/offerings table. More objects for spinning –spindles or distaffs– are located behind the woman. A female figure on a seal impression from the North-Syrian site of Emar (modern Meskene) is holding a spindle in the same way (Fig. 5.2). In this 14th century BC example an altar/banquet table is present as well.[199]

[188] See, in general, Haas 1994, 372–373.
[189] Cf. Miller 2002, page 423 in particular.
[190] See Beckman 1983, 118–119.
[191] See, for example, the so-called "Paškuwatti Ritual" (CTH 406). Refer to Miller 2010 for the latest discussion on it.
[192] E.g. the "First Soldiers' Oath". Cf. Oettinger 1967, 10–13.
[193] Cf., for instance, Kassian *et al.* 2002, 98–99.
[194] See recently Vigo 2010, 302, with previous bibliography.
[195] Refer to Haas 2003, 638–649.
[196] For these objects see section 2.2.
[197] See section 2.3.
[198] Teissier 1994, No. 348. For discussion see Bonatz 2000a, 79.
[199] Beyer 2001, F7. Its style is similar to that of Mittani seals with local (Syrian) influences. But the comparison with

Other seal impressions show female figures that appear to be holding spindles, although the damage does not permit us to be sure about the object represented. This is the case of a stamp seal impression from Ḫattuša (Fig. 5.3), on which one can recognise a seated woman raising a cup and a spindle, while in front of her stands an offerings table.[200]

An interesting comparison for these scenes can be found in the iconography of a *stele* from Yağrı.[201] Most scholars date this monument to the second half of the 2nd millennium BC, although some doubts persist.[202] The relief shows a banquet scene involving two figures, a man and woman seated at each side of a table: the man is poorly preserved, but one can see a raised arm holding a cup in a way identical to the woman on the other side, still clearly visible. This second figure was probably the most important and she is holding a mushroom-shaped item, likely to be a spindle, in her left hand.

In order to find further representations of spindles in the art of Anatolia, one has to look at the funerary memorial monuments dating back to the 1st millennium BC. The funerary art of this more recent period could have been influenced by that (unpreserved) produced in the 2nd millennium BC.[203] These Iron Age *stelae* represent lone women, couples or three people sculpted in relief. On some of these monuments, women have attributes such as spindles, spindle whorls and distaffs (Figs 5.4–5.5):[204] in some cases a single spindle with its whorl, in others spindle and distaff together.[205]

In all the representations, the spinning tools are always full of fibres (flax or wool) or yarn. Distaffs are represented as sticks; the fibres are

Fig. 5.1: Seal Impression. Kaneš (18th–17th centuries BC). Teissier 1994, No. 348.

Fig. 5.2: Seal Impression. Emar (14th century BC). Beyer 2001, No. F7.

Fig. 5.3: Seal Impression. Ḫattuša (16th century BC). Boehmer and Güterbock 1987, No. 145d.

the Kültepe impression recalls the Anatolian iconography.

[200] Boehmer and Güterbock 1987, No. 145.

[201] First published by Crowfoot 1899, 40–45. See also Garstang 1929, 147–148, fig. 10; Bittel 1976, 201, fig. 230; Bonatz 2000a, 52–53. Cf. Darga 1992, 191, fig. 195.

[202] The few Anatolian (Luwian) hieroglyphic signs are difficult to date. See the remarks by Meriggi (1975, 263, 264).

[203] Cf. Bonatz 2000b, 204 and note 44, 210. Orthmann 1971, 377–380.

[204] See Bonatz 2000a. *Stelae* with representations of spinning tools: C 21–25, 27, 33, 50–52, 59–61, 62, 68, 69.

[205] For a lone spindle see Bonatz 2000a, Pl. 12, C22, for spindle and distaff see Fig. 5.5.

Fig. 5.4: Funerary stele. Maraş (8th century BC). Bittel 1976, fig. 313.

Fig. 5.5: Funerary stele. Maraş (8th century BC). Bonatz 2000a, Pl. 21, C60.

wrapped tightly around them forming a sort of round ball.[206] Spindles look similar, but the shape of the wrapped thread is, as expected, fusiform.[207]

In most representations it is impossible to distinguish the spindle whorl, even though one can imagine its location on the lower part of the spindle.[208] When one finds spindles and distaffs together, the spindle always appears smaller, but when alone, it can be bigger.[209]

Even when together, these objects represent symbols not in use: women hold these in the same hand, as one can clearly observe on a funerary monument dating back to 9th–8th centuries BC. (Fig. 5.5). On this monument, the banquet scene involves a seated man and woman and another woman standing. The woman on the chair rests an arm on the other's shoulder and in the left hand she holds a spindle and distaff. The standing woman, who might be the daughter of the deceased couple, raises a mirror in her right hand and again a spindle and distaff in her left.[210]

One of the *stelae,* coming from Maraş and dating to 8th century BC, shows a lady sitting with a spindle in one hand as a scribe stands in front of her (Fig. 5.4).[211] This scene could be interpreted as a representation of a private moment: the lady of the house spinning.[212] As stated by Dominik Bonatz, the smaller figures, depicted standing by the deceased, should be identified as descendants or heirs and not as servants.[213] Comparisons are evident when one looks at a little stone relief from Susa, dated to

[206] Cf. Fig. 5.5 and Bonatz 2000a, Pl. 23, C68. Völling 2008, 95, figs 30–31.
[207] Cf. Bonatz 2000a, Pl. 12, C22; Teissier 1994, No. 348.
[208] On a stone relief from Susa the spindle whorl is clearly visible located at the superior edge. Cf. Völling 2008, 93, fig. 27. On this relief see also *infra*.
[209] Cf. Bonatz 2000a, Pl. 18, C51; Pl. 20, C59.
[210] Women holding spindles and distaffs in one hand were sculpted on Greek and Roman funerary monuments too. For a brief overview see Völling 2008, 95, note 378 and figures and Rova 2008. More details in Cottica and Rova 2006.
[211] Bonatz 2000a, Pl. 18, C51.
[212] Völling 2008, 93–94.
[213] Bonatz 2000b, 191.

8th century BC, in which a woman seats with a spindle in her hands, while an attendant stands beside her. She holds the tool carefully in front of her, close to a banquet table.[214] In both of these representations, it is uncertain whether this performance represents the quotidian action, a ritual or a symbol. Thus, in Anatolian art, one finds no definite representations of women spinning. An example from the neighbouring region is a well-known intarsia panel from Mari, dating back to the first half of the 3rd millennium BC, depicting a scene involving at least two spinning couples.[215] Following a common interpretation[216] the woman standing holds a distaff helping the seated and spinning woman (on her left). However, it is more likely that these women are not spinning but rather making skeins. The woman standing holds a big spindle, as the seated one unwinds yarn with both hands.[217]

Clearly and in conclusion, the spindle and distaff mark femininity in all the Hittite examples. Many works have already pointed to the interesting symbolic connotation of these instruments.[218] As outlined in the previous section, spindles and distaffs symbolise womanhood in many Hittite texts. Thanks to visual art, one can add that mirrors symbolise femininity too. Visual representations and archaeological data confirm connections among spindles, distaffs and mirrors (e.g. in the grave goods of Horoztepe and Alaca Höyük).[219] In visual art, these objects appear together in some burial *stelae*, dating to the 1st millennium BC.[220] These three items occasionally represent goddesses' *regalia*,[221] inviting an interpretation of these women as priestesses.[222] Ancient texts connect spinning to people's destinies and to particular goddesses involved in childbirth.[223]

The yarn has an evident connotation with the thread of life and as women create thread, they also create life in all its aspects. Spindle and distaff represented in art stress the femininity in two ways: first, they are symbols of textile economic activities typical of women who were the main manufacturers of textiles. Women spent their whole lives spinning, weaving and crafting clothes; this was true for every status. Second, spindle and distaff stress the most important role of females: the creation of life. This second point is particularly interesting because, as noted above, these symbols are often represented on funerary *stelae* of the Neo-Hittite period. Maybe the spinning tools carved on these monuments represents a hope for the afterlife because a woman can, in the same way, re-generate the life as she could create yarn (similar to the umbilical cord) and textiles.[224] In

---

[214] Cf. Völling 2008, 93. She proposes that the lady is spinning very delicate yarn (appropriate for embroidery).
[215] Parrot 1961–1962, 178, figs 7–8.
[216] Völling 2008, 85–86, with references.
[217] Breniquet (2008, 292; Breniquet 2010, 60) is in favour of this second interpretation.
[218] Recently, Rova 2008; Cottica and Rova 2006.
[219] For the first see Özgüç and Akok 1958, 44; Pl. VII, 1. For the second Košay 1951, tomb L.
[220] For example, Bonatz 2000a, Pl. 13, C27, Pl. 19, C53, Pl. 21, C60.
[221] Ninatta and Kulitta, servants of Šauška, holding mirrors on the relief of Yazilikaya (see Bittel 1975, Pl. 22). Kubaba on a Karkemiš relief holds a mirror and probably a distaff (see Bittel 1976, 254, fig. 289). With regard to this, the representation of a figure with a spindle in one hand and a mirror in the other on the Hasanlu gold bowl is particularly interesting. Cf. Winter 1989, 101, fig. 14. This female divine figure seated on a lion has been associated with Kubaba, but lions, mirrors and spindles are attributes of the goddess Ištar/Šauška too.
[222] Cf. Yakar and Taffet 2007, 782.
[223] On the metaphorical meaning of spinning and weaving in other cultures see recently the bibliography offered in Michel and Nosch 2010b, x, note 35. For Lamaštu amulets with a spindle see Wiggermann 2000; Farber 1980–1983.
[224] The connection between the thread of life and the umbilical cord is self-evident and has been identified by anthropologists and psychoanalysts in many cultures, ancient and modern. On this topic see the still very interesting paper by Róheim 1948. Although these suggestions are very intriguing, the application of these models of analysis in the field of the Ancient Near East requires further studies.

some way, these representations provide a glimpse into the activities of the past life and a hope for a new one. The archaeological data provided in previous paragraphs, by underlining the presence of a huge number of spindle whorls in funerary contexts, could confirm the two hypotheses.

In the case of spindle whorls, one deals with items that were certainly used in daily life. In that of metal spindles in rich graves, scholars are not so sure. Their inclusion in funerary deposits seems not to depend on their use, but because they recall crafting activities as well as the femininity of the buried person. Otherwise, they can represent the hope for the pursuing of creation activities in the future.[225] They could also be items not used for spinning, but to perform rituals.

Once spun, the yarn is ready to be woven. Iconography could help by enlightening us on the nature of ancient looms, hence providing us with information about textile production.

Illustrations of looms appear on early Mesopotamian seals and ceramic vessels and on Egyptian wall paintings and tomb models.[226] These provide important documentary evidence that confirms the archaeological record and contributes to our understanding of loom construction in the ancient Near East and Egypt. The specific situation for Hittite Anatolia is different. In the total lack of such representations, archaeological finds and comparisons with images coming from other areas and periods help to determine the nature of looms used in Anatolia.[227]

As already pointed out, while the quotidian weaving in Anatolia and in the ancient Near East was generally done by women, on the contrary, some stages of textile production were probably entrusted to men. This is because some processes were hard and dangerous for children who were certainly spending the day with their mothers.[228] Ritual weavers were also women, but the craftsmen entrusted by the palace to weave precious textiles appear to be mainly males.[229] In ancient Near Eastern iconography, although seldom, one finds male weavers or men involved in other phases of the textile production.[230] A procession of a ceremony involving the queen is reproduced on an interesting Urartian belt that presents a seated male beating a finished rug in a corner (Fig. 5.6).[231]

3. Textile Use in Anatolia of 2nd millennium BC and in Neighbouring Areas

3.1. Archaeological Finds

The study of textile remains is crucial for a comparative analysis linking archaeological, epigraphic and iconographical data.

Textile remains found in funerary and non-funerary contexts are considered here separately, focusing on those dating to the 2nd millennium BC, with references to previous periods.

The first example of non-funerary context dates to the Old Assyrian colony period in Anatolia (19th–18th centuries BC). A number of samples of fabric impressions were identified on the back

[225] Barber 1994, 207–210.

[226] For these Early Mesopotamian seals and seal impressions reproducing weaving and the preparation of the warp, see Amiet 1972, Nos. 673, 674. For representations of different kinds of loom see Breniquet 2008, figs 84–91, Breniquet 2010, 61–62 and Fig. 4.7. For Egyptian tomb models see El-Shahawy 2005, 136, No. 84, Pritchard 1969, No. 142. For Egyptian paintings reproducing looms see Pritchard 1969, No. 143.

[227] For the vertical loom in Anatolia see section 2.2. For an interesting overview on Early Bronze Age Mesopotamia, see Breniquet 2008 and Breniquet 2010, 58 and fig. 4.2. Examples dating to the mid. 1st millennium BC Aegean area: Völling 2008, 145, fig. 54; Burke 2010, 106, fig. 52.

[228] On characteristics of women's work, so to conciliate children care, see anthropologist J. Brown, quoted in Barber 1994, 29–30.

[229] See Vigo in section 2.3.

[230] In Amiet 1972, No. 674 weavers' sex is not clear. Other examples proposed by Breniquet (2010, 63) are still uncertain.

[231] Ziffer 2002, 647, fig. 4.

Fig. 5.6: Urartian belt. Detail of attendants of the Queen (8th century BC). Ziffer 2002, fig. 4.

of a number of *bullae* of Kaneš/Kültepe.[232] Although the context of the discovery is often uncertain, Veenhof suggests that the clay sealings could have been used to seal containers such as bags sacks or clothes, travelling along the trade routes between Aššur and Cappadocia.[233] A selection of these textile bags contained neither food nor other kinds of goods but rather tablets. This evidence adds another element to the interpretation of textiles use: bags, sacks or textile containers could have been used not only in the trade of goods but also in the transportation of tablets.[234]

As far as the fabric imprints on a number of seal impressions are concerned,[235] it may be suggested that the seals were rolled over pieces of fabric. The use of fabrics as support in various activities would then represent a new element overlooked so far by analysis based on textual data or archaeological investigations.[236]

Examples of textiles from Kaman-Kale Hüyük also date to the Old-Assyrian colony period in Cappadocia. They were found in Room 150 (Kaman Phase IIIc). Most of these charred cloth pieces consisted of bundles of threads;[237] but among them is a small fragment of fabric with decorative motifs. In the first case we deal with loose thread where warp and weft are not definable, while the second example quite clearly presents a weaving structure known as "Sumac-technique".[238]

[232] As confirmed by cloth impressions on the back of *bullae* Kt.87/k328, Kt.87/k329. See Özguç and Tunca 2001, Pl. 92. Völling 2008, 240 FO(59).
[233] Veenhof 1972.
[234] Veenhof 1972, 28; Veenhof 1997.
[235] Cf. several seal impressions on *cretulae* among those published by Özguç and Tunca 2001. For example, Pl. 78 (St. 46).
[236] Völling 2008, 240 FO(59).
[237] Fairbairn 2004, 109, Pl. 118, fig. 3.
[238] Fairbairn 2004, 109, Pl. 118, fig. 4. Völling 2008, 240–241 FO(60). For details on this technique, see Völling 2008, 293–294.

According to Fairbairn, given the context, the charred fragments may have been part of bags used to store grains or belonging to the clothing of the inhabitants.[239] The second example is the oldest proof of decorative technique on fabric. According to the excavator, it could belong to a textile imported from Assyria to Cappadocia.[240] This finding suggests that in Anatolia, at the beginning of the 2nd millennium BC, there existed richly decorated fabrics. They were produced through different weaving techniques, embroidered with golden threads, overlaid with beads or probably decorative plaques, all of which contributed to the creation of different motifs.[241]

The site of Acemhüyük also provides evidence of textiles in Anatolia at the beginning of 2nd millennium BC. Three stripes of one fabric, unfortunately extremely burnt, were discovered in the Sarikaya palace.[242] It is interesting to note that some of these pieces were decorated with faïence beads and golden threads. They were probably part of a garment, enriched with the first evidence of a technique of decoration similar to medieval brocade.

It is important here to include some Anatolian funerary contexts, even if they exceed the chronological span of our analysis. Their peculiarity is the presence of well preserved textile remains with evidence of decorations and traces of colours.

In a funerary deposition of Alişar Hüyük, dating to the mid. 3rd millennium BC (Burial e X14), archaeologists found some fragments of fabric stuck to skin and bones.[243] Microscope analysis has identified traces of dark brown and yellow colours, suggesting that this may not be a shroud but instead a garment with a specific meaning.[244] According to this interpretation, this garment could have actually been worn by the deceased or used to wrap the body.

The use of valuable fabrics, often dyed with a symbolic value in connection with the funerary context, implies that garments and cloths were generally considered precious goods, as well as bearers of meanings.

Another remarkable and recently published find comes from the Royal Tomb of Arslantepe, dating to the beginning of the 3rd millennium BC.[245] The Royal Tomb is located in an isolated area and consists of a circular pit with a cist grave surrounded by stones. A male body was buried in the cist with a rich funerary deposit (two necklaces, a calcite vessel and 14 ceramic pots or jars). The body and a selection of grave-goods were placed on a wooden surface upon which were identified many traces of fibres, so abundant that the whole platform might have been originally covered by a sheet. Textile fragments were discovered near the shoulder and the left tibia of the body, others underneath the two necklaces suggesting that this fabric might have been used as a shroud or a mortuary dress. The deceased was not only decorated with jewels but also with precious fabrics, which, according to the archaeologists, were wrapped around the body and the grave-goods.[246] The presence of two adolescents' skeletons on top of stones covering the cist indicates a high social status of the deceased. A boy and a girl lay in an unusual position, both wearing a copper pin, two spirals in the hair and a diadem. They were probably also wearing a garment and a veil, as

[239] Fairbairn 2004, 109, 114.
[240] Fairbairn 2004, 115.
[241] These embroidered clothes are not attested in Old Assyrian texts, although they are quoted in texts from Mari according to Rouault 1977a, No. 6; Rouault 1977b, 151 (for embroidery or decorative applications similar to sequins, ll. 40–46).
[242] Refer to Völling 2008, 241, with previous bibliography.
[243] Völling 2008, 238–239, FO(57).
[244] Fogelberg and Kendall 1937, 334–335 and fig. 60.
[245] Frangipane *et al.* 2009, 17–20, fig. 14–15.
[246] Frangipane *et al.* 2009, 18.

suggested by the cloth fragments under the boy's diadem and others around the pins.[247] Two more female skeletons were located at the feet of the first couple. According to the position in the grave, they appear to have been of a lower social status.

Up to now, there have been no textile remains recorded for 2nd millennium funerary contexts in Anatolia. For the neighbouring area, it is important to recall the cases of Jericho in the Palestinian region and Tell el Saʾidīyeh in Jordan.[248]

The discovery of the Royal Tomb in Qatna is crucial for the comparison of funerary contexts in the Syrian area.[249] The textile remains brought to light in the Qatna Tomb come from different contexts and are located in different areas of the burial complex.

Two main groups of fabric remains will be investigated here. The first group deals with the remains identified through microscopic analysis of sediment samples found in many spots in many areas within the Royal Tomb. Traces of textiles were recorded, for example, in main Chamber (1), in Chamber 3–4 (on the floor),[250] along with fragments that show traces of purple dye.[251] Belonging to this group are fabric remains found inside the sarcophagus in Chamber 4, on the wood platform in the North-East corner of Chamber 1 along with a number of fragments in advanced state of mineralization, which were found attached to beads and golden objects.[252]

The second group of textiles encompasses a relevant number of well-preserved pieces found in deposits on a table in Chamber 4.[253] These remains show different levels and folds in the fabric stratification. In particular, they showed many coloured fragments with refined decorations, indicating that weavers were highly skilled in their craft.[254] This decoration involves the overlay of fabrics.

The findings in the Qatna Tomb are absolutely striking in their state of preservation and in their manufacture. They emphasise the prestige and luxury of these funerary contexts.

3.2. Written Sources (Part Two: The Use of Textiles and Garments)

Textiles as finished products are listed among luxury goods in many Hittite text categories. Since it is impossible here to refer to a huge variety of clothes mentioned in the Hittite documentation, we limit our survey to significant samples in an interdisciplinary perspective.

Textiles and garments were exchanged between royal courts. In a letter sent by the Hittite king Šuppiluliuma I to the pharaoh Amenhotep IV, found in the el-Amarna archives, the sovereign of Ḫatti tried to come to an agreement with the newly enthroned king of Egypt. In order to ease the process, Šuppiluliuma sent to his "brother" wonderful golden statues, embellished with lapis lazuli. Among the magnificent luxury goods that symbolise a new friendship after the death of the previous pharaoh, *ḫuzzi*-cloths are listed.[255] In a similar way, the king Tušratta of Mittani, a neighbouring land locatable to the modern Khābūr valley (North-Eastern Syria), needed to enhance the agreement he

[247] Frangipane *et al.* 2009, 19 and fig. 19.
[248] Crowfoot 1960; Crowfoot 1965; Pritchard 1980.
[249] Al-Maqdissi *et al.* 2002, 189–218.
[250] Reifarth and Drewello 2011, 469–482.
[251] James *et al.* 2009, 1109–1118; Reifarth and Baccelli 2009, 216–219; James *et al.* 2011, 449–468.
[252] Reifarth and Drewello 2011, 478, Pl. 2.
[253] Dohmann-Pfälzner and Pfälzner 2011, 483–485.
[254] Reifarth 2011, 499.
[255] EA 41, 35–36. The *ḫuzzi*-cloth may refer to a precious Hurrian fabric.

came to with the father (Amenhotep III) of the heir to the throne (Amenhotep IV).[256] Hence, he sent to the pharaoh a Hurrian tunic (TÚG.GÚ.È.A *ḪURRI*) and a precious over-garment (^{TÚG}BAR.DUL).[257] The "Hurrian shirt/tunic" seems to be one of the most fashionable garments among the ancient Near Eastern sovereigns, as also testified by its occurrence in the Hittite palace inventories.[258] Among the subjugated persons that appear on the wall paintings of the Men-kheper-Re-seeb's tomb in Thebes (Egypt), two have been identified as the "Prince of Ḫatti" and the "Prince of Tunip", respectively.[259] According to Goetze the latter is wearing what can be considered a Hurrian shirt.[260] The ^{TÚG}BAR.DUL ("cloak"/"mantle"?) forms part of the gods' clothing set in Mesopotamian texts, also in the Akkadian form *kusītu*.[261] It is mentioned in a letter between the pharaoh and the king of Cyprus (Alašiya)[262] and in an Egyptian inventory of goods stored in the treasury, from the el-Amarna archive.[263] The strange form ^{TÚG}*kušiši*(-)DUL quoted in a Hittite palace inventory together with minas and shekels of gold with copper as tribute, may perhaps indicate that the Hittite $^{\text{(TÚG)}}$*kušiši-* is a loan word from the Akkadian *kusītu*.[264] The logogram $^{\text{[TÚG]}}$BAR.DUL is attested only twice in a fragmentary palace inventory.[265] Sifting through the Hittite documentation, we can assume that the same over garment is mentioned several times in different text categories by means of the logogram ^{TÚG}BAR."TE".[266] According to Goetze the Hittite word for ^{TÚG}BAR.DUL/"TE" should be a neuter gender noun because of *ku-e* ^{TÚG}BAR."TE"^{MEŠ} in KUB VII 8, col. III 16.[267] This can be the case of the *i*-stem noun *kušiši-* indeed. Unfortunately the alleged forms [^{TÚG}BAR."TE"$^{\text{ḪI.}}$]$^{\text{A?}}$*-aš* of KUB IX 27, col. I 12 and ^{TÚG}BAR. "TE"*-eš* of KUB XXXV 133, col. I 21, although not clear at all, raise some doubts. Apparently, the ^{TÚG}BAR."TE" and the *kušiši*-garments appear in similar contexts. The BAR."TE"^{MEŠ} are frequently mentioned in palace inventories that list precious garments assigned to individuals and palace officers[268] or as luxury incoming clothes from different places and persons as tributes.[269] They usually come in blue purple or green-blue purple colours.[270] They are almost always listed together with shoes (^{KUŠ}E. SIR$^{\text{ḪI.A}}$), leggings/gaiters or underclothes (TÚG GAD.DAM$^{\text{MEŠ/ḪI.A}}$ or †*kattama-*?[271]), shirts (TÚG. GÚ.È.A), belts/waist-bands (^{TÚG}E.ÍB)[272] and head-covers (^{TÚG}SAG.DUL), forming the main elements of a complete dress. This kind of over garment is spread over a patient during a ritual:

---

[256] EA 27, rev. 110–111.

[257] The name denotes a ready-to-wear garment: "garment (*túg*) which covers (*dul*) the (out)side (*bar*)". Perhaps the logogram defines a kind of mantle.

[258] See Siegelová 1986, 651. This "Hurrian shirt" must be a more ornate variation of the simple shirt (TÚG.GÚ.È.A). It may be embroidered or trimmed with gold or silver. A good quality (SIG$_5$) shirt seems not being an expensive item though (three shekels). Cf. "Hittite Laws": § 182. Hoffner 1997, 145–146.

[259] Cf. Pritchard 1969, 15, fig. 45; 255, No. 45. Cf. the following section (3.3.).

[260] Goetze 1955, 54. Cf. Pritchard 1951, 39, fig. B; 40. Pritchard 1969, fig. 45; 255, No. 45.

[261] CAD "K", 586 h, 587 3'.

[262] EA 34, 23.

[263] EA 14, col. III 27.

[264] Cf. already Goetze 1947a, 178–179; Goetze 1955, 57.

[265] KBo XVIII 175, col. VI 1-2. In the same text we surprisingly find also BAR."TE" ḪI.ḪI (col. II 5). The result of this provisional search is based on a CHD files survey (January 2013).

[266] For ^{TÚG}BAR.DUL$_{(1)}$ ≠ ^{TÚG}BAR."TE", See already Goetze 1955, 57. For an in depth discussion whether DUL and TE are really different signs and how the Hittite conceived of these logograms, see Weeden 2011, 170–171.

[267] Goetze 1955, 57, n. 80.

[268] E.g. KUB XLII 106.

[269] E.g. NBC 3842.

[270] Here we follow the colour designations recently sketched by Singer (2008, 23–24).

[271] As tentatively suggested by Weeden (2011, 226–227).

[272] Cf. Weeden 2011, 377, 470. See below for further interpretations.

> "*The cloaks* (TÚGBAR."TE"MEŠ) *or the tunics which are lying on the soldier bread he will spread out* (each) *night* [...] *Once more they spread a bed for him down in front of the table. They also spread out below for him the cloaks or tunics which have been lying upon the soldier bread. The patient lies down,* (to see) *if he will see in a dream the goddess* (Uliliyašši) *in her body; she will go to him and sleep with him.*"[273]

It is also part of the festive garments to dress up statues of gods;[274] it is even worn by the king-substitute during a ritual.[275] The BAR."TE"MEŠ are also "presented", together with the garments mentioned above, to determine the exact aspect of a situation which has caused a deity's anger.[276]

The *kušiši*-garments are used to spread paths for gods:

> "*For you* (plur.) *I have spread paths with a swath* (TÚGkurešnit)[277] *of a* k. (TÚGkušišiyaš)";[278] "*Over the paths* (made) *of fine oil and honey he spread out a piece of cloth/a swath* (TÚGkureššar) *from the soldier bread below, saying as follows: «O Storm-god of Kuliwišna, keep walking on a path* (*made*) *of a swath* (TÚGkurešnaš) *of a* k.-*cloth* (TÚGkušišiyaš)*! And for you, may your feet not trample brushes and stones! May* (*the path*) *be smooth under your feet!*"[279]

We find the *kušiši*-garment in a ritual against impurity that implies as *Materia Magica* soldier bread and other garments[280] and in a funerary ritual.[281]

From a passage of the prayer of the king Arnuwanda I and Ašmunikal to the Sun-goddess of Arinna about the ravages of the Kaška people we can infer that *kušiši*-garments, though scarcely attested among the bare lists of tributes, were probably offered to deities in temples:

> "*The lands that were supplying you, O gods of heaven, with offering bread, libations, and tribute, from some of them the priests, the priestesses, the holy priests, the anointed, the musicians, and the singers had gone, from others they carried off the tribute and the ritual objects of the gods. From others they carried off the sun-discs and the* lunulae *of silver, gold, bronze and copper, the fine garments, the festive ones*? (TÚG.ḪIAadupli),[282] *shirts/tunics of a* k. (kušišiyaš),[283] *the offering bread and the libations of the Sungoddess of Arinna.*"[284]

The *kušiši*-garment is poorly attested in the palace inventories, but it is always listed together with other festive-garments (TÚG NÍG.LÁMMEŠ) like head-bands (*lupan*(*n*)*i*-) and *kureššar*.[285] Beside the *kušiši*[286] these two items can form the royal dress of kings and queens.[287]

---

[273] CTH 406 (Ritual of Paškuwatti against Impotence [or Homosexuality]): *Excerpta* §§ 12; 17. We believe that cloaks are spread over "soldier bread" and then on the bed in which the patient lays in order to absorb virility and pass it to the patient himself.
[274] E.g. KBo XLVII 266+, col. I 7–8.
[275] KBo VII 21, 8'-10'.
[276] KUB XXII 70, obv. 10–12. Cf. Ünal 1978, 33; 84–85.
[277] For the word TÚG*kureššar*- see below.
[278] KUB XV 34, col. I 40–41.
[279] CTH 329/330. Cf. Groddek 2007, 332–333.
[280] E.g. KBo XXIX 202+KBo XXXVIII 219, col. III 1–8. Cf. Groddek 1999, 34–35. See also KBo VII 29.
[281] KUB XXX 28(+)XXXIX 23, obv. 24–27.
[282] Siegelová (1986, 706) considers *ADUPLI* listed in the palace inventories as an Akkadogram. *Contra* Starke 1990, 207–208: *aduplit*- ("festive garment") as loan word from Akk. *a*/*utuplu* (+°*it*) (KUB LVIII 33, col. III 26: *aduplita* nom./acc. Plur.); but see, for instance, *a-tu- up-li-aš*! of KBo XXXIX 217, 5'.
[283] Following Singer's translation (2002, 41), we prefer to interpret it as a genitive singular instead of a comm. gender accusative plural of *kušiši*-. *Contra* Tischler (HEG "K", 674). Note also that it lacks of determinative TÚG.
[284] KUB XVII 21++, col. II 14'–17'.
[285] E.g. KUB XLII 14++; KUB XLII 55; KUB XLII 56.
[286] See the attestations in Goetze 1947a, 177.
[287] E.g. KUB XLII 98, col. I 10–12.

The Hittite term $^{\text{(TÚG)}}$*kureššar-*, literally “cut of cloth”,[288] basically defines a piece of cloth used during rituals:

> E.g. “*She*? *sets* [*the b*(*asket*)] *of “drawing* [*the deity…*] *along the road”* [(*dow*)]*n.* [*…they wr*(*a*)*p*] *the red wool* […] *She*? *spreads a cut of cloth* [and then] *she speaks as follow,* [(*cal*)]*ls the deceased* [(*by name:*)] *«May these reeds be* [the br(idg)]e? *for you!»*”[289]

It can also indicate the veil worn by goddesses and queens in religious contexts. It is attested as a precious garment entrusted to high dignitaries of the Hittite court and itemized among other ritual clothes.[290] We believe that both the *kušiši*-garment (hence cloak/mantle?) and this female cloth are represented in the Hittite rock reliefs.[291]

Based on this preliminary investigation of the Hittite textiles terminology, we would also tentatively suggest that the Hittite logogram ^{TÚG}E.ÍB quoted several times in the inventories of incoming items,[292] could be represented as part of the dress of the well-known king/deity in the “King’s Gate” relief of Ḫattuša. Looking at the belt that fastens that trimmed kilt (^{TÚG}ÍB. LÁ? *MAŠLU*),[293] the association with ^{TÚG}E.ÍB.KUN seems plausible. Indeed this waist band/belt sometimes provided with a sort of tale (KUN) as the one in the relief, is occasionally mentioned in texts together with golden or bronze inlays and weapons as they were part of a special kit.[294]

As it has been outlined in this paper, there are no textile remains surviving from 2nd millennium funerary contexts in Anatolia. Nonetheless, Hittite funerary rituals refer to precious/festive garments (^{TÚG}NÍG.LÁM^{MEŠ}) offered to the statues which might represent the royal couple during the funeral:

> “One?] *man* [puts] *a bow* [⟨and⟩ arrows] ⟨in⟩ *his* (i.e. statue of the deceased?) *hand. But* [*if it is a wo*]*man* (i.e. if the queen has died) [he puts] *a distaff* [and spindle ⟨in⟩ her hand.] *And* [they give?] *to her precious/ festive garments.*”;[295]

or to preserve the purified bones of the deceased:

> “⟨They take⟩ *a silver* ḫu<p>par-*vessel* (weighing) *twenty minas and a half*(?), *filled with fine oil. They tak*[*e*] *out the bones with silver tongs*? *and put them into the fine oil in the* ḫu<p>par-*vessel. They take them out of the fine oil and lay them down on the linen* kazzarnul-*cloth. A fine cloth is laid under the linen cloth. When they finish gathering the bones, they wrap them in the linen and fine cloths.*”[296]

This brief and selective survey on Hittite clothing aimed not only to show the use of textiles in different contexts but also to propose some key elements for further comparisons with the Hittite artistic production.

3.3. Textiles Art Representations: κτῆμα ἀεὶ

Textiles crafted in Anatolia or in the neighbouring areas were certainly used in many different ways. They were common in everyday life in the form of bags, clothes, bandages, bed clothes, but also

[288] Cf. HED “K”, 262; EDHIL, 494.
[289] KUB LX 87, rev.? 3’-10’ with dupl. Cf. Kassian *et al.* 2002, 730–731.
[290] See the list of attestations in Siegelová 1986, 363–369. Cf. Goetze 1947a, 178.
[291] Refer to following section (3.3.). Cf. already Goetze 1947a, 178, n. 19; Goetze 1955, 57, n. 78.
[292] E.g. KUB XLII 48.
[293] Cf. Goetze 1955, 56 also for the particular use of the verb *:putal(l)iya/e-* “put on light clothes”. But see also CHD “P”, 401–402. For trimmed kilt(?) (^{TÚG}ÍB.LÁ? *MAŠLU*) together with waist bands/belts see KBo XVIII 181, obv. 5, 24.
[294] Compare the attestations in Goetze 1955, 55–56.
[295] KBo XXV 184, col. II 60–62. Cf. Kassian *et al.* 2002, 98–99.
[296] KUB XXX 15+, obv. 3–8. Cf. Kassian *et al.* 2002, 260–261.

tents. As known from texts quoted in the previous pages, textile products were often considered as luxury goods. Precious clothes were, hence, richly decorated and used for furniture, garments and gifts or given as tributes. Visual art is somewhat revealing on these uses of textiles in ancient Anatolia.

The most obvious use of fabric in representational art is the depiction of clothing. In the art of the ancient Near East clothes indicate civilization and power.[297]

Garments have the primary purpose to protect the body, but have also other important functions such as indicating social status. In all the time periods, the elite wear better quality clothing than lower classes. Unfortunately, common people rarely appear in art while the majority of information found in visual representations concerns garments of the elite or ritual clothes, as well as the wonderful pieces thought to be worn by gods.

The case of Hittite Anatolia is suggestive of this problem. Observing Hittite art we realize that the main purpose was always symbolic. Reliefs and seals dating to the 14th–13th centuries BC reproduced human figures identifiable with gods and goddesses, kings and their families.[298] The garments worn by Hittite kings and queens on monuments and seals seem to be highly representative, in order to communicate immediately to the observer the power of the figure in front of him.[299] Many Hittite reliefs represent the king. In some cases he dresses as a warrior, but most times he wears particular garments – a long tunic or mantle – which appear to be ceremonial. The attitude of the sovereign in this last case is similar to that of a priest (Fig. 5.7). The Hittite king is represented this way, for example, on two orthostats at Alaca Höyük, in two in Alalaḫ, on two reliefs at Yazılıkaya, on the Sirkeli relief and on some seal impressions found in the Hittite capital and in other sites.[300] In some other examples (such as the reliefs at Yazılıkaya, an ivory plaque from Megiddo, a gold and lapis lazuli tiny figure from Karkemiš and seal impressions from peripheral sites such as Emar), divine figures are dressed in the same way as the priest-king.[301]

The king wears a two-piece robe: a loose-fitting, short-sleeved garment that reaches to the feet, and a cloak with edges falling over both shoulders. In some cases there is a sort of pointed tail on the back.[302] The hem of this mantle is often trimmed, although details are not clear. The king dressed in this way usually bears a round cap. In the example from Alaca Höyük (Fig. 5.7) though, he has the head and the back covered by a sort of long veil, probably fixed by a metal band.[303]

---

[297] It is important to remind that Enkidu in the epic of Gilgameš is metaphorically dressed once civilized (Bier 1995, 1582). Nakedness was specifically used to indicate prisoners, disgraced and humble people (for examples in seals, cf. Otto 2000, No. 434) or in fertility contexts (on naked women as symbol of fertility, see Mazzoni 2002; Pruss 2002). Significant exceptions where the heroes are naked, have also been encountered in visual representation in Anatolia (E.g. Kültepe seal impressions show this motif. See Teissier 1994, 161, 163–165).

[298] Reliefs dating to the 1st millennium BC, as those observed for the representation of spindles and distaff in section 2.4, will not be analysed here. For detailed description, see Özgen 1985. Moreover, we refer only occasionally to the garments worn by gods, leaving this topic for a future study.

[299] On this topic, see recently Bonatz 2007.

[300] Reliefs. Alaca Höyük: fig. 7. Yazılıkaya: Ehringhaus 2005, 25, figs 38, 44. Sirkeli: Ehringhaus 2005, 198, figs 175–176. Seals. Ḫattuša: Herbordt 2005, Nos. 317, 494; Herbordt *et al.* 2011, Nos. 39–42, 53.

[301] Cf. for Yazılıkaya Bittel 1976, fig. 234. For the Hittite ivory plaque from Megiddo see Loud 1939, Pl. 10F. For Karkemiš golden divine figures, cf. Bittel 1976, fig. 242. For Emar seals, Beyer 2001, Nos. A1, A7, A10, etc. On the problem of the identification of the figure surmounted by a winged sun, see Mora 2004, 446–447 with previous bibliography. Examples of images of actual priests performing rituals are encountered among the extremely interesting reliefs in Alaca Höyük. Cf. Bittel 1976, figs 212, 220, 222.

[302] Cf. Ehringhaus 2005, 25, fig. 38.

[303] See recently Vigo 2010, 310–315.

Fig. 5.7: Offering scene. King and Queen. Relief from Alaca Höyük (14th century BC). Ehringhaus 2005, fig. 3.

The king is represented as a divine warrior on seal impressions and on reliefs, such as two relief blocks from Temple 5 (Fig. 5.8) and from Chamber 2 in Boğazköy, the Hittite capital Ḫattuša (Fig. 5.9), and a rock relief in Firaktin.[304] Princes are also represented as warriors, for example, on rock reliefs at Hanyeri and Hamide.[305] In all these cases the royal figures wear a short kilt. On their head, they can wear a high, conical hat with multiple horns, but also a rounded cap as in the cases of Hanyeri and Hamide. Details of the kilt are not always clear, as exemplified in Fig. 5.9. In Fig. 5.8 the kilt is one piece, trimmed at the lower edge and worn with a thick belt, while in Firaktin, Hanyeri and Hamide's examples edges appear overlapping at the front.

The martial kilt worn by the male figure on the so-called King's Gate at Boğazköy is very short.[306]

This type presents an elongated edge overlapping the actual kilt in the front. The decorative pattern of bands of diagonal hatches and volutes that probably represented an actual garment is remarkable (Fig. 5.10). According to Elizabeth Barber,[307] the cloth "must have been woven vertically on the loom – or the fringed edge woven separately and sewn on." Alternatively this could be a sort of belt/waist band as cautiously suggested above.

In Egyptian wall paintings representations of people identified as Hittites dressed in a short white kilt, or wearing a light tunic with a sort of kilt that probably was military attire are depicted

[304] Bittel 1976, fig. 198.

[305] Bittel 1976, figs 201, 202. Ehringhaus 2005: 70–80, 107–112. Bonatz 2007, 120–123. See, recently, different interpretations by Simon 2012.

[306] Bittel 1976, figs 267–268. Whether one should identify the figure as the divinized king or a god is a debated question. Similar kilts in Bittel 1976, figs 148, 262, 263.

[307] Barber 1991, 336–337.

Fig. 5.8: King Tuhdaliya as a warrior. Neve 1993, fig. 100.

Fig. 5.9: Warrior, tentatively identified as king Suppiluliuma II. Relief from Ḫattuša-Südburg, Kammer 2 (13th century BC). Ehringhaus 2005, fig. 54.

(Fig. 5.12).[308] On other Egyptian reliefs, Hittite soldiers wear long garments, wrapped around the body.[309]

On festive occasions Hittite men wore a longer tunic with long sleeves called Hurrian shirt, which is often mentioned in Hittite palace inventories, as reported above. A Hurrian shirt has been identified by Pritchard as the clothing of some figures (although indicated as Syrians) on Egyptian wall-paintings.[310] In these paintings the shirt is white, but decorated in blue and red along the edges and with a long line running down the front. Other peculiar clothes were those worn by musicians and acrobats on reliefs at Alaca Höyük (Fig. 5.11). They look like knee-length long-sleeved robes with trimmed edges overlapped and shut in the front by means of a bow-belt. Also the clothing worn by a hunter on another Alaca relief is knee length, long sleeved and open at the front.[311]

[308] For representation of Hittites wearing short white kilt see Pritchard 1969, fig 45. For light tunic richly decorated and kilt see Pritchard 1969, fig. 35.
[309] Cf. Pritchard 1969, figs 7, 322, 333.
[310] Pritchard 1951, 40. Pritchard 1969, figs 45, 46 (see also catalogue, 255, Nos. 45, 46). Cf. here section 3.2.
[311] Bittel 1976, fig. 225.

Fig. 5.10: Figure sculpted on "King's Gate" at Ḫattuša. Detail of the kilt. Bittel 1976, fig. 268.

Fig. 5.11: Musicians and acrobats. Relief from Alaca Höyük (14th century BC). Ehringhaus 2005, fig. 3.

Fig. 5.12: Hittite prisoner on faiance tile (1195–1164 BC). Pritchard 1969, no. 35.

Women in ceremonial attire wore a tunic and a mantle-like veil. This could have been of different colours, as exemplified in Figs 5.15–5.17. Others wore a two-piece cloth similar to that of a goddess as in Fig. 5.13, i.e. a short-sleeved tunic with a round neckline and a veil covering the head. The tunic is not belted at the waist and reaches down to the ankles. The veil falls down the back of the goddess, but the damages does not allow a reconstruction of the top of the head.

Goddesses are represented with a long pleated skirt. Something similar is also worn by the queen performing a ritual together with the king in his priestly attire.[312]

Goddesses and queens sometimes wore a high *pólos*.[313] In some examples, dating to the beginning of the 1st millennium BC, a long veil covers the *pólos*.[314] As suggested in the previous section this long veil could tentatively be identified with the well attested *kureššar*-garment.

We lack visual information about the colours of these garments. Of course, there are some indications of colours of textiles in the Hittite texts, but, even when it is possible to translate them, the subject is complicated by the difficulty on how these hues were perceived.[315]

Fig. 5.13: Seated goddess. Relief from Alaca Höyük (14th century BC). Detail. Bittel 1976, fig. 216.

Apart from the Egyptian wall paintings and faïence tiles, the only support in Anatolian visual art comes from the *Reliefkeramik*, dating back to 16th century BC.[316] The red-polished surface of these vases was decorated in relief with the additional use of some colours like dark brown and white/cream.[317] In some cases the decoration involved human figures. These are men and women participating in a rite, some sort of a sacred marriage ceremony. Men wear short white/cream tunics with long sleeves, but in some cases we observe a sort of back extension that looks like a swallowtail (Fig. 5.16).[318] More seldom we note men wearing long, long-sleeved tunics of dark brown colour (Fig. 5.14). Women are usually dressed in white or cream long tunics with long sleeves, but dark tunics are also attested (Figs 5.14–5.16). Both men and women wearing this kind of long tunics

[312] This is the case of the queen as represented on a block relief in Alaca Höyük, here Fig. 5.7.

[313] Ehringhaus 2005, 22–23, figs 32–33.

[314] Bittel 1976, 253, fig. 287; 255, fig. 289.

[315] See Vigo 2010, 298–302, with references and previous paragraph.

[316] For an overview see Özgüç 2002. On the Bitik vase, see Özgüç 1957; Bittel 1976, 145; on İnandık vase, see Özgüç 1988; on the recently discovered Hüseyindede vase, see Yıldırım 2009.

[317] These colours – reddish-brown, very dark brown and cream (sometimes going to yellow) – are the local traditional colours beginning with this very period. Cf. Bittel 1976, 145.

[318] Bittel 1976, 145, fig. 144. Özgüç 1988, Pl. I; Yıldırım 2009, Pls. 27–29, figs 8–12.

Fig. 5.14: Royal (?) couple involved in a ceremony. Fragment of the Bitik vase (16th century BC). Bittel 1976, fig. 140.

Fig. 5.15: Woman, musician. Detail of the frieze of the İnandık vase (16th century BC). Özgüç T. 1988, Pl. K, fig. 3.

also wear a thick reddish brown belt around their waists, although sometimes it is hidden as in the case of the couple involved in a sort of marriage scene on the Bitik vase (Fig. 5.14).[319] Women are sometimes depicted with a long veil covering their heads and whole bodies; this is the case of the female figures sitting on the bed on İnandık vase and Hüseyndede vase and on the seat on the Bitik vase. In other cases the veil, coloured in dark brown, looks lighter and follows the back of the figures to the feet (Fig. 5.15).[320]

As already stressed, the garments worn by these figures are indicated in white (in some cases fading to yellow), with few exceptions. In some examples decorative bands are marked, coloured in dark brown or in relief.[321]

Men, women, gods and goddesses represented in Hittite art wear a peculiar kind of shoes with the point turned upwards. Identified as ^{KUŠ}E. SIR$^{HI.A}$ by scholars, they were usually made of leather and not included in the present paper.

Other uses for textiles among the Hittites are, unfortunately, rarely represented in art. The sample provided here deals with interior furnishings, such as beds.

Hittite inventory texts list precious textiles. Among other luxury goods, these catalogues of gifts or tributes register *lakkušanzani*-linen, interpreted as a kind of bed cover,[322] or a sort of canopy for the bed.[323] In Hittite art, one cannot yet find any representation of canopy beds. Although model beds are quoted in texts for ritual purposes, the actual models have not been preserved and this piece of furniture is not usually represented in art.[324] The decoration of one of the friezes of the İnandık vase includes

[319] Cf. Bittel 1976, 143, fig. 140.
[320] Özgüç 1988, Pl. I, 4, Pl. K, 3.
[321] Özgüç 1988, Pl. K, 4.
[322] Siegelová 1986, 604.
[323] Košak 1982, 17; recently Vigo 2010, 297.
[324] In Mesopotamia bed models or other representations of beds in art exist, for example in scenes of sexual intercourse or in scenes involving death or the healing of a sick person. See Nevling Porter 2002.

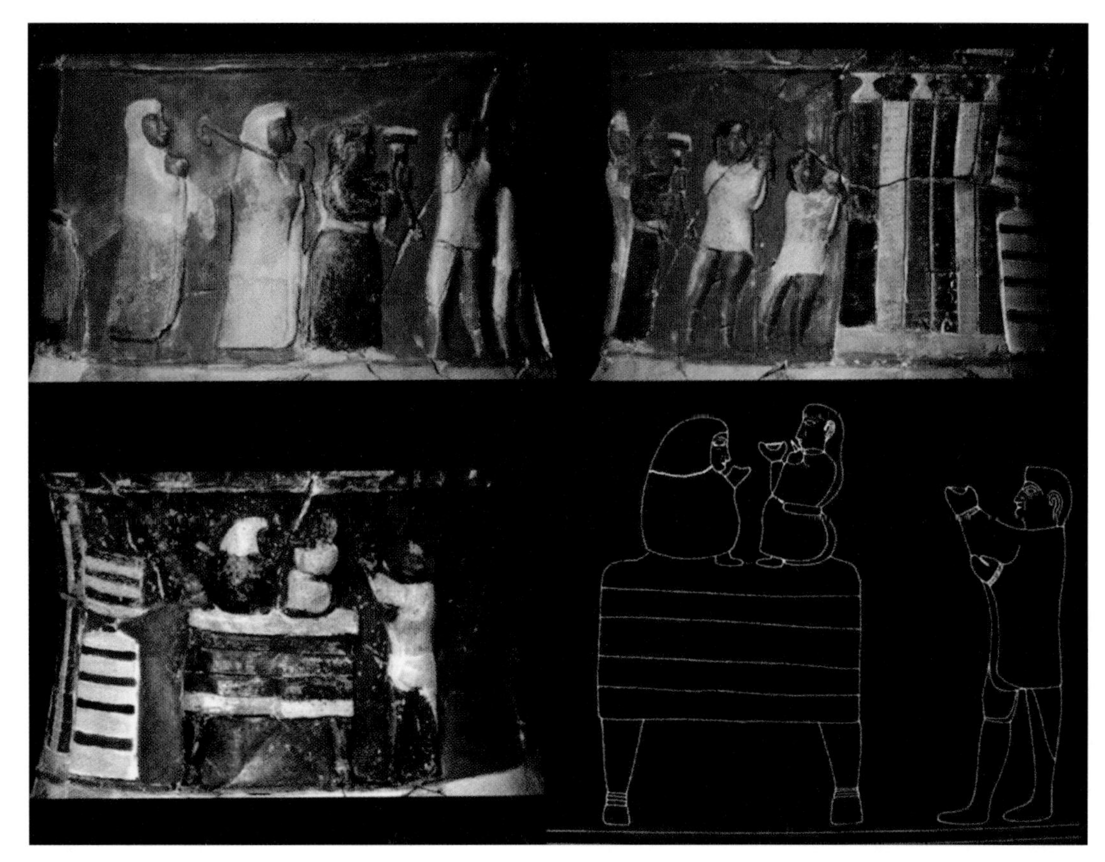

Fig. 5.16: Different scenes from the frieze of vase found in Hüseyndede (16th century BC). Yıldırım 2009, Pl. 28, fig. 11.

a rare image of a tall bed.[325] Either rich sheets or blankets cover the bed leaving the legs exposed. It appears basically white, decorated with two horizontal bands, one black and one brownish-red (the uncoloured surface of the vase).

The recent publication of the Hüseyndede vase adds a second artistic example of the same kind of bed (Fig. 5.16).[326] In this second example the bed is even higher, while the clothing is decorated with five bands: two white/cream and three very dark brown. The material of the sheets is unknown, although we cautiously opt for linen, considered more prestigious.[327]

Looking for comparisons in other representations from the ancient Near Eastern art, we encounter in Neo-Assyrian art some images of precious beds on which the king rests. Although these are not canopy beds, in some cases they were surmounted by precious tents that protected his majesty from the sun. Urartu art (1st millennium BC Eastern Anatolia), confirms the use of precious

[325] Scholars have interpreted the relief of a couple sitting on this bed as a sacred marriage ceremony (Özgüç 1988, 96, Pl. 51.1.). This interpretation has to be re-discussed in the light of the new evidence from Hüseyndede.
[326] Yıldırım 2009, Pl. 28, fig. 11.
[327] See the list of linen as bed clothes in Vigo 2010, 297 notes 68–69. On the prestigious use of linen attested in Mesopotamian texts see Waetzoldt *apud* Breniquet 2010, 54.

embroidered textiles. A beautiful metal belt depicts a veiled queen walking and performing some kind of ritual.[328] Two attendants protect her with a tasselled canopy. A second metal belt shows an almost identical procession,[329] but in this case there is also a nuptial bed. Two women raise sheets over it and allow interpreting the scene as a kind of marriage ceremony similar to that represented on the Old Hittite vases.

Artistic objects from other periods and geographical areas inform us that many other textiles were gifted as luxury goods. Neo-Assyrian reliefs and Urartian belts provide many examples of beautiful carpets and tapestries.[330] These objects were probably a work of art by themselves, as confirmed by the finding of a richly embroidered carpet in the so-called Pazyryk tomb (mid. 1st millennium BC).[331] Unfortunately, archaeological excavations in Anatolian and surrounding areas only reveal traces of these textile items.[332]

4. Concluding Remarks

The main purpose of the general overview of production and use of textiles in the 2nd millennium Anatolia presented here was to offer some elements for future comparative studies.

In doing so, we have tried to join the information provided by archaeological data, written documentation and iconography. The comparative approach of our research was successful in some cases, less so in others. As far as textile production of Hittite Anatolia is concerned, we have shown that weaving activities in domestic contexts are mostly drawn from the analysis of the archaeological data, but cannot be proved by written sources. On the contrary, the clay tablets found in the Hittite archives tell us something about the textile production in Anatolia, Syria and other neighbouring areas that did not leave archaeological traces. Pertaining to use, the presence of textile remains in funerary contexts of Hittite Anatolia can only be confirmed by Hittite funerary texts. Nevertheless, archaeology provides us with interesting finds from neighbouring areas and/or different periods (3rd millennium BC). Moreover, the multidisciplinary approach (iconography and philology) demonstrates that is sometimes possible to define the garments worn by Hittite elites, those exchanged between Near Eastern courts and the ready-to-wear dresses. The analysis of texts cannot be exhaustive in defining the spinning or weaving techniques of that time, because the clay tablets we have discussed were not meant to instruct anyone. The techniques mentioned are only incidentally preserved because they were part of a ritual.[333] The daily practices of these activities were carried out by skilled artisans and common people too (women in their domestic environments). In this context the related instructions were most likely transmitted orally. Perhaps the Hittite documentation offers examples of techniques that needed to be written, such as the glass-making instructions and the horse training. This is surely not the case of the Hittite *Textiltechnik*. The evidence considered here brings to light the lack of representations of spinning and weaving in Hittite Anatolia. One finds textile tools simply represented in ancient Anatolian art. Yet, when

[328] Ziffer 2002, 647 and fig. 4.
[329] Kellner 1991 No. 282.
[330] An example of a stone threshold from the Assyrian palace of Khorsabad imitating a carpet is A17598, Oriental Institute Museum, Chicago.
[331] For tapestries see Völling 2008, 173–181 and figures; now Smith 2013. For Assyrian textiles, Dalley 1991.
[332] See section 3.1.
[333] On the problem of the prescriptive *vs.* descriptive character of many "Hittite rituals"; the "Kizzuwatnean" ones in particular, see, above all, Miller 2004, 1–5; 476; 536–537.

observing these images closer, one can automatically interpret them differently. Iconography evolves to iconology: attempting to talk of quotidian life and local economy, one is forced to consider symbolism and religion. This lack of information does not mean insufficiency of Hittite visual representations. Rather, it means the absence of representation of daily life among the Hittites. The few narrative scenes recognizable on Hittite reliefs, for example, the ones found in Alaca Höyük, deal with rituals or sacred hunting, not with everyday life. Reliefs never depict battle and military triumph scenes. Hittite art was highly representative and served to perpetuate the high standing of the king. By focusing on religious aspects, this art reveals the piety of the king and his exclusive connection with gods. For this reason, the lack of spinning and weaving representations is not surprising, despite the fact that the crafting of textiles indeed comprised a very important component in the Hittite world.

The comparison among iconographic sources, archaeological remains and texts for the crafting of textiles in Hittite Anatolia encounters many problems. Some of the challenges in analysing these data include a difference in the quantity and quality along with the large chronological span.

As stated above, the few representations of spinning tools date to Middle and Late Bronze Age and to Iron Age Anatolia. Regarding the Early Bronze Age, the data is mainly derived from archaeological finds, and for the Hittite period mainly from texts. But, even if one compares this data in some way to "even out" the chronological span, there is a difference in meaning. The aim of the visual representation of spindles and distaffs (both on seals and on *stelae*) was clearly not to depict actual spinning. The same happens with texts. As pointed out, the texts that mention terms for spindle and distaff are not economic or administrative but mainly describe rituals. The comparison between epigraphy and art history is in this case possible and interesting, but does not properly concern the production of yarns. Spinning is a metaphor. The analysis of archaeological finds of spinning tools in Anatolia is partly headed in the same direction. Metal spindles were found in the contexts of graves. It becomes possible to compare archaeological finds and iconography of spindles. But were these tools really used? They were probably ritual items, used in life or forged just for funerary deposits.

A huge amount of spindle whorls also come from all principal Anatolian sites. The items are made mainly of clay and testify the regular, "real", spinning practice.

Fabrics and garments are listed in texts, but usually not described. They were found in excavations both in funerary and non-funerary contexts, but in a very poor state of preservation. Fabrics and garments are represented in visual art and provide important information about forms and tradition. The only garments depicted in art include those of religious contexts, similar to those worn by deities. Were these real garments? The appearance of some of the luxury clothes in art and in texts confirms that these garments were real. They were produced in Anatolia or in the surrounding regions – maybe in northern Syria – sent as tributes or gifts to the Hittite capital, treasured or committed to be refined and then worn during rituals and representative moments. These are the garments sculpted on block or rock reliefs, entrusted to eternity.

Because this field of research is very intriguing and liable to future expansion, archaeologists, art historians and philologists need the support of specialists from different disciplines. Scholars should combine linguistic analysis on textile terminology with knowledge provided by experimental archaeology in order to decode the terms of a technique, foreign to us today, and define the solid know-how's of crafts. Moreover, experts in topography, together with natural scientists, could help the archaeologists to define "textile topography".

Although the discrepancy between the North European tool-and-technique method and the South European historical method defines the framework of textile research around the world, research centres, such as the *Danish National Research Foundation's Centre for Textile Research* in Copenhagen, are developing projects that lead in this direction.[334] We hope that in the near future there will be other opportunities to merge different proficiencies and academic disciplines in order to obtain a more detailed and comprehensive picture of ancient textiles in addition to the one sketched here for the Hittite Anatolia.

Acknowledgements

Giulia Baccelli wrote 2.1., 2.2., 3.1.; Benedetta Bellucci wrote 2.4., 3.3., 4. (together with M.V.); Matteo Vigo wrote 1., 1.1., 2.3., 3.2., 4. (together with B.B.). The authors wish to thank C. Michel and Th. P. J. van den Hout for their helpful comments.

Abbreviations

AAS	Annales archéologiques arabes syriennes. Damascus.
ARET	Archivi Reali di Ebla: Testi. Roma.
ARM	Archives Royales de Mari. Paris.
BASOR	Bulletin of the American Schools of Oriental Research. Philadelphia – Boston.
Bo	Inventory numbers of Boğazköy tablets excavated 1906–1912.
BoḪa	Boğazköy – Ḫattuša. Ergebnisse der Ausgrabungen. Berlin.
CAD	R. D. Biggs, J. A. Brinkman, M. Civil, W. Farber, I. J. Gelb†, A. L. Oppenheim†, E. Reiner†, M. T. Roth and M. W. Stolper (eds), *The Assyrian Dictionary of the Oriental Institute of the University of Chicago*. Chicago 1956–2011.
CHD	H. G. Güterbock†, H. A. Hoffner Jr. and Th. P. J. van den Hout (eds), *The Hittite Dictionary of the University of Chicago*. Chicago 1980-.
CRAI	Comptes Rendus des Séances de l'Académie des Inscriptions et Belles Lettres. Paris.
CTH	E. Laroche, *Catalogue des Textes Hittites*. Paris 1971.
DBH	Dresdner Beiträge zur Hethitologie. Wiesbaden.
EA	Texts from el-Amarna, according to J. A. Knudtzon, *Die el-Amarna-Tafeln*, Vorderasiatische Bibliothek 20. Leipzig 1915.
EDHIL	A. Kloekhorst, *Etymological Dictionary of the Hittite Inherited Lexicon*, Leiden Indo-European Etymological Dictionary Series 5. Leiden – Boston 2008.
HED	J. Puhvel, *Hittite Etymological Dictionary*. Berlin – New York – Amsterdam 1984-.
HEG	J. Tischler, *Hethitisches Etymologische Glossar*, Innsbrucker Beiträge zur Sprachwissenschaft–Band 20. Innsbruck 1983-.
HW^2	J. Friedrich†, A. Kammenhuber†, (A. Hagenbuchner-Dresel *et alii*) (eds), *Hethitisches Wörterbuch.* (*Zweite, völlig neubearbeitete Auflage auf der Grundlage der edierten hethitischen Texte*). Heidelberg 1975-.

[334] http://ctr.hum.ku.dk/

HZL	C. Rüster and E. Neu, *Hethitisches Zeichenlexikon – Inventar und Interpretation der Keilschriftzeichen aus den Boğazköy-Texten*, StBoT-Beiheft 2. Wiesbaden 1989.
IBoT	Istanbul Arkeoloji Müzelerinde Bulunan Boğazköy Tabletleri(nden Seçme Metinler) – Istanbul 1944, 1947, 1954 – Ankara 1988.
JANER	Journal of Ancient Near Eastern Religions. Leiden – Boston.
JBL	Journal of Biblical Literature. Atlanta.
JCS	Journal of Cuneiform Studies. Baltimore.
JEOL	Jaarbericht "Ex Oriente Lux". Leiden.
KBo	Keilschrifttexte aus Boghazköi. Berlin 1916-.
KUB	Keilschrifturkunden aus Boghazköi. Berlin 1921-.
LSU	K. K. Riemschneider, *Die hethitischen Landschenkungsurkunden*, *MIO* 6 (1958), 321–381.
MDOG	Mitteilungen der Deutsche Orient-Gesellschaft zu Berlin. Berlin.
MIO	Mitteilungen des Instituts für Orientforschung. Berlin.
MSL	B. Landsberger, *Materialen zum Sumerischen Lexikon*. Roma 1937-.
MZL	R. Borger, *Mesopotamischen Zeichenlexikon*, Alter Orient und Altes Testament 305. Münster 2003.
NABU	Nouvelles Assyriologiques Brèves et Utilitaires. Paris.
NBC	Nies Babylonian Collection – Yale University.
OIP	Oriental Institute Publications. Chicago.
PIHANS	Publications de l'Institut historique-archéologique néerlandais de Stamboul. Leiden.
RGTC VI	G. F. del Monte and J. Tischler, *Die Orts- und Gewässernamen der hethitischen Texte*. Beihefte zum Tübinger Atlas des Vorderen Orients – Reihe B Geisteswissenschaften Nr. 7. Wiesbaden 1978.
RGTC VI/2	G. F. del Monte, *Die Orts- und Gewässernamen der hethitischen Texte: Supplement*. Beihefte zum Tübinger Atlas des Vorderen Orients – Reihe B Geisteswissenschaften Nr. 7/Supplement. Wiesbaden 1992.
RlA	Reallexikon der Assyriologie und Vorderasiatischen Archäologie, Berlin 1928-.
StBoT	Studien zu den Boğazköy-Texten. Wiesbaden.
TC	Tablettes cappadociennes du Louvre. Paris.
VBoT	Verstreute Boghazköi-Texte, Marburg a.d. Lahn.

Bibliography

Al-Maqdissi M., Dohmann-Pfälzner, H., Pfälzner, P. and Suleiman, A. 2003 Das königliche Hypogäum von Qatna. Bericht über die syrisch-deutsche Ausgrabung im November–December 2002. *MDOG* 135, 189–218.

Alp, S. 1968 *Zylinder- und Stempelsiegel aus Karahöyük bei Konya*. Ankara.

Amiet, P. 1972 *Glyptique Susienne, des origines à l'époque des Perses Achéménides*. Paris.

Andersson, E., Felluca, E., Nosch, M.-L. and Peyronel, L. 2010 New Perspectives on Bronze Age Textile Production in the Eastern Mediterranean. The First Results with Ebla as a Pilot Study. In P. Matthiae, F. Pinnock, L. Nigro and N. Marchetti (eds), *Proceedings of the 6th International Congress on the Archaeology of the Ancient Near East, Volume 2*, 159–176. Wiesbaden.

Archi, A. 1975 Città Sacre d'Asia Minore. Il problema dei laoi e l'antefatto ittita. *Parola del Passato* 164, 329–344.

Archi, A. 1985 *Testi amministrativi: assegnazione di tessuti (Archivio L. 2769)*, ARET 1. Roma.

Baccelli, G. (2011) Das Prestige und die Bedeutung von Textilien in West-Syrien in der II Jahr. Vor. Chr. Unpublished PhD thesis, Karls-Eberhards Universität Tübingen.
Balfanz, K. 1995 Eine spätbronzezeitliche Elfenbeinspindel aus Troja, VII A. *Studia Troica* 5, 107–116.
Barber, E. J. 1991 *Prehistoric Textiles: The development of the Cloth in the Neolithic and Bronze Ages with special reference to the Aegean*. Princeton.
Barber, E. J. 1994 *Women's Work. The First 20.000 Years. Women, Cloth, and Society in Early Times*. New York/London.
Beckman, G. 1983 *Hittite Birth Rituals – Second Revised Edition*, StBoT 29. Wiesbaden.
Beckman, G. 1996 *Hittite Diplomatic Texts*, Society of Biblical Literature: Writings from the Ancient World 7. Atlanta.
Bellinger, L. 1962 Textiles from Gordion. *Bullettin of the Needle and Bobbin Club* 46, 4–33.
Beyer, D. 2001 *Emar IV – Les sceaux*, Orbis Biblicus et Orientalis – Series Archaeologica 20. Fribourg (Switzerland)/ Göttingen.
Bier, C. 1995 Textile Arts in Ancient Western Asia. In J. Sasson (ed.), *Civilizations of the Ancient Near East*, 1567–1588. New York.
Biga, M. G. 1992 Les vêtements neufs de l'Empereur. *NABU* 19, 16–17.
Biga, M. G. 2010 Textiles in the Administrative Texts of the Royal Archives of Ebla (Syria 24th Century BC) with Particular Emphasis on Coloured Textiles. In C. Michel and M.-L. Nosch (eds), *Textile Terminologies in the Ancient Near East and Mediterranean from the Third to the First Millennia BC*. Ancient Textiles Series 8, 146–185. Oxford.
Biga, M. G. and Milano, L. 1984 *Testi amministrativi: assegnazioni di tessili (Archivio L.2769)*, ARET 4. Roma.
Bittel, K. 1975 *Das hethitische Felsheiligtum Yazılıkaya*. Berlin.
Bittel, K. 1976 *Les Hittites*. Paris.
Blegen, C. W. 1963 *Troy and Trojans*. New York.
Blegen, C. W., Caskey, J. L., Rawson, M. and Sperling, J. 1950 *Troy, Vol. I. General Introduction: The First and Second Settlements*. Princeton.
Boehmer, R. M. and Güterbock, H. G. 1987 *Glyptik aus dem Stadtgebiet von Boğazköy*, BoḪa 14. Berlin.
Bonatz, D. 2000a *Das syro-hethitische Grabdenkmal*. Mainz.
Bonatz, D. 2000b Syro-hittite Funerary Monuments. A Phenomenon of Tradition or Innovation? In G. Bunnens (ed.), *Essays on Syria in the Iron Age*, Ancient Near Eastern Studies Supplement 7, 189–210. Louvain-Paris-Sterling, Virginia.
Bonatz, D. 2007 The Divine Image of the King: Religious Representation of Political Power in the Hittite Empire. In M. Heinz and M. H. Feldman (eds), *Representation of Political Power. Case Histories from Times of Change and Dissolving Order in the Ancient Near East*, 111–136. Winona Lake.
Bordaz, L. A. 1980 *The Metal Artefacts from the Bronze Age. Excavations at Karataş.-Semayük. Turkey and their Significance in Anatolia, the Near East and the Aegean*. London.
Breniquet, C. 2008 *Essai sur le tissage en Mésopotamie: des premières communautés sédentaires au milieu du 3e millénaire avant J.-C.* Paris.
Breniquet, C. 2010 Weaving in Mesopotamia during the Bronze Age: Archaeology, techniques, iconography. In C. Michel and M.-L. Nosch (eds), *TTextile Terminologies in the Ancient Near East and Mediterranean from the Third to the First Millennia BC*. Ancient Textiles Series 8, 52–67. Oxford.
Burke, B. 2005 Textile Production at Gordion and the Phrygian Economy. In E. Kealhofer (ed.), *The Archaeology of Midas and the Phrygians: Recent Work at Gordion*, 69–81. Philadelphia.
Burke, B. 2010 *From Minos to Midas. Ancient Cloth Production in the Aegean and in Anatolia*, Ancient Textiles Series 7. Oxford.
Burnham, H. B. 1965 Çatal Höyük. The Textiles and the Twined Fabrics. *Anatolian Studies* 15, 169–174.
Calmayer, P. 1995 Möbel. B. Archäologisch. *RlA* 8, 334–337.
Christiansen, B. 2006 *Die Ritualtradition der Ambazzi. Eine philologische Bearbeitung und entstehungsgeschichtliche Analyse der Ritualtexte* CTH *391,* CTH *429, und* CTH *463*, StBoT 48. Wiesbaden.
Cottica , D. and Rova, E. 2006 Fuso e rocca: un percorso fra Occidente e Oriente alla ricerca delle origini di una simbologia. In D. Morandi Bonacossi, E. Rova, F. Veronese and P. Zanovello (eds), *Tra Oriente e Occidente. Studi in onore di Elena Di Filippo Balestrazzi*, 291–322. Padova.
Crowfoot, J. W. 1899 Exploration in Galatia Cis Halym, *The Journal of Hellenic Studies* 19, 34–318.

Crowfoot, E. 1960 Report on Textiles. In K. Kenyon (ed.), *Excavations at Jericho* 1, 519–526. London.
Crowfoot, E. 1965 Report on Textiles. In K. Kenyon (ed.), *Excavations at Jericho* 2, 662–663. London.
D'Alfonso, L. 2010 'Servant of the king, son of Ugarit, and servant of the servant of the king': RS 17.238 and the Hittites. In Y. Cohen, A. Gilan and J. L. Miller (eds), *Pax Hethitica – Studies on the Hittites and their Neighbours in Honour of Itamar Singer*, StBoT 51, 67–86. Wiesbaden.
Dalley, S. 1991 Ancient Assyrian Textiles and Origins of Carpet Designs. *Iran* 29, 117–132.
Darga, A. M. 1992, *Hitit Sanatı*. Istanbul.
Desroiers, S. 2010 Textile Terminologies and Classifications: Some Methodological and Chronological Aspects. In C. Michel and M.-L. Nosch (eds), *Textile Terminologies in the Ancient Near East and Mediterranean from the Third to the First Millennia BC.* Ancient Textiles Series 8, 23–51. Oxford.
Dohmann-Pfälzner H. and Pfälzner P. 2011 Archäologischer Kontext und Rekonstruktion des Bestattungstisches in Kammer 4. In P. Pfälzner (ed.), *Interdisziplinäre Untersuchungen zur Königsgruft in Qaṭna*, Qaṭna Studien 1, 483–485. Wiesbaden.
Ehringhaus, H. 2005 *Götter, Herrscher, Inschriften. Die Felsreliefs der hethitischen Grossreichszeit in der Türkei.* Mainz am Rhein.
Elliot, C. 1991 The Ground Stone Industry. In M. Yon (ed.), *Ras Shamra-Ougarit 6. Arts et industries de la pierre,* 9–99. Paris.
El-Shahawy, A. 2005 *The Egyptian Museum in Cairo: a walk through the alleys of ancient Egypt*. Cairo.
Emre, K. 1978 *Yanarlar. Afyon yöresinde bir hitit mezarligi- A Hittite Cemetery near Afyon*. Ankara.
Fairbairn, A. 2004 Archaeobotany at Kaman-Kalehöyük 2003. *Kaman-Kalehöyük Anatolian Archaeological Studies* 13, 107–120.
Farber, W. 1980–1983 Lamaštu. *RlA* 6, 439–446.
Fogelberg, J. M. and Kendall, A. I. 1937 Chalcolithic Textile Fragments. In H. H. von der Osten, J. A. Wilson and Th. G. Allen (eds), *The Alishar Höyük 1930–1932 III*, OIP 30, 334–335. Chicago.
Foster, B. R. 2010 Clothing in Sargonic Mesopotamia: Visual and written evidence. In C. Michel and M.-L. Nosch (eds), *TTextile Terminologies in the Ancient Near East and Mediterranean from the Third to the First Millennia BC.* Ancient Textiles Series 8, 110–145. Oxford.
Frangipane, M., Andersson, E., Laurito, R., Möller-Wiering, S., Nosch, M.-L., Rast-Eicher, A. and Wisti Lassen A. 2009 Arslantepe (Turkey): Textiles, Tools and Imprints of Fabrics from the 4th to the 2nd millennium BC. *Paléorient* 35/1, 5–30.
Garstang, J. 1929 *The Hittite Empire. Being a Survey of the History and Monuments of Hittite Asia Minor and Syria.* London.
Gachet-Bizollon, J. 2007 *Ras Shamra-Ougarit 16. Les Ivoires d'Ougarit*. Paris.
Gelb, I. J. 1955 *Old Akkadian Inscriptions in Chicago Natural History Museum. Texts of Legal and Business Interest*. Chicago.
Goetze, A. 1947a The Priestly Dress of the Hittite King. *JCS* 1, 176–185.
Goetze, A. 1947b Contributions to Hittite Lexicography. *JCS* 1, 307–320.
Goetze, A. 1955 Hittite dress. In H. Krahe (ed.), *Corolla Linguistica. Festschrift Ferdinand Sommer zum 80. Geburtstag am 4. Mai 1955*, 48–62. Wiesbaden.
Goetze, A. 1956 The Inventory IBoT I 31. *JCS* 10, 32–38.
Goldman, H. 1956 *Excavations at Gözlu Kule, Tarsus II*. Princeton.
Groddek, D. 1999 Fragmenta Hethitica dispersa VII/VIII. *Altorientalische Forschungen* 26/1, 33–52.
Groddek, D. 2005 *Hethitische Texte in Transkription. KUB 58*, DBH 18. Wiesbaden.
Groddek, D. 2007 Varia Mythologica. In D. Groddek and M. Zorman (eds), *Tabularia Hethaeorum. Hethitologische Beiträge Silvin Košak zum 65. Geburtstag*, DBH 25, 331–339, Wiesbaden.
Guy, P. L. O. 1938 *Megiddo Tombs*, OIP 33. Chicago.
Haas, V. 1994 *Geschichte der hethitischen Religion,* Handbuch der Orientalistik I/15. Leiden.
Haas, V. 2003 *Materia Magica et Medica Hethitica – Ein Beitrag zur Heilkunde im Alten Orient Vol. II.* Berlin/New York.
Hazenbos, J. 2003 *The Organization of the Anatolian Local Cults during the Thirteenth Century B.C. An Appraisal of the Hittite Cult Inventories*, Cuneiform Monographs 21. Leiden/Boston.

Herbordt, S. 2005 *Die Prinzen- und Beamtensiegel der hethitischen Grossreichszeit auf Tonbullen aus dem Nişantepe– Archiv in Hattusa*, BoḪa 19. Mainz.
Herbordt, S. *et al.* 2011, *Die Siegel der Grosskönige und Grossköniginnen auf Tonbullen aus dem Nişantepe – Archiv in Hattusa*, BoḪa 23. Mainz.
Hoffner, H. A. Jr. 1966 Symbols for Masculinity and Femininity. Their Use in Ancient Near Eastern Sympathetic Magic Rituals. *JBL* 85, 326–334.
Hoffner, H. A. Jr. 1997 *The Laws of the Hittites. A Critical Edition*. Leiden/New York/Köln.
Hoffner, H. A. Jr. and Melchert, H. C. 2008 *A Grammar of the Hittite Language*. Winona Lake.
van den Hout, Th. P. J. 1995 Tutḫalija IV, und die Ikonographie hethitischer Großkönige des 13. Jhs., *Bibliotheca Orientalis* 52/5–6, 546–571.
van den Hout, Th. P. J. 2010, [LÚ]DUB.SAR.GIŠ – "Clerk"? *Orientalia* 79, 255–267.
Houwink ten Cate, P. H. J. 1988 Brief Comments on the Hittite Cult Calendar: The Main Recension of the Outline of the *nuntarriyašḫaš* Festival, especially Days 8–12 and 15–22. In E. Neu, and Ch. Rüster, (eds), *Documentum Asiae Minoris Antiquae. Festschrift für Heinrich Otten zum 75. Geburtstag*, 167–194. Wiesbaden.
James, M. A., Reifarth, N., Mukherjee, A. J., Crump, M. P., Gates, P. J., Sandor, P., Robertson, F., Pfälzner, P. and Evershed, R. P. 2009 High prestige Royal Purple dyed textiles from the Bronze Age royal tomb at Qatna, Syria. *Antiquity* 83 (322), 1109–1118.
James M., Reifarth, N. and Eversched, R. P. 2011 Chemical Identification of Ancient Dyestuffs from Mineralised Textile Fragments from the Royal Tomb. In P. Pfälzner (ed.), *Interdisziplinäre Untersuchungen zur Königsgruft in Qaṭna*, Qaṭna Studien 1, 449–468. Wiesbaden.
Kassian, A., Korolev, A. and Sidel'tsev, A. 2002 *Hittite Funerary Rituals. šalliš waštaiš*, Alter Orient und Altes Testament 288. Münster.
Kellner, H. J. 1991 *Gürtelbleche aus Urartu*. Stuttgart.
Kemp, B. J. and Vogelsang-Eastwood, G. 2001 *The Ancient Textile Industry at Amarna*. Egyptian Exploration Society. London.
Klengel, H. 2008 Studien zur hethitischen Wirtschaft 4. Das Handwerk. Werkstoffe: Wolle und Leder, Holz und Rohr. *Altorientalische Forschungen* 35/1, 68–85.
Klengel, H. and Klengel, E. 2009 "Hurritische Hemden" in den keilschriftlichen Tradition. *Altorientalische Forschungen* 36/2, 205–208.
Klinger, J. 1992 Fremde und Außenseiter in Hatti. In V. Haas (ed.), *Außenseiter und Randgruppen. Beitrage zu einer Sozialgeschichte des Alten Orients. Xenia* 32, 187–212.
Košak, S. 1982 *Hittite Inventory Texts (*CTH *241–250)*, Texte der Hethiter 10. Heidelberg.
Košak S. 2002 *http://www.hethport.uni-wuerzburg.de/hetkonk/.*
Košay, H. 1951 *Les Fouilles d'Alacahöyük, Rapport préliminaire sur les travaux en 1937–39*. Ankara.
Košay, H. and Akok, M. 1966 *Ausgrabungen von Alaca Höyük. Vorbericht über die Forschungen und Entdeckungen von 1940–1948*. Ankara.
Kull, B. 1988 Textilherstellung (Spinnen und Weben) Textilgerät. In M. Korfmann (ed.), *Demircihüyük V, Die Mittelbronzezeitliche Siedlung*, 196–205. Mainz.
Lebrun, R. 1976 *Samuha, foyer religieux de l'empire hittite*, Louvain la-Neuve.
Leemans, W. F. 1960 *Legal and Administrative Documents of the Time of Hammurabi and Samsuiluna (Mainly from Lagaba)*, Studia ad Tabulas Cuneiformes a F. M. Th. De Liagre Böhl Collectas Pertinentia: 1/3. Leiden.
Lloyd, S. and Mellaart, J. 1965 *Beycesultan II*. London.
van Loon, M. N. 1978 *Korucutepe 2*. Amsterdam/New York/Oxford.
Loud, G. 1939 *The Megiddo Ivories*. Chicago.
Macqueen, J. G. 1975 *The Hittites and their Contemporaries in Asia Minor*. London.
Mallet, J. 1987 Le temple aux rhytons. In M. Yon (ed.) *Ras Shamra-Ougarit 3. Le Centre de la Ville, 38e-44e campagnes (1978–1984)*, 213–248. Paris.
Marazzi, M. 1982 "Costruiamo la reggia, 'fondiamo' la regalità": Note intorno ad un rituale antico-ittita (CTH 414). *Vicino Oriente* 5, 117–169.
Matthiae, P., Pinnock, F. and Scandone-Matthiae, G. 1995 *Ebla. Alle origini della civiltà urbana. Trent'anni di scavi in Siria dell'Università di Roma "La Sapienza"*. Roma.

Matthiae, P. 1998 Les fortifications de l'Ebla paléo-syrienne: Fouilles à Tell Mardikh, 1995–1997. CRAI 1998, 555–586.

Mazzoni, S. 2002 The squatting woman: Between Fertility and Eroticism. In S. Parpola and R. M. Whiting (eds), *Sex and Gender in the Ancient Near East. Proceedings of the 47th Rencontre Assyriologique Internationale, Helsinki, July 2–6, 2001*, 367–377. Helsinki.

Melchert, H. C. 1998 Once more Greek τολύπη, *Orpheus-JIEThS* 8, (2000 = Memorial Volume of V.I. Georgiev), 47–51.

Melchert, H. C. 1999 Hittite *karzan-* 'basket of wool'. In S. de Martino and F. Imparati (eds), *Studi e Testi II*, Eothen 10, 121–132. Firenze.

Melchert, H. C. 2001 A Hittite fertility rite? In G. Wilhelm (ed.), *Akten des IV. Internationalen Kongresses für Hethitologie. Würzburg, 4.–8. Oktober 1999*, StBoT 45, 404–409. Wiesbaden.

Melchert, H. C. 2003 Hittite *antaka-* "loins" and an Overlooked Myth about Fire. In G. M. Beckman and R. H. Beal and G. McMahon (eds), *Hittite Studies in Honor of Harry A. Hoffner Jr. on the Occasion of His 65th Birthday*, 281–287, Winona Lake.

Melchert, H. C. 2012 Hittite "Heteroclite" s-Stems, in ed. A. Cooper, J. Rau and M. Weiss (eds), *Multi Nominis Grammaticus. Studies in Classical and Indo-European linguistics in honor of Alan. J. Nussbaum on the occasion of his sixty-fifth birthday*, 175–184, Ann Arbor.

Mellaart, J. 1962 Excavation at Çatal Höyük. *Anatolian Studies* 12, 41–56.

Mellaart, J. and Murray, A. 1995 *Beycesultan. Vol. III. Part II. Late Bronze Age and Phrygian Pottery and Middle and Late Bronze Age Small Objects*. Ankara.

Mellink, M. 1956 *A Hittite Cemetery at Gordion*. Philadelphia.

Meriggi, P. 1975 *Manuale di Eteo-geroglifico. Parte II: Testi 2ª e 3ª Serie*. Roma.

Michel, C. 2001 *Correspondance des marchands de Kaneš au début du IIᵉ millénaire avant J.-C.*, Littérature anciennes du Proche-Orient 19. Paris.

Michel, C. and Nosch, M.-L. 2010a *Textile Terminologies in the Ancient Near East and Mediterranean from the Third to the First Millennia BC.* Ancient Textiles Series 8. Oxford.

Michel, C. and Nosch, M.-L. 2010b Textile terminologies. In C. Michel and M.-L. Nosch (eds), *Textile Terminologies in the Ancient Near East and Mediterranean from the Third to the First Millennia BC.* Ancient Textiles Series 8, vii–xviii. Oxford.

Michel, C. and Veenhof, K. R. 2010 The Textiles Traded by the Assyrians in Anatolia (19th–18th Centuries BC). In C. Michel and M.-L. Nosch (eds), *Textile Terminologies in the Ancient Near East and Mediterranean from the Third to the First Millennia BC.* Ancient Textiles Series 8, 209–269. Oxford.

van de Mieroop, M. 1987 *Crafts in the Early Isin Period. A Study of the Isin Craft Archive from the Reigns of Ishbi-Erra and Sh-Ilishu*, Orientalia Lovaniensia Analecta 24. Leuven.

Miller, J. L. 2002 The katra/i-women in the Kizzuwatnean Rituals from Ḫattuša. In S. Parpola and R. M. Whiting (eds), *Sex and Gender in the Ancient Near East Part II. Proceedings of the XLVIIᵉ Rencontre Assyriologique Internationale, Helsinki, July 2–6, 2001*, 423–431. Helsinki.

Miller, J. L. 2004 *Studies in the Origins, Development and Interpretation of the Kizzuwatna Rituals*, StBoT 46. Wiesbaden.

Miller, J. L. 2010 Paskuwatti Ritual: Remedy for Impotence or Antidote to Homosexuality? *JANER* 10/1, 83–89.

Mora, C. 2004 Sigilli e sigillature di Karkemiš in età imperiale ittita. I. I re, i dignitari, il (mio) Sole". *Orientalia* 73, 427–450.

Mora, C. 2006 Riscossione dei tributi e accumulo dei beni nell'Impero ittita. In M. Marazzi and M. Perna (eds), *Atti del Convegno «Fiscality in Mycenaean and Near Eastern Archives», Naples, 21–23 October 2004*, 133–146. Napoli.

Mora, C. 2007 I testi ittiti di inventario e gli 'archivi' di cretule. Alcune osservazioni e riflessioni. In D. Groddek and M. Zorman (eds), *Tabularia Hethaeorum. Hethitologische Beiträge Silvin Košak zum 65. Geburtstag*, DBH 25, 535–550.

Mora, C. and Vigo, M. 2012 Attività femminili a Ḫattuša. La testimonianza dei testi d'inventario e degli archivi di *cretulae*. In N. Bolatti-Guzzo, S. Festuccia and M. Marazzi (eds), *Centro Mediterraneo Preclassico. Studi e Ricerche III. Studi vari di egeistica, anatolistica e del mondo mediterraneo*, Serie Beni Culturali 20, 173–223. Napoli.

Nakamura, M. 2002 *Das hethitische nuntarriyašḫaš-Fest*, PIHANS 94.

Neve, P. 1993 *Ḫattuša. Stadt der Götter und Tempel*. Mainz am Rhein.

Nevling Porter, B. 2002 Beds, Sex, and Politics: The Return of Marduk's Bed to Babylon. In S. Parpola and R. M. Whiting (eds), *Sex and Gender in the Ancient Near East. Proceedings of the 47th Rencontre Assyriologique Internationale, Helsinki, July 2–6, 2001*, 523–535. Helsinki.
Oettinger, N. 1967 *Die Militärische Eide der Hethiter*, StBoT 22.
Ofitsch, M. 2001 Zu heth. *ḫueša*-: Semantik, Etymologie, kulturgeschichtliche Aspekte. In G. Wilhelm (ed.), *Akten des IV. Internationalen Kongresses für Hethitologie. Würzburg, 4.–8. Oktober 1999*, StBoT 45, 478–498.
Orthmann, W. 1971 *Untersuchungen zur späthethitischen Kunst*, Saarbrücker Beiträge zur Altertumskunde 8. Bonn.
von der Osten, H. H. 1937 *The Alishar Höyük, Seasons of 1930–32*, Vol. I–III, OIP 28–30. Chicago.
Otten, H. 1988 *Die Bronzetafel aus Boğazköy. Ein Staatsvertrag Tutḫalijas IV.*, StBoT-Beiheft 1.
Otto, A. 2000 *Die Entstehung und Entwicklung der Klassisch-Syrischen Glyptik*. Berlin/New York.
Özgen, İ. 1985 *A Study of Anatolian and East Greek Costume in the Iron Age*. University Microfilm International. Ann Arbor, Michigan.
Özgüç, N. and Tunca, Ö. 2001 *Kültepe-Kaneš. Mühürlü ve yazıtlı kil bullalar. Sealed and Inscribed Clay Bullae*. Ankara.
Özgüç, T. 1957 The Bitik Vase. *Anatolia* 2, 57–78.
Özgüç, T. 1982 *Maşat Höyük II. A Hittite Center Northeast of Boghazköy*. Ankara.
Özgüç, T. 1988 *İnandıktepe. An Important Cult Center in the Old Hittite Period*. Ankara.
Özgüç, T. 2002 Die Keramik der althethitischen Zeit. In AA.VV., *Die Hethiter und ihr Reich. Das Volk der 1000 Götter*, 248–255. Stuttgart.
Özgüç, T. and Akok, M. 1958 *Horoztepe – An Early Bronze Age Settlement and Cemetery*. Ankara.
Parrot, A. 1961–1962 La Douzième Campagne de Fouilles à Mari. *AAS* 11–12, 173–184.
Pasquali, J. 2005 Remarques comparatives sur la symbolique du vêtement à Ebla. In L. Kogan, N. Koslova, S. Loesov and S. Tishchenko (eds), *Memoriae Igor M. Diakonoff*, Babel und Bibel 2, 165–184. Winona Lake.
Pecchioli Daddi, F. 1982 *Mestieri, professioni e dignità nell'Anatolia ittita*, Incunabula Graeca 79. Roma.
Peyronel, L. 2004 *Gli strumenti di tessitura. Dall'età del Bronzo all'epoca persiana, Materiali e Studi archeologici di Ebla-IV, Roma.*
Pomponio, F. 2008 *Testi amministrativi: assegnazioni mensili di tessuti. Periodo di Arrugum*, ARET 15/1.
Pomponio, F. 2010 New Texts Regarding the Neo-Sumerian Textiles. In C. Michel and M.-L. Nosch (eds), *Textile Terminologies in the Ancient Near East and Mediterranean from the Third to the First Millennia BC.* Ancient Textiles Series 8, 186–200. Oxford.
Pritchard, J. B. 1951 Syrians as Pictured in the Paintings of the Theban Tombs. BASOR 122, 36–41.
Pritchard, J. B. 1969 *The Ancient Near East in Pictures Relating to the Old Testament. (Second Edition with Supplement)*. Princeton.
Pritchard, J. B. 1980 *The Cemetery at Tell el Sa'idīyeh, Jordan*. Philadelphia.
Pruss, A. 2002 The Use of Nude Female Figurines. In S. Parpola and R. M. Whiting (eds), *Sex and Gender in the Ancient Near East. Proceedings of the 47th Rencontre Assyriologique Internationale, Helsinki, July 2–6, 2001*, 537–545. Helsinki.
Reifarth, N. 2011 Die Textilien vom Bestattungstisch in Kammer 4. Vorbericht zu den mikrostratigraphischen und textiltechnologischen Untersuchungen. In P. Pfälzner (ed.), *Interdisziplinäre Untersuchungen zur Königsgruft in Qaṭna*, Qaṭna Studien 1, 499–526. Wiesbaden.
Reifarth, N. and Baccelli, G. 2009 Königsornat in Purpur und Gold- Die Textilfunde. In AA.VV, *Schätzte des Alten Syrien, Die Endeckung des Königsreichs Qatna*, 216–219. Stuttgart.
Reifarth, N. and Drewello, R. 2011 Textile Spuren in der Königsgruft. Vorbericht zu ersten Ergebnissen und dem Potential zukünftiger Forschungen. In P. Pfälzner (ed.), *Interdisziplinäre Untersuchungen zur Königsgruft in Qaṭna*, Qaṭna Studien 1, 469–482. Wiesbaden.
Riemschneider, K. K. 1958 Die hethitischen Landschenkungsurkunden. *MIO* 6, 321–381.
Riis, P. J. 1948 *Hama. Fouilles et recherches de la Fondation Carlsberg 1931–1938, II 3. Les cimetières à crémation.* Copenhagen.
Róheim, G. 1948 The Thread of Life. *Psychoanalytic Quarterly* 17, 471–486.
Rouault, O. 1977a *Mukannišum. L'administration et l'économie palatiales à Mari*, ARM 18.

Rouault, O. 1977b L'approvisionnement et la circulation de la laine à Mari. D'après une nouvelle lettre du roi à Mukannišum. *Iraq* 39, 147–153.
Rova, E. 2008 Mirror, Distaff, Pomegranate, and Poppy Capsule: on the Ambiguity of some Attributes of Women and Goddesses. In H. Kühne, R. M. Czichon and F. J. Kreppner (eds), *Proceedings of the 4th International Congress of the Archaeology of the Ancient Near East. 29 March–3 April 2004, Frei Universität Berlin*. Vol. I, 557–570. Berlin.
Rüster, Ch. and Wilhelm, G. 2012 *Landschenkungsurkunden hethitischer Könige*, StBoT-Beiheft 4.
Sallaberger, W. 2009 Von der Wollration zum Ehrenkleid. Textilen als Prestigegüter am Hof von Ebla. In B. Hildebrandt and C. Velt (eds), *Der Wert der Dinge-Güter im Prestigediskurs,* Münchner Studien zur Alten Welt 6, 241–278. München.
Sauvage, C. 2013 Spinning from old Threads: The Whorls from Ugarit at the Músee d'Archéologie Nationale (Saint-Germain-en-Laye) and at the Louvre. In M.-L. Nosch, H. Koefoed and E. Andersson-Strand (eds), *Textile Production and Consumption in the Ancient Near East: Archaeology, Iconography, Epigraphy*, Ancient Textiles Series 12, 189–214. Oxford.
Schaeffer, C. F. A. 1952 *Enkomi-Alasia. Nouvelles missions en Chypre 1946–1950*. Paris.
Shamir, O. 1996 Loom weights and Whorls. In D. T. Ariel and A. de Groot (eds), Excavations at the City of David, 1978–1985 IV. *Qedem* 35, 135–170.
Sharp Joukowsky, M. 1996 *Early Turkey. Anatolian Archaeology from Prehistory through the Lydian Period.* Dubuque, Iowa.
Siegelová, J. 1984 Gewinnung und Verarbeitung von Eisen im Hethitischen Reich im 2. Jahrtausend v.u.Z. *Annals Náprstek Museum* 12, 71–168.
Siegelová, J. 1986 *Hethitische Verwaltungspraxis im Lichte der Wirtschafts- und Inventardokumente*. Praha.
Siegelová, J. 1995 Möbel. A. II. Bei den Hethitern. *RlA* 8, 330–334.
Simon, Z. 2012 Hethitische Felsreliefs als Repräsentation der Macht – einige ikonographische Bemerkungen. In G. Wilhelm (ed.), *Organization, Representation, and Symbols of Power in the Ancient Near East, Proceedings of the 54th Rencontre Assyriologique Internationale at Würzburg 20–25 July 2008*, 699–714. Winona Lake.
Singer, I. 1984 The AGRIG in Hittite Texts. *Anatolian Studies* 34, 97–127.
Singer, I. 2002 *Hittite Prayers*. Atlanta.
Singer, I. 2007 Who were the Kaška? *Phasis* 10, 166–178.
Singer, I 2008 Purple dyers in Lazpa. In B. J. Collins, M. R. Bachvarova and I. C. Rutherford (eds), *Anatolian Interfaces. Hittite, Greeks and their Neighbours. Proceedings of an International Conference on Cross-Cultural Interaction, September 17–19, 2004 Emory University, Atlanta, GA*, 21–43. Oakville.
Smith, J. S. 2013 Tapestries in the Bronze and Early Iron Ages of the Anceint Near East. In M.-L. Nosch, H. Koefoed and E. Andersson-Strand (eds), *Textile Production and Consumption in the Ancient Near East: Archaeology, Iconography, Epigraphy*, Ancient Textiles Series 12, 161–188. Oxford.
Soysal, O. 2004 *Hattischer Wortschatz in hethitischer Textüberlieferung*, Handbuch der Orientalistik I/74. Leiden/ Boston.
Starke, F. 1990 *Untersuchungen zur Stammbildung des keilschriftluwischen Nomens*, StBoT 31.
Teissier, B. 1994 *Sealing and Seals on Texts from Kültepe Kārum Level 2*, PIHANS 70.
Ünal, A. 1978, *Ein Orakel Text über die Intrigen am hethitischen Hof (KUB XXIII 70 – Bo 2011)*, Texte der Hethiter 6. Heidelberg.
Veenhof, K. R. 1972 *Aspect of old Assyrian Trade and its Terminology*, Studia et Documenta ad Iura Orientis Antiqui pertinentia, Volumen X. Leiden.
Veenhof, K. R. 1997 "Modern" Features in Old Assyrian Trade, *Journal of the Economic and Social History of the Orient*, 40/4, 336–366.
Verderame, L. 2008 Il controllo dei manufatti tessili a Umma. In M. Perna and F. Pomponio (eds), *The Management of Agricultural Land and the Production of Textiles in the Mycenaean and Near Eastern Economy,* 101–133. Napoli.
Vigo, M. 2010 Linen in Hittite Inventory Texts. In C. Michel and M.-L. Nosch (eds), *Textile Terminologies in the Ancient Near East and Mediterranean from the Third to the First Millennia BC.* Ancient Textiles Series 8, 288–320. Oxford.
Völling, E. 2008 *Textiltechnik im Alten Orient: Rohstoffe und Herstellung*. Würzburg.

Waetzoldt, H. 1972 *Untersuchungen zur neosumerischen Textilindustrie*, Studi economici e tecnologici 1. Roma.
Weeden, M. 2011 *Hittite Logograms and Hittite Scholarship*, StBot 54.
Wiggermann, F. 2000 Lamaštu, Daughter of Anu, a Profile. In M. Stol (ed.), *Birth in Babylonia and the Bible, Its Mediterranean Setting*, Cuneiform Monographs 14, 217–252. Groningen.
Winter, I. 1989 The "Hasanlu Gold Bowl": Thirty Years Later. *Expedition* 31/2–3, 87–106.
Wisti Lassen, A. 2010a The Trade in Wool in Old Assyrian Anatolia. *JEOL* 42, 159–179.
Wisti Lassen, A. 2010b Tools, Procedures and Professions: A review of the Akkadian textile terminology. In C. Michel and M.-L. Nosch (eds), *Textile Terminologies in the Ancient Near East and Mediterranean from the Third to the First Millennia BC.* Ancient Textiles Series 8, 270–280. Oxford.
Wisti Lassen, A. 2013 Technology and Palace Economy in Middle Bronze Age Anatolia: the Case of the Crescent Shape Loom Weight. In M.-L. Nosch, H. Koefoed and E. Andersson-Strand (eds), *Textile Production and Consumption in the Ancient Near East: Archaeology, Iconography, Epigraphy*, Ancient Textiles Series 12, 78–92. Oxford.
Woolley, L. 1955 *Alalakh – An Account of the Excavations at Tell Atchana in the Hatay, 1937–1949,* Reports of the Research Committee of the Society of Antiquaries of London 18. Oxford.
Yakar, J. and Taffet, A. 2007 The Spiritual Connotations of the Spindle and Spinning: Selected Cases from Ancient Anatolia and Neighbouring Lands. In M. Alparslan, M. Doğan-Alparslan and H. Peker (eds), *Belkıs Dinçol ve Ali Dinçol'a Armağan. VITA. Festschrift in Honor of Belkıs Dinçol and Ali Dinçol*, 781–788. Istanbul.
Yıldırım, T. 2009 Hüseyindede: a Settlement in Northern Central Anatolia – Contributions to Old Hittite Art. In F. Pecchioli Daddi, G. Torri and C. Corti (eds), *Central-North Anatolia in the Hittite Period. New Perspectives in Light of Recent Research. Acts of the International Conference Held at the University of Florence (7–9 February 2007),* Studia Asiana 5, 235–246. Roma.
Yon, M., Caubet, A., Mallet, J., Lombard, P., Doumet, C. and Desfarges, P. 1983 Ras Shamra-Ougarit 1981–1983 (41e, 42e et 43e campagnes). *Syria* 60, 201–224.
Yon, M., Lombard P. and Renisio, M. 1987 L'organisation de l'habitat. Les maisons A, B et E. In M. Yon (ed.) *Ras Shamra-Ougarit 3. Le Centre de la Ville, 38e–44e campagnes (1978–1984)*, 11–128. Paris.
Zawadski, S. 2006 *Garments of the Gods. Studies on the Textile Industry and the Pantheon of Sippar according to the Texts from the Ebabbar Archive*, Orbis Biblicus et Orientalis 218. Freibourg/Göttingen.
Ziffer, I. 2002 Four New Belts from the Land of Ararat and the Feast of the Women in Esther 1:9. In S. Parpola and R. M. Whiting (eds), *Sex and Gender in the Ancient Near East. Proceedings of the 47th Rencontre Assyriologique Internationale, Helsinki, July 2–6, 2001*, 645–657. Helsinki.

6. Buttons, Pins, Clips and Belts… 'Inconspicuous' Dress Accessories from the Burial Context of the Mycenaean Period (16th–12th cent. BC)

Eleni Konstantinidi-Syvridi

During the Late Bronze Age, pictorial evidence from the Aegean seems to be restricted to representations of the formal or ritual costume. The mostly female figures on the frescoes from Thera, Knossos, Mycenae, and Tiryns, are considered to be related to religious ceremonies or feasts.[1] It is obvious however, that the typical Creto-Mycenaean costume depicted on the frescoes from the 16th to the 13th centuries BC, does not justify the plethora of clothing ornaments revealed in the burial and domestic contexts of the period. This seems to be the case already at the time of the Akrotiri frescoes which suggest that the jewellery in settlements does not correspond to that depicted in the wall-paintings.[2]

There were certainly other more practical dress types for everyday use; for the Mycenaean period, Linear B tablets document some types of costumes, the two most commonly mentioned being the short skirt, accompanied by an adjective meaning "flounced" and a short textile with sleeves (*pa-wo*), probably the one described in Homer (*Odyssey* V, 228–232) by the term "*pharos*", a kind of wrap-around cloak.[3] There is also the term *e-ra-pe-me-na* (*ραμμένα*) in Knossos tablet L 647,[4] referring to sewn textiles.

Turning to evidence from other cultures, we know that in Egypt, from as early as the 3rd millennium BC, clothes – generally made of linen and to a lesser extent of wool[5] – consisted of a simple short skirt for men which was wrapped around the hips and left the knees uncovered, and a long, narrow dress with straps for women.[6] Later, the length of the kilt reached the calf and a sleeveless shirt was added with an opening for the head cut at the centre. Circular capes and shawls were occasionally in use; other than that, Egyptian clothing did not change much over the centuries. Clothing ornaments were never in fashion but dresses were decorated with colour and embroidery.[7]

At the same time, in the north, Scandinavia has provided us with some of the oldest preserved woollen garments; in Borum Eshøj for example, a woman between 50 and 60 years old was buried

[1] Among others, Kritseli-Providi 1982; Warren 2005; Jones 2009.
[2] Televandou 1984; Vlachopoulos-Georma 2012, 38.
[3] Killen 1966, 109; Melena 1996, 170; Nosch 2012, 51.
[4] Tzachili 2000, 75, n. 14.
[5] Tzachili 2000, 73.
[6] Eman 1969, 202; Tzachili 2000, 75–76.
[7] Eman 1969, 227.

wearing a short tunic and a full-length skirt fastened with two belts, while in Egtved, a young woman aged between 18 and 20 years, wore a knee-length corded skirt and a short sleeved tunic.[8] Evidence from both Egypt and Europe indicates the existence of a variety of garments – some of them uniform – according to age, social or even marriage status of the individuals.

For the Aegean, since there are no actual remains of the textiles, apart from small bits and some indirect evidence,[9] it is not possible to reconstruct costume types other than the ones depicted on frescoes and figurines. Moreover, it is not known how popular the use of the burial shroud was, in geographical, chronological and social terms.[10] There are indications that, at least from a certain point onwards, people of a high social status were covered from head to toe in a shroud over their clothes, like the figures seen on the Tanagra clay *larnakes*.[11]

A closer look on the jewellery associated with the dress and clothing ornaments in particular, confirms a certain variety of Mycenaean costume types. Clothing ornaments of the period consist of discs or roundels, rosettes and cut-out plates in a variety of shapes, mainly of gold, with holes on the periphery for sewing onto the dress/shroud, pins, buttons and button-like objects, bands of fine gold foil or beadwork – which again were probably affixed on a lining cloth – belts and belt ornaments, and towards the end of the period, fibulae.[12]

As far as roundels and cut-outs are concerned, it is evident by the position in which most of them were discovered, that they probably decorated the selvage of the dress/shroud and the sleeves. Roundels and cut-outs first appear in Mochlos[13] and Platanos,[14] Crete, from the Early Minoan Period, while some gold discs decorated with dotted rosettes are also known from the Aigina Treasure.[15] Cut-out reliefs, mainly in the shape of a rosette, with eight, ten, twelve or sixteen petals, become popular between the 15th and 14th centuries BC, especially in the Argolid and Messenia. They usually bear holes on the periphery for the stitched attachment onto the cloth. In Tomb 4 at Selopoullo, Crete, more than a hundred rosettes of gold foil were concentrated around two burials,[16] in their majority near the upper part of the body, both above and below the skeletons; some of them were still placed on a line, along the left forearm, in a vertical position.

Rosettes and cut-outs are found in large quantities, furnishing wealthy burials, and seem to reflect a burial custom rather than fashion. Indeed, most examples could have decorated a kind of shroud, as indicated by both the finesse of the gold foil and their careless construction; often the perforation holes are neither symmetrical nor smoothed (Fig. 6.1). It has even been suggested that at least some of these ornaments had been worked in the tomb itself, at the time of the burial.[17] Due to the thinness of the foil, we are fortunate to have instances where the imprint of the textile

[8] Ehrenberg 1992, 119–121, figs 31–32.
[9] For an up-to-date bibliography on the Mycenaean textile industry see Nosch 2012, 43–55, esp. 43–44, n. 5; for remains of textiles in the Aegean, Barber 1992, 312–357, esp. 312–313 for Linear B evidence.
[10] Cavanaugh and Mee 1998, 109, n. 51.
[11] Spyropoulos 1972, 206–209, esp. 208c on X-ray technique.
[12] The article does not aim to present a catalogue of the jewellery available nor to repeat what is already known on the subject, but rather to suggest some ways of using clasps for fastenings, based on items kept in the Mycenaean Collection of the National Archaeological Museum.
[13] Xanthoudides 1924.
[14] Davaras 1975.
[15] Higgins 1957.
[16] Popham-Catling 1974, 203, 214, fig. 12.
[17] Protonotariou-Deilaki 1969, 105.

Fig. 6.1: Gold rosettes, NAM 2843. Mycenae Chamber Tombs.

it once covered is still preserved; such is the case from the Chamber tombs at Mycenae,[18] where a couple of rosettes bear a net or grid pattern on their surface (Fig. 6.2).

The most popular clothing ornaments by far are buttons. By this term, we usually refer to the conical, biconical or disc-shaped accessories, often of steatite, with vertical perforation,[19] although there are several other varieties. Buttons occur singly or in varying numbers and colours, both in burial and in domestic contexts; it is possible that they had multiple uses, as buttons, spindle whorls or even necklace beads (the finest examples);[20] however, when they are

Fig. 6.2: Detail of a gold rosette with imprint of net pattern, Mycenae Chamber Tombs.

[18] Xenaki-Sakellariou 1985, 187–188.

[19] Iakovides 1977, 113–119.

[20] Marinatos 1971, 217; Mylonas-Shear 1987, 132–133; Tournavitou 1995, 216–217; Konstantinidi 2001, 235.

Fig. 6.3: Selection of steatite shanked buttons, NAM 2359. Mycenae Chamber Tomb 2.

found in large numbers and have a careful fabrication and a finely polished surface, they should be considered as jewellery items or dress accessories (Fig. 6.3). Tomb 16 in Perati, Attica,[21] which held the intact burial of a young woman with eleven steatite buttons beneath the region of the knees remains unique (Fig. 6.4). The excavator himself suggested that they served as weights for a short skirt, in order to form pleats. The burial seems to have belonged to a wealthy lady, since she was also furnished with gold, silver and ivory items, along with a bronze mirror placed in front of the skull. The same type of buttons are depicted on a fresco from Xeste 3, hanging from the edges of female "overshirts", on top of the dress (Fig. 6.5). It would not be unreasonable to suggest therefore that, depending on the circumstances, the stone or clay items usually referred to as buttons, were in fact used as dress decoration in various ways – perhaps even as tassels from the edges of belts – and to a lesser extent, as necklace beads (the small delicate examples), weights or spindle whorls.

In Prosymna, Tomb II,[22] in the "Tomb of Clytemnestra"[23] and in two of the Mycenae Chamber tombs, apart from the type mentioned above, there are large quantities of round glass and ivory buttons with a flat base and a deep groove between base and the slightly convex top (Fig. 6.6):

[21] Iakovides 1969, A 258–9, tables 73d and 180a.
[22] Blegen 284–285, fig. 446, nos.1–4, 6–10.
[23] Wace *et al.* 1921–23, 371–2.

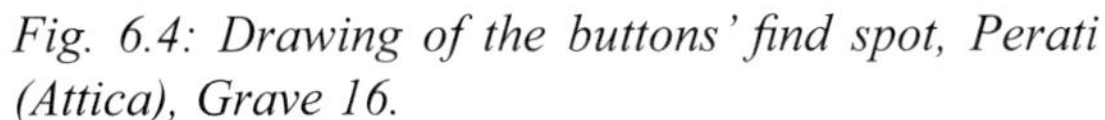

Fig. 6.4: Drawing of the buttons' find spot, Perati (Attica), Grave 16.

Fig. 6.5: by one of the crocus-gatherers, Xeste 3, Akrotiri (Thera).

Chamber tomb 15 produced around 67 such buttons of various sizes[24] and tomb 69 produced around 60 of them.[25] Unfortunately, it is not possible to attribute them to specific burials, but it seems that each garment they decorated held many of them. The grooved buttons must have been attached by means of a fabric loop, either in the middle of a strap (Fig. 6.7a) or on both sides of the costume, mainly for decoration; thus, the type of garment they decorated could be a kind of vest that would clasp in front with woollen or leather straps, similar to the vest worn by the faience Snake Goddess from the "Temple Repositories" at the Palace of Knossos.[26] Alternatively, they could be worn in a line, forming a fabric belt made by continuous loops (Fig. 6.7b).

Mycenae produced a number of other types of "buttons" as well, some of them quite interesting, as for example the one from Chamber tomb 28, a conical ivory button with several blind holes

[24] National Archaeological Museum, Athens, inv. no. 2272 (hereby NAM), d. 0.11–0.4cm, thickness 0.05–0.08cm. The buttons were found in the tomb's dromos that held six burials, Xenaki-Sakellariou 1985 (hereby *Ch,.T.*), 76–77, pl. 12.
[25] NAM 2951, d. 1.7–3.1cm., *Ch.T.*, 197, 201, pl. 87.
[26] Rethymiotakis 1998, 110.

Fig. 6.6: Selection of "grooved" buttons, NAM 2373. Mycenae Chamber Tombs.

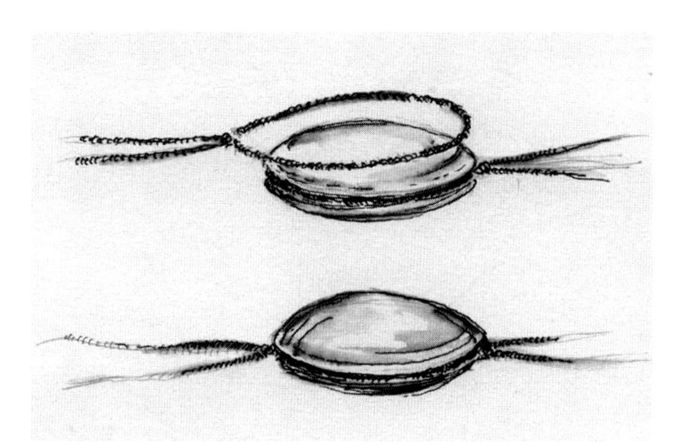

Fig. 6.7a: Suggested reconstruction of grooved buttons' clasp (Drawing by A. Goumas).

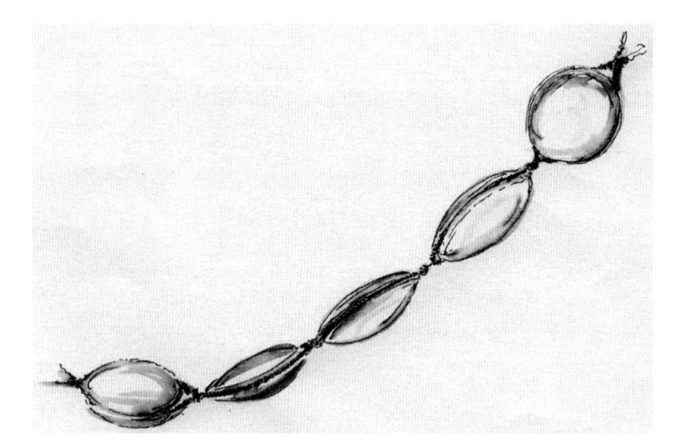

Fig. 6.7b: Suggested reconstruction of grooved buttons in a series of loops (Drawing by A. Goumas).

forming an arched perforation on the flat surface and a circle on the curved surface.[27] Another elaborate example is the hemispherical ivory button with incised decoration and applied miniature gold nails on the periphery (Figs 6.8, 6.9 right).[28] The back of the button has two parallel holes for the passing of a thread to be sewn onto a dress. In a similar way, another ivory button of the same type but with no decoration bears an almost hollow interior with six holes for the attachment on the dress (Figs 6.8, 6.9 left).[29] Both examples have a decorative groove on the base of the hemisphere. Finally, there is another ivory button of biconcave form, with several holes on only one surface, while the other is left plain (Figs 6.10, 6.11).[30]

Bands and belts were a popular accessory, used by both men and women. Pictorial evidence from Minoan Crete allows us to draw a typology on belts:[31] thus, Minoan belts are divided into cinched – of fabric or metal – concave fitted, single or double rolls. For the Mycenaean belts, however, some information is provided from a few terracotta figurines of the Phi and Psi types (Fig. 6.12) and a couple of ivory examples, namely the young girl from the Ivory Triad[32] and an ivory female figurine from Prosymna;[33] in those cases, a loose belt is clearly distinguishable, though its clasp cannot be identified (Fig. 6.13). From the burial context though, there are several examples of fragmentarily preserved bands made of fine gold foil, some with perforation holes on the periphery, probably meant to be sewn on a lining cloth (Fig. 6.14). Mycenae Chamber Tomb 15 yielded fragments of bands made of fine gold foil (thickness 0.02mm) that still preserve the imprint of the textile they once covered (Figs

[27] NAM 2404, Ch,T. 103, pl. 26.
[28] NAM 1001, Poursat 1977, 9, pl. I; Sakellarakis 1979, 69–70, figs 96, 97.
[29] NAM 507, Karo 1930, 110, pl. CI; Poursat 1977, 61.
[30] NAM 1991, Poursat 1977, 151, pl. XLVI.
[31] Verduci 2012, 641–642.
[32] Barber 2012, 26, Pl. VIIIf.
[33] Konstantinidi 2012, 267.

Fig. 6.8: Ivory buttons with attachment holes from Mycenae (NAM 507 and 1001).

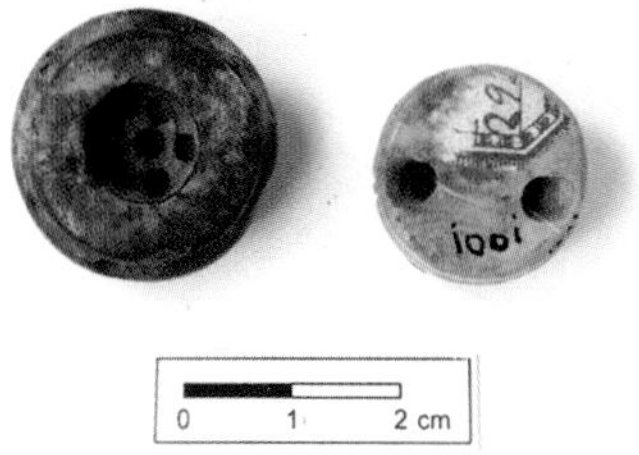

Fig. 6.9: The back of the ivory buttons from Mycenae (NAM 507 and 1001).

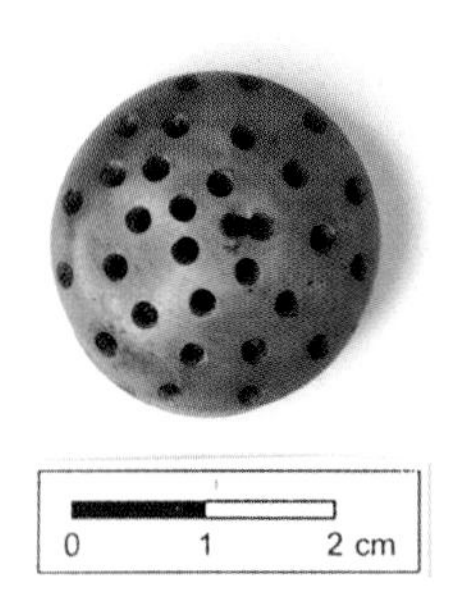

Fig. 6.10: Ivory button NAM 1991, Mycenae.

Fig. 6.11: Plain side of the ivory button NAM 1991, Mycenae.

6.15–6.17);[34] from the same tomb comes a bronze band, with apparently the same use as the gold ones, pierced with perforation holes throughout its periphery (Fig. 6.18).[35] An earlier context, Tomb Beta of Grave Circle B at Mycenae, contained the remains of a mature man in his early 30s, furnished with two gold bracelets, a knife and vases; a gold band (l. 0.395 m.) was found by the right side of the pelvis and above it;[36] the "belt" had no holes for attachment, so the original fabric backing – if there was any – would embrace the metal with the edges probably sewn. Linear B tablets may indicate the combination of hard materials on some of the textiles, as bronze and linen for example are recorded together,[37] which may have been used either for bands of that type or for some other type of garment.[38] In contrast to the roundels and cut-outs where fine plate is usually an indication of their burial use, metal bands were meant to be affixed on fabric.

---

[34] NAM 2301(3), *Ch.T.*, 77. Dim. of larger fragment 0.075 × 0.043m.
[35] NAM 2781, *Ch.T.*, 78, pl. 13.
[36] Mylonas 1973, 38, 42, pl. 28b.
[37] Ventris and Chadwick 1973, 320, 487–488 and chapter 6, n. 11.
[38] Barber 1992, 313, Tablet KN 1963.

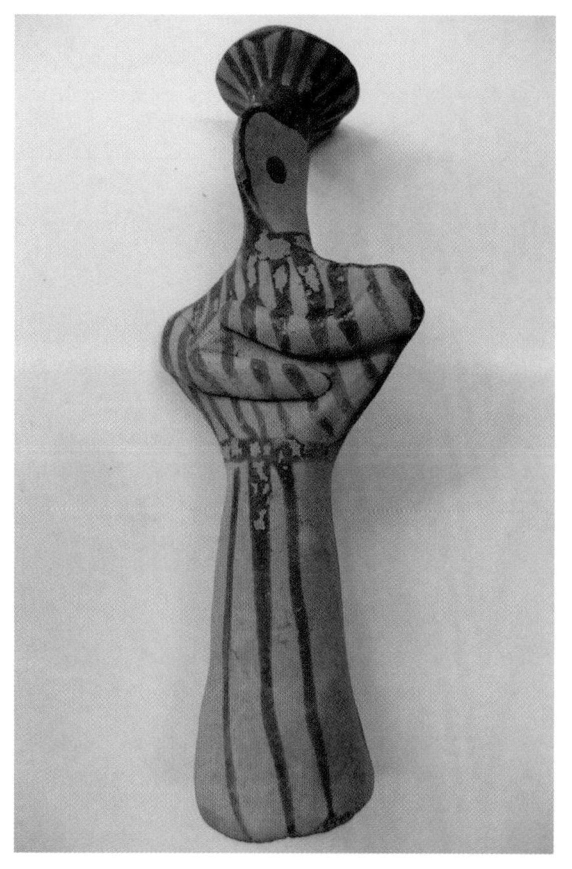

Fig. 6.12: Mycenaean terracotta figurine wearing the typical long dress with a belt.

Fig. 6.13: Detail of the skirt, showing the loose belt on ivory figurine NAM 6580, Prosymna.

Fig. 6.14: Gold belt with attachment holes on one side. NAM 2792 (7), Mycenae Chamber Tomb 58.

Fig. 6.15: Microscopic view of the gold foil with net pattern imprint NAM 2703 (3), Mycenae Chamber Tomb 15.

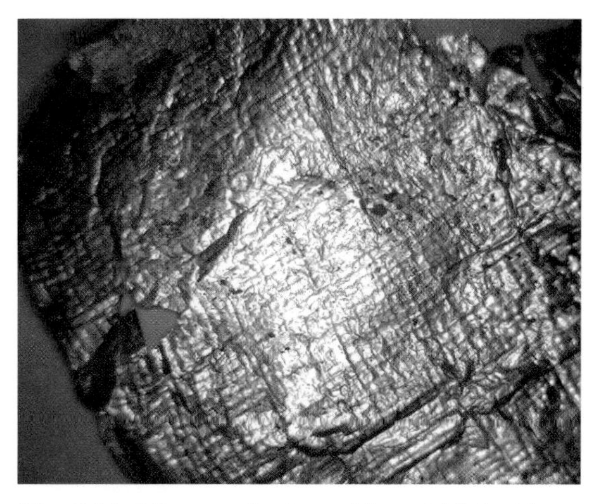

Fig. 6.16: Microscopic view of the gold foil with net pattern imprint NAM 2703 (3), Mycenae Chamber Tomb 15.

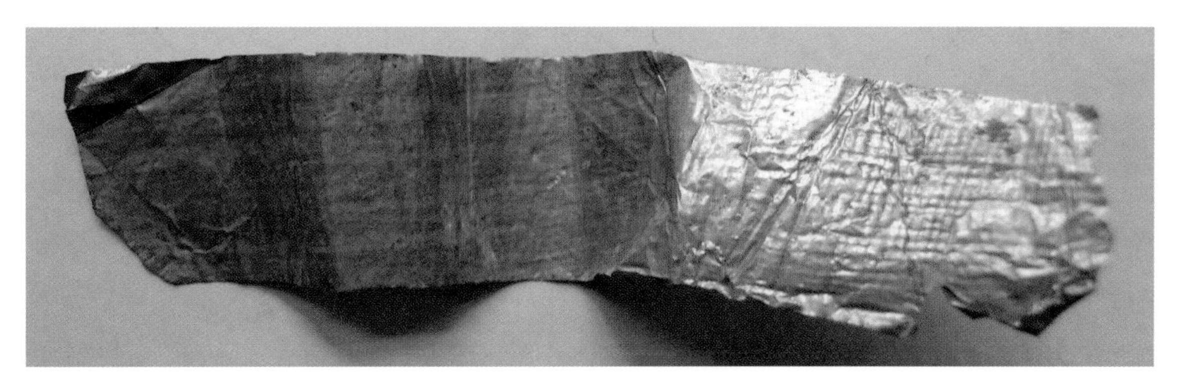

Fig. 6.17: Small elongated plate with a grid imprint on one edge. NAM 2703 (4), Mycenae Chamber Tomb 15.

Fig. 6.18: Bronze belt with perforation holes throughout the periphery, NAM 2781, Mycenae Chamber Tomb 15.

Fig. 6.19: Minute faience beads as preserved in a lump of earth. NAM 19308, Dendra.

Figs 6.20–6.22: (left to right) Microscopic views of the faience beads NAM 19308, showing the zigzag pattern and the variety of colors.

Apart from metal bands, there would have been beaded bands as well. Several scholars have recognised the use of sewn or embroidered jewellery and beads on textiles in Aegean representations.[39] From chamber tomb 2 at Dendra – Persson's "cenotaph" – comes the unique remnant of a beaded garment;[40] the excavator reports some 40,000 beads, most of them threaded after their excavation, while several lay still in the lump of earth, where they were incorporated after the decay of the textile (Fig. 6.19). Persson mentions five colours in this order, white, yellow, brown, black and blue, in a zigzag pattern. Microscopic views showed in addition two more shadings of blue, as well as black-spotted white (Figs 6.20–6.22).[41] Those tiny cylindrical beads with the extraordinary well

[39] For selected bibliography see Borgna 2012, esp. 339, n. 30; also, Shaw 2000.

[40] Persson 1931, 106.

[41] The views were taken by Dinolite Pro microscope with polarizing light and the enlargement varies from 50x to 130x.

Fig. 6.23: Detail of a "belt" from Grave Circle A, Mycenae with a clasp at one edge.

preserved colours would originally have decorated a garment or a belt, for which Persson suggests an Egyptian provenance.[42] Indeed, in Egypt such beads had been placed in graves since Predynastic times (*c.* 3000 BC), and by the time of the Dendra tombs (2nd half of 15th cent. BC), beads sewn on cloth or even woven into the material abound; there are indications by the textiles preserved that at least six different types of beading were in use.[43] Sometimes even, the Egyptians combined glass and faïence on the same garment. The Dendra example seems to consist entirely of faïence beads and it could have belonged to a belt or to an "apron", a kind of a kilt worn exclusively by men.[44]

From Grave Circle A, Mycenae, come two belt-like ornaments of gold with impressed decoration and an interesting clasp (Fig. 6.23). The two ornaments cannot have been used as actual belts, since they are made of a relatively thick plate decorated with circles which do not seem to have been folded, and their length is too large for practical use. However, the clasp accessory at the edges of the ornaments, a fusiform bar of gold with a notch in the middle, is also known from other instances and has passed unnoticed.[45] It recalls a simple form of clasp, which has been used mainly on the dogs' collars, where the other end of the belt/collar is formed as a loop – made of a variety of materials, like leather for instance – where the bar is vertically inserted (Fig. 6.24). A similar accessory of a bronze and silver alloy comes also from Grace Circle A (Fig. 6.25);[46] another one of gold, decorated with incised rings at the ends framing the notch, comes from Chamber tomb 102, Mycenae.[47] Two more ivory examples come from Mycenae[48] and another one of bronze with gold inlay decoration comes from Asine[49] chamber tomb I:1 (Fig. 6.26). All of them were found in tombs and they are most probably all that is left from the fabric belts they once decorated.

Pins, either as part of hair-decoration or as dress fasteners,[50] never become an indispensable accessory in the Mycenaean world, therefore evidence on their use is insufficient. So far, dress pins have been considered to fasten a kind of shawl or light jacket on the shoulders; when more than two, they could have been used to clasp a vest. The use of fibulae, the dress accessory that appears toward the end of the Mycenaean period, is more straightforward. They have all been found by the shoulders and have been clearly used for the fastening of a heavier cloth on the shoulders, like a *peplos* – a garment consisting of a couple of strips of woven material

[42] Persson 1931, 137–139.
[43] Nicholson and Shaw 2000, 280.
[44] Nicholson and Shaw 2000, 287.
[45] With the exception of Xenaki-Sakellariou who describes it as a belt clasp, see *Ch. T.*, 282, table 139 (NAM 4916.3, dim. 4 × 0.5cm.).
[46] NAM 863.
[47] NAM 4916 (3), *Ch.T.*, 282, pl. 139.
[48] NAM 1048 and 1049, Poursat 1977, 11, Pl. II.
[49] Frodin-Persson 1938, 371 (toggle-pin), fig. 241.
[50] Killian-Dirlmeier 1984, 37–65, tables 4, 5.

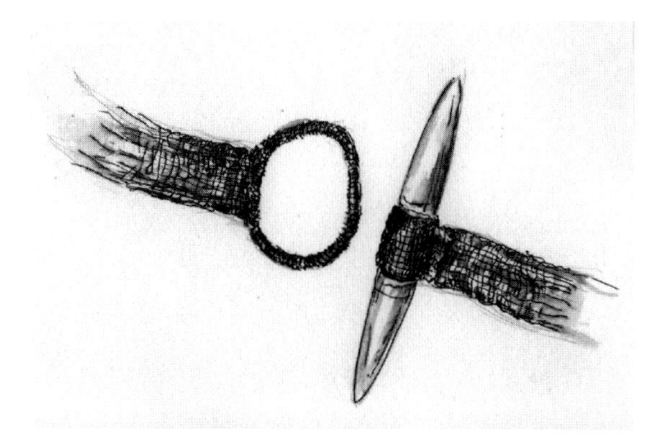

Fig. 6.24: Suggested reconstruction of the clasp (Drawing by A.Goumas).

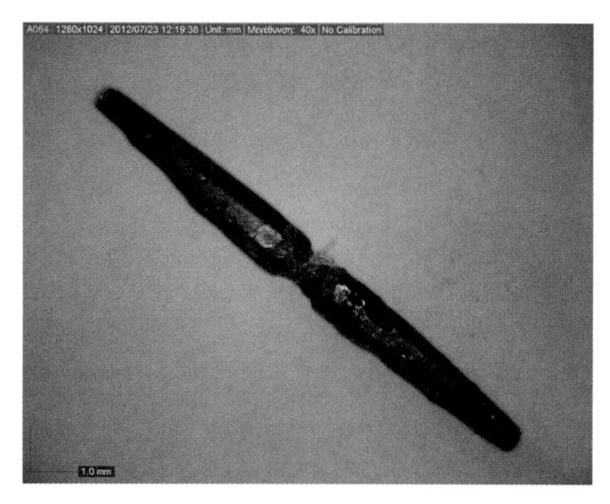

Fig. 6.25: Belt clasp NAM 863, Grave Circle A, Mycenae.

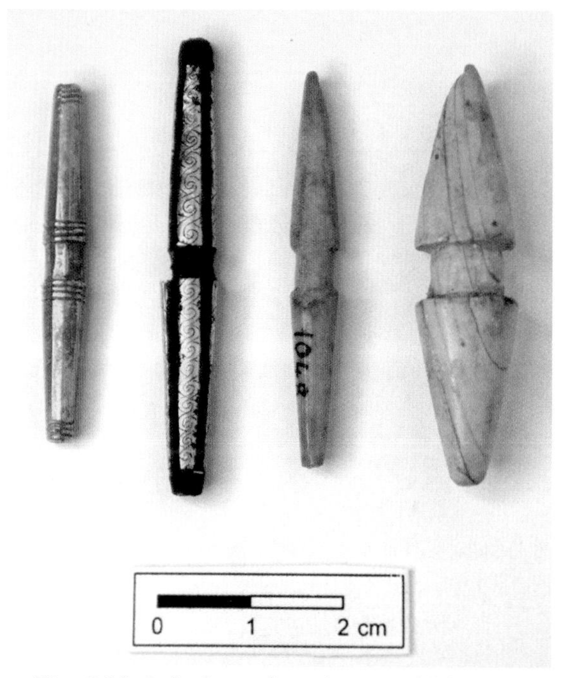

Fig. 6.26: Belt clasps from Asine and Mycenae.

sewn together down one side – that has been connected with a climatic change or a fashion coming from the north.[51]

Finally, there is another type of accessory, which may have been used to decorate the dress: according to the excavators, burials in Selopoullo Tomb 3 mentioned above, were originally wrapped in decorated shrouds and were provided with bronze utensils, weapons, as well as tubes and "tassels" of gold foil.[52] Gold tubes are also known from the chamber tombs of Mycenae[53] and their thickness justifies the passing of a thread (Fig. 6.27). A recent discussion by Barber[54] suggests the existence of a ritual dress type in the Aegean, the string skirt, worn by pre-pubescent girls, a tradition already known in several regions of the Balkans, Northern Europe and Eastern Russia.

Tassels have been discovered in Danish burials of the Bronze Age, while tubes of bronze leaf furnished several female burials, indicating the existence of a corded skirt. In a barrow, a woman buried in an oak coffin, held among other finds, about 125 bronze tubes lying in two rows with

[51] Konstantinidi 2001, 212.
[52] Popham-Catling, 1974, 195–258.
[53] Unnumbered in Sakellariou's catalogue.
[54] Barber 2012, 25–26.

Fig. 6.27: Gold tubes, Mycenae Chamber Tombs.

an interval at the middle of the body; in each tube were remnants of two parallel threads of wool, wound about with thinner woollen threads.[55] Although there seems to be no particular association between costume types of the two distant regions (Scandinavia and the Aegean), it is possible that all tassels and tubes had the same use, to decorate the fringe of the skirt or the lower garment.

Along with evidence from Linear B tablets, glyptic art too speaks in favor of the production of prestige clothing during the Late Bronze Age.[56] The creative imagination of the Mycenaean "fashion designers", reflected in the variety of decorative patterns known from frescoes and figurines of the period, would certainly also apply to the less elaborate costumes of practical use. Despite the fact that the geomorphological conditions of the Aegean did not allow the preservation of textiles, the clothing ornaments revealed in the burial context (certainly there are a lot more still remaining unidentified), indicate that the Mycenaeans had at their disposal a wide choice of costumes, both wrap-around and cut-to-shape (sewn), plain, embroidered or otherwise decorated. The type of the cloth, the length, as well as the decoration must have depended on the practical needs, but also on age, gender and social status of the peoples, as was the case in all contemporary societies.

Acknowledgements

I would like to extend my warmest thanks to Akis Goumas for the comprehensive drawings and to my colleagues from the National Archaeological Museum, Rhania Kapsokoli, Panayotis Lazaris, Eirini Miari, Katia Manteli and Katerina Kostanti for all their help.

[55] Broholm and Hald 1940, 148–151, figs 190–191.
[56] Crowley 2012, 236–237.

Bibliography

Barber, E. J. W. 1992 *Prehistoric Textiles, The Development of Cloth in the Neolithic and Bronze Ages*. Princeton.

Barber, E. J. W. 2012 Some Evidence for Traditional Ritual Costume in the BA Aegean. In R. Laffineur and M.-L. Nosch (eds) *KOSMOS. Jewellery, Adornment and Textiles in the Aegean Bronze Age, Proceedings of the 13th International Aegean Conference, University of Copenhagen, Danish National Research Foundation's Centre for Textile Research, 21–26 April 2010*. Leuven/Liège, 25–29.

Blegen, C. W. 1937 *Prosymna, the Helladic Settlement preceding the Argive Heraeum*, Vols. I, II. Cambridge.

Borgna, E. 2012 Remarks on Female Attire of Minoan and Mycenaean Clay Figures. In R. Laffineur and M.-L. Nosch (eds) *KOSMOS. Jewellery, Adornment and Textiles in the Aegean Bronze Age, Proceedings of the 13th International Aegean Conference, University of Copenhagen, Danish National Research Foundation's Centre for Textile Research, 21–26 April 2010*. Leuven-Liège, 335–341.

Broholm, H. C. and Hald, M. (eds) 1940 *Costumes of the Bronze Age in Denmark*. Copenhagen.

Cavanaugh, W. and Mee, C. 1998 *A Private Place: Death in Prehistoric Greece*, SIMA vol. CXXV. Jonsered.

Davaras, C. 1975 Early Minoan Jewellery from Mochlos. *BSA* 70, 101–114.

Ehrenberg, M. 1992 *Women in Prehistory*. London.

Erman, A. 1969 *Life in Ancient Egypt*. London.

Frödin, O. and Persson, A.W. 1938 *Asine, Results of the Swedish Excavations 1922–1930*. Stockholm.

Higgins, R. 1957 The Aegina Treasure Reconsidered. *BICS* 4, 27–41.

Iakovides, S. 1969–70 *Περατή, το Νεκροταφείο*. Athens.

Iakovides, S. 1977 On the Use of Mycenaean "buttons". *BSA* 72, 113–119.

Jones, B. 2009 New Reconstructions of the "Mykenaia" and a Seated Woman from Mycenae. *AJA* 113, 309–337.

Karo, G. 1930–33 *Die Schachtgraber von Mykenai*. Munich.

Kilian-Dirlmeier, I. 1984 Nadeln der frühelladischen bis archaischen Zeit von der Peloponnes. *Prähistorische Bronzefunde* XIII.8. München.

Killen, J. T. 1966 The Knossos Lc (cloth) Tablets. *BICS* 13, 105–111.

Konstantinidi, E. 2001 Jewellery Revealed in the Burial Contexts of the Aegean Bronze Age. *BAR IS* 912. Oxford.

Kritseli-Providi, I. 1982 *Τοιχογραφίες του Θρησκευτικού Κέντρου των Μυκηνών*. Athens.

Marinatos, S. 1973. Ανασκαφαί Θήρας V, *Praktika* (1971). Athens, 181–225.

Ruipérez, M. S. and Melena, J. L. 1996. *Οι Μυκήναιοι Έλληνες*. Athens.

Mylonas, G. 1973 *Ο Ταφικός Κύκλος Β των Μυκηνών*. Athens.

Mylonas-Shear, I. 1987 *The Panagia Houses at Mycenae*, University Museum Monograph 68. Philadelphia.

Nicholson, P. T. and Shaw, I. (eds) 2000 *Ancient Egyptian Materials and Technology*. Cambridge.

Nosch, M.-L. 2010 From Texts to Textiles in the Aegean Bronze Age. In R. Laffineur and M.-L. Nosch (eds) *KOSMOS. Jewellery, Adornment and Textiles in the Aegean Bronze Age, Proceedings of the 13th International Aegean Conference, University of Copenhagen, Danish National Research Foundation's Centre for Textile Research, 21–26 April 2010*. Leuven-Liège, 43–55.

Persson, A. 1931 *Royal Tombs near Dendra*. Lund.

Popham, M. R., Catling, E. A. and Catling, H. W. 1974 Sellopoulo Tombs 3 and 4, Two Late Minoan Graves near Knossos. *BSA* 69, 195–257.

Poursat, J. C. 1977 *Catalogue des ivoires mycéniens du Musée national d'Athenes*. Paris.

Protonotariou-Deilaki, Ev. 1969 *Ανασκαφή Θολωτού Μυκηναϊκού Τάφου (Καζάρμα)*, *AD* 24, B1, 104–105.

Rethymiotakis, G. 1998 *Ανθρωπομορφική Πηλοπλαστική στην Κρήτη. Από τη Νεοανακτορική έως την Υπομινωική Περίοδο*. Athens.

Sakellarakis, Y. 1979 *Το Ελεφαντόδοντο και η κατεργασία του στα Μυκηναϊκά χρόνια*. Athens.

Shaw, M. 2000 Anatomy and Execution of Complex Minoan Textile Patterns in the Procession Fresco from Knossos. In Karetsou, A. (ed.), *Κρήτη-Αίγυπτος. Πολιτιστικοί Δεσμοί Τριών Χιλιετιών*, Μελέτες. Athens, 52–63.

Spyropoulos, T. G. 1972 Mycenaean Tanagra: terracotta sarcophagi. *Archaeology* 25, 206–209.

Televandou, Chr. 1984 *Κοσμήματα Προϊστορικής Θήρας*. *AE*, 14–54.

Tournavitou, I. 1995 The Ivory Houses at Mycenae. *BSA* Supplement 24.

Tzachili, I. 2000 Αιγυπτιακές και Μινωικές Ενδυμασίες, In Karetsou, A. (ed.), *Κρήτη-Αίγυπτος. Πολιτιστικοί Δεσμοί Τριών Χιλιετιών*. Athens, 71–77.

Ventris, M. and Chadwick, J. 1973 *Documents in Mycenaean Greek* (2nd ed.) Cambridge.
Verduci, J. 2010 Wasp-waisted Minoans: Costume, Belts and Body Modification in the LBA Aegean. In R. Laffineur and M.-L. Nosch (eds) *KOSMOS. Jewellery, Adornment and Textiles in the Aegean Bronze Age, Proceedings of the 13th International Aegean Conference, University of Copenhagen, Danish National Research Foundation's Centre for Textile Research, 21–26 April 2010*. Leuven-Liège, 639–646.
Vlachopoulos, A. 2010 Georma, Jewellery and Adornment at Akrotiri, Thera. The Evidence from the Wall Paintings and the Finds. In R. Laffineur and M.-L. Nosch (eds) *KOSMOS. Jewellery, Adornment and Textiles in the Aegean Bronze Age, Proceedings of the 13th International Aegean Conference, University of Copenhagen, Danish National Research Foundation's Centre for Textile Research, 21–26 April 2010*. Leuven/Liège, 35–42.
Wace, A. J. B. *et al*., 1921–23 Excavations at Mycenae. *BSA* 25, 283–402.
Warren, P. 2005. Flowers for the goddess? New fragments of wall paintings from Knossos. In L. Morgan (ed.) *Aegean Wall Painting: A Tribute to Mark Cameron*, British School at Athens Studies 13. London, 131–148.
Xanthoudides, S. 1924 *The Vaulted Tombs of Messara*. London.
Xenaki-Sakellariou, A. 1985 *Οι Θαλαμωτοί Τάφοι των Μυκηνών, Ανασκαφής Χρ. Τσούντα (1887–1898)*. Paris.

7. Textile Semitic Loanwords in Mycenaean as *Wanderwörter*

Valentina Gasbarra

Language is the most direct means of expression and the most spontaneous reflex of the culture which it represents. In this sense, the decipherment of Linear B and the publication of Mycenaean archives have led us to examine how Mycenaean society was organized and, from a strictly linguistic point of view, the contacts and exchanges between the Mycenaean world and its immediate or more distant neighbours, as well as the connections with 1st millennium Greek forms.

Even though Mycenaean tablets consist exclusively in bureaucratic or administrative documents, they testify to the most fundamental linguistic categories of later Greek and allow us to follow and reconstruct the evolution of the language between the two stages under the phonological, morpho-syntactical and lexical profiles. This task does not occur without any surprises: as we can see, for example, by taking a glance at some morphological categories, such as the compounds. The Mycenaean lexicon displays a well consolidated tendency in replacing some terminological blanks with neologisms, which are often not yet included in the standard vocabulary, and for this reason present with a high degree of internal transparency and a clear recognizability in terms of constituents. On this subject, the other strategy available is the borrowing, and particularly, the borrowing of special terminology. This sector of research is not completely exhaustive at the present time,[1] although Mycenaean studies have known a significant impulse in recent years, thanks both to the interest of scholars and to the edition of corpora of documents, which have made a wide survey of the Mycenaean archives and have shown the spread of the language.

The infrequent loanwords in the Mycenaean Linear B archives belong mainly to the field of commercial exchange and they provide valuable evidence of Greek-Semitic interaction in the 2nd millennium BC. Let's start from the examples of textile terminology (e.g. those associated with fibre production, textile names, weaving and manufacture of garments, names of the workers employed in the textile industry) which are borrowed in the Mycenaean tablets: this contribution is aimed at elucidating the procedures and the categorization of linguistic borrowings, taking into account the typology of loanwords and the degree of the adaptation phenomena, such as the formation of compounds and derivatives modeled by using the morpho-syntactical structures of the Greek language. Another topic, which will be focused on, is the continuity between the semantic classes

[1] For a survey of studies about Greek and Semitic interference during the 2nd millennium BC, see: Vaniček 1878; Muss-Arnolt 1892; Lewy 1895; Grimme 1925; Mayer Modena 1960; Astour 1967; Masson 1967.

in the Semitic loanwords of the 2nd millennium BC and those of the later stage of the Greek language. Although the number of Semitic loanwords in the Mycenaean tablets is few, the terms that the Greek language continues borrowing from the Semitic languages are still related to the names of plants, metals, materials and garments and, mostly, to technical and commercial terminology. *A latere* of these considerations, the influence of the Anatolian languages must be underlined: the role of Hittite, particularly as the intermediary language from which Mycenaean Greek inherited some Semitic loanwords, will be also stressed.

The pre-Greek substrate and Greek in contact with other languages

The contacts between Greek and other languages, and the effects produced by these contacts, provide the most conspicuous evidence of the historicity of language. In ancient times, just as today, linguistic borrowing reflects judgements of cultural value and historical progress of a language is dependent on precisely such judgements.[2] The question of Greek in contact with other languages cannot be separated from the reflection on "common Greek" and on the substrate and contact languages on the Greek territory before the Hellenization. After the collapse of the Mycenaean kingdoms and the disintegration of the palatial societies, the linguistic outline of Greece was completely thrown. The so-called "Greek Dark Ages" (9th–8th centuries BC) corresponds with a social, economic and cultural withdrawal[3] as testified by the archaeological evidence, and with the consequent loss of the use of writing. The Dark Ages can therefore be considered as a formative period of the culture of archaic and classical Greece, at the end of which[4] the adoption of alphabetic writing inherited from the northern Semitic scripts (φοινικήια γράμματα 'Phoenician script') is one of the most important innovations.

The linguistic outline of Greece before the introduction of the Linear A and B writings is widely debated, and scholars are divided between those who believe that the Greek language has become dominant on a pre-existent Indo-European substrate,[5] and those who are inclined to believe in a so-called "Indo-Mediterranean substrate" with a very general and indefinite features, but with a clear and discriminating Non-Indo-European origin.[6]

These are both *a priori* assumptions and, as such, cannot be defended. The only fact that we can also evaluate is the presence of a number of words in the Greek lexicon which have no obvious connection in the cognate languages and are therefore suspected of being loans of "autochthonous" population inhabiting those areas before the arrival of the Indo-Europeans.[7] The generic notion of substrate has to be interpreted in a weak sense, as a sort of "inheritance" or, as "migrant words", which occur in other Mediterranean languages and for which no plausible etymology can be found (the most well known examples are: ἔλαιον 'olive oil' or οἶνος 'wine').[8]

On the other hand, the question of loanwords from pre-Greek languages is particularly complex in the evaluation of words like ξένος 'foreign', ἄναξ 'lord', βασιλεύς 'king', πόλεμος 'war', θεός 'god'

[2] For an exhaustive introduction of languages in contact with Greek, see Christidis 2007, 721–732.
[3] On the return to a self-sufficient economy and on the abandonment of complex and hierarchical social organizations as reaction caused by the collapse of the palatial administration, see Snodgrass 1987, 186–188.
[4] Amadasi Guzzo, 1991, 293–309.
[5] Palmer 1980, 4–9.
[6] Belardi 1954, 610; Silvestri 1974, 35–38.
[7] Szemerényi 1964, 404–405.
[8] On the names and on the "etymological reasons" of 'olive oil' and of 'wine' see Silvestri 2013, 335–340.

etc., with a high degree of specialization in the form and in the meaning and with no connection in the other Indo-European languages, although attested in the Mycenaean archives and, afterwards, in all 1st millennium Greek dialects.

It's not possible to go further: all the efforts directed to individualize a specific substrate (Aegean, Pelasgian, Asianic etc.), sometimes according to the testimony of ancient historians about the origin of their language,[9] have never met unanimous consensus among scholars, because they always show a lot of weaknesses on the phonetic, morphological or semantic point of view, since many of these terms may be borrowed from languages not yet directly attested.[10]

Mycenaean Contacts with the Near East

The Mycenaean palatial system required intensive exploitation of regional resources: sudden expansion of the power of a single palatial centre to control broader regional resources and production would have created new hierarchies of power, work and socio-political networks.[11] That organization required a high degree of specialization within specific industries (e.g. wool, flax and dye substances for cloth production; olive oil; perfumed substances and related pottery manufacture etc.). The long list of trades and occupations, which can be identified in the Linear B documents, implies the development of a specialization of labour, which goes far beyond that seen in Homer. Textile production in particular, is one of the most ancient human technologies, playing a crucial role in societies world-wide throughout our past and giving a clear measure of the level of technical know-how. Textile production reflects human interactions with the environment since the end of the Ice Age. Across the Mediterranean area, it testifies to cultural contacts and exchanges between the West and the Ancient Near East. The textile loanwords in the Mycenaean archives point primarily to extensive commercial relations with the Semitic East, but also to the high level of lexical (and, consequently, social and cultural) permeability between the Semitic and the Greek world. The study of textile terminology has a strong inter-disciplinary component, because it is closely connected with the study of material culture and techniques and with the role of textile production in ancient societies with its significance in the economy. The complex organization of production in Mycenaean times might in any case be inferred from the high level of trades. They can also be identified on similar tablets from the Eastern archives, in which craft production was of prime importance and, although textiles are largely invisible in the archaeological record due to their perishable nature, the presence of a linguistic term of a given procedure or tool implies its existence in the society where the language was spoken.[12]

When Linear B was first deciphered, it was immediately clear[13] that "the most useful and significant analogies" lay with the better documented and more fully understood societies of ancient Near East: often, though, the cryptic practices described in the Mycenaean tablets have been illuminated by the Near Eastern documentation.[14] The presence of foreign goods in Greece and

[9] Strabo (*Geogr.* 7.7.1) said: « Ἑκαταῖος μὲν οὖν ὁ Μιλήσιος περὶ τῆς Πελοποννήσου φησὶν διότι πρὸ τῶν Ἑλλήνων ᾤκησαν αὐτὴν βάρβαροι ».
[10] Duhoux 2007, 223–228.
[11] See Palaima 2004, 285.
[12] For a general introduction to ancient Near Eastern craft and technology, see Sasson 1995, vol. 1 chapter 7. For a recent review of the ancient Western and Near Eastern textile terminology, see Michel and Nosch 2010.
[13] Ventris and Chadwick 1956, 106; 113; 133; 135–136.
[14] For the analogies between Mycenaean and Near Eastern societies, see Shelmerdine 1998, 296–298.

Crete, confirmed by the presence of foreign words (names of spices, plants, metals and materials) in the Linear B written texts, testifies to trade contacts with the Semitic populations for most of the 2nd millennium BC. For example, many of the foreign references come from Pylos, where the tablets date to the last year of the palace administration at the end of LH IIIB. At this time, trade contacts with the Near East continued, though probably not on as large a scale as prevailed during LH IIIA2–IIIB1. On the other hand, the Pylos archives provide textual evidence for state-organized production of linen textiles and perfumed oil in industrial quantities, similar to the Knossos wool industry.

Mycenaean textile industry

Textile production is, however, labour intensive and involves many different processes: it implies specialization and division of tasks. Textile terminologies are closely associated with the study of material culture and techniques, and to the role of textile production in society and its significance in the economy. For this reason, the tablets record specific occupations such as spinners, weavers and fullers.[15] We have information on textiles from various Mycenaean centres (Thebes, Mycenae, Knossos, Pylos), but the most extensive documentation comes from Knossos,[16] where we can distinguish data about flocks, wool, production of clothes and names of textile workers, and from Pylos, where the production of flax and linen cloths is well documented. In the Mycenaean world, the textile industry – whose expertise already existed throughout the palatial territories, probably inherited from Minoan culture – was controlled and monitored by the palaces by supporting workers[17] and by controlling the quantities of raw materials from stage to stage until products were finished. The raw materials were distributed to textile workers with the expectation that set production targets would be met and finished textiles delivered back to the palace.[18] Distribution and requisition of raw materials to dependent workers is known as *ta-ra-si-ja* system,[19] and well documented in several areas of craft production. The palace control of goods and materials as well as the management of economic activities involved not only the textile industry, but also all the specialized industries evidenced in the Linear B texts[20] (furniture and woodworking, the manufacture of perfumed oil, bronze production, pottery, work with precious materials as gold, lapis-lazuli and ivory etc.).

As evidenced above, the main typologies of fibres testified in the Mycenaean archives are wool and flax, which is documented both as cultivated plant and as a fibre ready to be woven.[21] The cultivation of flax and the linen industry were wide spread in Greece, as it is shown by the terms designating production and manufacture of flax/linen articles in all periods of the Greek language and by place-names[22] derived from the term for "flax" etc. Cultivation and manufacture

[15] See Killen 1984; Palaima 1997; Nosch 2000.
[16] See Luján 2010.
[17] As Killen 1984 evidenced, groups of textile workers were supported by palatial food rations.
[18] Cline 2010, 436.
[19] See Nosch 2000; 2011.
[20] See Palaima 2003, 166.
[21] See Del Freo *et al.* 2010, 344–345, who identifies a regular distinction in the tablets between the plant and the fibre through two different syllabograms: *31=SA (LINUM), attested in Pylos and Knossos, and RI, attested in the Ma series of Pylos, in PY Mm 11 and in KN Nc 5100.
[22] For a review of ancient, koine, medieval and modern Greek terms and names for "flax", "linen" and their derivatives, see Georgacas 1959.

of flax (*linum usitatissimum*) are also well attested in the documents from Near Eastern archives, in which different kinds of cloths and different kinds of employment are regularly distinguished. This subtle distinction is not noticeable in the Mycenaean texts, which only make reference to a particular typology of linen in the Knossos tablet J 693 where the expression *ri-no re-po-to*, Gr. λίνον λεπτόν 'very fine linen', before *ki-to*, Gr. χιτών is attested. The other terms with a Semitic etymology, like βύσσος 'byssos' (Akk. *būṣu*; Ugar. and Phoen. *bṣ*; Hebr. *būṣ*) and σινδών 'fine woven cloth, fine linen garment' (Akk. *saddinu/šaddi(n)um*; Hebr. *sadīn*), that denote different and more valuable typologies of linen, appear in the Greek vocabulary exclusively from the 5th century BC. This late attestation – in a certain sense – confirms the pure nature of "loanwords by necessity",[23] connected with the need for naming new products obtained thanks to the improvement of cultivation and manufacturing techniques.

Textile terminology and Semitic loanwords in the Linear B texts

Lincar B records a very small number of names of garments, in strict connection with the flax industry. The Mycenaean documents record the word *ki-to*, Gr. χιτών 'chiton, tunic, designation of a garment without sleeves'. The term is *passim* attested in the Knossos archive and it represents a well-known example of a Semitic loanword, probably lent from the Akk. *kitû(m)*, and which can be compared with Ugar. and Phoen. *ktn*, Hebr. *kutonet*.[24]

Although the etymology of the word is widely debated,[25] the Akkadian term *kitû(m)*, on which the Greek χιτών is modelled, is probably inherited from the Sumerian GAD, GADA[26] 'linen, linen garment'.

The term *ki-to* (nom. sing.) is attested in the Knossos archive twice (KN Lc 563.B and L 693.1), as well as the forms *ki-to-ne* (nom. pl.) and *ki-to-na* (accus. sing.) attested respectively in KN L771.2 and KN Ld 785.2b, and the instrumental *ki-to-pi* in KN Ld 787.B. The term represents a good degree of adaptation into the Mycenaean lexicon, making a derivative and internally transparent adjective through the insertion of Greek affixes, like *e-pi-ki-to-ni-ja*,[27] Gr. ἐπιχιτωνία, an adjective that specifies a cloth which is 'worn over the *ki-to*'.

The fact, however, that the term is spread among numerous Indo-European and Non-Indo-European languages and cultures with the regular and very general meaning 'tunic, linen tunic', suggests a close relation with the category of "wandering words", words which have been borrowed from language to language, across a significant geographical area.

In studies of linguistic interference, it is important to record the distribution of words of foreign origin, making a clear distinction between those words which are widely attested in the host language and those of more limited occurrence. The early contacts between Greek and Semitic attested in the Mycenaean tablets belong mainly to the field of commercial exchange, for this reason the borrowed names with a Semitic etymology in the Linear B texts coherently exhibit this kind of behaviour. They also belong to the categories of plants/spices (e.g. Myc. *ku-mi-no-(a)*, Gr. κύμινον

[23] For a general introduction to lexical borrowings see Haspelmath and Tadmor 2009, particularly chapter 2, 35–54.
[24] For all the North-West Semitic attestations, see Hoftijzer and Jongeling 1995, s.v.
[25] See, lately, Vita 2010, 330.
[26] Cfr. AHw, CAD (vol. 8) and CDA, s.v.; Ellenbogen 1962, 96; Masson 1967, 29; GEW and EDG, s.v.
[27] In KN L 693 and, probably, in KN L 7514.

'cumin' to be compared with Akk. *kamūnu(m)*, Phoen. *kmn* and Hebr. *kammon;*[28] Myc. *sa-sa-ma*, Gr. σήσαμον 'sesame' to be compared with Akk. *šamaššammū(m)* and Ugar. and Phoen. *ššmn*;[29] Myc. *ku-pa-ro*, Gr. κύπαιρος 'cyperus', whose model, maybe, could be traced in Hebr. *koper*) and metals (e.g. Myc. *ku-ru-so*, Gr. χρυσός 'gold'). The word for 'gold' is widely attested in many Semitic languages, like Akk. *ḫurāṣu(m)*, Ugar. *ḫrṣ*, Phoen. *ḥrṣ*[30] and Hebr. *ḥārūṣ*, and its frequency demonstrates the importance of the gold trade in the ancient economy of Aegean and Near East. Beside the noun for 'gold', the Mycenaean archives record the material adjective *ku-ru-so* and *ku-ru-sa-pi*[31] (Hom. Gr. χρύσειος, χρύσεος, Aeol. χρύσιος 'golden, made of gold'), and a compound in *-wo-ko /worgos/*, *ku-ru-so-wo-ko* (PY An 207.10), Gr. *χρυσο-Fοργός 'gold-worker', which is inscribed in a large group (*c*. 40) of compounded substantives with verbal second member, usually indicating professions or functions, characterized by their internal recognizability.

In addition to the Semitic words just mentioned, two terms for precious materials can be added: Myc. *ku-wa-no*, Gr. κύανος 'lapis-lazuli' and Myc. *e-re-pa*, Gr. ἐλέφας, 'ivory'. *Ku-wa-no* and *e-re-pa* represent a different typology of loanwords,[32] because they have been inherited in Mycenaean Greek not directly from a Semitic language,[33] but through the intermediation of Hittite, as the Hittite forms *ku(wa)nna(n)–* and *laḫpa* – clearly demonstrate.

Some tentative conclusions

If the analysis conducted is correct, we can also assume that 2nd millennium Greek displays a small nucleus of terms with a Semitic etymology. These loanwords belong to the field of "special terminology" and they shed a light about contacts and exchanges between Mycenaean Greeks and their immediate or more distant neighbours in the Mediterranean basin during the Bronze Age. They also seem to confirm a high degree of continuity in the semantic classes, because the terms that the Greek language continues to borrow from Semitic languages during its history belong mainly to the field of trades and techniques. These later loanwords are evident in a specific proportion of the need for naming new activities and new objects. For example, the etymology of the Greek word μνᾶ (to be compared with Lat. *mina* and Skt. *manā́-*), which appears in Greek texts and inscriptions from the 6th century BC and which designates the name of a weight standard and a sum of money, can be traced in the Akkadian *manû(m)*[34] (Hebr. *mānē*, and Ugar. *mn*), the term for the verb 'to count' and for 'a mina-weight (*c*. 480 grams)'.

Similarly, the Greek word σίγλος/σίκλος (Lat. *siclus*) 'shekel', which represents both a coin and a unity of weight (but with a smaller geographical distribution than μνᾶ), can be considered a loanword from Akk. š*iqlu(m)* (Hebr. š*eqel*), the name of a weight and capacity measure.

A latere of these more general considerations, it is important to evaluate – although the terminology in Mycenaean archives is always profoundly fragmentary and scarce – the typology

[28] For Phoenician and Hebrew, see Hoftijzer and Jongeling 1995, s.v.
[29] See Hoftijzer and Jongeling 1995, s.v.
[30] See Hoftijzer and Jongeling 1995, s.v. *ḥrṣ*$_4$.
[31] *ku-ru-so* in PY Ta 714.2.2 and *ku-ru-sa-pi* in PY Ta 707.1; 714.3.
[32] For a further analysis of the role of Hittite as a bridge language between Indo-European and Non-Indo-European world, see Gasbarra and Pozza 2012, particularly paragraph 3.2.
[33] Cfr. the Akkadian terms *uqnû(m)* 'lapis-lazuli' and *alpu(m)* 'bull, ox'.
[34] The Akkadian *manū* has been generally interpreted as a loanword from Sumerian MANA, see AHw, CAD (vol. 10 part I) and CDA, s.v.

of linguistic interference we can analyze in the 2nd millennium BC Greek documentation. The analysis of loanwords, within the context in which they appear, suggests a close relation with the category of "wandering words" (*Wanderwörter*). This class of words is spread among numerous languages and cultures, usually in connection with trade, and it reveals a wide range of difficulty in establishing the etymology of the terms, or even their original source-language. The separation of *Wanderwörter* from loanwords is often ambiguous, and they may be considered a special class of loanwords, well distinguished from the category of *Lehnwort*.

In this sense, the textile terminology inherited from a Semitic source shows a coherent behaviour with all the terminology of plants, metals and materials in Mycenaean archives with a Non-Indo-European origin. These loanwords are also well adapted in the Mycenaean lexicon, as shown by the formation of derivative adjectives or compounds, and they represent a particular combination of endogenous and/or exogenous structures, creating new words well anchored to the sphere of technical pertinence and without any secondary semantic developments.

Acknowledgements

This paper is a product of the PRIN project "Linguistic representations of identity. Sociolinguistic models and historical linguistics" coordinated by Piera Molinelli (PRIN 2010/2011, prot. 2010HXPFF2, sponsored by the Italian Ministry of Education and Research). More specifically, the author works within the Research Unit at the University of Rome "La Sapienza", whose coordinator is Paolo Di Giovine.

Abbreviations

AHw von Soden, W. 1965–1981 *Akkadisches Handwörterbuch: unter Benutzung des lexikalischen Nachlasses von Bruno Meissner (1868–1947). Bearbeitet von Wolfram von Soden*, I–III. Wiesbaden.

CAD Gelb I. J. *et al.* 1956–2010 *The Assyrian Dictionary of the Oriental Institute of the University of Chicago*. Chicago.

CDA Black, J., George, A. and Postgate, N. (eds) 2000[2] *A Concise Dictionary of Akkadian*. Wiesbaden.

Bibliography

Amadasi Guzzo, M. G. 1991 The shadow line. In Baurain C. and Krings V. (eds), *Réflexions sur l'introduction de l'alphabet en Grèce, Phoinikeia Grammata: Lire et Écrire en Méditerranée,* Collection d'études classiques 6. Liège, 293–312.

Astour, M. C. 1967 *Hellenosemitica: an Ethnic and Cultural Study in West Semitic Impact on Mycenaean Greek*. Leiden.

Aura Jorro, F. 1985–1993 *Diccionario Micénico*. Madrid.

Belardi, W. 1954 Una nuova serie lessicale indomediterranea. *Rendiconti dell'Accademia Nazionale dei Lincei. Classe di Scienze Morali, Storiche e Filologiche*, Serie 8/9, 610–644.

Cline, E. H. 2010 *The Oxford Handbook of Bronze Age Aegean (ca. 3000–1000 BC)*. Oxford.

Christidis, A. F. (ed.) 2007 *A History of Ancient Greek from the Beginning to Late Antiquity*. Cambridge.

Del Freo, M., Nosch, M.-L. and Rougemont, F. 2010 The Terminology of Textiles in the Linear B Tablets including some Considerations on Linear A Logograms and Abbreviations. In C. Michel and M.-L. Nosch (eds), *Textile*

Terminologies in the Ancient Near East and Mediterranean from the Third to the First Millennia BC, Ancient Textiles Series 8. Oxford, 338–373.

Duhoux,Y. 2007 Pre-Greek Languages: Indirect Evidence. In A. F. Christidis (ed.), *A History of Ancient Greek: from the Beginnings to Late Antiquity*. Cambridge, 223–228.

EDG = Beekes, R. 2010 *Etymological Dictionary of Greek*. Leiden/Boston.

Ellenbogen, M. 1962 *Foreign Words in the Old Testament: their Origin and Etymology*. London.

Gasbarra, V. and Pozza, M. 2012 Fenomeni di interferenza greco-anatolica nel II millennio a.C.: l'ittito come mediatore tra mondo indoeuropeo e mondo non indoeuropeo. *AION* 1 (N.S.), 165–214.

Georgakas, D. J. 1959 Greek Terms for "Flax", "Linen" and their Derivatives; and the Problem of the Native Egyptian Phonological Influence on the Greek of Egypt. *Dumbarton Oaks Papers* 13, 253–269.

GEW = Frisk Hj. 1960–1972, *Griechisches etymologisches Wörterbuch*, I–III. Heidelberg.

Grimme, H. 1925 Hethitisches im Griechischen Wortschatze, *Glotta* 14, 13–26.

Haspelmath, M. and Tadmor, U. 2009 *Loanwords in the World's Languages. A Comparative Handbook*. Berlin.

Hoftijzer, J. and Jongeling, K. 1995 *Dictionary of the North West Semitic Inscriptions*, I–II. Leiden.

Killen, J. T. 1984 The Textile Industries at Pylos and Knossos. In T. G. Palaima and C. W. Shelmerdine (eds), *Pylos Comes Alive: Industry and Administration in a Mycenaean Palace. Papers of a Symposium*. New York, 49–63.

Lewy, H. 1895 *Die Semitischen Fremdwörter im Griechischen*. Berlin.

Luján, E. R. 2010 Mycenaean Textile Terminology at Work. The KN Lc(1)-Tablets and the Occupational Nouns of the Textile Industry. In C. Michel and M.-L. Nosch (eds), *Textile Terminologies in The Ancient Near East and Mediterranean from the Third to the First Millennia BC*, Ancient Textiles Series 8. Oxford, 374–387.

Mayer Modena, M. L. 1960 Gli imprestiti semitici in greco. *Rendiconti Istituto Lombardo di Lettere* 94, 311–351.

Masson, E. 1967 *Recherches sur les plus anciens emprunts sémitiques en grec*. Paris.

Michel, C. and Nosch, M.-L. (eds) 2010 *Textile Terminologies in the Ancient Near East and Mediterranean from the Third to the First Millennia BC*, Ancient Textiles Series 8. Oxford.

Morpurgo Davies, A. 1979 Terminology of Power and Terminology of Work in Greek and Linear B. In E. Risch and H. Mülestein (eds), *Colloquium Mycenaeum. Actes du sixième Colloque International sur les textes mycéniens et égéens tenu à Chaumont sur Neuchâtel (7–13 Septembre 1975)*. Genève, 87–108.

Muss-Arnolt, W. 1892 On Semitic Words in Greek and Latin. *Transactions of the American Philological Association* 23, 35–156.

Nosch, M.-L. 2000 Acquisition and Distribution: *ta-ra-si-ja* in the Mycenaean Textile Industry. In C. Gillis, C. Risberg and B. Sjöberg (eds) *Trade and Production in Premonetary Greece: Acquisition and Distribution of Raw Materials and Finished Products. Proceedings of the 6th International Workshop, Athens 1996*. Sweden/Åström, 43–61.

Nosch, M.-L. (2011) Production in the Palace of Knossos: Observations on the Lc(1) Textile Targets. *American Journal of Archeology* 115 (4), 495–505.

Palaima, T. G. 1997 Potter and Fuller: The Royal Craftsmen. In P. P. Betancourt and R. Laffineur (eds), *TEXNH: Craftsmen, Craftswomen and Craftsmanships in the Aegean Bronze Age. Proceedings of the 6th International Aegean Conference. Philadelphia, Temple University, 18–21 April 1996*, Aegeum 16. Liège/Austin, 407–412.

Palaima, T. G. 2003 'Archives' and 'Scribes' and Information Hierarchy in Mycenaean Greek Linear B Records. In M. Brosius (ed.), *Ancient Archives and Archival Traditions. Concepts of Record-Keeping in the Ancient World*. Oxford, 153–194.

Palaima, T. G. 2004 Mycenaean Accounting Methods and Systems and Their Place within Mycenaean Palatial Civilization. In M. Hudson and C. Wunsch (eds), *Creating Economic Order. Record-Keeping, Standardization, and the Development of Accounting in the Ancient Near East*. Bethesda/Maryland, 269–302.

Palmer, L. R. 1980 *The Greek Language*. London.

Sasson, J. (ed.) 1995 *Civilizations of the Ancient Near East*, I–IV. London.

Shelmerdine, C. W. 1998 Where do we go from here? And how can the Linear B tablets help us get there? In E. H. Cline and D. Harris-Cline (eds), *The Aegean and the Orient in the Second Millennium, Proceedings of the 50th Anniversary Symposium, University of Cincinnati, 18–20 April 1997*, Aegeum 18. Liège, 291–298.

Silvestri, D. 1974 *La nozione di indomediterraneo in linguistica storica*. Napoli.

Silvestri, D. 2013 Interferenze linguistiche nell'Egeo tra preistoria e protostoria. In L. Lorenzetti and M. Mancini (eds), *Le lingue del Mediterraneo antico. Culture, mutamenti, contatti*. Roma, 333–375.

Snodgrass, A. M. 1987 *An Archaeology of Greece. The present State and future Scope of a Discipline*. Berkeley/ Los Angeles.

Szemerényi, O. 1964 On Reconstructing the Mediterranean Substrata. *Romance Philology* 17, 404–418.

Vaniček, A. 1878 *Fremdwörter im Griechischen und Lateinischen*. Leipzig.

Ventris, M. and Chadwick, J. (1956) *Documents in Mycenaean Greek*. Cambridge.

Vigo, M. 2010 Linen in Hittite Inventory Texts. In C. Michel and M.-L. Nosch (eds), *Textile Terminologies in The Ancient Near East and Mediterranean from the Third to the First Millennia BC*, Ancient Textiles Series 8. Oxford, 290–322.

Vita, J. P. 2010 Textile Terminology in the Ugaritic Texts. In C. Michel and M.-L. Nosch (eds), *Textile Terminologies in the Ancient Near East and Mediterranean from the Third to the First Millennia BC*, Ancient Textiles Series 8. Oxford, 323–337.

Wisti Lassen, A. (2010) Tools, Procedures and Professions: A review of Akkadian Textile Terminology. In C. Michel and M.-L. Nosch. (eds), *Textile Terminologies in the Ancient Near East and Mediterranean from the Third to the First Millennia BC*, Ancient Textiles Series 8. Oxford, 272–282.

8. Constructing Masculinities through Textile Production in the Ancient Near East

Agnès Garcia-Ventura

Throughout history, and in a multitude of geographical settings, the production of textiles has been associated primarily with women. This is the case both in real life and in symbolic contexts: for example, the spindle and the distaff are instruments that are traditionally connected with women. Likewise women and goddesses are described as weavers of lives and destinies. In the ancient Near East, the Aegean, Italy and Egypt there is a large body of evidence for the close relationship between women and textile production in the primary sources – images, archaeological remains, texts – over a period of several centuries.[1] As a result, scholars have tended to analyse how textile production shaped the construction of different models of femininity, and have largely disregarded the notion of men playing any role in the process. This happens in part due to a tendency that affects not only studies of textile production but studies of other areas as well: women and femininities tend to be more discussed and problematized than men and masculinities. Given this imbalance, in this chapter I will focus on the construction of masculinities and will use the construction of femininities only as support material.

With regard to the chronological framework, I take as my starting point the administrative texts dealing with textile production from Ur III (*c.* 2100–2000 BC). For Ur III it is assumed that spinners were women, while those responsible of finishing tasks were men; for other productive phases such as plucking wool or weaving, the sexual division of labour is not so clear cut. I will discuss whether the sexual division of labour that characterized Ur III also applies to other periods of ancient Near Eastern history or to certain literary texts or visual sources ranging from the mid-4th millennium to the 1st millennium BC. As support materials I also use sources from ancient Egypt.

I will concentrate on certain specific stages of textile production:[2] spinning, and finishing. As spinning is the stage most closely associated with women, as stated above, I propose to analyse the contexts in which men appear linked to spinning or to spinning tools. The finishing stage is usually associated with males, and so provides an ideal context for scrutinizing how masculinities

[1] For the link between women and textile production in Antiquity see, among others, Barber 1994; Keith 1998, 499; Andò 2005; González Marcén and Picazo 2005, 141–144; Bevan 2006, 61; Larsson Lovén 1998 and 2013.

[2] For a recent review of the *chaîne opératoire* in ancient Near Eastern textile production of using a multidisciplinary approach, see Andersson 2012.

were shaped. In any case it is instructive to analyse the extent to which masculinities were defined *per se* or in opposition to attributes linked to femininity. In this regard I also explore which of these attributes were portrayed as positive or as negative.

The main theoretical framework for the analysis is provided by gender studies. Gender studies have concentrated from their very beginnings on 'women' as a category of analysis. One of the first goals was to establish that this category, 'women', was a social construct rather than something given. Simone de Beauvoir's *Le deuxième sexe*, published in 1949, contained the famous adage "one is not born, but rather becomes, a woman". But it was only in the 1970s that scholars began to consider masculinities as a social construct. The group *Achilles Heel*, promoted by Victor J. Seidler and others, was a pioneer in this area.[3] But despite the passage of time and the quantity and quality of publications dealing with the study of masculinities,[4] the analysis of this issue in certain contexts (the ancient Near East, for example) is almost non-existent.[5] A good summary of the situation and the reasons for the differences in the treatment of men and women as subjects and objects of study has been formulated by David Morgan 1993:

> "[...] women tend to be more embodied and men less embodied in social scientific, popular and feminist writings and representations, various reasons might readily be provided for such a bias. Very generally, it may be seen as part of a wider problem which has only recently begun to be rectified, namely one where women are more likely to be problematized than men. [...] Further, a greater tendency to write and speak of women and their bodies may be seen as reflecting the well-known ideological equation between women/men and nature/culture".[6]

In this chapter I maintain the man-woman contrast and the association with masculinities and femininities. This dichotomy as a category of analysis and this association have both been questioned from *queer* studies and post-feminist perspectives.[7] Combining these critiques, one of the most prominent proposals was published in 1998 by Judith Halberstam in her monograph entitled *Female Masculinity*. Halberstam dealt with masculinity as something produced by women, not as something linked exclusively to male bodies. Halberstam contended that women were able to construct masculinities and that masculinities constructed by women were distinct from masculinities constructed by men: women were not simply imitating men, but proposing new models. Despite the great interest of this interpretation and although I will use it as the basis for certain interpretations, I will concentrate only on masculinities constructed by male bodies.

---

[3] One of the first results of this group was the volume published 1978 by Victor J. Seidler (editor) under the title *The Achilles Heel Reader: Men, Sexual Politics and Socialism*.

[4] As publications are numerous, here I only quote three of them where previous references can be found and that serve as introduction to the topic for a reader with interest on masculinities and Antiquity. See the volume edited by Foxhall and Salmon 1998 as one of the first compilations with ancient history case studies with a special focus on masculinities. For an introduction to the study of masculinities in archaeology, with previous references, see Alberti 2006. For a manual on the study of masculinities in general (not specific neither for archaeology nor for Antiquity), see Whitehead and Barrett 2001.

[5] For some exceptions to this situation see Winter 1996, Asher-Greve 2008 and Suter 2012. The latter is especially interesting as it includes an introduction about the concepts of masculinity, femininity and their application to the ancient Near Eastern sources (Suter 2012, 433–437). Also in an attempt to deal with the construction of masculinities, the author presented a poster at the 57th *Rencontre Assyriologique Internationale* in Leiden, July 2012 together with J. Vidal. The poster was entitled "Ugaritic army: professional soldiers and the militia" and included a section entitled "Constructing masculinities".

[6] Morgan 1993, 70.

[7] Judith Butler's *Gender Trouble* (1990) is the pioneering work in this direction. It has been the basis for most of the proposals in this area published since then.

Spinning: real and symbolic duties

Spinning is a stage of textile production that is linked almost invariably to women.[8] Margarita Gleba[9] suggests two main reasons for this. On the one hand, spinning is more time-consuming than weaving. On the other hand, the spindle was more visible and easy to show in public; one could walk around while holding it, whereas the loom was heavier. For the case of the ancient Near East, even in Kassite Babylonia (*c.* 1500–1155 BC), where textile production was mainly carried out by male workers, the texts indicate that spinning was an exclusively female task.[10]

If we move to the Ur III period, when most workers in the textile sector were women, women seem to have done all the spinning.[11] In Ur III administrative texts we find specific Sumerian occupational terms[12] for workers devoted to weaving (u š - b a r) or to finishing (a z l a g_2), but only sporadically we find the term for those spinning yarns (g u).[13] Administrative texts recording raw materials and working days linked to spinning appear only occasionally and register the Sumerian word g e m e_2, which could be translated as 'low rank female worker'. As this term is attested in multiple contexts and not only linked to spinning, it gives us information about the rank and sex of the workers, but not about their occupation; the latter information is inferred by the context.[14]

Moreover, women are associated with spinning through countless symbolic and linguistic usages of terms related to this activity. In fact, this association is not only present in the administrative texts that record the institutional textile production from different periods; Sumerian literature contains frequent references to hair clasps (^{giš}k i r i d) and spindles (^{giš}b a l a) as objects associated with femininity, while weapons (^{giš}t u k u l) are associated with masculinity. Some of the best and most widespread examples of the validity of these associations are birth incantations. They are attested since mid-3rd millennium to mid-2nd millennium BC in multiple versions and copies, some of them being written in Sumerian and others in Akkadian language.[15]

One of the earlier birth incantations preserved is from Fara (*c.* 2600–2500 BCE). Below is Marten Stol's translation into English of the excerpt alluding to the abovementioned attributes:[16]

> "If it is female, let her bring out of the spindle and the pin;
> if it is a male, let her bring out of it the throwing stick and the weapon".

Another well-known incantation text from Ur III (UM 29-15-367) has been quoted, transliterated and translated several times in specialized literature, what allows us to match different versions.[17]

---

[8] For examples of the link between women and spinning in several ancient contexts other than the Near East, see Andò 2005, 45–53 and Gleba 2011. For the ancient Near East, see the entries for 'Spinnen' in the *Reallexikon der Assyriologie* (Völling 2011; Waetzoldt 2011b).

[9] Gleba 2011, 26.

[10] Sassmannshausen 2001, 90.

[11] Waetzoldt 1972, 120–122.

[12] Transliteration follows the Assyriological form (spaced for Sumerian, italicised for Akkadian and Hittite).

[13] Waetzoldt 1972, 88; Waetzoldt 2011a, 406, footnote 3.

[14] For examples of Ur III texts dealing with spinning, all accompanied by translation into English, see Firth and Nosch 2012.

[15] For a good overview of diverse types of incantations from different periods, presented in transliteration, translation and commentary, see Cunningham 1997. On symbols of masculinity and femininity see especially Cunningham 1997, 33, 74–75.

[16] Stol 2000, 60 (with previous references). For a detailed study of this incantation with some parallels, see Krebernik 1984, 36–47 ("Beschwörung 6"). For a recent quotation of this incantation paying special attention to the symbols of masculinity and femininity, see Suter 2012, 438.

[17] See Stol 2000, 61, footnote 80 for previous references. For a pioneer reference on the topic of the birth incantations and the attributes and the publication of this text, see van Dijk 1975, 57 and 61 (transliteration and translation of the text into English respectively). Van Dijk matches this text with an Old-Babylonian one (VAT 8539) also analysed by the

Below I quote again Stol's translation into English as the most recent one:[18]

> "If it is a male, let him take a weapon, an axe, the force of his manliness.
> If it is a female, let the spindle and the pin be in her hand."

For the gender issue I discuss here the use of "manliness" in Stol's translation is especially interesting. Likewise Jan van Dijk[19] translates "virilité" in his French version. However Graham Cunningham[20] chooses "heroism". The Sumerian a$_2$ nam-ur-sag of line 46 is translated literally as "valorous arm" and nam-ur-sag alone as "heroism". It is clear that the expression refers to a quality associated to masculinity, but it seems better to preserve the translation closer to the Sumerian term. Indeed, as has been discussed in some analyses, some Sumerian terms translated as "masculinity" or "manliness" are not faithful enough as they add a nuance in a specific direction.[21] At this point, then, I prefer "heroism" as proposed by Cunningham.

Moving now to the spheres of gods and goddesses, Enki[22] attributes the spindle and the hair clasps to the goddess Inanna as symbols of femininity.[23] This is highly significant because Inanna, despite being a female from the point of view of biological sex, is not a prototypical example of femininity, since she is also associated with war and weapons. In like manner, it is no coincidence that Inanna's male devotees wore women's clothes and make-up; in other words, they transgressed the expected appearance of prototypical masculinity and adopted that of prototypical femininity as well. This impression is reinforced by the fact that they carried both spindles and swords, again attributes of femininity and masculinity respectively – opposites brought together by the gender ambiguous Inanna.[24]

Some accounts linked to Inanna also mention that the goddess had the power to turn men into women and women into men. In other words, she was able to materialize the metaphor of the "world turned upside down" mentioned above. One example is the hymn to the goddess Inanna for Išme-Dagan, an Assyrian king from the 18th century BC (hymn Išme-Dagan K). The following excerpt mentions the alluded metaphor and the exchange of attributes linked to femininity and masculinity:

> "Inana was entrusted by Enlil and Ninlil with the capacity to gladden the heart of those who revere her in their established residences, but not to soothe the mood of those who do not revere her in their well-built houses; to turn a man into a woman and a woman into a man, to change one into the other, to make young women dress as men on their right side, to make young men dress as women on their left side, to put spindles into the hands of men ......, and to give weapons to the women."[25]

---

author in a previous paper in 1972. Michel 2004, 409, footnote 52 quotes this text too matching it to an Old Assyrian one (kt 90/k, 178). Michel notes that birth incantations use symbolic pairs associated to male and female infants: weaponry versus textile tools respectively, as discussed in this contribution, or savage ram and cow, as is the case in her text.

[18] Stol 2000, 61.

[19] Van Dijk 1975, 65.

[20] Cunningham 1997, 72.

[21] On the Sumerian nam-šul-la and the lack of a Sumerian word for "masculinity", see Suter 2012, 435, footnote 16. On the Sumerian nam-guruš and what lies behind its diverse translations as "manly" or "manliness" see Garcia-Ventura 2014, 14.

[22] Enki is one of the main gods of the Mesopotamian pantheon. He is associated with wisdom, arts and creation.

[23] Waetzoldt 2011b, 2.

[24] Teppo 2008, 78–79.

[25] English translation from the "Electronic Text Corpus of Sumerian Literature", onwards *etcsl*. For the Išme-Dagan K hymn, see http://etcsl.orinst.ox.ac.uk/cgi-bin/etcsl.cgi?text=t.2.5.4.11#. This composition has been also quoted by Suter 2012, 435, footnote 13 as an example of the symbols attributed to femininity and masculinity in Sumerian literature.

These transgressions here are perceived as positive thanks to the intervention of the goddess, but in other contexts they are described as negative and interpreted as a punishment. As an example, I quote an excerpt from the Hittite text "The first soldier's oath",[26] translated into English by Billie Jean Collins in 1997:

> "They bring a woman's garment, a distaff and a spindle and they break an arrow (*lit. reed*). You say to them as follows: "What are these? Are they not the dresses of a woman? We are holding them for the oath-taking. He who transgresses these oaths and takes part in evil against the king, queen and princess[27] may these oath deities make (that) man (into) a woman. May they make his troops women. Let them dress them as women. Let them put a scarf on them. Let them break the bows, arrows, and weapons in their hands and let them place the distaff and spindle in their hands (instead)".[28]

This text shows how elements that were common in everyday life were symbolically transformed in certain Hittite rituals, in this case the attributes linked to sex roles.[29] It is stated that women were associated with the spindle and the spinning whorl, while men were associated with weapons.[30] As far as the translation is concerned, I should highlight a discrepancy between the English version published by A. Goetze[31] and the more recent versions by B. J. Collins[32] again in English and by J. V. García Trabazo[33] in Spanish. This discrepancy has a strong bearing on the main gender issue discussed here: where Goetze translates "mirror",[34] Collins and García Trabazo translate "distaff" for the Hittite ^{GIŠ}*ḫulāli*. In fact all these elements – distaff, spindle and mirror – are regularly associated with femininity, but in this case, in view of the presence of the determinative GIŠ (wood) accompanying the substantive, it seems that "distaff" is a more plausible translation than "mirror".[35]

This confusion is frequent not only in texts, but also in iconography. At the 4th ICAANE meeting in Berlin in 2004, Elena Rova presented a paper showing images of women holding something that might have been either a mirror or a tool used for spinning. Rova proposes that this ambiguity was probably deliberate as both these attributes were linked to femininity. She also suggests criteria that would help to distinguish between the different tools.[36]

---

[26] CTH 427, lines 42–53 (= *Catalogue des textes hittites*, E. Laroche 1971). Concerning the dating of the text, I quote Collins 1997, 1.66: "The language of the composition indicates that it was composed in the Middle Hittite period (late 15th century BC), although the copies that survive were inscribed in the Empire period".

[27] Goetze 1950, 354 translated it as "the king (and) the queen (and) the princess", as did Collins 1997, 1.66 some decades later. However, checking the transliteration is possible that the Sumerian logograms (DUMU.MEŠ LUGAL) suggest that this refers to the king's offspring rather than to the princess. The most recent translation into Spanish by García Trabazo 2002, 533 supports this possibility too: "al rey (y) la reina (y) a los hijos del rey". In any case it seems that both options are plausible as the Sumerogram DUMU itself has no grammatical gender. It could be specified attaching the words male/female or masculine/feminine, but it does not happen always as we see in this example. As an example of the use of DUMU (son) or ŠEŠ (brother) attached to a feminine anthroponym, see the case of the princess Enanedu as referred by Lion 2009, 166–169.

[28] Collins 1997, 1.66, epigraph 9.

[29] González Salazar 2004, 152.

[30] The text is also quoted by Hoffner 1966, 331–332, who compares it with another Hittite text where this link between men, women and these attributes is clear. In addition, other Hittite texts associated to funerary rituals reinforce the link between women and goddesses and spinning tools like distaff and spindle. To this respect see Rova 2008, 559–560.

[31] Goetze 1950 (text included in the classical edition of texts from the ancient Near East by J. B. Pritchard).

[32] Collins 1997.

[33] García Trabazo 2002, 533.

[34] Goetze 1950, 354.

[35] Kloekhorst 2008, 357.

[36] Rova 2008.

In addition to the Hittite text and the incantations quoted above, the association of women and goddesses with spinning tools is common in countless contexts. In Ugaritic literary texts,[37] for example, the goddess Athiratu is portrayed holding a spindle.[38] A Phoenician inscription from the 1st millennium BC from Karatepe[39] confirms the association of women with spindles as the allegory of safety in the country: a woman depicted strolling peacefully while spinning with the spindle, without being disturbed by anyone. Another reference is the Bible, which again links spindles and distaffs to femininity (Prov 31, 19).[40] As late as the first half of the 20th century AD in the region of the Argolid (Greece), herdsmen used to manufacture a wooden distaff as a gift for their fiancées.[41] Modern-day English also retains expressions like "the distaff side" referring to the feminine side of the family, and "the spear side" referring to the masculine side.[42] These examples suggest that the link between women, femininity and spinning tools is practically universal. As a result, it is interesting to identify the contexts in which males rather than females are associated with these tools.

In some texts is attested the expression "the spindle man". As in the Hittite text above, this link also bears a negative connotation. It seems that "the spindle man" is a negative reference, as it is used to describe the effeminate behaviour of certain males.[43] Therefore, the association of spinning tools with "real men" and not to Inanna devotees or characters of mythical stories would have been perceived as dysfunctional, a symptom of femininity. For all these reasons, I think it is possible to identify the construction of masculinities and femininities as associated with certain tools, artefacts and tasks. Reinforcing this link, certain magical texts show a reversal of attributes clearly linked to a prototypical masculinity or to a prototypical femininity.[44]

Another text, in this case a Sumerian proverb, presents a man associated with a spindle as being unfortunate. The character in question is a carpenter. Comparing the line with the situations mentioned above, perhaps here the misfortune is not attributed to the femininity associated with the spindle, but to the difference in status of those who produce the tools compared with those who use them.[45] Below is the published translation by Bendt Alster:[46]

> "A disgraced scribe becomes an incantation priest.
> A disgraced singer becomes a piper.
> A disgraced lamentation priest becomes a flutist.[47]

---

[37] Like KTU 1.4:II.3–4, among others.

[38] Marsman 2003, 421–422; cf. Hoffner 1966, 330–331 for examples of other Ugaritic texts.

[39] KAI 26 A, col. II, lin. 5–8. The first publication of this inscription, corresponding to the numbering of the text here quoted, was by Donner and Röllig 1964: *Kanaanäische und Aramäische Inschriften* (=KAI). Bron 1979 also refers to this excerpt of the inscription.

[40] Hoffner 1966, 329 also quotes II Sam 3, 29 as an example of this link. However this quotation seems to be erroneous, as it has nothing to do with the issue under discussion.

[41] Bouza 1976.

[42] These and similar definitions can be found in the *Oxford English Dictionary* (2nd edition, Clarendon Press, 1989). For "distaff side" see vol. 4, s.v. § 4, p. 849. For "spear side" see vol. 16, s.v. § 10a, p. 146.

[43] Bottéro and Petschow 1972–1975, 465.

[44] Hoffner 1966, 328.

[45] A hypothesis defended by Gordon 1959, 213.

[46] Alster 1997, 55 (proverb 2.54, vol. 1). Cf. Gordon 1959, 211. See Alster 1997, 365 (vol. 2) p. 365 for a comment on similarities and differences among Alster's more recent version and Gordon's former translation into English. Alster 1997, 365 (vol. 2) describes this literary composition as "a sententious, short poem listing jobs remaining for professionals who have lost their professional skills."

[47] There are some discrepancies in these two lines as translated by Alster 1997, 55 or Gordon 1959, 211 and by *etcsl* (http://etcsl.orinst.ox.ac.uk/cgi-bin/etcsl.cgi?text=t.6.1.02#). The latter translates them as follows: "A disgraced singer

A disgraced merchant becomes a twister (?).
A disgraced carpenter becomes a man of the spindle.
A disgraced smith becomes a man of the sickle.
A disgraced mason becomes a «clay dragger» (?)."

With regard to what the images suggest regarding the sexual division of labour in Mesopotamia in different periods,[48] most have been interpreted as proof that the spinners were mainly women. Indeed, most of the human figures are interpreted as women because they wear something resembling a pony tail. Recently, Julia M. Asher-Greve revised some Late Uruk seals (mid-4th millennium BC) which she had previously studied for her PhD.[49] She suggests that features such as position or activity may identify certain pony-tailed figures as females, and others as males.[50] On the other hand, Susan Pollock and Reinhard Bernbeck, despite their suggestive classification of figures represented in Protodynastic seals as men, women, pony-tailed or ambiguous,[51] do not hesitate in identifying pony-tailed figures carrying out textile activities as women.[52] In any case, what both proposals stress is that some representations are ambiguous as regards sex and that we have to question which secondary attributes (in this case, hair style) we should consider in order to identify males and females. These secondary sexual attributes change over time and we may reach mistaken conclusions if we apply these criteria uncritically. Identifying ambiguity and considering that it may have been a deliberate strategy for representing human bodies opens up new areas of interpretation.[53]

Another example demonstrating that our association of women with spinning might be based on preconceptions and not on actual data comes from archaeology. In some archaeological digs, grave goods have been associated with men or with women before any analysis of the bones, just because it is assumed that women use certain implements and men others. Núria Rafel[54] has drawn attention to this misconception in certain Iberian funerary contexts (*c.* 8th–3rd centuries BC), as has M. Carmen Vida Navarro[55] in the case of some 1st millennium BC tombs at Pontecagnano, in southern Italy. In neither context were the expected pairings – man-weapon versus woman-spindle whorl – systematically reproduced, as only in some cases they were verified. Sometimes both kinds of grave goods were found in the same tomb, or sometimes spindle whorls appeared in masculine tombs. In Pontecagnano, Vida Navarro questions the association of specific typologies of *fibulae* to men and to women, as sometimes they are carried out before the analysis of the bones; so, as suggested above, they are based on preconceptions on style, gender and stereotypes, not on data. Despite bone analysis are not 100% reliable, at least they add more data to be assessed.

becomes a flute-player. A disgraced lamentation priest becomes a piper". This discrepancy would be the consequence of the usual controversy when translating terms that name musical instruments.

[48] See Breniquet 2008, 287–290 for a collection of representations of spinning. To compare and complete these analyses, some contemporary examples are interesting, such as the spinning scenes from Egypt (Newberry 1893, vol. 2, print 13; Newberry 1893, vol. 2, print 4; Winlock 1955, prints 26 and 27) or from the Aegean (Barber 1994, 82 and 220).

[49] Asher-Greve 1985.

[50] Asher-Greve 2008.

[51] Pollock and Bernbeck 2000, 155.

[52] Pollock and Bernbeck 2000, 159.

[53] On sexual ambiguity see Garcia-Ventura 2012, 508; Garcia-Ventura and López-Bertran 2013. On ambiguity related to the objects and characters represented, in this case concerning music and weaving, see Breniquet 2011. In this paper the author deals with an Old-Babylonian terracotta relief (AO 12454) usually interpreted as representing a harpist. Breniquet 2011, 287–290 suggests that the presumed harp could be a waist loom too.

[54] Rafel 2007.

[55] Vida Navarro 1992.

Elisabeth Völling[56] also alerts us to the risks of preconceived associations regarding the use of specific raw materials for manufacturing spinning tools. At the *Vorderasiatisches Museum* at Berlin (VAM), there is an onyx artefact classified as a sceptre. In spite of the value of the raw material, Völling proposes that the artefact might in fact be a spindle with a spindle whorl. This example and the ones above show that our preconceptions may lead us to link spinning exclusively with women or to disregard archaeological remains that may have been used in spinning. If we can leave these assumptions aside, we will be able to appreciate the involvement of men – and not only of women – in the process of spinning, and the positive or negative connotations of this trade.

Finishing textiles: setting or challenging hegemonic masculinities?

Moving now to the last stage of textile production, finishing, it includes a variety of techniques such as cropping, scouring, bleaching, laundering, pleating, smoothing or fulling, among others. In Ur III Mesopotamia the main duties of those responsible for finishing clothes were fulling and cleaning.[57]

The images related to finishing available to us from Mesopotamia are not comparable to the ones related to spinning.[58] Obviously, though, the near absence of images of a stage of production does not mean that it did not exist. Some images show evidence of folding, which is one of the tasks included in finishing. Looking at these images it is difficult to determine whether the figures carrying out this activity are male or female.[59] Fortunately some paintings and bas-reliefs from Egyptian tombs are more illuminating, as they clearly depict men carrying out the finishing stage.[60]

In this case, then, most of the evidence about finishing is found in texts. Ur III texts give a detailed register of these workers with their personal names, duties, and allotments in some cases. The Sumerian word for those responsible of finishing tasks is a z l a g$_2$,[61] a term almost always preceded by the determinative l u$_2$, translated as "person" or "human being", but also as "man". It is no accident that the deeply patriarchal Mesopotamian society used the same word for "man" and for "person". The same happens in many other languages where "man" is used for both men and women and as a broad term not specifically linked to men or to masculinity. Nevertheless, we must be careful when translating l u$_2$ in some contexts, as Asher-Greve and A. Lawrence Asher have pointed out: "persistent mistranslation in many texts of the word l u$_2$ as "man" probably contributed to scholarly neglect of women".[62] In spite of these considerations, it seems clear that the evidence (e.g. personal names) suggests that all or almost all a z l a g$_2$ in Ur III were men.[63] In addition, Hartmut Waetzoldt quotes a Garšana text recording an exceptional situation in which

[56] Völling 1998.

[57] Waetzoldt 1972, 155. I will use finishing with this meaning, referring to these two activities. When not, I specify to which tasks I am referring.

[58] Breniquet 2008, 313.

[59] Breniquet 2008, 317.

[60] Some good examples are found in registers of the images from the tombs of Khety (Newberry 1893, vol. 2, print 13) and Baqt (Newberry 1893, vol. 2, print 4), both from Beni Hasan, from the 11th Dynasty (Middle Kingdom of Egypt, threshold from the 3rd to the 2nd millennium BC).

[61] a z l a g$_2$ usually written as LU2.TUG2 in lexical lists, is equivalent to the Akkadian *ašlāku*. For references and sources from different periods where the term is attested with its variatons, see Waetzoldt 1972, 153 and CAD A. vol. 2. s.v., 445–447 (= *Chicago Assyrian Dictionary*).

[62] Asher-Greve and Asher 1998, 40.

[63] See Waetzoldt 1972, 153–154 as a classic reference. For a more recent study (concentrating not only on those responsible of the finishing of textiles, but on some of their duties as well) including previous references, see Verderame 2008.

female workers were asked to help these professionals because the men were unable to deal with the amount of work to be done.[64] Waetzoldt shows that some years later, the team was enlarged to meet the demand.

Below is a transliteration and translation of a text from Girsu (text L. 2628), from the third reign year of the king Ibbi-Suen,[65] as an example of one of the Ur III texts in which those doing finishing tasks as well as female weavers were mentioned:

obverse	*obverse*
1. 1 (u) 7 (aš) 3 (bariga) še gur	5280 sila of barley
2. še-ba geme$_2$ uš-bar /Gir$_2$-su^{ki}	as barley ration
	for the female weavers (from) Girsu
3. 2 (aš) Lugal-KA-gi-na$^!$	600 (sila of barley for) Lugal inimgina
4. 2 (aš) Lugal-u$_2$-šim-e	600 (sila of barley for) Lugal-ušime
reverse	*reverse*
5. 4 (bariga) Lu$_2$-^{d}Ba-u$_2$/dumu Ur-^{d}E$_2^?$-ša-GIŠGAL	240 sila (for) Lu-Ba'u, Ur ešgal's son:
6. ^{lu2}azlag$_2$-m[e]	they are cleaners.
7. ki A-a-kal-l[a]-ta	From A(ya)kalla
8. nam-zi-tar-ra	Namzitara
9. šu ba-ti	received
10. mu si-mu-ru-um/^{ki} ba-hul	year: Simurrum was destroyed

Since in Ur III the term ^{lu2}azlag$_2$ was associated exclusively (or almost exclusively) with men, I contend that the link between men and finishing constituted one of the multiple strategies used to construct masculinities. However, scrutinizing the sources becomes clear that finishing and spinning and its relationship with masculinity and femininity respectively are not comparable. Indeed spinning and spinning tools were associated with women and were considered as attributes that identified a pattern of femininity in multiple contexts other than textile production, as detailed above. It does not happen with finishing and masculinity, an association only present when dealing directly with textile production. Despite that, in what follows, I will show that, from my point of view, is possible to analyse certain aspects of the construction of masculinities through the finishing of textiles.

Finishing tasks were included in what was known as the *ramo de agua* – the area of textile production requiring the use of water – in textile production during the industrial revolution in Catalonia, Spain's most flourishing region. Those workers occupied with activities involving water (including dyeing too, in this particular case) were traditionally men, whereas the previous stages such as spinning and weaving were carried out by women. In her study of this sector, Virginia Domínguez[66] shows that 90% of workers in the *ramo del agua* were men. The reason for this sexual division of labour present in these different locations and eras lies in the fact that cleaners and dyers worked outside with toxic, heavy and corrosive substances, which were considered characteristically male tasks.[67]

[64] Waetzoldt 2011a, 407–408.
[65] Lafont and Yildiz 1996, text 2628 = TCTI 2, 2628.
[66] Domínguez 1999, 14.
[67] See Murdock and Provost 1973 as a classic reference on this issue.

Concentrating now on laundry, one of the tasks in the finishing phase, it has been regendered many times. Some Neo-Babylonian (*c.* 626–539 BC) texts portray it as a male task when performed in the framework of public institutions like temples.[68] Customers of these laundries would have been gods (in a symbolic sense) and elites, as registered in the texts. At this point, we may wonder why these elites selected certain clothes to be cleaned at these laundries and not at home. One possible explanation is that there was a specialization in the cleaning of expensive or special clothes,[69] just as today we wash everyday clothes at home but use laundry services for the special and delicate ones.

Another facet of this process of regendering of laundry that deserves mention is its depiction in the English-language press of the late 19th and early 20th centuries.[70] Washing clothes was mainly a female task, but with the introduction of technological improvements it became progressively appropriated by men. In fact this trend is identified by George P. Murdock and Caterina Provost as one of the factors that determine the sexual division of labour.[71] While the involvement of men in the arena of 'public' or commercial laundry was described in press reports or advertisements, domestic washing of clothes, associated to women, was ignored and, if mentioned at all, was merely ridiculed. With this example in mind, perhaps the situation reflected in the ancient Near Eastern sources is similar; perhaps laundry is linked with the masculine sphere because the records deal with commercial laundry, but they tell us nothing of the washing of clothes at home, a task most probably performed by women.

If we accept this hypothesis, then there is at least one exception to the rule. In some Ugaritic literary texts both men and women are described washing clothes.[72] Even the goddess Athiratu appears washing her own clothes[73] and the clothes of a man named Dani'ilu are washed by his son.[74] These examples suggest the interaction of different categories in the distribution of duties related to laundry. Perhaps there was sexual division of labour, since it appears that both women and men washed their own clothes, and perhaps age was a factor that determined who washed whose clothes. Finally, perhaps gender affected the way hierarchies were perceived: we find a goddess washing her own clothes, but not a god.

Moving on now to how literary texts describe the men who carried out these finishing tasks, the text entitled "At the cleaners"[75] is particularly interesting. This is a humorous Old-Babylonian (*c.* 1900–1600 BC) text[76] which satirises the occupation and duties of a laundrymen. In the story a customer arrives at the laundry and orders the washing of some clothes, giving strict instructions. The laundryman listens to all instructions and the payment proposed, and then declines the offer;

[68] Waerzeggers 2006, 94.
[69] Waerzeggers 2006, 95.
[70] Mohun 2003.
[71] Murdock and Provost 1973.
[72] Marsman 2003, 421.
[73] KTU 1.4:II.5–9.
[74] KTU 1.17:I.33.
[75] In all publications of the text quoted here in following notes the title attributed to it is "At the cleaners". There is only the exception of Reiner 1995 who preferred "At the fullers".
[76] Cf. "At the cleaners" with this other Old-Babylonian text published in transliteration, translation and comment by Sylvie Lackenbacher 1982, text AO 7026. Unlike the one discussed here, this other text registers some technical details concerning the duties of laundrymen such as types of fabrics, their weights, etc. For this reason it would be interesting to compare them, as they are to some extent complementary. In her paper, Lackenbacher alludes to the satire quoted here and contends that the most suitable title for the composition would be "Dialogue du blanchisseur avec son client" (Lackenbacher 1982, 144).

he suggests that the customer should do it himself because no other laundrymen will accept the order either. This last detail gives us an insight into their collective perception of themselves as members of a kind of "guild", if we may use this term in an anachronistic way.

The text, found at Ur (U.7793), was first published by Cyril J. Gadd. Gadd published it in a transliterated and translated form[77] in a paper in the journal *Iraq* in 1963, where he announced that a copy of the text was to be published in the second part of the sixth volume of the *Ur Excavation Texts* series. More than 20 years passed before a new study of the text was published, a revised transliteration and translation by Alasdair Livingstone, in 1988. Since then, several publications have been made of the text: collating it, that is to say, revising the reading of certain lines,[78] analysing the content and comparing it to other literary traditions,[79] updating the translation,[80] or both, presenting new analysis and new translations.[81] Taking into account all these different transliterations, comments and translations, below I quote Nathan Wasserman's translation as the most recent one, which incorporates previous updates:

> Come fuller, let me instruct you, treat my garment! [82]
> What I instruct you, do not lay aside,
> Your own (ideas), you should not do!
> As for the hem of the garment, you will lay down the selvage,
> You will stitch the outer side to the inside,
> You will pick up the thread of the (shorter) border.
> You will soak the delicate part (of the cloth) in beer,
> You will strain it through a sieve.
> You will loosen the hem with selvage.
> You will spray it with clear water,
> You will wipe it like a *kimdum* cloth, and you will...:
> To the weft yarns you will [*brush?*] so that the warp yarns...
> You will... *in* a *barrier*(?)/*basin*(?),
> ... you will *mix* alkali with gypsum (to prepare fuller's earth?).
> You will [*beat*? or: *press*?] it on/with/under a stone.
> ... in a vessel.
> In case you have *applied* a (laundry) mark, (then) you must... and you will have to comb (the fabric).
> You will tap (the garment) repeatedly with an *e'ru*-wood stick (to felt smooth the fabric).
> Y[ou will arrange] the fringe on the washer's stool.
> You will [sew/repair] the work, the (damaged) warp, with a needle.
> *Rev.*

---

[77] Below are some excerpts of the story following the first version published by C. J. Gadd (1963, 184–185; copy of the text in UET [= *Ur Excavation Texts*] 6/2, 414) for comparison with the English version quoted here: "1. Come now, Cleaner, let me give you an order – clean my suit. 2. The order which I give you don't lay aside, 3. that (process) of your own don't do. 4. The hem and the coat you will lay down, 5. the front you will beat inwards, 6. the bits you will pick off [...] 26. you will bring (the finished work) to (my) house, and a seach of barley will be poured into your lap. 27. The cleaner answers him: By Ea, master of craftsmanship, who preserves me, 28. not excepting me (to anybody), what you are talking is stuff and non[sense] [...] 31. The order you are giving me, to repeat (and) say over, 32. to speak and to recite, I haven't the power. [...] 35. the big job which you have in hand do it by yourself. [...]".

[78] George 1993.

[79] Reiner 1995.

[80] Foster 2005, 151–152, in his last review of his anthology of literary Akkadian texts.

[81] Wasserman 2013.

[82] Gadd (1963, 184) and Livingstone (1988, 177) and Wasserman (2013, 275) translate it in the singular as "suit" (Gadd) or "garment" (Livingstone and Wasserman), while Foster (2005, 151) proposes the plural "clothes", as does the revised version published at CDLI (P274721, see *Cuneiform Digital Library Initiative* website: http://www.cdli.ucla.edu/).

You will spread and flatten the hem.
You will dry (the garment) in the break of evening,
so that the fabric will not dry (and wrinkle).
(Afterwards) you will place it in a box (and) in a chest.
It had better be smooth! Bring (it) to me; I will make you very happy – promptly!
You will bring (the garment) to the house, (one) will pour a *seah* of barley into your lap.
The fuller answers: By the name of Ea, the lord of wisdom who keeps me alive!
Drop it! Not me! What you are saying – only my creditor and my tax collector
have the nerve (to talk) like you!
Nobody's hands could manage this work!
What you have instructed me I cannot repeat, utter or reiterate!
Come upstream of the city, in the environs of the city-
let me show you a washing-place! And then (you could)
set yourself (to do) the great work you have in your hands!
The meal time should not pass[83] – come in and stay and
unravel the cleaner's many threads!
If you don't calm yourself down
there will be no fuller who will bother for you.
You will be mocked. Your heart will burn,
and you will cause a rash (?) to appear on your body.
Its lines, their number (is) 41 (sic).

This exceptional text is one of the few examples of a humorous and ironic text written in Akkadian.[84] It is interesting as it portrays the cleaner from an entirely different point of view from the one present in the Ur III administrative texts. While the administrative texts record information of relevance to the institutions (wages, duties, names, work teams), this text deals with the perception of the cleaner as a professional independent worker (or at least not totally dependent on or dictated to by the institutions). It seems that this cleaner is working by himself and has the power to decide whether or not to accept an assignment. In this regard, then, the organization of work resembles the pattern attested in Old-Babylonian documents more closely than that found in the Ur III texts.

In the Egyptian tradition we find another ironic text, comparable to the one just described. But, unlike the Akkadian text, the Egyptian one (an excerpt from a text from the 12th Dynasty (beginning of the 2nd millennium BC) mocks the laundrymen and their working conditions rather than the customers. The text is known by Egyptologists as the "Satire of Trades"[85] or more formally the "Teaching of Duaf's son Khety".[86] In the text, a father trying to persuade his son to learn to be a scribe, speaks contemptuously of other occupations,[87] two of which, mat-weaving and laundering, are related to textile production in a broad sense. Below I quote the excerpts as translated by Miriam Lichtheim and Richard B. Parkinson respectively:

[83] Foster (2005, 152), translates this line in a completely different way: "Don't miss your chance, seize the day!" According to Foster, this line insists on the idea of *carpe diem* as the laundryman advises his customer to make the most of the moment, reinforcing the text's comical effect (see especially Foster 2005, 152, footnote 3).
[84] Despite this is the most common interpretation, Wasserman (2013, 259) in his recent study of the text proposes that probably the main aim was didactical, instead of satirical, as all instructions are listed in detail.
[85] On this informal name, see Lichtheim 1975, 184.
[86] The text is complete at *Papyrus Sallier*, II, 8.2.
[87] On the purpose of the satire see, among others, Lichtheim 1975, 184–185; Parkinson 1991, 72–83; Roccati 2000, 6.

> "The weaver[88] in the workshop, he is worse off than a woman; with knees against his chest, he cannot breathe air. If he skips a day of weaving, he is beaten fifty strokes; he gives food to the doorkeeper, to let him see the light of day".[89]

> "And the washerman washes on the shore, and nearby is the crocodile. «Father, I shall leave the flowing (?) water», say his son and daughter, «for a trade that one can be content in, more so than any other trade», while his food is mixed with shit. There is no part of him clean, while he puts himself amongst the skirts of a woman who is in her period(?); he weeps, spending the day at the washing board. He is told: «Dirty clothes! Bring yourself over here», and the (river-)edge overflows with them".[90]

Interestingly, in both cases the disadvantages of the occupations of the mat-weaver and the laundryman are highlighted through an explicit comparison with women: in the first case, the disadvantage is discomfort, and in the second it is dirt. In both cases these negative aspects are linked to women described as squatting (i.e., discomfort) or as menstruating (i.e., dirt). In this case, then, masculinity is not constructed through the exaltation of positive aspects or through the usual link between men and weapons, but in contrast to femininity; this is an alternative, negatively perceived masculinity constructed referring to the negative characteristics associated with the female sex.

Conclusion

Literary and administrative texts, images and certain archaeological remains shed interesting light on the sexual division of labour in the societies of Antiquity. In addition, as I contend here, they allow us to analyse how masculinities and femininities were constructed. At this second level of analysis it is possible not only to determine who was doing what, but to envisage how society perceived certain trades, which attributes were considered appropriate for males and for females, and which were perceived as positive or negative.

In the case of spinning I would like to highlight two points. First, spinning was normally performed by women, but we should be careful to avoid preconceptions in our analysis of the sources. Second, even if we accept that women were almost exclusively the spinners, there are certain contexts in which men are mentioned. Analysing these contexts sheds light on certain strategies used to construct femininities and masculinities. One conclusion of this analysis is that men were linked to spinning mainly in symbolic or ritual contexts. A second conclusion is that this link is sometimes used to ridicule men, as is perceived as a threat to the construction of a hegemonic masculinity.

Similarly, we also find some descriptions of men working as laundrymen. Though this occupation is associated predominantly with males, in some cases it is described, again, as deconstructing hegemonic masculinity. In other words, the fact that men were connected with laundry tasks in many contexts does not lead to an automatic construction of a hegemonic masculinity through this occupation; in fact, such a conception is strongly challenged by the Egyptian "Satire of Trades" quoted above. These sources, then, enable us to identify diverse strategies used to construct diverse masculinities. They bear witness to the lack of uniformity in the construction of these patterns, even when concentrating on the same tasks in similar contexts.

In both arenas, spinning and finishing textiles, the association of female attributes with men carrying out these trades had clearly negative connotations. Probably this is an indication of the

[88] This weaver refers to mat-weaver, as notices Lichtheim 1975, 192, footnote 11.
[89] Lichtheim 1975, 188.
[90] Parkinson 1991, 75.

importance of sexual division, both symbolically and from the point of view of social prestige. When analysing administrative texts, factors such as age, hierarchy or speciality appear to be as influential as gender. However, in some other written sources gender appears as the main structuring factor. It has been suggested that this situation was accentuated between the 3rd and the 1st millennia BC in the ancient Near East;[91] during these two millennia women lost legal capacities, their visibility in public arenas, and social prestige. The materials analysed here do not necessarily support this proposal, but it may be a fruitful avenue for future interdisciplinary work to pursue, and sources related to textile production may well provide valuable insights.

Acknowledgements

This contribution was prepared during my research fellowship at the CRC 933 "Material Text Cultures. Materiality and Presence of Writing in Non-Typographic Societies" at the Ruprecht-Karls-Universität Heidelberg. I thank Elsa Chesa for advice about published translations of the "Satire of Trades" and Érica Couto for advice about birth incantations. Likewise I thank the editors of the volume for their insightful comments. Obviously any remaining errors or omissions are my own responsibility.

Bibliography

Alberti, B. 2006 Archaeology, Men, and Masculinities. In S. M. Nelson (ed.), *Handbook of Gender in Archaeology*, Lanham, 401–434.

Alster, B. 1997 *Proverbs of Ancient Sumer. The World's Earliest Proverb Collections*. Bethesda, Maryland.

Andersson Strand, E. 2012 The Textile *chaîne opératoire*: Using a multidisciplinary approach to textile archaeology with a focus on the Ancient Near East, *Paléorient* 38, 1–2, 21–40.

Andò, V. 2005 *L'ape che tesse. Saperi femminili nella Grecia antica*. Roma.

Asher-Greve, J. M. 1985 *Frauen in altsumerischer Zeit*. Malibu, California.

Asher-Greve, J. M. 2006 "Golden Age" of Women? Status and Gender in Third Millenium Sumerian and Akkadian Art. In S. Schroer (ed.), *Images and Gender: Contributions to the Hermeneutics of Reading Ancient Art*, Freiburg/Göttingen, 41–81.

Asher-Greve, J. M. 2008 Images of Men, Gender Regimes, and Social Stratification in the Late Uruk Period. In D. Bolger (ed.), *Gender through time in the Ancient Near East*, Lanham/New York/Toronto/Plymouth, 119–171.

Asher-Greve, J. M. and Asher, A. L. 1998 From Thales to Foucault... and back to Sumer. In J. Prosecký (ed.), *Intellectual Life of the Ancient Near East. Papers presented at the 43rd Rencontre Assyriologique Internationale. Prague, July 1–5, 1996*, Prague, 29–40.

Barber, E. J. W. 1994 *Women's Work the First 20,000 Years. Women, Cloth and Society in Early Times*. New York/London.

Bevan, L. 2006 *Worshippers and Warriors. Reconstructing Gender and Gender Relations in the Prehistoric Rock Art of Naquane National Park, Valcamonica, Brescia, Northern Italy*. Oxford.

Bottéro, J. and Petschow, H. 1972–1975 Homosexualität. In *Reallexikon der Assyriologie und Vorderasiatischen Archäologie* 4, Berlin/New York, 459–468.

Bouza Koster, J. 1976 From Spindle to Loom: Weaving in the Southern Argolid, *Expedition* 19, 1, 29–39.

Breniquet, C. 2008 *Essai sur le tissage en Mésopotamie. Des premières communautés sédentaires au milieu du IIIe millénaire avant J-C*. Paris.

Breniquet, C. 2011 Une plaquette «au harpiste» d'Eshnunna. In F. Wateau (ed.), *Profils d'objets. Approches d'anthropologues et d'archéologues*, Paris, 283–296.

[91] Asher-Greve 2006; Suter 2008.

Bron, F. 1979 *Recherches sur les inscriptions phéniciennes de Karatepe*. Genève/Paris.

Butler, J. 1990 *Gender Trouble. Feminism and the Subversion of Identity*. London/New York.

Collins, B. J. 1997 The First Soldier's Oath. In W. W. Hallo (ed.), *The Context of Scripture. Volume I: Canonical Compositions from the Biblical World*, Leiden/New York/Köln, 1.66.

Cunningham, G. 1997 *'Deliver Me from Evil'. Mesopotamian Incantations 2500–1500 BC* (=Studia Pohl 17), Roma.

Domínguez Álvarez, V. 1999 Dona i treball tèxtil. Sabadell 1900–1960. In *Dona i treball tèxtil. Sabadell 1900–1960*, Sabadell, 9–33.

Donner, H. and Röllig, W. 1964 *Kanaanäische und Aramäische Inschriften*. Wiesbaden.

Firth, R. and Nosch, M.-L. 2012 Spinning and Weaving Wool in Ur III Administrative Texts, *Journal of Cuneiform Studies* 64, 67–84.

Foster, B. R. 2005 *Before the Muses. An Anthology of Akkadian Literature*. Bethesda, Maryland.

Foxhall, L. and Salmon, J. 1998 *Thinking men: masculinity and its self-representation in the classical tradition*. London.

Gadd, C. J. 1963 Two Sketches from the Life at Ur, *Iraq* 25, 177–188.

Gadd, C. J. and Kramer, S. N. 1966 *Ur Excavations Texts VI: Literary and Religious Texts, Second Part* (=UET 6/2), London – Philadelphia.

García Trabazo, J. V. 2002 *Textos religiosos hititas. Mitos, plegarias y rituales*. Madrid.

Garcia-Ventura, A. 2012 From engendering to ungendering: revisiting the analyses of Ancient Near Eastern scenes of textile production. In R. Matthews and J. E. Curtis (eds), *Proceedings of the 7th International Congress on the Archaeology of the Ancient Near East (ICAANE. British Museum and University College London, April 2010)*, Wiesbaden, 505–515.

Garcia-Ventura, A. 2014 Ur III Biopolitics. Reflections on the Relationship between War and Work Force Management. In D. Nadali and J. Vidal (eds), *The Other Face of the Battle. The Impact of War on Civilians in the Ancient Near East* (=AOAT 413), Münster, 7–23.

Garcia-Ventura, A. and López-Bertran, M. 2013 Figurines & Rituals. Discussing Embodiment Theories and Gender Studies. In C. Ambos and L. Verderame (eds), *Approaching Rituals in Ancient Cultures*, Rivista degli Studi Orientali Nuova Serie, Volume LXXXVI, Roma, 117–143.

George, A. R. 1993 Ninurta-Paqidat's dog bite, and notes on other comic tales, *Iraq* 55, 63–75.

Gleba, M. 2011 The "Distaff Side" of Early Iron Age Aristocratic Identity in Italy. In M. Gleba and H. W. Horsnaes (eds), *Communicating Identity in Italic Iron Age Communities*, Oxford, 26–32.

Goetze, A. 1950 Hittite Rituals, Incantations, and Description of Festival. In J. B. Pritchard (ed.), *Ancient Near Eastern Texts Relating to the Old Testament*, Princeton/New Jersey, 346–365.

González Marcén, P. and Picazo, M. 2005 Arqueología de la vida cotidiana. In M. Sánchez Romero (ed.), *Arqueología y género*, Granada, 141–158.

González Salazar, J. M. 2004 Rituales mágico-religiosos hititas relacionados con las actividades militares del reino de Hatti (II milenio a.C.). In J. Fernández Jurado, C. García Sanz and P. Rufete Tomico (eds), *Actas del III congreso español de Antiguo Oriente Próximo. Huelva, del 30 de Septiembre al 3 de Octubre de 2003*, Huelva, 147–157.

Gordon, E. I. 1959 *Sumerian Proverbs. Glimpses of everyday life in Ancient Mesopotamia*. Philadelphia.

Halberstam, J. 1998 *Female Masculinity*. Durham.

Hoffner, H. A. 1966 Symbols for Masculinity and Feminity: Their Use in Ancient Near Eastern Sympathetic Magic Rituals, *Journal of Biblical Literature* 85, 3, 326–334.

Keith, K. 1998 Spindle Whorls, Gender, and Ethnicicty at Late Chalcolithic Hacinabi Tepe, *Journal of Field Archaeology* 25, 497–515.

Kloekhorst, A. 2008 *Etymological Dictionary of the Hittite Inherited Lexikon*. Leiden/Boston.

Krebernik, M. 1984 *Die Beschwörungen aus Fara und Ebla. Untersuchungen zur ältesten keilschriftlichen Beschwörungsliteratur* (=TSO 2), Hildesheim/Zürich/New York.

Lackenbacher, S. 1982 Un texte vieux-babylonien sur la finition des textiles, *Syria* 59, 129–149.

Lafont, B. and Yildiz, F. 1996 *Tablettes cunéiformes de Tello au Musée d'Istanbul. Datant de l'époque de la IIIe Dynastie d'Ur (ITT II/1,2544–2819, 3158–4342, 4708–4713)*. Leiden.

Larsson Lovén, L. 1998 *Lanam fecit*. Woolworking and female virtue. In L. Larsson Lovén and A. Strömberg (eds), *Aspects of Women in Antiquity*, Jonsered, 85–93.

Larsson Lovén, L. 2013 Textile production, female work and social values in Athenian vase painting. In A.-L. Schallin (ed.), *Perspectives on Ancient Greece. Papers in celebration of the 60th anniversary of the Swedish Institute at Athens*, Stockholm, 135–151.

Lichtheim, M. 1975 *Ancient Egyptian Literature. Volume I: The Old and Middle Kingdoms*. Berkeley.

Lion, B. 2009 Sexe et genre (2). Des prêtresses fils de roi. In F. Briquel-Chatonnet, S. Farès, B. Lion and C. Michel (eds), *Femmes, cultures et sociétés dans les civilisations méditerranéennes et proche-orientales de l'Antiquité*, Lyon, 165–182.

Livingstone, A. 1988 "At the Cleaners" and Notes on Humorous Literature. In G. Mauer and U. Magen (eds), *Ad bene fideliter seminandum. Festgabe für Karlheinz Deller zum 21. Februar 1987* (=AOAT 220), 175–187. Neukirchen-Vluyn.

Marsman, H. J. 2003 *Women in Ugarit & Israel. Their Social & Religious Position in the Context of the Ancient Near East*. Leiden/Boston.

Michel, C. 2004 Deux incantations Paléo-Assyriennes. Une nouvelle incantation pour acompagner la naissance. In J. G. Dercksen (ed.), *Assyria and Beyond. Studies presented to Mogens Trolle Larsen* (=PIHANS 100), Leiden, 395–420.

Mohun, A. P. 2003 Industrial Genders: Home/Factory. In N. E. Lerman, R. Oldenziel and A. P. Mohun (eds), *Gender & Technology. A Reader*, Baltimore, Maryland, 153–176.

Morgan, D. 1993 You too can have a Body Like Mine: Reflections on the Male Body and Masculinities. In S. Scott and D. Morgan (eds), *Body Matters*, London, 69–88.

Murdock, G. P. and Provost, C. 1973 Factors in the Division of Labor by Sex: a Cross-cultural Analysis, *Ethnology* 12, 2, 203–25.

Newberry, P. E. 1893 *Beni Hasan*. London.

Parkinson, R. B. 1991 *Voices from Ancient Egypt. An Anthology of Middle Kingdom Writings*. London.

Pollock, S. and Bernbeck, R. 2000 And They Said, Let Us Make Gods in Our Image: Gendered Ideologies in Ancient Mesopotamia. In A. E. Rautman (ed.), *Reading the Body: Representations and Remains in the Archaeological Record*, Philadelphia, 150–164.

Rafel Fontanals, N. 2007 El textil como indicador de género en el registro funerario ibérico. In P. González Marcén, C. Masvidal Fernández, S. Montón-Subías and M. Picazo Gurina (eds), *Interpreting household practices: reflections on the social and cultural roles of maintenance activities*, Barcelona, 115–146.

Reiner, E. 1995 At the Fuller's. In M. Dietrich and O. Loretz (eds), *Vom Alten Orient zum Alten Testament. Festschrift für Wolfram Freiherrn von Soden zum 85. Geburtstag am 19. Juni 1993* (=AOAT 240), Neukirchen-Vluyn, 407–411.

Roccati, A. 2000 Réflexions sur la Satire des Métiers, *Bulletin de la société française d'Égyptologie* 148, 5–17.

Rova, E. 2008 Mirror, Distaff, Pomegranate, and Poppy Capsule: on the Ambiguity of some Attributes of Women and Goddesses. In H. Kühne, M. Czichon and F. Janoscha Knepper (eds), *Proceedings of the 4th International Congress of the Archaeology of the Ancient Near East*, Wiesbaden, 557–570.

Sassmannshausen, L. 2001 *Beiträge zur Verwaltung und Gesellschaft Babyloniens in der Kassitenzeit*. Mainz am Rhein.

Stol, M. 2000 *Birth in Babylonia and the Bible: its Mediterranean setting* (=CM 14), Groningen.

Suter, C. E. 2008 Who are the Women in Mesopotamian Art from ca. 2334–1763 BCE? *KASKAL. Rivista di storia, ambienti e culture del Vicino Oriente Antico* 5, 1–55.

Suter, C. E. 2012 The Royal Body and Masculinity in Early Mesopotamia. In A. Berjelung, J. Dietrich and J. F. Quack (ed.), *Menschenbilder und Körperkonzepte im Alten Israel, in Ägypten und im Alten Orient*, Tübingen, 433–458.

Teppo, S. 2008 Sacred Marriage and the Devotees of Ištar. In M. Nissinen and R. Uro (eds), *Sacred Marriages. The Divine-Human Sexual Metaphor from Sumer to Early Christianity*, Winona Lake, Indiana, 75–92.

van Dijk, J. 1972 Une variante du thème de « l'Esclave de la Lune », *Orientalia* 41, 339–348.

van Dijk, J. 1975 Incantations accompagnant la naissance de l'homme, *Orientalia* 44, 52–79.

Verderame, L. 2008 Il controllo dei manufatti tessili a Umma. In M. Perna and F. Pomponio (eds), *The Management of Agricultural Land and the Production of Textiles in the Mycenaean and Near Eastern Economies*, Paris, 111–133.

Vida Navarro, M. C. 1992 Warriors and Weavers: sex and gender in Early Iron Age graves from Pontecagnano, *Journal of the Accordia Research Center* 3, 67–100.

Völling, E. 1998 Bemerkungen zu einem Onyxfund aus Babylon, *Mitteilungen der Deutschen Orient-Gesellschaft zu Berlin* 130, 197–221.
Völling, E. 2011 Spinnen. B. Archäologisch. In *Reallexikon der Assyriologie und Vorderasiatischen Archäologie* 13, Berlin/Boston, 3–5.
Waerzeggers, C. 2006 Neo-Babylonian Laundry, *Revue d'Assyriologie* 100, 83–96.
Waetzoldt, H. 1972 *Untersuchungen zur neusumerischen Textilindustrie*. Roma.
Waetzoldt, H. 2011a Die Textilproduktion von Garšana. In D. I. Owen (ed.), *Garšana Studies*, Bethesda, Maryland, 405–454.
Waetzoldt, H. 2011b Spinnen. A. Philologisch. In *Reallexikon der Assyriologie* 13, Berlin/Boston, 1–3.
Wasserman, N. 2013 Treating Garments in the Old Babylonian Period: "At the Cleaners" in a Comparative View, *Iraq* 75, 255–277.
Whitehead, S. M. and Barrett, F. J. 2001 *The Masculinities Reader*. Cambridge.
Winlock, H. E. 1955 *Models of Daily Life in Ancient Egypt*. Cambridge/Massachusetts.
Winter, I. 1996 Sex, Rethoric, and the Public Monument: The Alluring Body of Naram-Sîn of Agade. In N. Boymel Kampen (ed.), *Sexuality in Ancient Art*, Cambridge, 11–26.

9. Spindles and Distaffs: Late Bronze and Early Iron Age Eastern Mediterranean Use of Solid and Tapered Ivory/Bone Shafts

Caroline Sauvage

Based on complete archaeological examples preserved in Ugarit and Delos, this chapter will investigate the interpretation of Late Bronze Age ivory pomegranate-knobbed[1] and whorled shafts as versatile three-piece spinning kits that could have been used alternatively as spindles (shaft + whorl) or distaffs (shaft + pomegranate knob).[2] Constitutive parts of such kits, i.e. ivory shafts, pomegranate knobs and spindle-whorls, have been found in the Late Bronze Age and Iron Age eastern Mediterranean in domestic, religious and funerary contexts in the Levant, Cyprus and the Aegean. If the identification of mounted whorls on ivory shafts has always been straightforward, solid ivory shafts and pomegranate knobs have not yet been systematically explored in relation to textile industry. Indeed, such knobbed shafts have been variously interpreted as sceptres, kohl rods, objects of prestige, or feminine symbols; while ivory shafts can be interpreted as kohl rods, cosmetic boxes fastening systems, or are simply characterized as "rods". The aim of this article is therefore to explore the use and function of the rod components of ivory/bone spinning kits in the Late Bronze Age and in the Iron Age. The careful study of deposition contexts and eventual association to textile tools of each type of artefacts should allow for a better understanding of these objects, and for pinpointing their use in relation to textile industry in the eastern Mediterranean.

On the use of spindles

Spinning fibres involves simultaneously three processes: drawing out (or drafting), twisting and winding the yarn.[3] These are typically achieved by using a spindle, which allows the thread to stay under constant tension, and thus avoids the newly formed thread from tangling or untwisting until further attention (i.e. plying) is given to it.[4] Not only do spindles prevent the thread from un-spinning, but they also allow faster and easier work, and permit control over the thickness and uniformity of the yarn. "A stick or a rock will do",[5] and will absorb enough rotation power to allow the spinner

[1] The identification of a pomegranate is subjected to caution, as it may also have been the representation of a poppy capsule; see for instance Smith 2002, 97–100.
[2] Sauvage 2012, 207.
[3] Hochberg 1977, 18; Barber 1991, 41.
[4] Barber 1991, 42.
[5] Barber 1991, 42. See also Hochberg 1977, 21–23; Montell 1941, 113–114, fig. 1.

to free one of his/her hand for drafting and extending the rotation as needed.[6] While sticks are great bobbins, stones are better flywheels and, because of their weight and density, rotate faster. Therefore combining a shaft with a weight (i.e. spindle-whorl) is more efficient.[7] Shaft and whorl can be made out of different materials (wood, reed, bone/ivory, metal, glass/faïence) and assembled together diversely.[8]

Two main spinning techniques co-existed in the ancient Near East and eastern Mediterranean: the low-whorl technique, and the high-whorl technique. On a low-whorl spindle the whorl is attached to the shaft near the bottom, and the rotation movement could be induced by "a flick of the thumb and fingers".[9] On such spindles, the thread passes underneath the whorl, then around the spindle and finally passes back to the top of the spindle (Fig. 9.1). It causes the thread to frequently come in contact with the extremities and down-facing end of the whorl. This technique was attested in Bronze Age Anatolia, Cyprus and the Aegean.[10] According to Barber, Anatolia actually used more of a "middle" whorl technique, as exemplified by the third millennium silver and gold or electrum "spindle" from tomb L at Alaca Höyük, or by the metal spindles from Horoztepe. She assimilates the middle-whorl technique with a low-whorl technique.[11] The high-whorl technique was attested in Egypt, Mesopotamia and Ugarit,[12] and required the rotation to be set with the hand palm: Egyptians typically rolled their spindles up or down the leg with one hand.[13] On these high-whorl spindles, only the maximum diameter area of the whorl would feel constant pressure from the thread (Fig. 9.1).[14] In both techniques, the spindle may be supported or hang from the thread (drop-spindle), see Fig. 9.20,[15] but the position of the whorl on a spindle is said to be culturally determined.[16] Low- and high-whorl spindles will produce two different type of treads: a so-called "S" spun fibre will be made by a high-whorl spindle (as for instance all the flax made in Egypt), while a low-whorl spindle would produce a "Z" spun thread.[17]

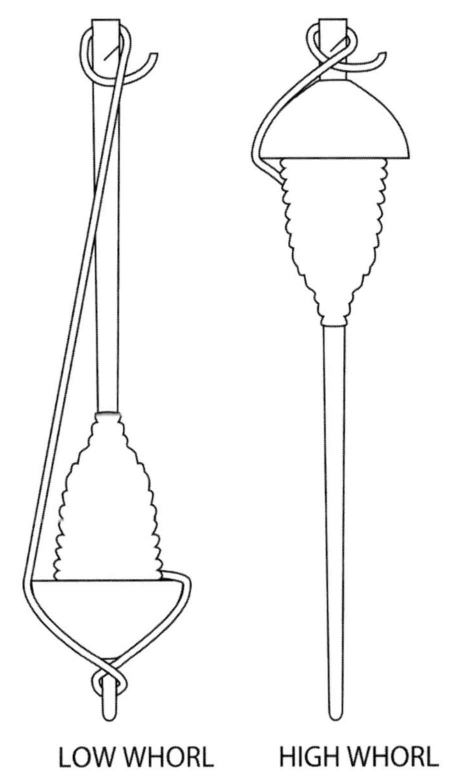

Fig. 9.1: Low- and high-whorls spindles showing point of contact of thread with whorls, after Crewe 1998, fig. 8.4 (© L. Crewe).

[6] Barber 1991, 42–43.
[7] Hochberg 1979b, 25; Barber 1991, 43.
[8] For instance, wood spindles (*giš-bala)* are attested in Ebla; Anderson, Felluca, Nosch, Peyronel 2010, 161.
[9] Barber 1991, 43.
[10] Barber 1991, 54–55; Frankel and Webb 1996, 193–194; Crewe 1998.
[11] Barber 1991, 60–61.
[12] Barber 1991, 56–58; Sauvage 2012, 197; Sauvage and Hawley 2013.
[13] Barber 1991, 43.
[14] Crewe 1998, 61.
[15] Barber 1991, 43. When the spindle is supported, its end may be on the floor, on the spinner's leg or in a cup.
[16] Crowfoot 1931, 34; Barber 1991, 53; Crewe 1998, 7; Crewe 2002, 218.
[17] Barber 1991, 66; Breniquet 2008, 110–112.

The length of the wool fibres could dictate the specific use of spindles as for instance short goat hairs could be spun using a hand-held spindle, while longer fibres such as sheep's wool are easier to spin with a suspended or supported spindle,[18] because long(er)-staple wool require the spinner to have both hands free to draw out the fibres.[19] When it comes to drafting fibres, the weight of the spindle itself is important and certainly contributes to it.[20] Therefore the heavier a spindle is, the bigger the tension, and the faster the fibres will be drawn out of the distaff. Such remark has of course implications for the choice of a spindle according to the type of fibres that one spinner wishes to work with. For instance, Barber pointed out that the short, fine, and slippery cotton fibres would draw out too fast with a light drop-spindle and that they therefore require a light-weight supported spindle.[21] The total weight of the spindle also impacts the thickness of thread that will be obtained: a lighter spindle makes a finer thread, while a heavier spindle will produce a thicker thread. Therefore, we could postulate, that once the spindle has been chosen for a specific type of fibres, the thickness of the thread to be produced would be monitored by selecting whorls according to their weight. Thus, with the same spindle, for a fine wool thread obtained from short fine wool, a 8g whorl can be used, while, a 33g whorl will certainly produce a thicker thread.[22]

Context and distribution of bone/ivory spindles and shaft

The present catalogue is not exhaustive and only takes into account the objects with a known context, and whose assemblage can be reconstructed. It derives from more substantial studies of ivories, pomegranates, and textile tools.[23]

Spinning kits and spindles

Ugarit

At Ugarit and Minet el-Beida, four spindles have been found. Spindle RS 4.221[A]) – Louvre AO 15757 was found in *dépôt 43* at Minet el-Beida (Fig. 9.2).[24] Its preserved length is 22cm, it has a diameter that varies from 0.85 and 1.27cm, and a dome-shaped whorl (ø 3.1cm, H.1.35cm, ø perf. 1.1cm) inserted at about on third of its preserved length. Its total weight is 30.6g. It was found with another "spindle", AO 15758 – RS 4.221B, that could be best understood as a spinning kit.[25] This preserved shaft is topped by a pomegranate knob and has a thin, almost flat, whorl inserted near in its middle. It is possible that a missing part was attached on the lower end of the shaft, opposed to the knob[26] (Fig. 9.3). The maximum diameter of this shaft is 1.35cm, its preserved length is 22.1cm, while its whorl is 4.04cm in diameter, 0.42cm thick and has a perforation of 1.2cm.[27] It weighs 44.9g. According to the excavation notebooks and inventory, the same deposit also yielded

[18] Barber 1991, 43.
[19] Barber 1991, 44.
[20] Barber 1991, 43, 52.
[21] Barber 1991, 43. Moreover, cotton fibres have to be spun with a bowl.
[22] Ryder 1968, 81; Barber 1991, 52. See the illustration of the different thickness of thread produced with different weigh range of whorls; Andersson, Nosch, Wisti Lassen 2007, 10, fig. 8.
[23] For instance see Gachet-Bizollon 2007; Ward 2003, Daviau 2002.
[24] Gachet-Bizollon 2007, no. 136.
[25] Sauvage 2012, 203–207.
[26] Gachet-Bizollon 2007, no. 137, p. 121–123, 260.
[27] Gachet-Bizollon gives no information on the weight.

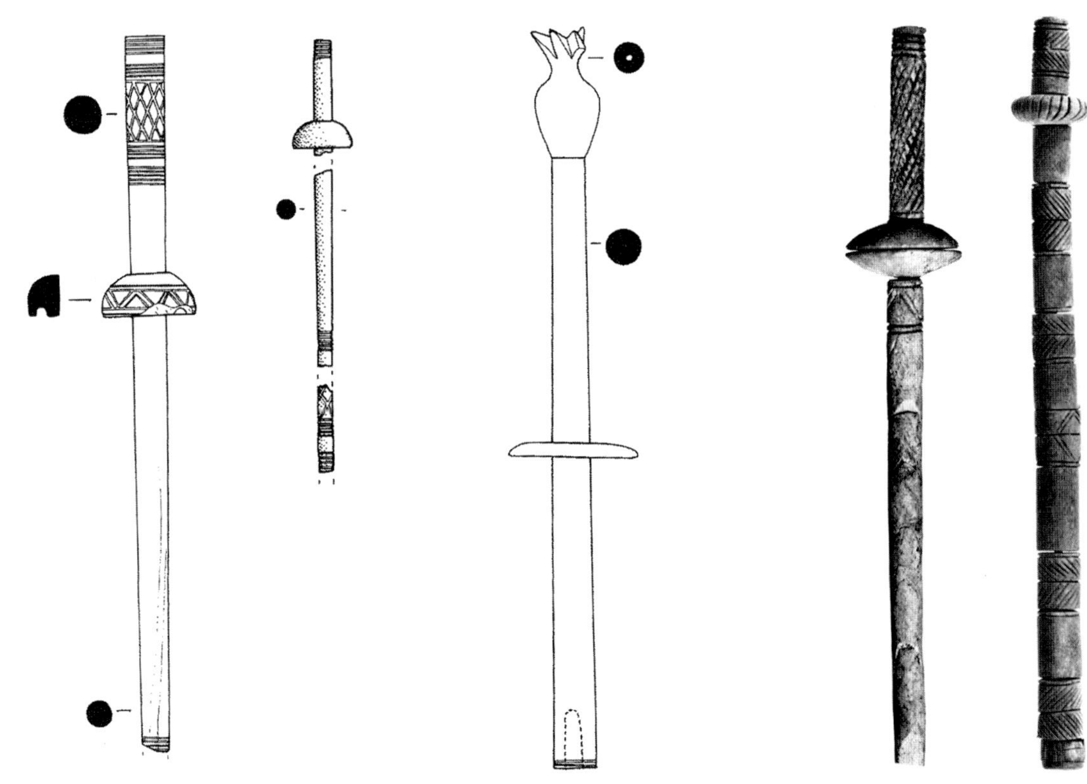

Fig. 9.2: Spindles from Ugarit RS 4.221[A] (AO 15 757) and RS 34.210; after Gachet-Bizollon 2007, nos. 136 and 139 (courtesy of J. Gachet-Bizollon, © Mission de Ras Shamra).

Fig. 9.3: Spinning kit from Minet el-Beida AO 15758 – RS 4.221B, after Gachet-Bizollon 2007, no. 137 (courtesy of J. Gachet-Bizollon, © Mission de Ras Shamra).

*Fig. 9.4: Spindles from Megiddo. On the left, M3568 (L. 15*cm*) from the upper level of tomb 1122; after Guy 1938, pl. 84.1. On the right, B433a (L. 20.2*cm*) from tomb 3018F; after Loud 1949, pl. 197.2 (courtesy of the Oriental Institute of the University of Chicago).*

another spindle, at least two groups of ivory/bone whorls and several whorls made of serpentite, ivory and faience. It is possible that this deposit corresponds to a tomb not seen by Schaeffer during the excavations.[28] A third, small broken spindle RS 34.210 preserved at the Lattaquia museum was found at Ras Shamra in room BD of the *maison aux albâtres*, located in block 1 of the *quartier résidentiel* (Fig. 9.2).[29] Its preserved length is 13.3cm, and its diameter is 0.5cm. Its whorl is 1.9cm in diameter and 0.8cm in thickness and its perforation is probably of 0.5cm. From the same room comes a bone/ivory dome-shaped whorl (ø 3.27, H. 0.8cm).[30]

Megiddo

At Megiddo, two spindles were found in tombs, while a third one possibly comes from a domestic context. From the upper level of tomb 1122 comes one bone spindle (M 3568) with two spindles-

[28] For the re-evaluation of the deposit and the list of the material discovered there, see Sauvage *forthcoming*.

[29] Gachet-Bizollon 2007, no.139.

[30] Gachet-Bizollon 2007, no. 563. Gachet-Bizollon classifies this object as a button.

whorls facing each other and attached to the two-part shaft by a pin (Fig. 9.4).[31] The smallest (lower) end of the spindle is broken (L. 15cm, ø 0.6–0.7cm). The top part of the shaft is decorated with horizontal lines and lattice pattern, while the top part of the piece, under the whorls, is decorated with horizontal and oblique lines. From the same layer, come three bone shafts (M 3569), one of which has both of its extremities dug by a mortise (L. 7, 7.1 and 10cm, ø 0.75–0.9cm).[32] The top layer of this tomb also yielded a 14.8cm long bronze "pin" and several spindle-whorls: seven dome shaped bone whorls, five conical with splayed edges bone whorls and one dome-shaped steatite whorl.[33]

In tomb 3018F (st. IX), a spindle made of several short ivory cylinders was found (Fig. 9.4). An ovoid whorl decorated with deeply incised radiuses or grooves was sandwiched in between the cylinders, the whole shaft being originally held together by an inner pin (inv. B 433a, L. 20.2cm, ø 1–1.2cm, whorl: ø 2.2cm, H. 0.8cm).[34] This 'articulated' spindle is the only example of its kind, and we can wonder whether it was practical to use, as we can easily imagine the rotation of the cylinders on the shaft if they were not firmly secured by the pin.

Finally, a three-part shaft spindle (inv. M 3530, total L. 25.2cm, shafts L. 7.2–9–7.4cm, ø 0.8–1.2cm, whorls ø 3cm, H. 07 and 0.8cm), mounted with two spindle-whorls, comes from the LB 1 room 1140 of square U17 at Megiddo.[35] The whorls are located at one third of the shaft length, closer to the thinner end. I have no knowledge of publication mentioning this context.

Artemision at Delos

An ivory knobbed and whorled solid shaft (i.e. spinning kit) comes from the Artemision at Delos (shaft: L. 22.5cm, ø 1–0.7cm; whorl: ø 3.5cm, thickness 0.7cm). The whorl is set at 2cm from its largest extremity, opposing the pomegranate (Fig. 9.5).[36] It was found in a *favissa* under temple E,[37] and was buried with several ivory, bone, faience and metallic objects, including ivory whorl inv. B. 7121 (ø 3.7cm; H. 0.6cm),[38] pierced silver disc inv. B. 7174 (ø 2.8cm; H. 0.1cm),[39] faience whorls inv. B 7163 (ø 1.7cm),[40] and B. 7164 (ø 1.7cm; ø perf 1cm),[41] and conical stone whorl inv. B 7193 (ø 3cm; H. 2.5cm).[42]

Perati

A fragmentary ivory spindle (Δ 108) comes from tomb 65 at Perati (Fig. 9.6). The shaft is 19.9cm long and its diameter varies from extremity to extremity from 0.8 to 0.4cm. A dome shape whorl

[31] Guy 1938, 170, pl. 84.1; Gachet-Bizollon 2007, 125 no. 9. The mounting of the whorls was reconstructed from LB1 spindle M 3530 from loc. 1140, see below.
[32] Guy 1938, pl. 84.2; Gachet-Bizollon 2007, 125 no. 11.
[33] Guy 1938, pl. 84.3–15.
[34] Loud 1948, pl. 197.2.
[35] Lamon and Shipon 1939, pl. 95.38; Gachet-Bizollon 2007, 125 no. 10.
[36] Gallet de Santerre and Tréheux 1947–48, 198–199, no. 36, fig. 16; Gachet-Bizollon 2007, 115.
[37] Gallet de Santerre and Tréheux 1947–48, 251–252.
[38] Gallet de Santerre and Tréheux 1947–48, 199, no. 37, fig. 17.
[39] Gallet de Santerre and Tréheux 1947–48, 221, no. 75, fig. 25. Parallels in silver and in ivory are known from the Artemision at Ephesus, where they are interpreted as top-whorls for hairpins (Hogarth 1908, 119, pl. XII.24 and pl. XXXIII.16).
[40] Gallet de Santerre and Tréheux 1947–48, 220, no. 65, pl. XXXVIII.7.
[41] Gallet de Santerre and Tréheux 1947–48, 220, no. 66, pl. XXXVIII.8.
[42] Gallet de Santerre and Tréheux 1947–48, 239, no. 91, fig. 32.

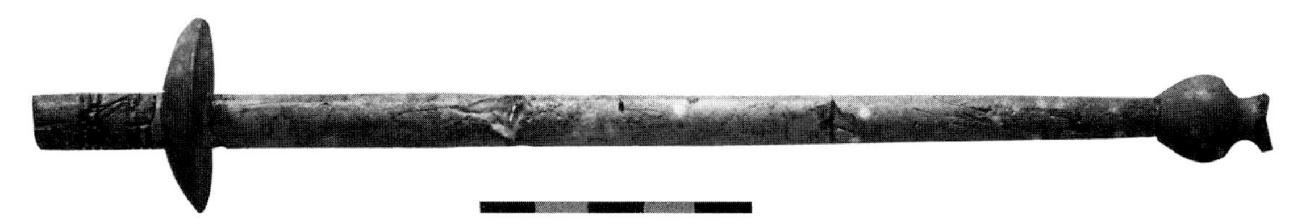

Fig. 9.5: Ivory shaft with whorl and knob from the Artemision at Delos; after Gallet de Santerre and Tréheux 1947–48, 198–199, fig. 16 (© EfA).

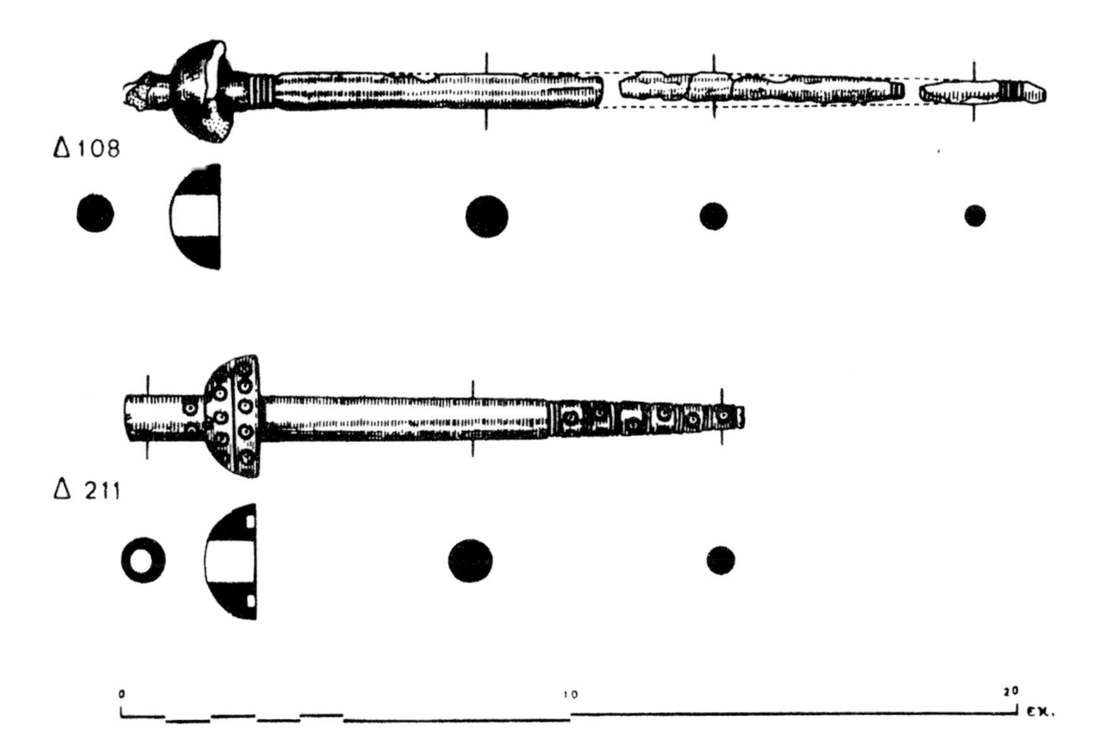

Fig. 9.6: Ivory spindles from tombs 65 and 152 at Perati; after Iakovidis 1978, fig. 117, 96 (courtesy of S. Iakovidis).

was inserted on its larger extremity. Another ivory dome-shaped whorl was found in the same tomb (Δ 112, ø 3cm; H. 0.7cm, ø perf. 0.6cm).[43] Conical stone "whorls" also come from the same tomb.[44]

From tomb 152 comes a complete ivory spindle (Δ 211, L. 13.1cm, ø 0.45–0.9cm, ø whorl 2.4cm, H. whorl 1.1cm) decorated with horizontal lines as well as pointed circles on the smallest part of the shaft (Fig. 9.6).[45] Pointed circles are also present on the shaft, below the whorl (for a

[43] Iakovidis 1969, pl. 23b; 1969A 70–78.
[44] Iakovidis 1969, pl. 22a–b.
[45] Iakovidis 1969, pl. 15a. See Iakovidis 1969 A, 52–56 for the tomb.

low-whorl spindle), and they are also used to decorate the dome part of the whorl. The dome-shaped whorl is inserted on the largest end of the shaft, at about 1.6cm from the extremity. Both ends are cut flat, and the one on the largest extremity is hollow.[46] Another fragmentary solid shaft (Δ 212, L. 7.3cm, ø 1.1–0.9cm) decorated with horizontal lines, pointed circles and conical stone whorls (L 311–312) comes from the same tomb.[47]

Conclusion

Late Bronze Age eastern Mediterranean complete spinning kits and spindles come from domestic (Ras Shamra), funerary (possibly *dépôt 43* at Minet el-Beida and Perati) and religious (Delos) contexts, therefore pointing to their effective use as spinning tools. They are associated with other textile tools such as spindle-whorls. In the Aegean, conical whorls have been interpreted by Iakovidis as "buttons" (i.e. skirt weights), and not as spindle-whorls,[48] but this interpretation has recently been challenged, and their function as spindle-whorls cannot be totally ruled out.[49] If these objects are buttons, then spindle Δ 211 from tomb 152 at Perati was associated with another ivory shaft but not with spindle-whorls.

Two main modules appear to have been used: a long spindle with a shaft of 20 to 23cm and a shorter one of about 13cm long and of a lesser diameter attested at Perati and Ugarit. The whorl is always inserted on the larger part of the spindle, which can also be hollowed by a mortise.

According to Barber, spindle Δ 211 from Perati tomb 152 is too short to have been rolled down the thigh or turned in the hand, and she thinks it was certainly best used as a drop-spindle.[50] This spindle is similar to RS 34.210 from Ugarit (L. 13.3, ø 0.5cm),[51] but none present a hook or attachment device. It is however possible that a hook was inserted in the mortise on the top of the objects. It would also have been possible to use these spindles as supported spindles.

Pomegranate shafts and knobs

Ugarit

In Ugarit, beside the pomegranate knobbed shaft RS 4.221[B] previously mentioned (Fig. 9.3), several pomegranate knobs were found in the city.[52] Most of them come from Schaeffer's excavations and therefore their context and assemblage are not always clear nor fully published. The proposed table (Table 9.1) is based on published data, and it is likely that it will be possible to complete and enhance it with further studies.[53]

When the pomegranate knobs contextual assemblage is known or possible to reconstruct, they are associated with spindle-whorls and/or loom-weights and to ivory shafts as in House E in *centre de la ville*.

---

[46] Iakovidis 1980, p. 95.
[47] Iakovidis 1969, pl. 15a–b.
[48] Iakovidis 1977, esp. pl. 24–25.
[49] Andersson and Nosch 2003, 202–203; Rahmstorf 2008, 296; Burke 2010, 102–103; Andersson, Mårtensson and Nosch 2011, p. 411.
[50] Barber 1991, p. 63.
[51] Gachet-Bizollon 2007, cat. 139.
[52] See Gachet-Bizollon 2007, cat. 249–264.
[53] See the ongoing study by V. Matoïan and J.-P. Vita; Matoïan and Vita 2009 (2010), esp. tables p. 483–485.

Table 9.1: Table of pomegranate knobs for Ugarit with their archaeological context, catalogue after Gachet-Bizollon 2007).

Gachet-Bizollon's catalogue no. and inventory no.	Type of object	Context	Associated textile tools
251. RS 96.[4016]	Incomplete H. 4.2, ø 1.6, H. calyx 2.2	Minet el-Beida. 1932 Trench 25.IV, pt. 207? (= atop tomb VI).	Uncertain association: Ivory shaft found in tomb VI, (Gachet-Bizollon 2007, cat. 190).
252. RS 4.91	Complete H. 4.4, ø 1.9, H. calyx 2.2	Minet el-Beida. 1932 Trench 25.IV; maybe in or near tomb VI as it appears between objects from tomb VI in the artefact register. It may also come from elsewhere.	Uncertain association: Ivory shaft found in tomb VI, (Gachet-Bizollon 2007, cat. 190).
249. RS 2.[053]	Complete. H. 3.8, ø pericarp 2.1, ø calyx 1.7	Ras Shamra. 1930 *Maison du grand prètre*, pt. 37, 1m.	?
250. RS 3.435	Complete?	Ras Shamra. Acropolis, *bibliothèque*, trench B6, 1m.	?
253. RS 8.30	Complete? H. 4.2	Ras Shamra. 1936 Acropolis *chantier 1, tranchée coudée* pt. 62, 1.60m.	?
254. RS 9.281	Complete. H. 4.1, ø 0.9	Ras Shamra. 1937 Northwest of the tell. Pt S 434, area of *les* écuries *et du temple hourrite*.	?
259. RS 22.109	Complete. H. 3.4, ø pericarp 1.7–1.9, ø calyx 2	Ras Shamra. 1959 Northwest of the tell, *tranchée terrasse* pt. 2378, 1.80m.	?
255. RS 11.[1002]	Complete. H. 4.5, ø pericarp 1.9, ø calyx 2.1	Ras Shamra. 1939 *Ville basse est*, tomb LXXXI (SM no. 24).	3 whorls and fragments Gachet-Bizollon no. 557 (Gachet-Bizollon identifies the objects as buttons).
256. RS 14.136	Incomplete. H. preserved 2.9, ø 2.7	Ras Shamra. 1950 *Ville basse ouest*.	?
257. RS 17.163	Incomplete. H. 3.1, ø 1.6	Ras Shamra. 1953 Near the royal palace, pt. 750 2.60, in the street, outside of loc. 49.	?
258. RS 21.14[B]	Incomplete. H. preserved. 1.5, ø 1.3	Ras Shamra. 1958 *Quartier résidentiel*, block 3, Rapanu's house, room 5, near the staircase of the tomb's dromos (tomb II – SM no. 301).	?
260. RS 23.635	Complete. H. 3.8, ø pericarp 1.9, ø calyx 2.1	Ras Shamra. 1960 *Ville sud*, pt. 3176, 1.20m.	?
261. RS 24.112	Incomplete. H. 3.6, ø 1.7	Ras Shamra. 1961 *Ville sud*, pt. 3374, 3.40m.	?
262. RS 25.366	Incomplete. H. 4.5, ø pericarp 2.1, ø calyx 2.1	Ras Shamra. 1962 *Tranchée sud acropole*, pt. 5118, 229E, 1.25 m.	?
263. RS 81.545	Complete. H. 2, ø pericarp 1.5, ø calyx 1.6	Ras Shamra. 1981 *Centre de la ville*, house E, loc. 1201 (small room).	1 stone spindle-whorl RS 81.514, (Elliott 1991, 42). 2 ivory shafts RS 81.596 and 81.547 (Gachet-Bizollon 2007, nos. 168, 234).
264. RS 81.3040	Incomplete. H. preserved 3.4, ø 1.9	Ras Shamra. 1981 *Centre de la ville*, house E, tanour of 1209 (entrance).	1 loom-weight (Matoïan and Vita 2009, 484). 1 shaft RS 81.3041, (Gachet-Bizollon 2007 no. 236).

Lachish

In Lachish, two ivory pomegranate rods (inv. 2772 and 2774) were found in the Canaanite temple, along with three solid ivory shafts (see below).[54] The knobbed shafts and solid shafts all come from a cache located in the southeast corner of structure III (loc. D.III, 181), which also contained amongst other ivory objects a comb, a box and a disc.[55] These probably belonged to discarded objects from the temple and could have been either cult material or offerings. Shaft inv. 2772 (L. 25.2cm, ø 0.5–0.8cm) is decorated with horizontal lines and lattice patterns at both ends. Its pomegranate knob (H. 5.2cm, H. calyx 2.8cm, ø 1.8cm), inserted on its smaller extremity, is large and the persistent calyx topping the fruit are straight and as long as, if not longer that the pericarp. The second shaft (L. 26cm, ø 0.7–0.8cm) is also decorated with horizontal lines and lattice pattern on both ends and is topped on its smaller end by a shorter pomegranate knob (H. 2.4cm, H. calyx 0.6cm, ø 1.4cm). A few spindle-whorls also come from the temple but none were associated with the ivory shafts.[56] The context of the shaft, in a secondary deposition context in a pit, cannot give viable information regarding its association to textile tools.

Another pomegranate shaft comes from tomb 216 (inv. 4653, L. 25.2cm, ø 1.1–1.4cm, H. knob 4.2.cm, ø pericarp 1.8cm)[57] and was found with a conoid spindle-whorl with thin edges splaying out (ø 2.2cm, ø perf. 0.3cm).[58] The perforation of the whorl is smaller than the diameter of the ivory shaft.

Finally, a pomegranate knob comes from level VII, square R10 (inv. a17), but it has no known association to ivory shafts or textile tools.[59]

Kition

Two knobbed pomegranate shafts were found in the upper burial of tomb 9 at Kition. Rod 132, made of elephant ivory is incomplete (L. 23.2cm), its lower extremity being broken. The shaft is slightly tapered towards the knobbed extremity, where it is decorated with horizontal lines and scale pattern. Its "lower" and larger end presents three perforations (Fig. 9.7).[60] Rod 60–62 is complete (L. 23.6cm) but its knob is damaged. The knob is inserted on its tapered end, while the opposite and larger end of the shaft is cut flat. The shaft is decorated with horizontal, and diagonal lines as well as with lattice pattern.[61] Both rods were found with several fragmentary solid ivory rods and six spindle whorls (three ivory, two bone and one steatite).[62]

Three ivory pomegranate (or 'poppy') knobs were found in the Kition temples, one comes from floor IIIA of courtyard C at Kition-Kathari (no. 5268) and was found with two loom-weights,[63] the

[54] For the pomegranate rods, see Tufnell, Inge and Harding 1940, pl. XX nos. 25, 26; for the ivory shafts without knob, see Tufnell, Inge and Harding 1940, pl. XX nos. 23, 27, 28. See also Gachet-Bizollon 2007, 125 nos. 1, 2 and 5.

[55] Tufnell, Inge and Harding 1940, 59, nos. 1, 2, 4, 10–18, 20, 21, 24–31, pl. XV–XX.

[56] Tufnell, Inge and Harding 1940, pl. XXIX. 29–33.

[57] Tufnell 1958, no. 4653, fig. 28.7 and 54.2; Gachet-Bizollon 2007, 125 no. 6.

[58] Tufnell 1958, no. 4649, fig. 54.1. This whorl, because of its dimensions, could have been a spindle-whorl or a bead; Sauvage 2012, 201.

[59] Loud 1948, pl. 197.20.

[60] Karageorghis 1974, no.132, pl. LXXXVII, CLXX, 69, 91; Gachet-Bizollon 2007, 126, no. 64. For a better illustration of the perforation see Smith 2009, p. 98, III.11.

[61] Karageorghis 1974, nos. 60–62, pl. LXXXVII, CLXX, 66, 91; Gachet-Bizollon 2007, 126, no. 63.

[62] Karageorghis 1974, nos. 58, 236, 240, 106, 107, 35, pl. LXXXVII, CLXX, CLXXI.

[63] Karageorghis and Demas 1985, 248, pl. CXCI; Smith 2002, 97–98, fig. III.11a. Eleven loom-weights were also found in between floor III and IIIA.

Fig. 9.7: Pomegranate knobbed shaft 132 from Kition tomb 9, upper burial, after Smith 2009, 98, fig. III.11 (©J. Smith).

Fig. 9.8: Pomegranate knobbed shaft E.003.241 from Swedish Tomb 3, disturbed layer, at Enkomi (courtesy of the Medelhavsmuseet).

second one comes from floor I in room 12 (no. 555) and was not found with textile tools,[64] and the third one comes from well 1 of temple 1 (no. 1982) and was found with beads.[65]

Enkomi

From Disturbed layers at Enkomi Swedish tomb 3, come two ivory pomegranate knobbed shafts.[66] They were found on the floor of the tomb. Both rods are 24.4cm long and have a large pomegranate knob, with a straight and long persistent calyx as long as, if not longer than the pericarp. The knob is inserted on their smaller end, while the larger end is decorated with incised lattice pattern and is cut flat. Shaft E.003.240 has an ovoid shape, while it overall tappers towards the pomegranate.[67] Its total weight is 41g. Shaft E.003.241 weighs 36g and has a cylindrical and slightly tapered shaft towards the knob (Fig. 9.8).[68]

From the same tomb comes a conical stone loom-weight as well as a complete unperforated bone spatula and a fragmentary perforated one.[69] Both of these bone tools could have been used in the textile industry.

Palaeopaphos

From tomb 119 at Palaeopaphos-Eliomylia, comes an almost complete hippo ivory rod (L. 20.3),[70] its lower extremity is tapered and its upper end cut into a peg.[71] A pomegranate ivory knob was

[64] Karageorghis and Demas 1985, 157.
[65] Karageorghis and Demas 1985, 247, pl. CCXXXIX.
[66] Gjerstad 1934, LXXVIII, fig. 1 nos. 240–241 pl. CUI, 4, p. 483; Gachet-Bizollon 2007, 126 nos. 48–49.
[67] See the Medelhavsmuseet online database: http://collections.smvk.se/pls/mm/rigby.VisaObjekt?pin_masidn=3200208 accessed 10/06/2012.
[68] See the Medelhavsmuseet online database: http://collections.smvk.se/pls/mm/rigby.VisaObjekt?pin_masidn=3200212 accessed 10/06/2012.
[69] Gjerstad 1934, 478 no. 32; LXXVI fig. 3: object on the left and top right.
[70] Hippo ivory comes from hippopotamus teeth; see Caubet and Pauplin 1987.
[71] Karageoghis and Michaelidès 1990, 80 no. 27A, pl. LXXXIII, LXXXVIII; Gachet-Bizollon 2007, 127 no. 67.

found in the same tomb (H. 1.8cm, ø 1.8cm)[72] with two other fragmentary ivory rods (L. 3.1 and 3.9cm).[73] No textile tools were found in this tomb.

Conclusion

Except from the tomb at Palaepaphos and two examples found in the Kition temple area, all the pomegranate shafts or knobs with known contexts and reconstructed assemblage are found with spindle-whorls and/or loom-weights. They can be found in funerary, domestic as well as religious contexts, attesting to their effective use in households and symbolic importance. One rod from Kition tomb 9 was perforated, maybe to allow suspension of the rod. However, one perforation would have been enough, and it is difficult to explain the three successive ones.

The repetitive association of such tapered knobbed shafts with textile tools may confirm their use in the textile industry, probably as distaffs as previously proposed.[74] The versatile character of the shaft, used either with a whorl as a spindle or with a knob as a distaff finds another confirmation with the perforation of the shaft from tomb 9 at Kition. When used as a spindle, the perforated side would have corresponded to its top, the thinnest and tapered end would then have been the bottom of the spindle.

Solid ivory shafts

Ugarit

In Ugarit, 78 bone/ivory shafts were catalogued by Gachet-Bizollon.[75] The large majority of them come from Schaeffer's excavations, and have either no context or when a find spot is known, their context and associated assemblage have not yet been fully studied or published. In the following table (Table 9.2), I compiled shafts with a known find spot and associated known textile tools.[76]

When it is possible to reconstruct a context, it appears that most ivory shafts were associated with other textile tools such as spindle-whorls, loom-weights or pomegranate knobs, in domestic, religious and funerary contexts. In domestic and religious contexts, when no textile tools were found in the same room, such specific tools were however found within the same building, allowing us to infer a somehow looser relationship, such as in the *temple aux rhytons* and in house D in the *centre de la ville*. The only instance where no textile tools were found is in tomb II (SM 139) located on the acropolis of Ugarit. This tomb was looted in antiquity and almost all of the recovered material came from its dromos.

Kazel

At tell Kazel, two 13th c. BC fragmentary bone shafts come from a domestic context (building I, room IC, level 5) and were associated with 14 bone spindle-whorls.[77] One of the shafts has its extremity preserved and decorated with horizontal lines. The group of whorls exhibits size and shape

[72] Karageoghis and Michaelidès 1990, 80 no. 27B, pl. LXXXIII, LXXXVIII. Its mortise is larger than the peg of the preserved rod.

[73] Karageoghis and Michaelidès 1990, 80 nos. 27E-D, pl. LXXXIII, LXXXVIII.

[74] Sauvage 2012.

[75] Gachet-Bizollon 2007, cat. 170–248.

[76] An ongoing archaeological study of textile tools by V. Matoïan will certainly shed more light on most of these hardly known contexts; see Matoïan and Vita 2009 (2010), esp. tables p. 483–485.

[77] Badre *et al.* 1994, 312, fig. 43c; Gachet-Bizollon 2007, 125 nos. 40–41.

variation, while all of the centred perforations do not have a diameter larger than 4mm.[78] From the same house, but from a different room and level (building I, room IE, level 6) comes a complete bone shaft (L. 19,5cm; ø 1,1cm).[79] One extremity is rounded and decorated with horizontal lines and a lattice pattern. The other extremity is stepped-down into a thin peg.

Dan

At tel Dan, four bones or ivory rods were found in collective tomb 387, the so-called "Mycenaean tomb".[80] Bone rod 229 is incomplete (L. 19.5cm, ø 0.8cm), its preserved end is stepped-down as well as rounded. It was found in cluster A, alongside duck cosmetic box 201 and has therefore, on typological and contextual basis, to be identified as a kohl stick (Fig. 9.9).[81] A total of ten whorls were found in cluster A in the western side of the tomb. They were mixed with bones inlays from a box (210). According to the excavator, the whorls from cluster A were probably contained in box 210.[82] Incomplete bone shaft 227 (15.4cm long, diameter 0.8cm) was also found in cluster A. From the same cluster, come two bone "needles"[83] 224a and b, not depicted on the plan.

Ivory rod 230 is 4.6cm long and has a diameter of 0.8cm. It is decorated with horizontal and zig-zag lines as well as a lattice pattern. One of its ends is smooth while the other is drilled and was certainly designed to host a peg.[84] It may have been the end extension part of a shaft. It was found in cluster B located in the south-eastern corner of the tomb and characterized by a group of about 100 whorls found 20cm above the pavement, near pyxis 208 and the skull of a 30 year-old male. If it is the object represented on the plan under vase 244, it was then surrounded by whorls. From the same area, near box 205, comes bone rod 231. It has a preserved length of 15.5cm and a diameter of 0.6 to 0.8cm and exhibits on its preserved and smaller end horizontal lines and a lattice pattern. The object is not represented on the plan and it is therefore difficult to know if it was associated with the ivory boxes 208 or 205 or with the whorls.

In this tomb, more than one hundred and ninety-three objects described as whorls or buttons and beads made of stone, ivory, bone, glass and faience were found.[85] The bone whorls were grouped in two main clusters A and B, while faience whorls (419–425) were grouped near the southern wall. The rest of the whorls were found at various levels and locations. The publication provides a table of all the whorls/beads/buttons found in the tombs including diameter, height and sometimes weight, but it lacks diameter of perforation,[86] and weight information for the objects identified as "buttons".[87] These "buttons" are dome shaped and made of stone, bone, deer antler, ivory or faience. They exhibit shape and decoration parallels with whorls from Ugarit.

In cluster A, one such bone button (no. 397: ø 2.8cm; H. 0.6cm; ø perf. 0.5cm) was attached to the remains of box 210 possibly indicating that some of these were used as buttons or decorative

[78] The picture of the whorls has no scale; Badre *et al.* 1994, fig. 43c.
[79] Badre *et al.* 1994, 320, fig. 46; Gachet-Bizollon 2007, 125 no. 42.
[80] Ben-Dov 2002 151, inv. 228a–e, 229–231. Gachet-Bizollon catalogued eight rods (2007, 125–126 nos. 36–39).
[81] Ben-Dov 2002, 151, 155 and fig. 2.119.
[82] Ben-Dov 2002, 157.
[83] Ben-Dov 2002, 224a–b, fig. 2.117. These objects have a rounded-pointy end, which may not have been sharp enough to pierce through fabrics. The other end is pierced by a hole. They are maybe to be compared to "styli" used in tapestry.
[84] Ben-Dov 2002, 151–152, fig. 2.120.
[85] Ben-Dov 2002, 157.
[86] It can however be deduced from the drawing for some objects.
[87] Buttons "are decorated with an incised pattern of lines emanating from the centre of the item or with incised circles or semicircles" Ben-Dov 2002, 160.

Table 9.2: Table of the ivory and bone shafts from Ugarit with a known archaeological context and their textile tools association, catalogue of the shafts after Gachet-Bizollon 2007, 119–121.

Gachet-Bizollon's catalogue and Inventory number	Type of object	Context	Associated textile tools
177. RS 1.[119]	Incomplete? Horizontal and oblique lines and lattice pattern?	Minet el-Beida. 1929 Tomb III (SM 1005).	1 bone/ivory spindle-whorl, (Gachet-Bizollon no. 140). 1 ivory spindle-whorl, (Gachet-Bizollon no. 520).
146. RS 1.[111]	Complete. Horizontal lines L. 3.9, ø 1.9 Mortise		
238. RS 1.[120]	Incomplete? Two shafts? L. 8 and 3.8, ø 0.18 and 0.12		
222. RS 1.[109]	Incomplete. Plain. L. 7.5, ø 1-1.2		
190. RS 4.77	Complete. Horizontal lines and scale pattern. Small mortise drilled on the tapered extremity of the shaft. L. 22, ø 1–1.4; w. 33.6g.	Minet el-Beida. 1932 Tomb VI (SM 1007).	1 pomegranate knob possibly found atop the tomb (Gachet-Bizollon 2007, cat 251). 1 pomegranate knob found in or near the tomb (Gachet-Bizollon 2007, cat 252).
151. RS 18.208[B]	Fragmentary, horizontal lines L. 5.2, ø 1.2	Ras Shamra. Royal palace. courtyard V or nearby.	1 ivory spindle-whorl from the area of courtyard V (Gachet-Bizollon no. 142).
197. RS 18.208[A]	Incomplete. Horizontal lines and scale pattern L. 5.3, ø 1–1.2	Ras Shamra. 1954. Royal palace. pt. 1434, 1.60 m, courtyard V or staircase 80, or 1431?	1 ivory spindle-whorl from the area of courtyard V (Gachet-Bizollon no. 142).
201. RS 23.715	Incomplete. Horizontal lines and scale pattern L. 3.5, ø 0.9–1.2	Ras Shamra. 1960. *Ville sud.* Block IV, house B, loc. 9. Pt. 2861, 0.9m.	2 bone/ivory spindle-whorls, (Gachet-Bizollon no. 530 from house B courtyard 7, pt. 2860, pt. 3059).
226. RS 22.[463]	Incomplete. Plain. L. 6, ø 1.5–1.6	Ras Shamra. 1959. *Ville sud.* Block II, house A, tomb 2650 (SM no. 502).	1 bone/ivory spindle-whorl, (Gachet-Bizollon no. 529).
165. RS 75.99	Incomplete. Horizontal lines L. 2, ø 0.9	Ras Shamra. 1975. North-west of the tell. *Résidence nord*, "1975–1976" A/151/SO19.	1 bone/ivory spindle-whorl from A/151/SE/11, (Gachet-Bizollon no. 516). 1 bone/ivory spindle-whorl from A/151/SE/23, (Gachet-Bizollon no. 517).
213. RS 75.118	Complete. Horizontal lines and scale pattern L. 4.8, ø 1 Mortise	Ras Shamra. 1975. North-west of the tell. *Residence nord* "1975–1976" A/6c/SE2.	1 ivory spindle-whorl, (Gachet-Bizollon no. 565 A/6c/SO1).
167. RS 81.3	Incomplete. Horizontal lines a: L. 4.6, ø 1.5 b: L. 7.5, ø 1.3, peg (ø 0.66)	Ras Shamra. 1981. *Centre de la ville.* Loc 1051 (dead end of street 1228).	3 stone spindle-whorls RS 81.612, RS 81.849, and RS 84.78 in street 1228, (Elliott 1991, 42). 1 stone loom-weight RS 79.56 (Elliott 1991, 40).
169. RS 88.385 + 88.606	Incomplete. Horizontal lines L. 4 and 2, ø 0.8 to 1.1	Ras Shamra. 1981. *Centre de la ville.* House D, room 1307.	None in the same room, but 4 stone spindle-whorls in the house: RS 81.501, RS 81.3119, 81.3072, RS 81.3642 (Elliott 1991, 42; Matoïan and Vita 2009, 484)

168. RS 81.596	Incomplete. Horizontal lines L. 8.3, ø 1.4 Peg (ø 0.7)	Ras Shamra. 1981. *Centre de la ville.* House E, room 1201.	1 stone spindle-whorl RS 81.514 (Elliott 1991, 42; Matoïan and Vita 2009, 484). 1 pomegranate knob RS 81.545 (Gachet-Bizollon 2007 no. 263).
234. RS 81.547	Incomplete. Plain. L. 8.1, ø 1.1–1.3 Mortise.		
236. RS 81.3041	Incomplete. Plain. L. 5.4, ø 0.9–1	Ras Shamra. 1981. *Centre de la ville.* House E, loc. 1209.	1 loom-weight (Matoïan and Vita 2009, 484). 1 pomegranate knob, RS 81.3040 (Gachet-Bizollon 2007, no. 264).
221. RS 81.505	Incomplete. Horizontal lines and scale pattern L. 3.6, ø 1.9	Ras Shamra. 1981. *Centre de la ville.* House F, loc. 1221/1222. 1222 may have been an area for washing or dying textiles (O. Callot, see ref. in Matoïan and Vita 2009, 484)	5 stone spindle-whorls RS 81.624,-627, RS 81.800, room 1222 (Elliott 1991, 42; Matoïan and Vita 2009, 484). 1 stone spindle-whorl room 1221, RS 81.654, (Elliott 1991, 42; Matoïan and Vita 2009, 484).
214. RS 79.5063	Incomplete. Horizontal lines and scale pattern L. 2.7, ø 1.4	Ras Shamra. 1979. *Centre de la ville.* *Temple aux rhytons*, loc. 36.	None in this room, but 4 spindle-whorls and 2 loom-weight upstairs (Matoïan and Vita 2009, 484). See also below.
215. RS 81.3026	Incomplete. Horizontal lines and scale pattern L. 5.9, ø 1.6 Mortise	Ras Shamra. 1981. *Centre de la ville.* Pit 1237 of the looter of the *temple aux rhytons.*	2 stone loom-weights RS 81.620, RS 81.3081, (Elliott 1991, 40).
220. RS 78.111	Incomplete. Horizontal lines and scale pattern L. 4.2, ø 0.9		
235. RS 81.3037	Incomplete. Plain. L. 5.4, ø 0.9-1		
237. RS 86.5062	Incomplete. Plain. L. 5.8, ø 0.1 × 0.11 and 0.6 × 0.8	Ras Shamra. 1986. *Centre de la ville* Room 81, annex of the *temple aux rhytons.*	1 stone spindle-whorl (fill above room 81), RS 80.5152, (Elliott 1991 42). 1 stone spindle-whorl RS 86.5011, (Elliott 1991, 43).
186. RS 25.[578]	Incomplete. Horizontal and oblique lines. L. 5.4, ø 0.7	Ras Shamra. 1962. Trench *sud acropole.* Tomb 5048 (SM 627).	1 ivory spindle-whorl, (Gachet-Bizollon no. 538).
219. RS 24.203	Complete. Horizontal lines and scale pattern L. 4.4, ø 0.8	Ras Shamra. Trench *sud-acropole.* pt. 3568, 0.75m (Z 129).	1 ivory spindle-whorl, (Gachet-Bizollon no. 562, near pt. 5017 129E).
225. RS 11.[1003]	Incomplete. Plain. L. 6.5, ø 0.9	Ras Shamra. 1939. *Ville basse est.* Tomb LXXXI (SM no. 24).	1 ivory spindle-whorl, (Gachet-Bizollon no. 522). 2 ivory spindle-whorls, (Gachet-Bizollon no. 554). At least one stone spindle-whorl (Matoïan and Vita 2009, 485).
240. RS 5.103	Incomplete?	Ras Shamra. 1933. Acropolis. Dromos of tomb II (SM no. 139).	No known textile tools. Looted tomb.
241. RS 5.149	Incomplete?		

inlays.[88] Four biconical beads were found inside box 210.[89] Buttons 401 (bone: ø 3.2cm; H. 1cm; w. 85.3g; ø perf. 0.5cm), 402 (deer antler: ø 2cm; H. 0.4cm; w. 9.4g), 404 (antler: ø 2.1cm; H. 0.8cm; w. 35.3g), and 405 (stone: ø 1.5cm; H. 0.55cm; w. 10.2g), were also part of cluster A.

The shafts found in this tomb were therefore deposited with or near textile tools, such as in cluster A where an incomplete rod was found with two needles and a box containing 10 whorl-like objects. The box was decorated by at least one. Their shape similarity to whorls from Ugarit found mounted on spindles as well as their discovery place, in cluster B, near bone rods 227 and 229, allow us to infer an association to textile tools, even if their documented diameter of perforation (0.5cm) is smaller than that of the shafts (0.8cm). It is also highly possible that box 210 found decorated by and used as a container for whorl-like objects could have been used to store the several spindle-whorls used by one of the deceased. In such a case, the "buttons" on the outside of the box would illustrate and display the content of this spindle-whorl storage box.[90]

Megiddo

At LB II Megiddo, several solid ivory shafts were found in tombs 877 B1, 40 and 989 C1. In tomb 877B1, a complete solid ivory shaft (M 2433, L. 23.6cm, ø 0.33–0.86cm)[91] decorated with horizontal lines and lattice pattern was recovered (Fig. 9.10). One of its extremities is tapered, while the other is cut flat. It was found with another fragmentary solid bone shaft (M 2435, L. 3.7cm, ø 0.55cm)[92] and eight spindle-whorls, one of them M 2828, in ivory, has a perforation large enough to be inserted onto one of the shafts (ø whorl 2.5, ø perf. 0.6cm).[93]

In tomb 989C1, a complete and solid bone shaft decorated with horizontal lines was found. Both ends were maybe cut flat, while its smaller end was maybe hollowed by a mortise (M 2856, L. 21.3cm, ø 0.38–0.54).[94] Another fragmentary solid bone shaft comes from the same tomb and is decorated with groups of horizontal lines near its preserved end (M 2853, L. 11cm, ø 0.7cm).[95] Seven spindle-whorls were also found in this tomb as well as a bone knob resembling a pomegranate (M 2836, H. 2.5cm, ø at base 0.8cm).[96] This knob could possibly have been inserted onto one of the bone shafts.

From tomb 40 comes a complete solid ivory shaft made of two rods originally attached by a pin or a peg. The larger end of the shaft is cut flat, while the other is stepped-down. It is decorated with horizontal lines and lattice pattern (inv. x 738, L. 22.5cm, ø 0.5–0.7cm, ø stepped end 0.3cm).[97] From the same tomb comes a fragmentary plain bone shaft (inv. x 632, L. 5.6cm, ø 0.6–0.7cm).[98]

[88] It is possible that 398 (bone: ø 2.9cm; H. 0.5cm; ø perf. 0.3cm) was also attached to the box with a bone pin; Ben-Dov 2002, 246.

[89] 410 (ivory, biconical: ø 1.6cm; H. 0.8cm; w. 29.8 g), 411 (ivory: ø 1.8cm; H. 0.75cm; w. 28g; ø perf. 0.25cm; uncentred perforation), 412 (bone, biconical: ø 1.9cm; H. 1cm; w. 38.6g), 413 (hematite, biconical: ø 2cm; H. 0.8cm).

[90] This box would, in such a case, have been part of the spinning kit of the deceased. Recent examples of Chancay from Peru (1300–1400 AD) show that an individual could own and use as many as 12 loose whorls, 57 spindles with whorls and 11 spindles without whorls; Liu 1978, 98.

[91] Guy 1938, pl. 95.50; Gachet-Bizollon 2007, 125 no. 13.

[92] Guy 1938, pl. 95.49; Gachet-Bizollon 2007, 125 no. 12.

[93] Guy 1938, pl. 95.41–48.

[94] Guy 1938, pl. 100.30; Gachet-Bizollon 2007, 125 no. 15.

[95] Guy 1938, pl. 100.29; Gachet-Bizollon 2007, 125 no. 14.

[96] Guy 1938, pl. 100.22–28.

[97] Guy 1938, pl. 156.13; Gachet-Bizollon 2007, 125 no. 16.

[98] Guy 1938, pl. 156.12.

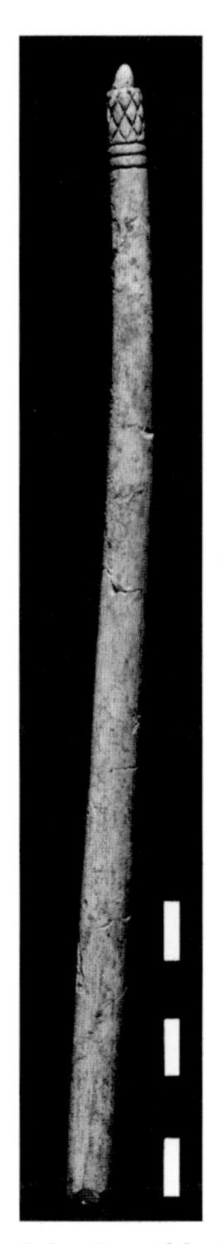

Fig. 9.9: Possible kohl rod from the so-called Mycenaean tomb at tel Dan (L. 19.5cm), after Ben-Dov 2002, fig. 2.119 (courtesy of Israel Antiquities Authority).

Fig. 9.10: Bone shaft (L. 23.6cm) from Megiddo tomb 877B1, after Guy 1938, pl. 95.50 (courtesy of the Oriental Institute of the University of Chicago).

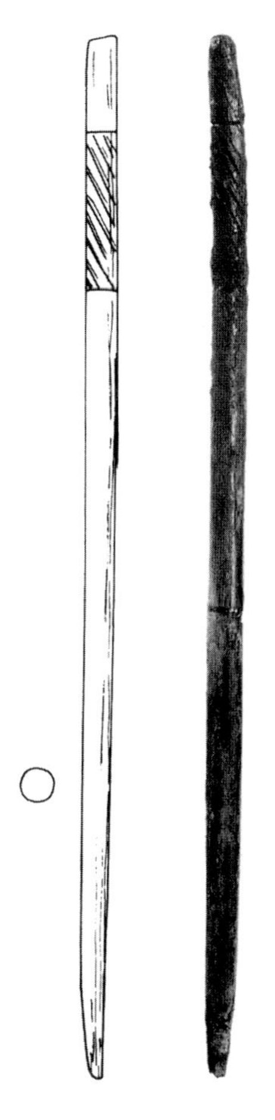

Fig. 9.11: Bone shaft (L. 20cm) from tell Deir 'Alla, (a) after Franken 1992, fig. 4–5.18, (b) after Kooij van der and Ibrahim 1989, no. 13, 101 (courtesy of J. van der Kooij, © Deir 'Alla Excavation).

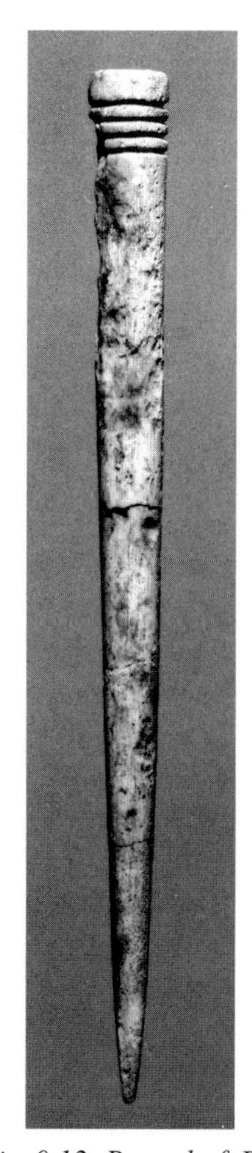

Fig. 9.12: Bone shaft BM 1872.0315.82 (11.7cm) from a tomb at Ialysos (Rhodes), © The Trustees of the British Museum.

Several bone shafts have been found in domestic contexts at Megiddo in st. VI, VIIA, VIIB, VIII and X. Most, if not all, of the ones from level VII were found in the same context as textile tools such as spindle-whorls and loom-weights. Two almost complete bone shafts (M 5673a–b)[99] from stratum VIIA in the northern quadrant of loc. 1771 were found with two paste beads and several beads (M 5648, M 6268), but no measurements nor illustrations are available, and therefore it is hard to say if these faience beads could have been whorls.[100] All the other shafts have a clear

[99] Loud 1948, pl. 197, 10–11.
[100] Loud 1948, 153.

correlation to textile tools: fragmentary bone shafts M 6056 and M 6130[101] from loc. 1835 northern quadrant (stratum VIIA) were found with two bone whorls (M 5069a–b),[102] two steatite whorls (M 6131),[103] one bronze needle (M 6129)[104] and a bone oval knob (M 6135).[105] The two bone shafts from loc. 1814 (stratum VIIA) were found with a clay (loom-?) weight (M 6163).[106] Bone shaft (M 5985)[107] from loc. 1825 southern quadrant (st. VIIA) was found with two whorls (M 5987 and M 6184) and another whorl (M 6152) comes from the eastern quadrant of this locus.[108] And the two shafts from the eastern quadrant of loc. 1831 (st. VIIB) were found with two bone whorls (M 6029, M 6031),[109] a limestone whorl, another whorl of unknown material, a clay disk, and a stone ring (M 6032, M 6033, M 6234, M 6232).[110] Another bone shaft (M 6028) also comes from the same locus.[111] Two bone shafts from layer VI were also associated with textile tools:[112] shafts M 5776–77 found on one side of loc.1769 (st. VI) were associated with two bone whorls,[113] a bronze needle and two ivory pin-heads located inside the locus (M. 5669, M 5667).[114] One shaft was recorded in layer VIII from square N15, but no textile tools were recorded in the find registry.[115] Seemingly, shaft b137 from layer X (loc. 2032 eastern quadrant) has no known association to textile tools.[116]

Eight shafts were found in the so-called treasury at Megiddo[117] within an assemblage of more than 382 ivories in a context that suggests hording of bits and pieces of ivory at the end of the Late Bronze Age.[118] Therefore their assemblage and eventual association with tools is not informative regarding to the use of shafts as textile tools.

It is reasonable to say that in most cases, solid ivory or bone shafts from Megiddo were found in contexts suggesting a possible association to textile industry.

Tell Deir ʻAlla

At Tell Deir ʻAlla, a 20cm long bone shaft with one tapered end, an almost rounded top and decorated with incised oblique lines in between horizontal lines was found in room E2 (Fig. 9.11).[119] The top horizontal line is deeper and may have been used as a groove to secure the thread. A conical and fragmentary knob was found in the same room and it is possible that it was a pomegranate or bud-shaped knob (H. 2.5cm) not previously discussed. In the same room, 8 spindle-whorls were

[101] Loud 1948, 156, pl. 197.6.
[102] Loud 1948, 156, pl. 172.33–34.
[103] Loud 1948, 156, not illustrated.
[104] Loud 1948, 156, not illustrated.
[105] Loud 1948, 156, pl. 197.21.
[106] Loud 1948, 155, not illustrated. It was found in the western quadrant of the locus.
[107] Loud 1948, 155, pl. 197.9.
[108] Loud 1948, 155, not illustrated.
[109] Loud 1948, 156, pl. 172.29.
[110] Loud 1948, 156, not illustrated.
[111] Loud 1948, 156, pl. 197.5.
[112] Shaft d712 could have been a handle. Its shape is different from all the other tapered solid ivory shafts described in this article; Loud 1948, 187, pl. 197.12.
[113] Loud 1948, 152, pl. 197.13, bone whorls are not illustrated.
[114] Loud 1948, 152, none of these objects is illustrated.
[115] Loud 1948, 147, pl. 197.3.
[116] Loud 1948, 158, pl. 197.1.
[117] Loud 1939, 20, pl. 55.286, 56.294–298, 57.299 and 303; Gachet-Bizollon 2007, 126, nos. 17–24. Whorls were part of the recovered material, see pl. 15.78–83, 95 and 100–103.
[118] Feldman 2009, esp. 188–189.
[119] Kooij van der and Ibrahim 1989, 58, cat 13, 92; Franken 1992, 42–43, fig. 4–5.18.

found with a carnelian 'bead', which could have been used as a whorl.[120] This room located east of the sanctuary's cella was occupied during the last phase of the Late Bronze Age and destroyed by an earthquake followed by a fire.[121] According to the excavators, its northern part was lost by erosion, it belonged to a house and was used as a shrine.[122]

Lachish

In Lachish, three ivory shafts were found with the two pomegranate rods previously cited, and therefore come from an ivory cache in the temple. Rods inv. 2971 (L. 19cm, ø 1–1.2cm) and inv. 2776 (L. 13.6cm, ø 0.7–1cm) were decorated with horizontal lines and lattice pattern, while inv. 2775 (L. 13.2cm, ø 0.6–0.8cm) is probably plain.[123] A solid tapered shaft decorated with groups of horizontal lines was found in pit tomb 501 (inv. 3400, L. 23.7cm, ø 0.75–1.37cm),[124] along with paste and carnelian discs.[125]

Kition

In Kition, several ivory rods come from the upper burial of tomb 9. Two incomplete solid ones have a broken end and a decorated cut-flat extremity: no. 119 (L. 15.5cm) is decorated with lines and scale pattern and no. 75 (L. 9.5cm) is decorated with lines and lattice pattern.[126] A third fragmentary rod has both extremities sheered off diagonally and is decorated with lines (L. 7cm).[127] A solid and almost complete shaft (L. 30.8cm) is decorated with lines and scale patterns on both extremities, one of which is broken.[128] These rods were found along with the two pomegranate rods nos. 60–62 and 6 spindle whorls.[129] All the whorls display a smaller perforation than the diameter of the rods. From the same tomb, but from a different burial (lower burial) comes ivory shaft no. 139 (L. 25cm) decorated with scale pattern and lines on the preserved flat end, the other end is broken away.[130]

Enkomi

At Enkomi, a complete solid ivory shaft comes from the early burial of Swedish tomb 6 chamber A.[131] Shaft 101, found on the floor, is 23.2cm long, is cut flat on its larger extremity (ø about 1cm) and has one pointed end (ø about 0.65cm before the point). It is decorated with horizontal lines on its larger end, while a single and deep incision is present near its point, at 1cm from its extremity. It may also be perforated at about 4cm from its larger extremity, but it is difficult to see it clearly on the published illustration. On the floor of this burial, the excavator found three stone spindle-whorls. Perforation would have allowed the use of the shaft wide side up. From Swedish tomb 17 (second group) comes an incomplete ivory shaft (L. 8cm) found on the floor of the tomb. Its flat extremity is preserved and decorated with lines.[132] No textile tools were found in this tomb.

[120] Franken 1992, 42–43, fig. 4–5.6–13 and 16 (top).
[121] Franken 1992, 7–8, 37.
[122] Franken 1992, 38.
[123] Tufnell, Inge and Harding 1940, pl. XX.23, 27–28.
[124] Starkey 1935, 202, pl. XVI.3; Tufnell 1958, pl. 28.15; Gachet-Bizollon 2007, 125 no. 8.
[125] Tufnell 1958, similar to pl. 29.1, 6
[126] Karageorghis 1974, nos. 75 and 119, pl. LXXXVII, CLXX, 66 and 91; Gachet-Bizollon 2007, 126 nos. 61–62.
[127] Karageorghis 1974, no. 111, pl. LXXXVII, CLXX.
[128] Karageorghis 1974, no. 248, pl. LXXXVII, CLXX, 76 and 91; Gachet-Bizollon 2007, 126 no. 65.
[129] See above.
[130] Karageorghis 1974, nos. 139, 56, pl. CXLIX; Gachet-Bizollon 2007, 126 no. 66.
[131] Gjerstad 1934, 496, pl. LXXIX, fig. 3, no. 101; Gachet-Bizollon 2007, 126 no. 46.
[132] Gjerstad 1934, 545, pl. LXXXVII, fig. 2, no. 81.

Two fragmentary ivory or bone rods (nos. 4926A and 4926B) come from French tomb 5.[133] Shaft 4926B (L. 21.5cm) was found, according to Schaeffer, under the left knee of a female individual located in the centre of the upper level. In the same level, the excavator found an ivory spindle-whorl (inv. 4916).[134] Rod 4926A comes from the top layer of the southern area of the tomb and was found near the head of an individual.[135] It is almost complete (L. 25cm) and has a diameter of 1cm. The same strata also yielded bone or ivory spindle-whorl inv. 4.994 no. 297.[136] From the same tomb (third layer) come two pieces of a solid ivory shaft whose preserved extremity is dug into a mortise (inv. 5025 a+b: L. 7.8 + 8.6cm; ø 1.5cm),[137] and a flat disc with a bronze peg covered by a gold nail (ø 3.6cm; H. 1.5cm) which could have been used on the shaft.[138] French tomb 5 also yielded several spindle-whorls such as inv. 4550 no. 160,[139] inv. 4551 no. 42,[140] inv. 4521 no. 37,[141] inv. 4916 no. 218,[142] inv. 4920[143] and 5028 no. 313[144] as well as a flat pierced ivory disc inv. 5053 no. 319 (ø 4.5cm; H. 0.6cm; ø perf. 0.66cm).[145] The bone/ivory disc, meant to be attached to an ivory shaft by a rivet,[146] can be interpreted as a low-whorl spindle because it would be impossible to attach a hook on it, but as Barber pointed out, it can also be interpreted as a distaff, similar to those from the third millennium BC found at Kish and Abu Salabikh,[147] and similar to the one from Lindos (see below, Fig. 9.14). It could also possibly be interpreted as a high-whorl hand-held supported spindle, meant to be rolled along the thigh. Alternatively, since flat ivory discs are not commonly associated with textile tools, it could also have had a completely different function.

One complete and solid shaft comes from British tomb 86. BM 1969,0701.56 is 9.7cm long and has a diameter of 1cm, its decorated extremity with lines and lattice pattern is cut flat, while a mortise is dug into its other end.[148] No textile tools were recorded in this tomb.

Fragmentary rod BM 1969, 0701.54 decorated with horizontal lines comes from British tomb 84 (L. 7.8cm; ø 1.1cm) and is sheered off diagonally at both ends.[149] From the same tomb, comes a bone oval "pin" (1897, 0401.1570; L. 10.7cm; W 0.8cm; Th. 0.7cm).[150]

A solid and complete ivory rod (BM 1897,0401.885; L. 15.9; ø 0.8cm) was found in British tomb 24.[151] It has a stepped-down and tapered end (H. 1.1cm; ø 0.43–0.38cm), while its other end is cut flat and decorated with deeply carved horizontal lines. Three dome-shaped spindle-whorls

---

[133] Schaeffer 1952, 181, inv. 4926A no. 193 pl. XLII; Schaeffer 1952, 185, inv. 4926B, fig. 75.207 and fig. 81.11/207; also Gachet-Bizollon 2007, 126 no. 53.
[134] Schaeffer 1952, 186 no. 218.
[135] Schaeffer 1952, pl. XXXV no.193. Spindle-whorls are not represented on the plan.
[136] Schaeffer 1952, 183 fig. 75; Caubet 1987, 32 no. 44.
[137] Caubet 1987, 32 no. 42; Gachet-Bizollon 2007, 126 no. 58.
[138] Schaeffer 1952, 194–195, no. 335 inv. 5025, fig. 84.2.
[139] Schaeffer 1952, 177.
[140] Schaeffer 1952, 166; Caubet 1987, 32 no. 45.
[141] Schaeffer 1952, 165, fig. 81; Caubet 1987, 32 no. 46.
[142] Schaeffer 1952, 186.
[143] Caubet 1987, 32 no. 47.
[144] Schaeffer 1952, 187; Caubet 1987, 32 no. 48.
[145] Schaeffer 1952, 189; Caubet 1987, 28 no. 7.
[146] Schaeffer 1952, 194–195, fig. 82; Barber 1991, 63.
[147] Barber 1991, 63, 58, fig. 2.19.
[148] BM online database; Crewe 2009, 86.34; maybe Gachet-Bizollon 2007, 125 no. 45.
[149] BM online database; Crewe 2009 84.22.
[150] BM online database; Crewe 2009 84.21.
[151] BM online database; Crewe 2009, 24.35.

were found in the same tomb: ivory whorl 1969,0701.30 (ø 1.9cm; H. 0.4cm; ø perf. 0.4cm; w. 1.02g),[152] chlorite whorl 1969,0701.31 (ø 3.5cm; H. 0.8cm; ø perf. 0.5cm; w. 12.2g),[153] and chlorite whorl 1969,0701.32 (ø 3.7cm; H. 0.5cm; ø perf. 0.5cm; w. 10.7g).[154] It is possible that the stepped and tapered end of the rod was designed to host any of these spindle-whorls, especially if some padding was used between the spindle and the whorl.[155] It is also possible that it was design to fit into another ivory rod, or to accommodate a pomegranate knob.

At least five bone/ivory solid shafts were found in domestic contexts at Enkomi and come from the Cypriot and French excavations.

A complete ivory shaft inv. 3384 (L 13.3cm), decorated with lines and lattice pattern, comes from Dikaios level IIB–IIIA (13th c. BC), area III. I–K 1–2 east (-14.10).[156] One of its ends is dug into a mortise. No textile tools are registered nearby. Complete shaft inv. 1263 (L. 10.2) comes from level IIIA, area III, room I, G Δ 38–40 east, almost on floor II (-14.46, level IIIB).[157] Two terracotta loom-weights are recorded in the same room (inv. 4359/4 one is from level IIIA and the second may be from level IIIB (?), but was found at the same altitude).[158] From area I room 12, Δ-E 16–18 south, almost on floor II (level IIIB) comes a fragmentary ivory rod decorated with groups of horizontal lines on each end and scale pattern (L. 10.5cm).[159] No textile tools were recorded in the same context.

From the French excavations, come three ivory shafts. The first one, inv. 1958.154, is 7.5cm long and has a diameter of 1.2cm.[160] It was found in Chantier Est under point 206, and is decorated with horizontal lines and scale patterns at both ends. From the same area, under point 206, on floor II at a depth of 1m,[161] comes ivory spindle-whorl 1958.152 (ø 3.1cm; H. 0.7cm)[162] and loom-weight 1958.163 (under point 206 at 1m).[163] The second shaft, inv. 1958.210, is broken but almost complete (L. 22.5cm, ø 1cm). It is decorated with a scale pattern on its larger end, while the thinner one is plain.[164] It was found in sounding XLI, pt. 25 at 1.35m. Two terracotta spindle-whorls were found in the same context (biconical whorl inv. 1958.103, ø 3.5cm; dome-shaped whorl 1958.214, ø 2.7cm, H. 0.8).[165] A third fragmentary and unillustrated shaft was found at point top. 232 (inv. 1959.18, 1.20m depth). It is 16.5cm long, 1cm large and is decorated with horizontal lines at both ends.[166] Two biconical steatite whorls are also recorded in the same context (ø 1.9cm, H. 2.3cm; ø 2.2cm, H. 2cm).[167]

---

[152] BM 1969,0701.30: BM online database; Crewe 2009, 24.32.
[153] BM 1969,0701.30: BM online database; Crewe 2009, 24.6.
[154] BM 1969,0701.30: BM online database; Crewe 2009, 24.7.
[155] See for instance the ivory spindle-whorls BM 1897,0401.1370 and 1897,0401.1371 found with the broken tip of the spindle in its hole; BM online database; Crewe 2009, U.205, U.206. However, the shape of these whorls is different and recalls more a disc than a spindle-whorl. It is possible that these object were not whorls but decorative buttons attached to boxes or furniture.
[156] Dikaios 1969–1971, vol. I, 255; vol. II, 662; vol. IIIa, pl. 128.47, pl. 156.42; Gachet-Bizollon 2007, 126 no. 50.
[157] Dikaios 1969–1971, vol. I, 277; vol. II, 680; vol. IIIa, pl. 132.7; Gachet-Bizollon 2007, 126 no. 51.
[158] Dikaios 1969–1971, vol. II, 707 and 752.
[159] Dikaios 1969–1971, vol. l, 293; vol. 2, 717; vol. IIIa, pl. 135.47 and 168.38; Gachet-Bizollon 2007, 126 no. 52.
[160] Courtois 1984, 57 no. 520, fig. 18.2; Gachet-Bizollon 2007, 126 no. 54.
[161] Depth is from the surface of the excavation and not from the point.
[162] Courtois 1984, 59 no. 560, fig. 18.23.
[163] Courtois 1984, 67 no. 622, fig. 20.23.
[164] Courtois 1984, 57 no. 521, fig. 18.3; Gachet-Bizollon 2007, 126 no.55.
[165] Courtois 1984, 71 no. 705, fig. 22.37 and 71 no. 713, fig. 22.39.
[166] Courtois 1984, 57 no. 522.
[167] Courtois 1984, 143 no. 1169–1170, fig. 43.40.

Only two ivory/bone shafts from domestic contexts were not found in contexts associated with textile tools, while three were. Shafts from Swedish tomb 17 and British tomb 86 were not found with textile tools. The finds of decorated ivory shafts in domestic contexts at Enkomi along with textile tools point to their effective use by the inhabitants.

Aegean

In the Aegean, solid ivory/bone shafts were found in limited numbers and come only from five sites: Perati, Asine, Phylakopi on Melos,[168] Ialysos and from the Cave of Zeus at Mount Ida.[169] However, only Perati and Asine have well published contexts allowing for a reconstruction of the objects' assemblage.

Several solid ivory shafts were found in three Perati tombs. Two fragment of ivory shafts Δ 43 (elephant ivory, L. 6.2, ø 0.3cm) and Δ 44 (L. 2.2, ø 0.2cm) were found in tomb 13.[170] From the same tomb, come several conical whorls, either spindle-whorls or buttons.[171] In tomb 16, fragments of an elephant ivory shaft(s) Δ 58 (L. of individual pieces between 0.8 and 3.2cm, ø 0.2–0.4cm) were found.[172]. Tomb 16 also contained several conical-shaped whorls found in a row above and beneath the tibias of a woman, pointing to a possible use as dress weight, attached to the hem of her skirt,[173] however, the function of these objects as spindle-whorls cannot be ruled out.[174] Two pieces of a fragmentary elephant ivory shaft Δ 125–126 (L. 4.6 and 2cm, ø 0.25–0.36 and 0.2–0.25cm) come from tomb 75, where they were found with a hippo ivory dome-shaped whorl (ø 1.2cm, ø perf. 0.3cm).[175] Shaft Δ 125 is perforated near its largest extremity.

A fragmentary short bone shaft was found in LH chamber tomb I:2 at Asine.[176] The preserved shaft is entirely decorated with a scale pattern. One of its extremities ends in a pin, while the other is broken (L.8.5cm, ø 2cm). Publication of the tomb also mentions seven bone buttons with a "shallow groove along the edge (ø 1.5–3cm)",[177] and a conical steatite whorl (H. 1cm, ø 1.5cm).[178]

A solid tapered ivory/bone shaft comes from a 14th c. BC tomb at Ialysos in Rhodes (British Museum 1872.0315.82).[179] The shaft is 11.7cm long and has a diameter of 1cm. Its widest extremity is cut flat and is decorated with four deep horizontal grooves (Fig. 9.12).

Conclusion

Solid ivory/bone shafts with a tapered end are mainly found with textile tools, suggesting their recurrent use in textile industry. In Ugarit, they are associated with spindle-whorls and/or loom-weights and in the case of tomb VI at Minet el-Beida, the shaft was maybe associated with pomegranate knobs. As far as we can deduce from the published data, Ugaritic domestic contexts show that ivory shafts regularly come from rooms or houses with spindle-whorls and loom-weights.

[168] Bone pin (L. 17cm, ø 0.25–0.4cm) with two incised lines on its top. No context is known; Atkinson *et al.* 1904, 192 pl. 40.9.
[169] Solid and fragmentary 13.3cm long ivory shaft; Heraklion museum inv. 69; Kunze 1935–1936, 232, pl. 86.21.
[170] Iakovidis 1969, pl. 86a; 1969A 285–295.
[171] Iakovidis 1969, pl. 87a–b (Δ48, Δ34, L56–58 and L67–69).
[172] Iakovidis 1969, pl. 75a; 1969A 254–258.
[173] Iakovidis 1969, pl. 74b (L83–73); Iakovidis 1977, 115, 117 fig. 2. A similar arrangement was observed in a tomb from Nauplia. Iakovidis argues that spindle-whorls are larger and flatter than the "buttons/*conuli*"; Iakovidis 1969 A 56, 76; 1969 B 351f.
[174] See for instance Andresson and Nosch 2003, 202–203.
[175] Iakovidis 1969, pl. 30c; 1969A 93.
[176] Frödin and Persson 1938, 388 no. 4, fig. 252.
[177] Frödin and Persson 1938, 388 no. 5, not illustrated.
[178] Frödin and Persson 1938, 390 no. 6, fig. 252.
[179] Schofield 2007, fig. 83.

In some cases, ivory/bone shafts were found with more than a dozen spindle-whorls, as attested in tell Kazel and at tel Dan, reinforcing our impression that they were almost always associated to textile tools.

In rare instances, such as in looted tomb II on the Ugarit acropolis, British tomb 86[180] and the rich Swedish tomb 17 at Enkomi,[181] no positive association could be established. If it is true that records can always be partially preserved, especially when dealing with domestic contexts, looted tombs, or tombs excavated in the 19th c. (British tomb 86 at Enkomi), it certainly is not the only plausible explanation, especially when no textile tools were recorded in intact Swedish tomb 17 at Enkomi.[182] It is, of course, possible to argue that ivory shafts could have been used as hand-held spindles, sometimes without whorls (see below), or could have been used with wooden whorls. However, bone rod 229 from the Mycenaean tomb at tel Dan provides a possible alternative function for the rods. In this case, it is identified as a kohl stick. If the tip of this rod had been broken, it would have been similar to most ivory/bone shafts recorded in the Levant and Cyprus.

I would therefore propose to identify most ivory shafts as possible textile tools (spindles or distaffs) when they are associated with other textile tools such as spindle-whorls, loom-weights, needles or pomegranate knobs. If their association to textile tools can certainly be hypothesised, their precise function is not so easy to gasp. Indeed, only in some instance (Minet el-Beida tomb III, the royal palace at Ugarit, Megiddo tomb 877B1 and tomb 16 at Perati) are bone/ivory rods found with spindle-whorls whose perforation diameter would have permitted their insertion onto the shaft, therefore turning the shaft into a spindle. In all other instances, the diameter of perforation of the spindle-whorls was smaller than the diameter of the shaft. These shafts could therefore have been used as hand-held distaffs. However, if the shaft was used as a spindle, it could have been used either as a hand-help spindle without whorl, or as a spindle with a disappeared wooden whorl.[183] In any case, the versatile possible uses of the shaft were probably attractive for their users. Few shafts (Tell Kazel, Megiddo tomb 40 and Enkomi tomb 24) exhibit a stepped-down extremity that could have allowed (1) the insertion of a pomegranate knob, (2) the insertion of a whorl with a lesser perforation diameter, (3) its insertion into another shaft, or (4) a different use. It is less reasonable to think that such pegs could have been designed for inserting spindle-whorls because of the recurrent position of whorls on larger ends of spindles, while most of the stepped-down shafts end are dug on the finest extremity. Therefore the insertion of a knob would be plausible, even if none was found with these rods. Typologically, the stepped-down end can also be roughly reminiscent of the rounded end kohl stick from tel Dan.

If ivory/bone rods were part of spinning kits, they were however not the norm and were certainly restricted to a few wealthy users. Indeed, if spindle-whorls are not frequent in LBA tombs in Cyprus,[184] bone/ivory rods are even less common and do not appear in all the tombs where spindle-whorls were found. For instance, at Enkomi, they were found in rich tombs, such as British tombs 24 and 84, each containing 54 and 59g of gold, in French tomb 5 (20g of gold), and Swedish tomb 6 (5g).[185]

[180] The content of the tomb was mostly smashed, Tatton-Brown 2003, 48.
[181] Keswani 2004, 125.
[182] For a table of the contents of the tomb, see Gjerstad 1934, 546 and Keswani 2004, 233.
[183] Wooden whorls are attested in Egypt, but also in the Near East with rare examples from Alalakh (see for instance BM 136477b; BM 136600 and BM 136477c).
[184] See Smith 2002, 283–284.
[185] See Keswani 2004, 231–239.

When incomplete or coming from contexts where no textile tools are to be found, another function has to be proposed, such as kohl sticks, chest/lid/drawer closing mechanisms,[186] or even throw-sticks used for playing games.[187]

Iron age objects

Hama

In Hama, several bone shafts were found in the mid-8th c. BC destruction layer of the city. In building V, a "small palace located outside of the royal zone",[188] shafts come from rooms L, B, A, E, F. An almost complete bone rod (inv. 8A431, L. 18.3cm, ø 0.7cm) comes from room L (Fig. 9.13). Its largest extremity is cut flat, sheered off diagonally and decorated with horizontal lines and zig-zag lines. Its tapering end is missing.[189] The same room yielded a loom-weight.[190] Seven fragments of a solid bone shaft were found in room B (L. 0.5 to 3.5cm, ø 0.4–0.9cm), one of its ends is carved into a rounded shape, and decorations of lines and herringbone pattern are present on the pieces.[191] In the same room, the excavators found 7 small flat whorls (ø about 1.3cm), 3 dome-shaped bone spindle-whorls (ø 1.2 to 1.9cm), and a bone button or spool.[192] In room A, three fragments of the same (?) shaft (L. 3.1 to 8.6cm, ø 0.9cm) were found along with another fragmentary shaft (L.2.41, ø 1cm).[193] In room E, a fragmentary shaft (L. 2.9, ø 0.5cm)[194] was found with two fragmentary clay spools.[195] Finally, another fragmentary plain shaft was found in room F (L. 6.5cm).[196] It is likely that building V hosted a bone/ivory workshop or was a storage place for bone/ivory objects,[197] however, this building also yielded a considerable number of textile related tools: 1 loom-weight,[198] 35 clay spools,[199] and 17 spindle-whorls,[200] and it is likely that it also hosted activities related to textile production.

From Building I, room C, comes a fragmentary bone shaft tapering towards its ends.[201] The preserved end is decorated with lattice pattern in between deep horizontal incisions and below a conical tip. From different rooms within the same building come a stone (ø 3.7cm, H. 2cm) and a bone spindle-whorl (ø 2cm, H. 0.9cm).[202]

Another bone shaft was found in building IV, room A. The incomplete plain shaft is broken at both ends and tapers into a fine point (L. 18cm, ø 0.6cm).[203] Seven bone whorls were found in the

---

[186] For a proposed restitution of such mechanism on a box from Kamid el-Loz, see Hachmann 1983, 103, fig. 51.
[187] See for instance Louvre E3674, E 3675 and E 3676 from a New Kingdome Theban Tomb, Bardies-Fronty and Dunn-Vaturi 2012, no. 6 p. 48.
[188] Riis and Buhl 1990, 14, 26.
[189] Riis and Buhl 1990, 207 no. 736, fig. 96.736.
[190] Riis and Buhl 1990, 207 no. 731.
[191] Riis and Buhl 1990, 207 no. 737, fig. 96.737.
[192] Riis and Buhl 1990, 208 no. 745, fig. 97.745; 212, no. 780, fig. 98.780; 212, no. 786, fig. 98.786.
[193] Riis and Buhl 1990, 207 no. 738, fig. 96.738; 208 no. 741.
[194] Riis and Buhl 1990, 208 no. 742.
[195] Riis and Buhl 1990, 207 no. 734.
[196] Riis and Buhl 1990, 208 no. 743.
[197] Riis and Buhl 1990, 207.
[198] Riis and Buhl 1990, 207 no. 731.
[199] Riis and Buhl 1990, 207 no. 732, 734, 735.
[200] Riis and Buhl 1990, 208–212 no. 745, 749, 752, 758, 767, 768, 777, 780.
[201] Riis and Buhl 1990, 208 no. 740, fig. 97.740.
[202] Riis and Buhl 1990, 210 no. 765, fig. 97.765; 210 no. 770, fig. 97.770.
[203] Riis and Buhl 1990, 208 no. 744, fig. 97.744.

same room (ø 1.2–1.5cm, H. 1cm). According to the excavators, their lack of polish may indicate that they were not used as spindle-whorls but rather as buttons or knobs.[204] None of the shafts found in the city was equipped with mortise or peg.[205]

Several solid shafts were found in the town cemetery. Publication allows the reconstruction of the assemblage, as summarized in Table 9.3, while the sex of the occupants was established according to the grave goods.[206]

Several bone shafts were found in the cremation cemeteries from periods I to IV, but only a few are illustrated in the publication.[207] Most, if not all of them were between 21.2 and 24.8cm, but one was shorter (14.5cm from G XXX8) and may be fragmentary. The shafts are generally tapered and in some instances carved into a small rounded bud.[208] The publication of the finds lacks a comprehensive catalogue and it is therefore difficult to interpret each group. Seemingly, it is difficult to link the presence/absence of spindle-whorls to specific shafts, but it is reasonable to say that none of the shafts from period III and IV were associated to bone or stone spindle-whorls. Such a pattern could reflect an 8th c. BC change of consumption habit and could signal the use of these shafts for another purpose, or a change in spinning habits. It is also possible, but not likely unless made of wood, that all of the spindle-whorls were destroyed during the cremation. Most of the period IV shafts are characterized by the carving of their thinner end into a small flower bud (Fig. 9.16).

Sarepta

At Sarepta, a pin (L. 11.7cm) tapered to a point at one end and carved in the shape of a pomegranate at the other end was found in area II-A-5 in level 2-1. A broken solid bone shaft (inv. 3031, L.15.5cm, ø max 1cm) decorated with horizontal lines and zig-zag pattern was found in area II-B-4 in a trench above E wall (possibly level 3), no other textile tool was recorded in this strata.[209]

Hazor

In Hazor, a plain solid bone rod tapered at both ends (L. 13cm, ø shaft 0.6.5, ø ends 0.35–4cm) was found in stratum IV (*ca.* 700 BC), loc 3116 (inv. B 572/1).[210] From the same locus, but possibly from St. V, comes a spindle-whorl (inv. B 969/1, ø 1.9cm, ø perf. 0.5cm).[211]

Kinneret (Tell el-'Orēme)

At Iron Age II Kinneret, a broken and undecorated bone shaft with a pointed end (L. 5 and 9.6cm, ø 0.6–0.8cm) was found in street 520, st. II.[212] The same locus also yielded a limestone dome-shaped spindle-whorl (ø 3.4cm, H. 2cm, ø perf. 0.8–1cm).[213] Another pointed end of a bone shaft was found in loc. 529, str. IIA.[214] At the same depth and from the same locus come an oblong terracotta

[204] Riis and Buhl 1990, 212 no. 781, fig. 98.741.
[205] Riis and Buhl 1990, 205.
[206] Riis 1948, 31.
[207] Riis 1948, fig. 217.
[208] Riis 1948, 173.
[209] Pritchard 1988, 111, 218, fig. 30:8.
[210] Yadin *et al.* 1960 pl. CV.27.
[211] Yadin *et al.* 1960 pl. CLXVI.14.
[212] Fritz 1990, 358–359, pl. 112.12.
[213] Fritz 1990, 346–347, pl. 106.15.
[214] Fritz 1990, 358–359, pl. 112.11.

Table 9.3. Context and association of the shafts in the cremation necropolis at Hama, after Riis 1948.

Hama Period	Tomb	Characteristic of the shaft	Associated Textile Tools
Period I **1200–(1075)** **B.C.E**	G IV 64 – woman	5E241	None
	G IV 76 – woman	Two fragments L. 10.1; L. 7.4 is tapered and one end is carved into a flower. One end is stepped into a peg Groups of horizontal lines	2 fragments of a thin dome-shaped spindle-whorl? (Riis 1948, 171.B) 1 stone dome-shaped spindle-whorl (Riis 1948, 172.D) 1 ovoid spindle-whorl (Riis 1948, 172.G)
	G IV 110 – woman	Groups of horizontal lines	None
	G IV 177 – woman	Mortise at one end	1 spindle-whorl
	G IV 257 – woman	1 shaft and fragments of shaft Groups of horizontal lines	1 dome-shaped spindle-whorl (Riis 1948, 172.D)
	G IV 289 – woman	Fragments	1 disc-shaped with circular groove spindle-whorl (Riis 1948, 172.E)
	G VII *ad* 1-20	2 fragments One end is stepped into a peg Groups of horizontal lines	1 conical spindle-whorl (Riis 1948, 172.C) and fragments of spindle-whorls
	G VIII 398 – woman	1 shaft and fragments of a shaft one end cut flat or rounded; mortise at one end tapered, end carved into a flower, L. 4.2	1 conical spindle-whorl (Riis 1948, 172.C)
	G VIII 436 – woman	Mortise at one end L. 11.9, thinnest end is rounded	1 ovoid spindle-whorl (Riis 1948, 172.G)
	G VIII 458 – woman	Smallest end carved into a flower/pomegranate Groups of horizontal lines	None
	G VIII *ad* 468 – woman		4 spindle-whorls: thin dome-shaped (Riis 1948, 171.B) conical with flat edges (Riis 1948, 172.I)
	G VIII 503 – woman	Both ends cut flat or rounded	1 disc-shaped with circular groove spindle-whorl (Riis 1948, 172.E)
	G VIII 537 – woman	At least two shafts Mortise at one end L. at least 16cm Groups of horizontal lines	2 spindle-whorls and fragments conical with flat edges (Riis 1948, 172.I) 1 bone spatula
	G VIII 572 – woman	fragments Mortise at one end Groups of horizontal lines	Fragments dome-shaped with circular groove spindle-whorl (Riis 1948, 172.F)
	G VIII 586 – woman	Both ends cut flat or rounded	1 spindle-whorl
	G VIII *ad* 586–653 (650?) Fill belonged to 650?	Fragments	4 stone spindle-whorls 2 bone spindle-whorls At least one disc-shaped with circular groove (Riis 1948, 172.E)
	G XII 138 – woman		2 spindle-whorls: Thin dome-shaped (Riis 1948, 171.B) Dome-shaped (Riis 1948, 172.D)
	G XII 142 – woman	5 fragments One end is stepped into a peg, mortise at the other Groups of horizontal lines	1 bone spindle-whorl 3 stone spindle-whorls: At least one thin dome-shaped (Riis 1948, 171.B) and 1 conical with flat edges (Riis 1948, 172.I) One flower/pomegranate knob
	G XIV 3 – woman	Fragments	None
	G XIV 4 – woman	Fragments Groups of horizontal lines	1 stone disc-shaped spindle-whorl (Riis 1948, 171.A)

Period I or II	G IV (k) in the fill		None
Period II (1075)–(925) BC	G IV 136 – woman	Fragments	None
	G VIII 172 – woman	Fragments Groups of horizontal lines	None
	G XII 58 – woman	Fragments Groups of horizontal lines	2 spindle-whorls disc-shaped (Riis 1948, 171.A) dome-shaped (Riis 1948, 172.D)
Period III (925)–800 BC	G VIII 1 – woman	Fragment	None
	G VIII 246 – woman	Fragments Groups of horizontal lines	None
Period IV 800–720 BC	G IX 36 – woman	Fragments	None
	G IX 95 – woman	Fragment	None
	G IX 145 – woman	Fragment Smallest end carved into a flower/pomegranate	None
	G IX 160 – woman	1 shaft Smallest end carved into a flower/pomegranate Groups of horizontal lines	None
	G IX 162 – woman	Fragment Groups of horizontal lines	None
	G IX 270 – woman	1 shaft Smallest end carved into a flower/pomegranate Groups of horizontal lines	None
	G XXX 8 – woman	1 shaft L. 14.5cm Smallest end carved into a flower/pomegranate Groups of horizontal lines	None

spindle-whorl with a hourglass perforation (ø 1.8cm, H. 2.4, ø perf. 0.4–0.6cm)[215] and a perforated clay disc with a hourglass perforation (ø 3.2cm, H. 1.2cm, ø perf. 0.8–1.2cm).[216]

Achziv

In Achziv, pomegranate knobs were found in six of the published tombs, while other examples come from unpublished tombs.[217] Solids ivory/bone shafts were also found in the tombs.

In tomb T.C.4, (Fig. 9.17) a pomegranate knob (no. 6512; H. 1.9cm; w. 1.3cm; ø of perforation 0.5cm) was associated with an ivory spindle-whorl decorated with spokes, reminiscent of Late Bronze Age Levantine tradition (no. 5-6998, H. 0.6cm; ø 2.9cm; ø of perforation 0.45cm) and a solid ivory shaft with a cut flat end (no. 6906/1; L. 4.9cm; ø 0.9cm).[218]

Tomb T.C.2 (Fig. 9.18) yielded one pomegranate knob perforated from end to end (no. 1404; H. 1.5cm; ø 1.34cm; ø of perforation 0.4cm), four conical ivory/bone spindle-whorls (no. 1450: H. 0.56cm, ø 0.94cm, ø of perforation 0.2cm; no. 1465/1: H. 0.54cm, ø 1.5cm, ø of perforation 0.26cm; no. 1092/2: H. 0.5cm, ø 1.5cm, ø of perforation 0.2cm; no. 1443: H. 0.85cm, ø 1cm, ø

[215] Fritz 1990, 346–347, pl. 106.9.
[216] Fritz 1990, 344–345, pl. 105.7.
[217] In unpublished tomb 979 excavated by M. Prausnitz two pomegranate knobbed shafts, one pomegranate knob, four fragmentary solid shafts and dozens of spindle whorls were found together; personal communication S. Wolff, publication of the tomb forthcoming.
[218] Mazar 2001, 19, 44–45, figs. 6–18; photographs 16–44.

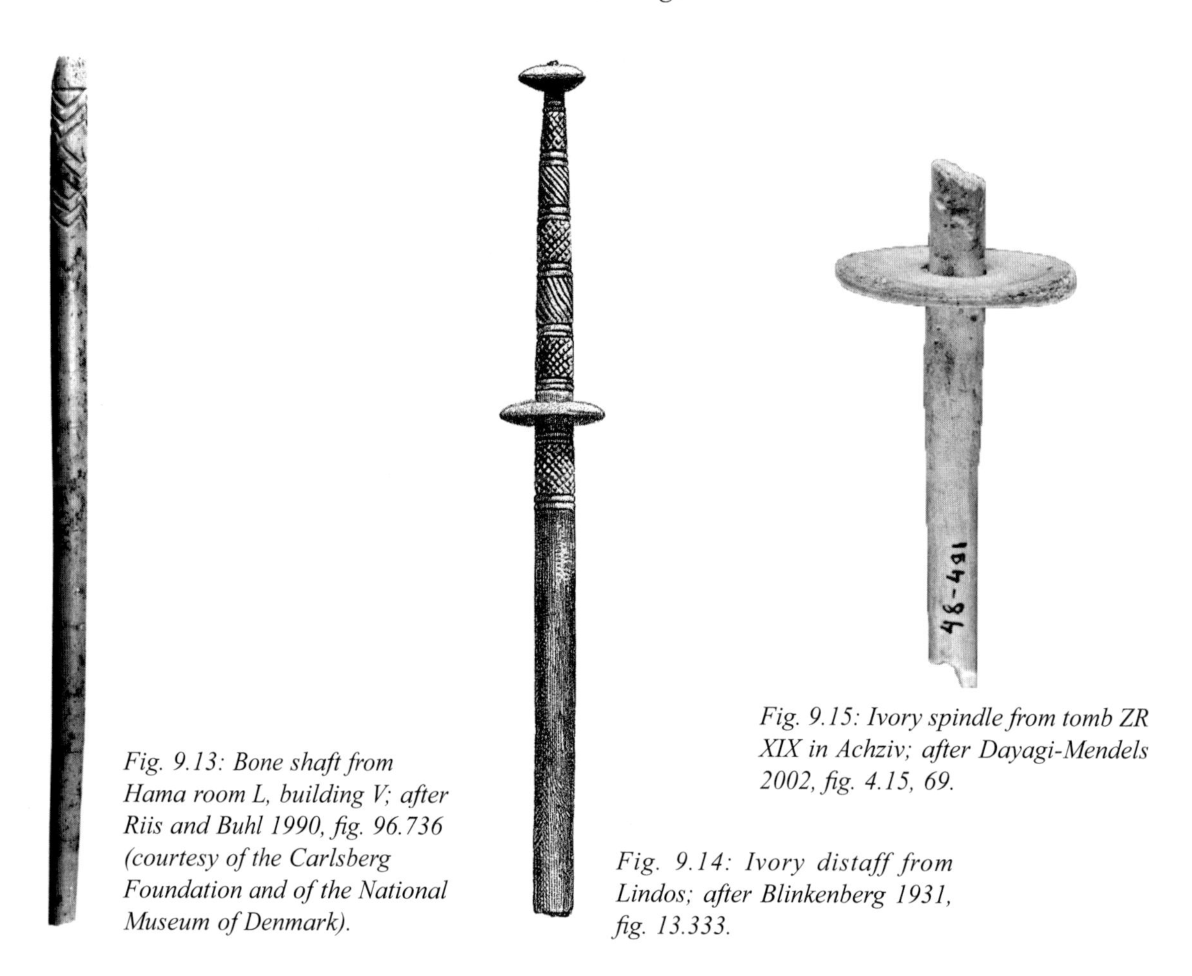

Fig. 9.13: Bone shaft from Hama room L, building V; after Riis and Buhl 1990, fig. 96.736 (courtesy of the Carlsberg Foundation and of the National Museum of Denmark).

Fig. 9.14: Ivory distaff from Lindos; after Blinkenberg 1931, fig. 13.333.

Fig. 9.15: Ivory spindle from tomb ZR XIX in Achziv; after Dayagi-Mendels 2002, fig. 4.15, 69.

Fig. 9.16: Bone shaft from Hama G VIII 458, after Riis 1948, fig. 217.A (courtesy of the Carlsberg Foundation and of the National Museum of Denmark).

of perforation 0.25cm), one slightly tapered solid shaft (no. 1402: L. 5.6cm; ø 0.3–0.4cm) and two fragmented (?) perforated solid shafts with deep oblique incisions on the shaft and a cut flat extremity near the perforation (no. 1371/3: L. 2.5cm, ø 0.5cm, ø of perforation 0.3cm, incision at 1.5cm from the centre of the perforation; no. 1371/2: L. 2.9cm, ø 0.6cm, ø of perforation 0.3cm, incision at 1.5cm from the centre of the perforation).[219]

In 9th–8th c. BCE Tomb Z V, a fragmentary pomegranate knob (no. 48-22, H. 1.5cm, w. 1.4cm) was found associated with 2 ivory spindle whorls (no. 48-24: fragmentary whorl: ø 1.1cm, ø of perforation 0.45; complete one: ø 1cm, ø of perforation 0.4cm).[220]

[219] Mazar 2001, 51; 66–67.

[220] Both whorls are identified as beads in the publication Dayagi-Mendels 2002, 17.

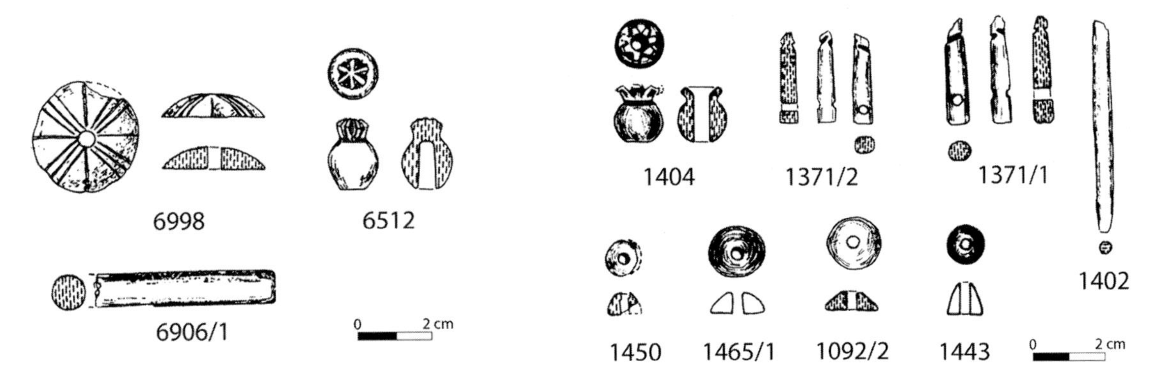

Fig. 9.17: Textile tools from tomb T.C.4 in Achziv; after Mazar 2001, fig. 18, 45.

Fig. 9.18: Textile tools from tomb T.C.2 in Achziv; after Mazar 2001, fig. 25, 67.

Tomb Z XI dated to the 10th–9th c. BCE yielded one knob in the shape of a pomegranate or poppy seed-pod as well as spindle whorls.[221] One bone/ivory dome-shaped spindle-whorl (no. 48-94: H. 0.6cm, ø 2.3cm, ø of perforation 0.85cm) has a perforation large enough to fit onto a solid ivory shaft. Two conical bone/ivory spindle whorls were also found in this tomb (no. 48-99/1: fragmentary, H. 1cm, ø 1.3cm (?), ø of perforation 0.2–0.4cm (?); no. 48-93: H. 1cm, ø 1.1cm) along with other smaller whorls that could be beads. This tomb contained one of the earliest and richest assemblages of the cemetery.[222]

In 10th–8th c. BCE tomb Z XX, two pomegranate knobs, a fragmentary solid ivory rod and eight spindle-whorls were found together.[223] Pomegranate knob no. 48-123 is the smaller (1.7cm high, 1.3cm wide). Knob no. 48-124 is perforated lengthwise and although its dimensions are almost similar to no. 48-123 (H. 2cm, w. 1.5cm), its exocarp is larger and its persistent calyx smaller than no. 48-123. Diameter of perforation of these knobs is unknown. A solid rod, on which the knobs could have apparently fitted is 1.9cm long and has a diameter of 0.5cm. One large ivory/bone whorl (no. 48-120, H. 2cm, ø 2.3cm, ø of perforation 0.8cm) could have been fitted on a solid ivory shaft. Another large ivory whorl which comes from the same context (no. 48-121, H. 1.3cm, ø 3.4cm) is not illustrated, and therefore its diameter of perforation is unknown. A group of six spindle-whorls is illustrated in the publication (no. 48-127). Their diameter varies from 1 to 1.2cm, their height from 0.5 to 0.7cm and their diameter of perforation from 0.2 to 0.3cm.

A broken ivory spindle was found in tomb ZR XIX. Fragmentary spindle no. 48491 has a flat whorl inserted near its larger part (Fig. 9.15).[224] The shaft, preserved to a length of 6.7cm is slightly tapered. Its larger extremity seems to have been diagonally cut flat (unless it is broken) and may be perforated, although it is not clear from the picture. The diameter of the shaft varies from 0.5 to 0.7cm in diameter. The whorl inserted on the shaft has a diameter of 3cm and is 0.2cm high. The tomb dates to the 10–8th c. BCE and was re-used in the Roman period.[225]

[221] Dayagi-Mendels 2002, 24–25.
[222] Dayagi-Mendels 2002, 24.
[223] Dayagi-Mendels 2002, 31–32, fig. 3.16, 35.
[224] Dayagi-Mendels 2002, 60, fig. 4.15:11.
[225] Dayagi-Mendels 2002, 68.

Table 9.4: Solid shafts found in the Achziv tombs.

Tomb	Characteristic of the shaft	Associated Textile Tools
T.A. 68	Solid plain fragmentary shaft. Slightly tapered, one preserved extremity is rounded. L. 6.25cm; ø 0.5–0.75cm.	Four ivory/bone spindle-whorls: Dome-shaped whorls 2582/3: H. 0.75cm; ø 2cm, ø of perf. 0.5cm. Dome-shaped whorl 2582/1: H. 0.5cm; ø 1.25cm, ø of perf. 0.25cm Conical whorl 2582/2: H. 0.6cm; ø 1cm, ø of perf. 0.25cm Dome-shaped whorl 2582/4: H. 0.6cm; ø 1.25cm, ø of perf. 0.35cm (Mazar 2001, 120–121).
T.A. 73	Solid plain ivory shaft 3026/2. L. 5.5.cm; ø 0.5cm	None (Mazar 2001, 98-99).
T.A. 76	Solid plain fragmentary shaft 5749/2, Slightly tapered. L. 4cm; ø 0.75–1cm.	Two ivory/bone spindle whorls: 5735: H. 0.3cm; ø 1cm, ø of perf. 0.15cm. 5737: H. 0.3cm, ø 1.25cm.; fragmentary. (Mazar 2001, 82).
T.A. 79	Solid plain fragmentary shaft 6162. Slightly tapered, one end cut flat. L. 5.5cm; ø 0.75–1cm. Solid plain fragmentary shaft 4537/7. L. 0.75cm; ø 0.55cm	three ivory/bone spindle-whorls: conical whorl with deep spokes 6166/2: H. 0.75cm; ø 1cm, ø of perf. 0.4cm. dome-shaped whorl with three deep spokes 6164/5: H. 0.75cm; ø 1.25cm, ø of perf. 0.15cm. conical spindle-whorl 6166/1: H. 0.75cm; ø 0.75cm, ø of perf. 0.25cm. (Mazar 2001, 92).
T.A. 80	Solid plain fragmentary shaft 6545: L. 2.25cm; ø 0.5cm.	Three conical bone/ivory spindle-whorls: 6548/1: H. 0.75cm; ø 1.25cm, ø of perf. 0.4cm. 6548/2: H. 0.75cm; ø 1.25cm, ø of perf. 0.4cm. 6548/3: H. 0.5cm; ø 1.25cm, ø of perf. 0.5cm. (Mazar 2001, 94).
Z XVIII	Solid tapered fragmentary shaft 48-112. L.11.3cm; ø 0.5–0.65cm. the largest extremity is rounded and may be perforated.	One ivory conical (?) spindle-whorl no. 48-111. H. 1.8cm; ø 1.2cm, ø of perf. 0.35cm. (Dayagi-Mendels 2002, 29).

Finally, one ivory pomegranate knob and several spindle-whorls were found in tomb ZR XXXVI.[226] This tomb dates to the 10th–down to the 7th c. BCE.[227] The pomegranate knob no. 46-671 is 1.4cm high and 1.1.cm wide. It was found with remains of its ivory shaft inserted in its perforation, and it seems, according to the picture, that the diameter of the shaft was 0.5cm. Its persistent calyx are really open and extend almost horizontally. Two large whorls or possible loom-weights come from the same tomb. No. 44-696 is made of clay and its diameter of perforation (0.8cm) would have allowed it to fit onto a rod (H. 1.5cm; ø 3.5cm). The second whorl no. 48-697 has a perforation of 0.6cm, but it is not centered and therefore the use of this object as an effective spindle-whorls has to be ruled out. Two lentoid and three dome-shaped bone spindle whorls also come from the same tomb. Their diameter varies from 0.7 to 2.3cm, their height from 0.5 to 0.8cm and their diameter of perforation from 0.23 to 0.4cm.

Solids shafts were found in tombs T.A. 68, 73, 76, 79 and 80, as well as in tomb Z. XVIII. They were always associated to spindle-whorls (Table 9.4).

[226] Dayagi-Mendels 2002, 98–102.
[227] Dayagi-Mendels 2002, 90.

The so-called Phoenician cemetery at Achziv contains textile tools akin to Late Bronze Age examples. When solid ivory shafts are found, they are always associated with spindle-whorls and/or pomegranate knobs. Pomegranate knobs are also always associated with textile tools.

Megiddo

From Megiddo st. IV (11th–9th c. BC) comes a fragmentary (?) bone rod with one end cut flat and decoration of horizontal and oblique lines at both preserved ends (M 1274, L. 9.9cm, ø 0.6–0.75cm). It was found in locus 404, in the northern "stable" compound.[228] No textile tools were recorded in this locus.[229] In the filling of st. IV, in loc. 1482 (a building with administrative offices and living quarters), a complete (?) bone rod was found (M 5176, L. 6.9cm, ø 0.6–0.75cm).[230] One of its extremities is possibly cut flat, while the other is carved into a pomegranate (H. 1cm, H. calyx 0.4cm, ø 0.9cm). An ivory whorl (M 4494, ø 5.2cm, ø perf. 1.1cm) was found, well stratified, in the same building, and could have been secured onto the shaft.[231]

At Megiddo level III (8th-mid–7th c. BC), a complete (?) ivory rod cut flat at both end and incised at one end with lines (M 4835, L. 13.35cm, ø 0.9–1cm) was found with a bone whorl (M 4393, ø 2.4cm, H. 1.6cm, ø pert. 0.9cm) in loc 1486.[232] In the large storage pit 1414 (7m deep, 11 m large) a fragmentary bone rod had its upper end carved into a schematic pomegranate while its lower end is broken.[233] From the same context come two bone spatulas (M 4453, M 4480),[234] probably used in tapestry weaving.

Beth Shan

At Beth Shan a solid ivory shaft (inv. 25-9-32, L. 15.3cm, ø 0.6–0.8cm) was found in block A west (locus 1002, upper level V).[235] Both ends are broken, the finer one is decorated with horizontal lines, lattice pattern and oblique lines, while the larger extremity is decorated with horizontal lines and lattice patterns. An alabaster whorl (inv. 25-9-2, ø 2.7cm, ø perf. 0.8cm) comes from the same locus.[236] The perforation of the whorl would have allowed it to fit onto the ivory shaft.

Tall Jawa

At tall Jawa (Jordan), three bone/ivory rods were found in an Iron II domestic context, in stratum VIII of building 300 (field E). Two plain solid shafts (TJ 1530 a+b, TJ 1603)[237] come from room 306 (B), where they were associated with a ceramic spindle-whorl (TJ 1689).[238] A bone rod (TJ 2203) decorated with a herringbone pattern and horizontal lines (L. 9cm; ø 0.95cm)[239] was found in room 307 (A) with two broken ceramic spindle-whorls (TJ 932, TJ 2200) and a loom-weight (TJ 906).[240]

[228] Lamon and Shipon 1939, 39, pl. 96.12.
[229] Lamon and Shipon 1939, 142.
[230] Lamon and Shipon 1939, 27, pl. 96.21.
[231] Lamon and Shipon 1939, 143, pl. 95.12.
[232] Lamon and Shipon 1939, 133, pl. 96.10 and 94.17.
[233] Lamon and Shipon 1939, 66–68, pl. 96.20.
[234] Lamon and Shipon 1939, 129, pl. 95.54–55.
[235] James 1966, 157 fig. 114.1. Locus 1002 corresponds to a later room built in the doorway between 1001 and 1004 (James 1966, 49).
[236] James 1966, fig. 114.4.
[237] Daviau 2002, 182–0, fig 2.141.1–2, E65.28/88–29/92; TJ 1530 a+b bone L.14.35; ø 0.70; TJ 1603: bone L.12.25, ø. 0.7–0.35cm.
[238] Daviau 2003, 236.
[239] Daviau 2002, 182, fig. 2.143:1; E64:62/85.
[240] Daviau 2003, 280.

Jerusalem

From Jerusalem come several bone and ivory rods, all found out of context, in dumps, rubble or on the surface. It is therefore impossible to examine their context for textile tool association.[241]

Palaeopaphos-Skales

In CG I tomb 76 at Palaeopaphos-Skales, three fragmentary bone rods, two decorated with groups of horizontal lines (L. 4.3, 3.8 and 2.5cm),[242] were found with a stone spindle-whorl (ø 2.5cm, H. 1.8cm).[243]

Lindos

An ivory/bone rod with two whorls, one secured as a knob on the tapered extremity of the shaft, and one almost in the middle 7.5cm from the top one, have been interpreted as a distaff by C. Blinkenberg (Fig. 9.14).[244] The preserved length of the object is 19.3cm, and it is 0.9cm in diameter. The shaft bears engraved decoration of oblique grooves and cross-hatching, on its upper part from the top whorl, down to a few centimetres below the low whorl. Its identification as a distaff is based on parallels with depictions from Greek vases and on parallels from archaeological objects from Etruria.[245] If we consider the upper whorl as a knob 'topping' the shaft, then this object bears striking resemblance to the whorled pomegranate knobbed shaft RS 4.221B - AO 15758 from Ugarit (Fig. 9.3).

The acropolis at Lindos also yielded a fragmentary ivory shaft decorated with parallel lines and oblique lines forming lozenges. It is 8.8cm long and has a diameter of 1.1cm.[246] Another solid shaft tapered at both ends and decorated with parallel horizontal lines separated by two diagonal bands filled with horizontal lines. It is 17.7cm long and has a diameter of 0.9cm.[247]

Conclusion

Typologically, Iron Age solid shafts present a strong continuity with their Late Bronze Age counterparts: dimensions of the rods as well as decoration patterns of engraved lines akin to Late Bronze Age examples, suggest similar uses. Such supposition is confirmed by their contextual deposition: they are, for instance, found with spindle-whorls and loom-weights in domestic contexts at tall Jawa. In Hama, these shafts were also found in a building certainly associated with textile manufacture, and especially with clay spools. However, based on the evidence recovered from the cremation cemetery period IV at Hama, the 8th c. BC solid shafts whose finer end is carved into a small bud or pomegranate may have had a different function since they are never associated to textile tools. From domestic contexts of the same period, all of the solid shafts but one have their tapered end broken. They are all associated with textile tools, included the one terminated by a small bud from room B. Removable ivory/bone pomegranate knobs are absent from the Iron Age

[241] See Ariel 1990, 140–141: BI 170 comes from an earth layer; BI 172 and 174 come from rubble, BI 173, 177, 178, 180, 182 and 183 from dumps and BI 179 was found on the surface. An inscribed pomegranate also has to be mentioned, although it has no known context, see Avigad 1990; Avigad 1994; Anon 1992; Goren *et al.* 2005; Lemaire 2006;

[242] Karageorghis 1983, 218 no. 47, pl. CXLIV.47, fig. CXLIII.47.

[243] Karageorghis 1983, 215 no. 5, pl. CXLIV.5, fig. CXLIII.5.

[244] Blinkenberg 1931, pl. 13.333, col. 135 no. 333.

[245] Blinkenberg 1931, col. 133–134.

[246] Blinkenberg 1931, pl. 13.334; col. 135, no. 334.

[247] Blinkenberg 1931, pl. 13.335; col. 135, no. 335.

records, and it is not certain that the few mid-8th c. BC shafts from Hama with their finest end carved into a tiny bud/pomegranate could be compared to these. Indeed such tiny rounded buds almost resemble the kohl stick identified in the Mycenaean tomb at tel Dan. Pins or knobbed shafts with a large and decorated pomegranate-like head were found in the archaic Artemisia at Ephesus.[248] These can be compared to the LBA and EIA pomegranate shafts, but they are also typologically very different and I will therefore not discuss them in this article. The Lindos "distaff" is strongly reminiscent of the Ugarit and Delos spinning kits, and this object could therefore be either a distaff or an Iron Age spinning kit.

Discussion

This survey of the context and assemblage of Late Bronze Age ivory spindles, knobbed shafts and solid shafts, shows that when it is possible to reconstruct their surrounding assemblage, the large majority of such rods were found along with textile tools such as spindle-whorls, loom-weights and bone spatulas. If there was no doubt about the function of spindles (whorl + shaft), the pomegranate knobbed shafts and the plain shafts function needed to be surveyed. The examples that we reviewed showed, as already noticed by Gachet-Bizollon,[249] that there are two different types of objects: one-piece solid spindles, slightly tapered, with a hemispherical whorl inserted on their larger part, which can be either short (about 12–15cm) or long (about 20–25cm). The second type of spindle is made of at least two solid pieces assembled together, and generally found without whorl. The examples from Megiddo tombs could also suggest that the whorls were – or could have been – mounted on a peg located in-between the different parts of the shaft.[250] Such a possibility could explain the number of spindle-whorls with a lesser diameter of perforation than the solid shafts found within the same contexts. When whorls are to be found on shafts, they are always positioned on the larger end, while knobs, when present, appear on the smaller end, regardless of the place of discovery, in areas using either low- or high-whorl spindles. The two pomegranate knobs and two solid shafts recovered in the Uluburun shipwreck[251] demonstrate that these objects circulated in the eastern Mediterranean. Similarity of shapes, decoration and whorl/knobs implantation in different cultural spinning habits tend to demonstrate that they were all produced within one coherent area, and then exported to the other. According to the amount of evidence found in the Levant, and especially at Ugarit, it is likely that such rods, and knobs were produced in the Levant, maybe in the northern part of the region.[252]

Such shafts and spindles were of course not the only spindles that existed, and we have to imagine that finer ones made of metal, wood, or reed existed.[253] It is also highly possible that other types of bone/ivory spindles were common, as exemplified by the three duck head elephant ivory 'pins' found in the royal burial ('treasury') at Kamid el-loz: the deep horizontal lines as well as

---

248 Hogarth 1908, pl. XXXIII.
249 Gachet-Bizollon 2007, 121–125.
250 Gachet-Bizollon 2007, 116. Gachet-Bizollon also proposes another type, topped by a pomegranate whorl (type 2), see Gachet-Bizollon 2007, 115–116.
251 Pulak 1992, 1; Gachet-Bizollon 2007, 127 no. 68–70.
252 Hypothesis already proposed by Gachet-Bizollon 2007, 121.
253 At Megiddo and tall Jawa, the ratio of spindles to spindle-whorls is about 2%, suggesting that perishable materials were also used; Daviau et al. 2002, 182.

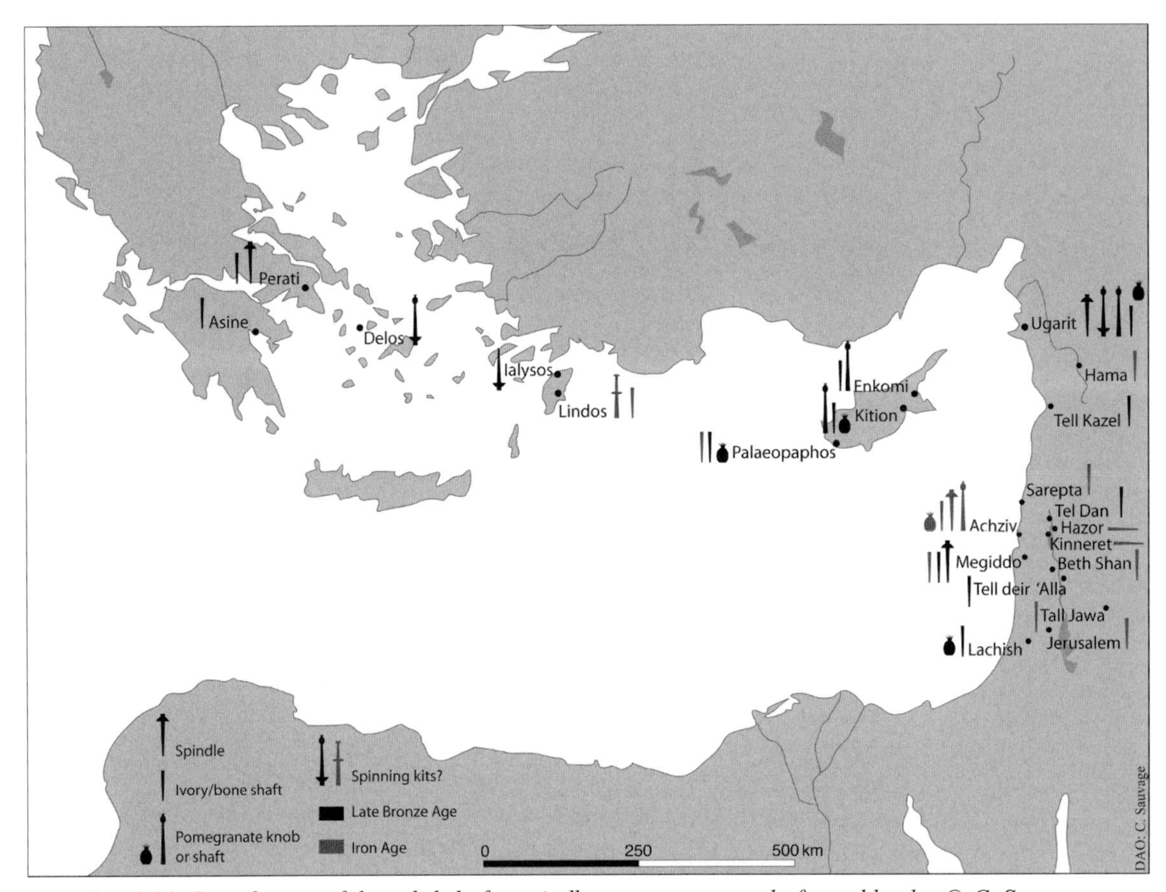

Fig. 9.19: Distribution of the solid shafts, spindles, pomegranate shafts and knobs, © C. Sauvage.

the diagonal groove on the top of the 19.1cm long KL 78:513 (ø 0.2–0.6cm)[254] could indicate their use as a suspended spindle.

If the proposed point of manufacture for most of the pomegranate knobbed solid shafts is correct, then they originated in an area where high-whorl spindles were used, showing that, when used as spindles, their largest, often cut flat end was atop, while the tapered end was facing downwards.

Often, high-whorls spindles are used as drop-spindles, and spin while suspended from the forming yarn. They are set in motion by rolling them against the thigh with the palm of the hand. Such method to initiate rotation, requires the rod to be long enough to accommodate: (1) the palm of the spinner; and (2) winding the newly spun yarn, and therefore necessitate a long(er) shaft than low-whorl spindles. High-whorl suspended spindles can also be set in motion by twirling them with the fingers, allowing a slower spin and therefore a slower drafting of the fibres and a less twisted yarn.[255] Suspended spindles produce a fine and even thread[256] because the weight of the

[254] Duck-heads "pins" inv. 78:520 (preserved L. 7.9cm), 78:512 (L.22.3cm), 78:513 (L.19.1cm); Miron 1990, cat. 515–517, 119, pl. 44.2–4. The tomb contained also spindle-whorls (see pl. 51.8–10 and 52.6–13).
[255] Hochberg 1979a, 29.
[256] Crowfoot 1931, 20.

spindle hanging in the air further drafts or stretches the fibres that the spinner is drawing, creating what is called a double drafting (from both ends).[257] According to Hochberg, this method is used for fast spinning, for yarns requiring a hard-twist and for plying.[258] A distaff is often used, but may be absent. This technique is efficient for long staple fibres, but the drafting created by the weight of the suspended spindle is too important for short staple or fine fibres.[259] Suspended spindles often have a thread notch or groove incised on their top, a metal hook, or can be perforated. In the archaeological records, none of the rods present oblique grooves nor hooks that would suggest their use as a suspended spindle. However, two, maybe three examples in Kition, Perati, and Enkomi are perforated near the largest end of the shaft, suggesting that, in some cases, the solid ivory/bone shafts were used as drop-spindles. It would also have been possible to attach the yarn onto the rod by half-hitch, "made by looping the yarn around the thumb".[260]

Given the length size of most of the solid ivory/bone shafts (i.e. 20–25cm), the absence of groove/notch, and their tapered end, it is also possible to suggest that they were used with techniques that actually do not require grooves nor hook, pointing to their use as either hand-held spindles, or supported spindles.

Hand-held spindles can be used with or without whorls (Fig. 9.20). They are typically rotated by hand: "the rotations is achieved by a combined movement of the whole hand and the muscles of the palm, the fingers playing little or no part."[261] It is possible that some solid ivory rods were used as hand-held spindles without whorls.[262] According to Hochberg, spinning with a stick is the easiest technique to learn.[263] When a whorl is used, this technique is extremely efficient to control wool, especially short-stapled wool.[264] Drafting and twisting are simultaneous, and this method also allows the doubling of the yarn.[265] Crowfoot distinguished two types of supported spindles: type A: "supported by resting lengthwise on the right thigh. Spindle usually large, chiefly used for wool" and type B: generally small spindles, standing erected on the ground, in a shell, bowl or cup. In recent ethnographic surveys, type B spindles were generally used for cotton.[266] A type A supported spindle can be used for any quality of wool, short or long. Drafting and twisting are separate: the spinner can draft with both hands while the spindle rests on his/hers thigh. A type B would allow the spinner to free one hand, so drafting and twisting are simultaneous, and no rotation interruption is required.[267] Supported spindles allow the spinner to spin fine threads made of short-stapled wool because only a light tension is applied to the forming thread.

I mentioned earlier that tension is critical in spinning, and that the tension applied to a fibre will determine the thickness of the yarn spun, but also the type of fibres that one can wish to work with. For instance, according to Hochberg, heavy spindles, of 100–150g, may be use to spin long staple wool.[268] Barber also notes that for heavy thread of long flax and for plying wool yarn, heavier

[257] Crowfoot 1931, 20.
[258] Hochberg 1979a, 29.
[259] Crewe 2002, 218.
[260] Ryder 1968, 79.
[261] Crowfoot 1931, 10.
[262] See for instance the techniques described by Liu (1978, p. 99).
[263] Hochberg 1977, 24. Contra Ryder (1968, 79) argued that it is not easy to use a spindle without whorl.
[264] Crowfoot 1931, 12–13.
[265] Crowfoot 1931, 13.
[266] Crowfoot 1931, 17.
[267] Crowfoot 1931, 19.
[268] Hochberg 1979b, 21.

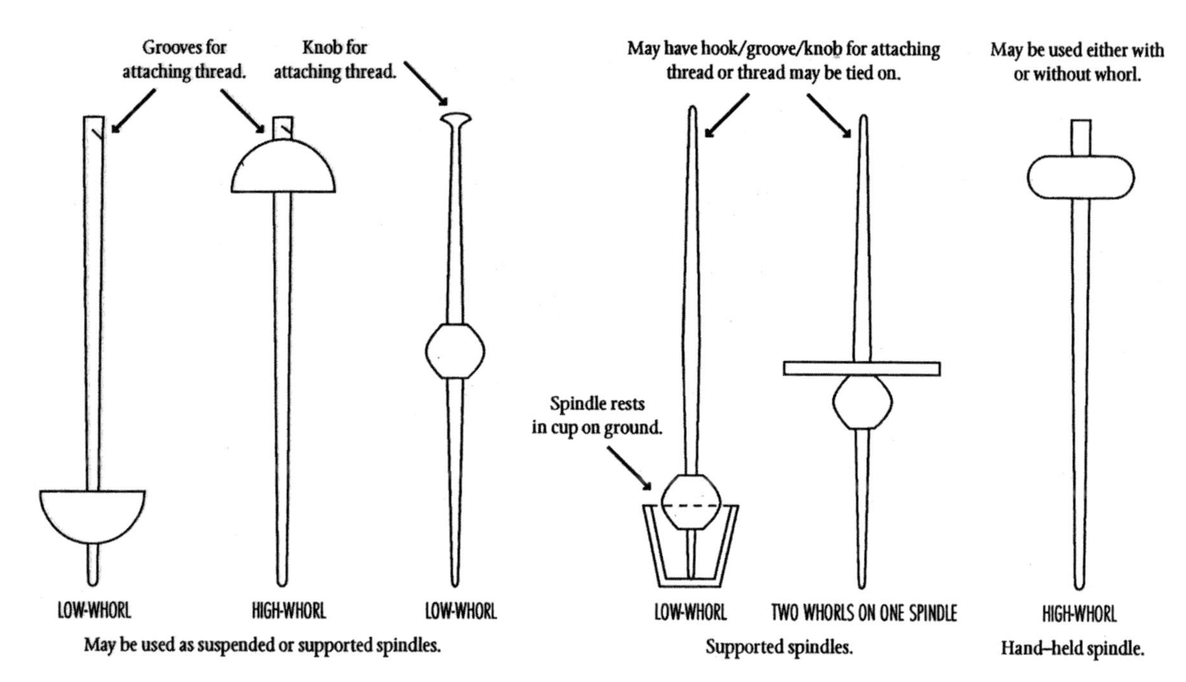

Fig. 9.20: Different types of spindles with possible variation for the position of the whorl, after Crewe 1998, 6, fig. 2.1 (© L. Crewe).

spindles may be used. But for short staple wool, flax tow or cotton, a light spindle is mandatory.[269] When data are available for complete spindles, solid shafts, or pomegranate knobbed shafts, their weight range between 30 and 45g. Therefore these rods, when used as spindles, corresponded to light spindles designed for delicate and possibly short staple fibres. They would certainly have produced fine threads. It is unlikely that the Late Bronze Age examples were used to spin cotton since the first appearance of this term (*kitinnû*) dates to the 9th c. BCE in southern Babylonia.[270] Possibly extra-fine fibres, carefully prepared were used with these luxurious spindles. Wool fibre variations are common in different breeds, but also within the same breed, depending on the age and sex of the animals, but also on the body part where the wool is collected. For instance, sheep wool from lamb, ewe, ram or wether will be of different quality, while wool from the thigh is coarser and longer than wool from the shoulder.[271] Experimental spinning with 4g whorls show that carefully prepared wool is necessary, allowing the spinner to work with a soft, fine and washed product. Such wool quality requires the wool to be brushed to remove as much underwool as possible,[272] and is more time consuming: experiments show that it takes about 9 hours to prepare 66g of wool after washing and drying, when it takes 6 hours to prepare wool for spinning with heavier spindle whorls.[273] Thread

[269] Barber 1991, 52.
[270] Zawadzki 2006, 28. There, cotton was at first rare and expensive, and was used for the garment of the gods. The situation changed under Nabonidus, when cotton fabrics were given to temple personnel, suggesting its popularity as well as its wider availability (Zawadzki 2006, 28–29).
[271] Andersson Strand 2010, 11.
[272] Mårtensson, Andersson, Nosch, Batzer, 2006.
[273] Andersson *et al.* 2008, 173; Andersson Strand 2010, 13. For the preparation time of less fine fibres, see Nosch 2012, 48.

spun with light whorls contains less fibre than when spun with heavier whorls (or spindles),[274] and can be considered as indicative of high-quality thread, certainly used for high-quality textiles.[275] Spinning with a light spindle therefore requires more preparation time, but is also more time consuming for the spinner and demands greater skills.[276] I would therefore postulate that the final product was a premium thread, in accordance with the quality and luxury of the constitutive material of the spindle.

The small-size spindles of about 12–13cm long attested in Ugarit, Perati and Ialysos are however too short to be hand-held spindles. Because of their small rod, the rotation of these spindles should have been activated with a twirl of the fingers, and used either as type B supported spindles, or as drop-spindles. No hook, groove or perforation can confirm their use as suspended spindles. These short spindles would have been even lighter than the long ones previously discussed, and would have created a finer thread.

After discussing in length the possible ways to use the ivory/bone rods as spindles, another question needs to be asked: how were they used in areas where low-whorl spindles are attested, knowing that the use of high- or low-whorl spindles is culturally determined? I already pointed out that the whorls, when found on the rods (i.e. spindles), are always positioned near the largest extremity, while when a knob is attached (i.e. distaff), it is always secured on the tapered end. It is therefore possible that these ivory/bone spindles were: (1) all used the same way in the eastern Mediterranean; (2) used upside-down in the Aegean and Cyprus; or (3) exchanged and deposited in tombs or temples for their value as prestigious material, while reminiscent of a familiar object.[277]

The domestic context of the solid bone/ivory shafts from tell Kazel, Megiddo, Ugarit and Enkomi points to an effective use of these objects in the Levant, their probable area of manufacture, but also in Cyprus, said to use low-whorl spindles.[278] Most of these shafts were, however, found in tombs, and if it is possible that they had been part of the deceased possessions during life, they may also have had the function of reflecting the status of the individual, family or social group, and/or were manufactured especially for the burial.[279] Textile tools made of prestigious material are certainly no exception and could have been made for tomb assemblage. For instance, this may be the case for the gold spindles from shaft grave circle A tomb III at Mycenae,[280] if one could argue that Helen of Troy is said to have been spinning purple dyed-wool with a gold spindle,[281] and that Herodotus' story of Evelthon of Salamis giving Pheretime of Cyrene a golden spindle and distaff as well as wool, suggests these are appropriate gifts for a women.[282] Late Bronze Age circulation of precious spindles (metal, stone[283] and bone/ivory/horn) as official gifts between kingdoms is attested by EA 25 (70–72)[284] and could also point to the effective use of these spindles during their

---

[274] Andersson *et al.* 2008; Andersson Strand 2010, 13.

[275] Andersson and Nosch 2003, 199.

[276] Nosch 2012, 48. The output of thread per hour is about 50m of yarn per hour when spun on an 18g whorl, and 35m of yarn per hour when spun on a 4g whorl.

[277] Sauvage 2012, 208.

[278] Frankel and Webb 1996, 193–194; Crewe 1998, 8.

[279] Smith 2002, 283.

[280] Karo 1930, 57, pl. XVII.93–95 and 106; Burke 2010, 103.

[281] Homer, *Il.* IV, 113–154.

[282] Herodotus *Histories* IV 162.

[283] For onyx textile tools; see Völling, 1998.

[284] List of the gifts of Tushratta of Mitanni to Amenhotep IV. Moran 1992, 79. EA 25. l. 70–72 "[x spindles of gol]d, 8 shekels in weight. 26 spindles of silver, 10 shekels in weight. [x spindles of …]. 10 spindles of lapis lazuli. 16 spindles of al[abas]ter. [x spindles of …] … 11 spindles of … […] stone. 33 spindles of horn."

owner's life. In the case of ivory spindles, both *scenarii* (prized possessions during life, and display objects made for the tomb assemblage) are possible in areas where they are found in both domestic and funerary contexts. It could, for instance, be the explanation for the 'articulated' spindle found in tomb 3018F at Megiddo (Fig. 9.4). However, in the Aegean, since they are only found in funerary and religious contexts, it is likely that they were deposited for their value, and perhaps not actually used by the deceased. It is possible that the function of the object, even if "useless" by Aegean people, could still have been considered as "accurate" during the gathering of funerary material, because of the resemblance of these with actual useful spindles. Textile tools were also most likely related to the religious sphere[285] and therefore their deposition in tombs may reflect more than the owner's identity as a spinner, and may, for instance, correspond to an evocation of the life cycle, as Breniquet suggested.[286]

Would it have been possible for Cypriots (or Aegean peoples) to use the ivory spindles up-side-down? If used in such a manner, then the largest part of the shaft, whose extremity is often flat, would have been the bottom part, while the smallest pointed and tapered extremity would have been the top. The lack of groove or knob on this part of the spindle suggests the use of the object as hand-held spindle or supported spindle and not as a suspended spindle. In the case of a supported spindle, the flat bottom, would have rendered difficult the rotation of the object, the large diameter and the flat extremity probably causing too much friction to obtain a real spin with a desirable moment of inertia. The only possible use of these spindles while up-side-down would then be as hand-held spindles.

The perforation on the larger end of rod 132 found in Kition tomb 9 (Fig. 9.7), and on shaft Δ 125 at Perati would also confirm that these rods were used (if used) with their larger end on top. But it is possible that these perforations originated at the place of manufacture and not at the place of deposition, and they cannot be taken for the effective use of the objects in Cyprus and the Aegean. Alternatively, the presence of Canaanites living, for instance, in Cyprus could explain this phenomenon,[287] but the archaeological records do not allow us to identify the people buried with ivory/bone rods at Kition or Enkomi as foreigners.

Low-whorl spindle evidence in Cyprus comes from an Early Cypriot III clay model found in Vounous tomb 29 (Fig. 9.21 right). This object has been discussed at length and the perforation on its smaller and tapered end suffices to eliminate doubts about its orientation, and places the tapered end atop.[288] However, the determination of the orientation for the ECI clay model from Vounous tomb 92 is only determined by analogy (Fig. 9.21 left). This model exhibits striking resemblance with the ivory/bone spindles discussed in this article, especially complete spindles RS 4.421[A], RS 34.210 and M. 3568 found at Ugarit and Megiddo (see Figs 9.2 and 9.4). If this locally-made clay model actually depicts a spindle resembling the bone/ivory ones, it can only mean that such spindles, maybe made of wood or other perishable material, were actually used by the locals, before the Late Bronze Age.[289] Without perforation on this second clay model, whose largest extremity

[285] Spindles, pomegranate knobbed shafts and rods were found in Lachish, Ugarit, Delos and the cave of Zeus at mount Ida. Similar 'symbolic' gifts are attested in Ebla, where two bronze spindles and a small whorl were found in a *favissa* inside the Ishtar precinct; Peyronel 2007, 26; Anderson, Felluca, Nosch, Peyronel 2010, 162–164.

[286] see Breniquet 2008, 359–380.

[287] See for instance textual evidence reviewed in Yon 2007, 17–24.

[288] See Frankel and Webb 1996, 193–194; Crewe 1998, 8.

[289] Several scholars noted the novelty of such solid ivory spindles in the Late Bronze Age; Guy 1938, 170; Barber 1991, 62. But it is possible that similarly shaped spindles, made of wood, were already used long before the Late Bronze Age.

is missing, I am again tempted to orient the tapered end downwards and to postulate for its use as supported or hand-held spindle. Since the top of this object is missing, it is difficult to assert how many centimetres are missing, and if this spindle should rather be identified as a high-whorl spindle such as the examples from Figs 9.2 and 9. 4 or as a mid-/low-whorl spindle. Several whorls from Late Bronze Age Cyprus display grooves or notches on their domes, pointing towards the use of high-whorls spindles.[290] Therefore, the presence of such models raises the question of whether the ancient Cypriots could simultaneously have been using both techniques, maybe one for spinning and the other for plying. If such was the case, Bronze Age Cypriot spinners could be compared to modern Tunisian weavers who use both high- and low-whorl spindles, to achieve different strengths of threads. [291] We also know that today modern Cypriot spinners use high-whorl spindles, and if we accept the hypothesis that Bronze Age Cyprus was maybe using both methods, then the technique of Cypriot spinners would not have dramatically shifted through time, but would possibly have selected one process over the other.

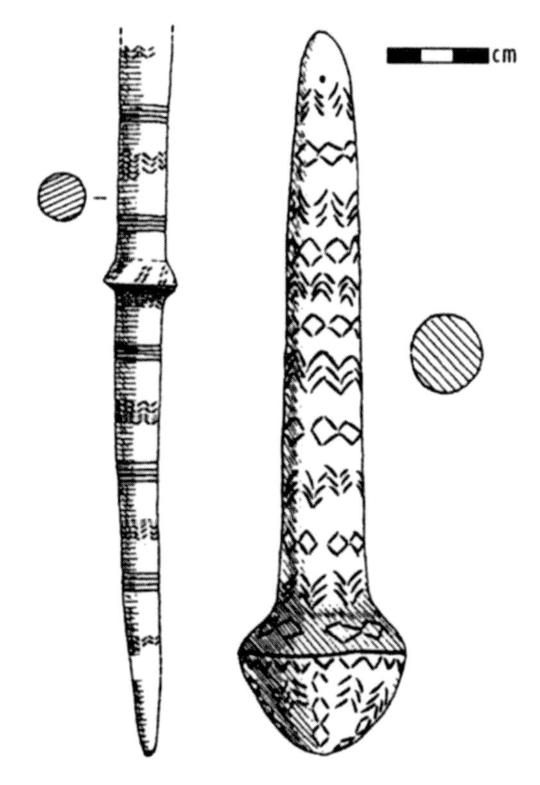

Fig. 9.21: Spindles models from Vounous. The left on dates to ECI and the right one to EC III, after Crewe 1998, 8 fig. 2.2; Stewart 1962, fig. 90.6; Dikaios 1940, pl. LVI (courtesy of the Medelhavsmuseet and of L. Crewe).

Conclusion

Two modules of bone/ivory spindles existed in the eastern Mediterranean Late Bronze Age. A short one of about 12–13cm long is attested in the Aegean and the Levant, while the majority of the attestations are spindles of a larger dimension, about 20–25cm long. These spindles are found in domestic, religious and funerary contexts and are always associated with other textile tools such as spindle-whorls, other spindles and bone/ivory shafts. Such spindles were probably produced in the northern Levant, an area known to use high-whorl spindles. The whorl is always inserted on the larger end of the shaft, the tapered end being therefore the bottom of the spindle.

The ivory/bone ones being a more expensive version of an older prototype.

[290] For instance, at Enkomi, whorls no. 1506 and 1507 in Dikaios 1969, vol. IIIA, pl. 178.1–2. For the use of grooved whorls as high-whorls, see Breniquet 2008, 119. Late Bronze Age Cyprus was certainly producing linen (IBoT I 31 obv. 2–4 (CTH 241.1) mentions "linen of Alashiya"; Beckman 1996, 33) and wool (A kind of wool 'from Cyprus' or of 'Cypriote type' is attested in Linear B texts; Bennet 1996, 57–58). Remains of fabric found on Cypriot objects were identified as S-spun linen fabrics (Åström 1965, 112–113: textile comes from the MC III Tomb 7 at Paleoskoutella; and textile impressions were found on a Cypro-Archaic II (600–475 BC) jug; Pieridou 1967, esp. 26). If they were locally produced (and not imported from Egypt), then a spinning technique similar to the Egyptian one (i.e. high-whorl spindles) should have been used in Cyprus to produce a S-spun thread.

[291] Barber 1991, 43; Crowfoot 1931, 31.

Except for three examples, all pomegranate knobs of knobbed shafts were found in domestic, religious and funerary contexts with other textiles tools (spindles, loom-weights and spindle-whorls), suggesting an association with textile industry. Based on the well-preserved examples from Ugarit and Delos, their use as distaff is likely, the knob being then placed on the thinner end of the shaft. Such an interpretation would also correspond with later Greek representations of distaffs, having in some cases a rounded knob visible above the fibres (Fig. 9.22).

Tapered bone/ivory shafts can thus have had a versatile use in textile industry, either as spindles or as distaffs. When found in archaeological contexts with no knob or spindle-whorls mounted on them, these shafts are mostly found with textile tools. Without being able to totally rule out the possible different functions of ivory/bone solid and tapered shafts, this contextual survey allow us to say that they were certainly used in the textile industry because of their frequent association with textile tools such as spindles, spindle-whorls, loom-weights, needles, etc. For the rare occasions where they are not associated to textile tools and not perforated, it seems that another function needs to be proposed. The several possible uses of these shafts are of course not exclusive from each other, pointing again towards the versatility of these objects.

The spindle-whorls often found with these shafts generally have a diameter of perforation smaller than the diameter of the shaft, therefore preventing their use onto these shafts. This may be an indication that ivory/bone shafts were mostly used as distaffs, un-whorled supported spindles, with wooden whorls or that the whorls were pinned in between constitutive parts of the shaft, as

Fig. 9.22: Black figure vase from the acropolis at Lindos. A women is spinning with a suspended low-whorl spindle and is drafting fibers from a distaff topped with a rounded knob; after Blinkenberg 1931, pl.129.2691+2692.

demonstrated by the Megiddo example. The small number of bone/ivory whorls that could have fit on the tapered shafts may also point to a restricted use of the shafts as spindles. As demonstrated earlier, these ivory spindles were really light and were therefore most likely used to spin high-quality, labour intensive, and costly threads, certainly made of premium, fine, and well sorted fibres. The Iron Age solid ivory/bone shafts show strong continuity in dimension, decoration, contexts and textile tools associations suggesting similar uses in the Levant and therefore a durable knowledge.

The ivory/bone solid tapered shafts, when used as spindles, would have been impossible to use up-side-down, with their thinner extremity on top if it was not perforated. In the Late Bronze Age Aegean, where they are not found in domestic contexts, is likely that such spindles were deposited in the Perati tombs as valued objects, reminiscent of familiar textile tools. Their occurrence in domestic contexts in Cyprus, at Enkomi and Kition, points towards the effective use of these rods by the inhabitants. Their use as distaffs is evident through the presence of numerous pomegranate knobs, but the presence of spindle-whorls with a perforation large enough to fit onto the shafts at both sites, and an EC model from Vounous similar in shape, and decoration to the ivory/bone spindles may point towards the simultaneous use of high-whorl and low-whorls in Bronze Age Cyprus.

Acknowledgements

I wish to thank the editors of the volume for their solicitation to contribute and for their useful comments. I am grateful to S. Cluzan (Musée du Louvre, département des Antiquités Orientales, Paris) and K. Göransson (Medelhavsmuseet, Stockholm) for providing me with weight measurement of spindles and ivory shafts from Ugarit and Enkomi. Charles Arnold (British Museum) kindly confirmed that the Ialysos spindle is not perforated. I am also grateful to S. Wolff for pointing out the Achziv material. I also wish to thank V. Matoïan for her comments on the Ugarit material.

Abbreviations

BCH	*Bulletin de correspondance hellénique*
BSA	*(Annual of the) British School of Archaeology at Athens*
INA	Institute of Nautical Archaeology
PEFQ	*Palestine Exploration Fund Quarterly*
OpAth	*Opuscula Atheniensia*
RDAC	*Report of the Department of Antiquities, Cyprus*
RSO	Ras Shamra-Ougarit

Bibliography

Andersson, E. Felluca, E. Nosch, M.-L. Peyronel, L. 2010 New Perspectives on Bronze Age Textile Production in the Eastern Mediterranean. The first results with Ebla as a Pilot Study. In P. Matthiae, F. Pinnock, L. Nigro and N. Marchetti, *Proceedings of the 6th International Congress on the Archaeology of the Ancient Near East, May 5th–10th 2008, Sapienza Università di Roma, Volume 1*, 159–176.

Andersson, E. Mårtensson, L. Nosch, M.-L. and Rahmstorf, L. 2008 New Research on Bronze Age Textile Production. *BICS* 51: 171–174.

Andersson, E Mårtensson, L. and Nosch M.-L. 2011 Textile Production in Late Bronze Age Khania. Evidence from the Greek-Swedish Excavations at the Agia Aikaterini Square, Kastelli. In Proceedings of the 10th International Cretological Congress (held in Khania, 2006), A3, 407-420.

Andersson, E. and Nosch M.-L. 2003 With a Little Help from my Friends: Investigating Mycenaean Textiles with help from Scandinavian Experimental Archaeology. In K. P. Foster and R. Laffineur, Metron, *Measuring the Aegean Bronze Age, Proceedings of the 9th International Aegean Conference, New Haven, Yale University, 18–21 April 2002*, 197–205.

Andersson Strand, E. 2010 The Basics of Textile Tools and Textile Technology: from Fibre to Fabric. In C. Michel and M.-L. Nosch, *Textile Terminologies in the Ancient Near East and Mediterranean from the Third to the First Millennia BC*, 10–22.

Andersson Strand, E. and Nosch, M.-L. 2007 *Technical Textile Tools Report. Methodological Introduction, Tools and Textiles* – Texts and Contexts Research Program, The Danish National Research Foundation's, Centre for Textile Research, University of Copenhagen.

Anon,1992. The Pomegranate Sceptre Head – from the Temple of the Lord or from a temple of Asherah? *Biblical Archaeology Review,* 18(3): 42–5.

Ariel, D. T. 1990 *Excavation at the City of David 1978–1985*, vol. II, Qedem 30, Jerusalem.

Åström, P. 1965 Remains of Ancient Cloth from Cyprus. *OpAth* V: 111–114.

Åström, P. 1972 *Swedish Cyprus Expdition vol. IV, 1D*. Lund.

Atkinson, T. D. *et al.* 1904 *Excavations at Phylakopi in Melos: conducted by the British School at Athens*, London.

Avigad, N. 1990. The Inscribed Pomegranate from the 'House of the Lord' *The Biblical Archaeologist* 53.3: 157–166

Avigad, N. 1994. The Inscribed Pomegranate from the 'House of the Lord'. *Israel Museum Journal*, 8: 7–16.

Badre, L., Gubel, E., Capet, E. and Panayot, N. 1994 Tell Kazel (Syrie). Rapport préliminaire sur les 4e–8e campagnes de fouilles (1988–1992). *Syria* 71: 259–346.

Barber, E. J. W. 1991 *Prehistoric Textiles: the Development of Cloth in the Neolithic and Bronze Age, with Special Reference to the Aegean*. Princeton.

Bardès-Fronty, I. and Dunn-Vaturi, A.-E. 2012 *Art du jeu. Jeu dans l'art, de Babylone à l'occident medieval*. Paris.

Beckman, G. 1996 Hittite Documents from Hattusa. In A. B. Knapp, P. W. Wallace and A. G. Orphanides, *Sources for the history of Cyprus vol. 2 Near Eastern and Aegean texts from the third to the 1st millennium B.C.*, Altamont, 31–35.

Ben-Dov, R. 2002 The Late Bronze Age "Mycenaen" Tomb. In A. Biran and R. Ben-Dov, *Dan II. A Chronicle of the Excavation and the Late Bronze Age "Mycenaen" Tomb*, 33–248, Jerusalem.

Bennet, J. 1996 Linear B Texts from the Bronze Age Aegean. In A. B. Knapp, P. W. Wallace and A. G. Orphanides, *Sources for the History of Cyprus vol. 2 Near Eastern and Aegean texts from the third to the 1st millennium B.C.*, Altamont, 51–58.

Blinkenberg, C. 1931 *Lindos, fouilles de l'acropole 1902–1914. I Les petits objets. Texte et planches*.

Breniquet, C. 2008 *Essai sur le tissage en Mésopotamie, des premières communautés sédentaires au milieu du IIIe millénaire avant J.-C.* Travaux de la Maison René-Ginouvès.

Burke, B. 2010 *From Minos to Midas. Ancient Cloth Production in the Aegean and in Anatolia*, Oxford.

Caubet, A. 1987 Enkomi (Fouilles Schaeffer 1934–1966): Inventaire complémentaire. *RDAC* 1987: 23–48.

Caubet, A. and Poplin, F. 1987 Les objets de matière dure animale : étude du matériau. In M. Yon (ed.) *Le centre de la ville, 38e–44e campagne (1978–1984)*, RSO III, 273–306.

Courtois, J. C. 1984 *Alasia III, Les objets des niveaux stratifiés d'Enkomi (Fouilles C. F. A. Schaeffer 1947–1970)*.

Crewe, L. 1998 *Spindle Whorls: a study of form, function and decoration in prehistoric Bronze Age Cyprus*, Studies in Mediterranean Archaeology and Literature. Pocket-books 149.

Crewe, L. 2002 Spindle-whorls and Loomweights. In G. W. Clarke and J. P. Connor *et al.* (eds), *Jebel Khalid on the Euphrates: Report on the Excavations 1986–1996. Volume 1*, 216–243.

Crewe, L., Catling, H. and Kiely, T. (ed.), 2009 *Enkomi*. London.

Crowfoot, G. M. 1931 *Methods of hand spinning in Egypt and the Sudan*, Bankfield Museum Notes ser. 2, no 12.

Daviau P. M. 2002 *Excavations at Tall Jawa, Jordan. Volume II. The Iron Age Artefacts*, Brill.

Daviau, P. M. 2003 *Excavations at Tall Jawa, Jordan. Volume I. The Iron Age Town*, Brill.

Dayagi-Mendels, M. 2002. *The Akhziv Cemeteries. The Ben-Dor Excavations 1941–1944. Israel Antiquity Authority Reports,* 15. Jerusalem.

Dikaios, P. 1940 The Excavations at *Vounous-Bellapais* in Cyprus, 1931–2. *Archaeologia* 88: 1–174.

Dikaios, P. 1969–1971 *Enkomi: excavations 1948–1958*. Mainz am Rhein.

Feldman, M. H. 2009 Hoarded Treasures: The Megiddo Ivories and the End of the Bronze Age. *Levant* 41: 175–1994.

Frankel, D. and Webb, J. M. 1996 *Marki Alonia. An early and middle Bronze Age Town in Cyprus. Excavations 1990–1994*, Jonsered.
Franken, H. J. 1992 *Excavations at Tell Deir 'Alla. The Late Bronze Age Sanctuary*. Leuven.
Fritz, V. 1990 *Kinneret. Ergebnisse der Ausgrabungen auf dem Tell el-'Oreme am See Gennesaret 1982–1985*, Wiesbaden.
Frödin, O. and Persson, A. W. 1938 *Asine. Results of the Swedish Excavations 1922–1930*, Stockholm.
Gachet-Bizollon, J. 2007 *Les ivories d'Ougarit et l'art des ivoiriers du Levant au Bronze Récent,* RSO XVI.
Gallet de Santerre, H. and Tréheux, J. 1947–48 Rapport sur le dépôt égéen et géométrique de l'Artémision à Délos. *BCH* 71–72, 148–254, pl. XIX–XLVI.
Gjerstadt, E. 1934 *The Swedish Cyprus Expedition. Finds and Results of the Excavation in Cyprus 1927–1931*, Stockholm.
Goren, Y., Ahituv, S., Ayalon, A., Bar-Matthews, M., Dahari, U., Dayagi-Mendels, M., Demsky, A. and Levin, N. 2005, A Re-examination of the Inscribed Pomegranate from the Israel Museum, *Israel Exploration Journal* 55:3–20.
Guy, P. L. O.1938 *Megiddo Tombs*, OIP 33.
Hachmann, R. 1983 *Frühe Phöniker im Libanon. 20 Jahre deutsche Ausgrabugen in Kāmid el-Lōz*. Mainz am Rhein.
Hochberg, B. 1977 *Handspindles*, Santa Cruz.
Hochberg, B. 1979a The High Whorl Spindle. *Shuttle Spindle & Dyepot* 1979: 27–29.
Hochberg, B. 1979b *Spin Span Spun: Fact and folklore for Spinners*, Santa Cruz.
Hogarth, D. G. 1908 *The British Museum Excavations at Ephesus, the Archaic Artemisia*. London.
Iakovidis, S. 1969 *Perati, to Nekrotapherion*, Athens.
Iakovidis, S. 1977 On the Use of Mycenaean 'Buttons'. *BSA* 72: 113–119.
Iakovidis, S. 1980 *Excavations of the Necropolis at Perati*, Institute of Archaeology, University of California Los Angeles, Occasional Paper 8.
James, F. 1966 *The Iron Age at Beth Shan, A study of Levels VI–IV*, Philadelphia.
Karageorghis, V 1983 *Palaepaphos-Skales, An Iron Age Cemetery in Cyprus*, Nicosia.
Karageorghis, V. 1985 *Excavations at Kition V, The Pre-Phoenician Levels Part II,* Nicosia.
Karageorghis, V. and Demas, M. 1985 *Excavations at Kition V, The Pre-Phoenician Levels Areas I and II. Part I,* Nicosia.
Karageorghis, V. and Demas, M. 1974 *Excavations at Kition I, The Tombs*, Nicosia.
Karageorghis, V. and Michaelides, D. 1990 *Tombs at Palaeopaphos 1. Teratsoudhia 2. Eliomylia*. Nicosia.
Karo, G. 1930 *Schachtgräber von Mykenai*. München.
Keswani, P. 2004 *Mortuary Ritual and Society in Bronze Age Cyprus*, London.
Kooij van der, G. and Ibrahim, M. M. 1989 *Picking up Threads ... A Continuing review of excavations at Deir Alla, Jordan*, Leiden.
Kunze, E. 1935–1936 Orientalische Schnitzereien aus Kreta. *Athenische Mitteilungen* 60–61: 218–233.
Lamon, R. S. and Shipton, G. M. 1939 *Megiddo I, Seasons of 1925–1934, Strata I–V*. Oriental Institute Publications 42. Chicago.
Lemaire, A. 2006, A Re-examination of the Inscribed Pomegranate: A Rejoinder, *Israel Exploration Journal,* 56:2: 167–177.
Liu, R. 1978 Spindle-Whorls: Part 1. Some Comments and Speculations. *The Bead Journal* 3: 87–103.
Loud, G. 1939 *The Megiddo Ivories*, OIP 52.
Loud, G. 1949 *Megiddo II. Seasons of 1935–1939*, OIP 62.
Mårtensson, L., Andersson, E., Nosch, M.-L. and Batzer, A. 2006b. *Technical Report Experimental Archaeology Part 2:2 Whorl or Bead?*, Tools and Textiles – Texts and Contexts Research Programme, The Danish National Research Foundation's, Centre for Textile Research, University of Copenhagen.
Matoïan, V. and Vita, J.-P. 2009 (2010) Les textiles à Ougarit, Perspectives de la recherche. *Ugarit Forschungen* 41: 467–504.
Mazar, E. 2001 *The Phoenicians in Achziv, the Southern Cemetery. Jerome L. Joss Expedition. Final Report of the Excavations 1988–1990*. Barcelona.
Miron, R. 1990 *Kamid el-Loz. 10. Das 'Schatzhaus' im Palastbereich. Die Funde*, Bonn.

Montell, G. 1941 Spinning Tools and Methods in Asia. In V. Sylwan *Woolen textile of the Lou-Lan People*, Stockholm, 109–125.

Moran, W. L. 1992 *The Amarna Letters*, Baltimore/London.

Nosch, M.-L. 2012 From Texts to Textile in the Aegean Bronze Age. In M.-L. Nosch and Laffineur R. (eds) *Kosmos Jewellery, Adornment and Textiles in the Aegean Bronze Age, Proceeding of the 13th International Aegean Conference, University of Copenhagen, National Research Foundation's Centre for Textile Research 21–26 April 2010*, Leuven/Liege, 43–53.

Peyronel, L. 2007 Spinning and Weaving at Tell Mardikh-Ebla (Syria): Some Observations on Spindle Whorls and Loom-Weights from the Bronze and Iron Ages. In C. Gillis and M.-L. Nosch (eds), *Ancient Textiles. Production, Craft and Society. Proceedings of the First International Conference on Ancient Textiles, Held at Lund, Sweden, and Copenhagen, Denmark, on March 19th–23rd, 2003*, Oxford, 26–35.

Pieridou, A. 1967 Pieces of Cloth from Early and Middle Cypriote Periods. *RDAC* 1967: 25–29.

Pritchard, J. B. 1988 *Sarepta IV. The Objects from Area II, X*, Beyrouth.

Pulak, C. 1992 The Shipwreck at Ulu Burun, Turkey: 1992 Excavation Campaign. *The INA Quarterly* 19.4: 4–11.

Rahmstorf, L. 2008 *Tiryns Forshungen und Berrichte XVI: Kleinfunde aus Tiryns: Terrakotta, Stein, Bein und Glas/Fayence vornehmlich aus der Spätbronzezeit*, Wiesbaden.

Riis, P. J. 1948 *Hama, fouilles et recherches 1931–1938 II.3 Les cimetières à crémation*. Copenhagen.

Riis, P. J. and Buhl, M.-L. 1990 *Hama. Fouilles et recherches de la fondation Carlsberg 1931–1938 II2, Objets de la période dite syro-hittite (âge du Fer)*, Copenhagen.

Ryder, M. L. 1968 The Origin of Spinning. *Textile History* 1: 73–82.

Sauvage, C. and Hawley, R. 2013 Une fusaïole inscrite au MAN. In V. Matoïan and M. Al-Maqdissi (eds), *Études Ougaritiques III*, Leuven, 365–394.

Sauvage, C. 2012 Spinning from old Threads: the Whorls from Ugarit at the Musée d'archéologie Nationale of Saint Germain en Laye and at the Louvre. In H. Koefoed (ed.), *Textile Production and Consumption in the Ancient Near East. Archaeology, Epigraphy, Iconography*, Oxford, 187–212.

Sauvage, C. *forth.* Tranchée 2 V dépôt 43, Minet el-Beida 1932. In C. Sauvage and C. Lorre (eds), *La collection d'Ougarit de C.F.A. Schaeffer au Musée d'Archéologie Nationale de Saint-Germain-en-Laye*.

Schaeffer, C. F. A. 1952 *Enkomi-Alasia I*. Paris.

Schofield, L. 2007. *The Mycenaeans*. Los Angeles.

Smith, J. 2002 Changes in the Workplace: Women and Textile Production on Late Bronze Age Cyprus. In D. Bolger and N. Serwint (eds), *Engendering Aphrodite: Women and Society in Ancient Cyprus.* CAARI Monographs 3. Boston: American Schools of Oriental Research Archaeological Reports, 281–312.

Smith, J. 2009 *Art and Society in Cyprus from the Bronze Age into the Iron Age*. Cambridge.

Starkey, J. L. 1935 Excavations at Tell Kuweir 1934–1935. *PEFQ:* 198–206.

Stewart, J. R. 1962 *The Swedish Cyprus Expedition Volume IV; Part 1A. The Early Cypriote Bronze Age*. 205–401. Lund.

Tatton Brown, V. 2003 Enkomi: the notebook in the British Museum. *Cahier du Centre d'Études Chypriotes 31*: 9–65.

Tufnell, O. 1958 *Lachish (tell ed-Duweir) IV. The Bronze Age*, London–Oxford–Toronto.

Tufnell, O., Inge, C. H. and Harding, L. 1940 *Lachish II (Tell ed Duweir): The Fosse Temple*. London/New York/ Toronto.

Völling, E. Bemerkungen Zu Einem Onyxfund Aus Babylon. *Mitteilungen der Deutschen Orient-Gesellschaft* 130: 197–221.

Ward, C. 2003 Pomegranates in Eastern Mediterranean Contexts during the Late Bronze Age. *World Archaeology* 34.3: 529–541.

Yadin, Y., Aharoni, Y., Amiran, R., Dothan, T., Dunayevsky, I. and Perrot, J. 1960 *Hazor II. An Account of the Second Season of Excavations.* Jerusalem.

Yon, M. 2007 Au Roi d'Alasia, Mon Père. *Cahiers du Centre d'Études Chypriotes* 37:15–39.

Zawadzki, S. 2006 *Garments of the Gods, Studies on the Textile Industry and the Pantheon of Sippar according to the texts from the Ebabbar Archives*, OBO 218, Academic Press Fribourg.

10. Golden Decorations in Assyrian Textiles: An Interdisciplinary Approach

Salvatore Gaspa

The intriguing description which Oppenheim made about the ornaments decorating the garments of the Mesopotamian gods' statues in 1949[1] led the reader to the world of textile decoration and to the care for divine paraphernalia in Babylonian cultic practice. As already observed by Fales and Postgate in their introduction to the edition of a group of administrative documents from the archive of Nineveh (modern Kuyunjik),[2] Oppenheim discussed the Babylonian practice of adorning the garments of divine statues without mentioning attestations of designations for such decorative elements from Neo-Assyrian textual sources. However, he tried to corroborate his assumptions by citing numerous attestations from Neo-Assyrian monumental art. In addition, the discovery of the tombs of queens in Kalḫu (modern Nimrud) and their valuable contents represents another important piece of evidence for the use of decorative elements in the fabrication of luxury garments of the first millennium BC Mesopotamia which cannot be ignored by scholars of ancient textiles. In order to update Oppenheim's considerations, the following remarks will attempt to give a more complete analysis of dress decorations in first millennium BC Mesopotamia. This will be made through an interdisciplinary approach combining textual data with contemporary archaeological and iconographical evidence. Combining words and *realia* is in most cases an insoluble problem for the identification of the items mentioned in ancient texts. Fortunately, the findings in the burials of the Assyrian queens represent a turning point for textile research and their treasury of textile-related data may now help us to ground the study of the first millennium BC metal appliqués on firmer foundations.

Golden Decorations for Textiles in First Millennium BC Mesopotamia: A Discussion on the Neo-Assyrian Evidence in Light of the Textual Sources

The life of the royal courts and the cultic ceremonies in the main temples in first millennium BC Mesopotamia oriented the local textile manufacture towards the production of finely elaborated items of clothing for the members of the ruling *élite* as well as for the gods' statues. In fact, the dressing of the gods, which were represented in the shrines by their statues, was a fundamental part

[1] Oppenheim 1949, 172–193.
[2] Fales and Postgate 1992, xxv.

of the regular service that temple personnel had to perform for the gods: accordingly, divine statues had to be properly washed, fed, dressed, and entertained in order to get the gods' benevolence and gifts. The extremely elaborate decoration of these luxury garments, which required the collaborative work of specialized goldsmiths and tailors, became an important sector in the palace- and temple-oriented textile economies of the Near Eastern states. The adornment of garments with golden appliqués is especially attested in Assyria and in Babylonia. Texts from the archive of the Eanna temple in Uruk (second half of seventh–mid sixth century BC), some of which were not known to Oppenheim when he wrote his paper, inform us that the vestments for the goddesses Ištar, Nanāya, and Bēltu-ša-Rēš were densely covered with hundreds of gold appliqués in the shape of rosettes (*aiaru*), stars (*kakkabu*), *ḫašû*-elements, *tenšû*-elements, and lions (*nēšu*).[3] The Assyrian counterparts of these Babylonian dress-ornaments have been recognized in some decorative elements of gold which are recorded in some of the Neo-Assyrian administrative lists of metal objects found in Nineveh and published in 1992 in a volume of the *State Archives of Assyria* series.[4] Although the fragmentary status of the texts and the concise style of the Assyrian bureaucracy do not give us details about the items which were adorned by the decorative elements, it is clear that some of the attested ornamental elements were used in textile decorations. These metallic elements are indicated by the words *takkussu* and *buṭu*[...]. A third element is only attested in the logographic form SIG.LU.KUR GAR-*nu* and no corresponding Akkadian syllabic writing is known at present. As alternative readings, Fales and Postgate suggest *pik-lu-lat* and SIG UDU KUR, both to be rejected.[5] Another possibility is to read the occurrence as *sik-lu-nat*; a plural form *siklunāt* would fit well to the quantity of the listed items (4 *sik-lu-nat*). The only possible term referring to textiles which comes to my mind is the word *sikulittu*, which is attested in Nuzi texts as a qualification of chairs and beds.[6] Is the form *siklunāt* in someway linked to the word *sikulittu*? The use of this item in connection with chairs and beds seems to be perfectly in line with what we know about the *dappastu*, as we will see in detail below.

What is important to note is that *takkussu* and *buṭu*[...] occur together and this confirms the hypothesis that they complemented each other, thus representing a possible counterpart of the rosettes and the *tenšû*s of the Neo-Babylonian garments. The first designation, *takkussu*, has been interpreted as denoting a tube or pipe,[7] while the interpretation of the second word is problematic, since in all the known attestations the last signs of the term are broken. Is the occurrence *buṭu*[...] to be referred to the word *buṭuttu*, "terebinth nut"? Beads used in jewellery were often named according to their appearance in ancient Mesopotamia.[8] Perhaps, the *buṭ*[*uttu*?] was a type of bead imitating

[3] Beaulieu 2003, 21–25. On the textile production and the management of garments for divine statues in first millennium BC Babylonia, see Zawadzki 2006.

[4] Fales and Postgate 1992, xv.

[5] Fales and Postgate 1992, 86. The word *pikallullu* refers to the vent for an oven (AHw 863a and CAD P 371a), while UDU KUR is the logographic form of *immer šadê*, "mountain sheep". Both the meanings do not seem to fit to the context concerning the description of a textile.

[6] See CAD S 261a. The word is listed as *zikulittu* in AHw 1527b and CDA 447b. No plural form of the term seems to be attested at present. In addition, mention should be made of a term of possible Hurrian origin which occurs in connection to textiles: Alalakh 362:6 ˹6˺ TÚG.*ši-ik-la(-)te-na* (in list of textiles). See CAD Š/II 436a s.v. *šiklu*.

[7] AHw 1307a; CAD T 78b; CDA 395a; AEAD 121a.

[8] See, e.g., the following names of beads, attested in Neo-Babylonian texts: *binītu*, "fish-roe-shaped bead", *erimmatu*, "egg-shaped bead", *nurmû*, "pomegranate-shaped bead", *zēr qiššê*, "melon-seed-shaped bead". See Beaulieu 2003, 13–14. For designations of beads referring to fruits and seeds in Mari texts, see Arkhipov 2012, 47–48 (*kisibirrum*, "coriander-shaped bead"), 49 (*murdinnum*, "bramble-shaped bead"), 52 (*nurmûm*, "pomegranate-shaped bead"), 52 (*papparḫum*,

the shape of the terebinth-nuts. New attestations of the word are needed to confirm or reject this hypothesis. In the texts, *takkussu* and *buṭu*[...] are associated with the textile known as *dappastu*.[9] One text mentions a red woollen *dappastu* with 382 tubes, 432 *buṭu*[...]-elements, and four SIG!.LU.KUR GAR-*nu*,[10] while another one has a *dappastu* with four SIG.LU.KUR! GAR-*nu*?, 136 tub[es ...], and 136 *buṭ*[*u*...]-elements.[11] In both cases, additional quantities of tubes are listed, in someway associated to the same textile product: respectively, 100 tubes for the first *dappastu*[12] and 404 tubes for the second one.[13] It is interesting to note that the number of SIG.LU.KUR-ornaments does not change, while the amounts of tubes and *buṭu*[...]s are variable. In one of the texts, the weight of (all?) the elements adorning a *dappastu* is given: 11 minas 13½ shekels (*c.* 11.33 or 5.66kg).[14] This weight shows that one *dappastu* with all these elements must have been very heavy. It is clear that the production of this type of textile and of all these precious metal appliqués was very expensive and involved the most skilled tailors and goldsmiths of the empire.

To come back to the decorative elements characterizing the *dappastu*, the fact that in both the *dappastu*s occurring in the administrative lists four SIG.LU.KUR GAR-*nu* (or *sik-lu-nat* GAR-*nu*) are mentioned seems to suggest that the elements in question had to do with the four sides or the four angles of the textile. The *dappastu* has been interpreted as a blanket or bedcover[15] and as a rug.[16] Accordingly, a square-shaped textile seems to be the best candidate for the item in question. An exemplar of blanket is provided by the iconographical evidence of the first millennium BC Assyria: it is depicted in the scene of the garden banquet of Assurbanipal and the queen in a relief from Room S' of the North Palace in Nineveh (*c.* 645 BC).[17] This blanket, whose use is associated with the king's couch, is bordered by a decorated band and, presumably, also by four angular tassels (Fig. 10.1).[18]

According to Neo-Assyrian texts, the *dappastu* came in two types, the woollen variety[19] and the linen variety.[20] The woollen variety could be red[21] or black.[22] From a list of grave goods for a king we learn that the "front part" of the *dappastu*, perhaps to be intended as the upper and visible part of it,[23] could be black.[24] The connection of this textile with beds is corroborated by the fact

purslane(?)-shaped bead), 54 (*šarûrum*, "melon(?)-shaped bead"), 56 (*uḫennum*, "fresh date-shaped bead"), 56 (*uṭṭeṭum*, "grain-shaped bead"), 57 (*zēr šakirêm*, "henbane-seed-shaped bead").

[9] This textile product is attested in CTN 2, 1:3'; 152:5; 154 r.3'; Kwasman 2009, 114, K 6323+ ii 1; ND 2307 e.24 (*Iraq* 16 [1954], 37, pl. VI); ND 2311:7 (*Iraq* 23 [1961], 20, pl. X); ND 2691:8 (*Iraq* 23 [1961], 44, pl. XXIII); ND 2758:7' (*Iraq* 23 [1961], 48, pl. XXVI); SAA 7, 64 r. i 7; 66 r. i' 1', 6'; 96:3'; 97:9'; 105:4'; 115 i 11; 117 r.3; 168:5'; SAA 16, 53:9; StAT 3, 1 r.18; TH 52:6; 64:3–4.

[10] SAA 7, 64 r. i' 7'–10'.

[11] SAA 7, 66 r. i' 1'–4'.

[12] SAA 7, 64 r. i' 11'.

[13] SAA 7, 66 r. i' 5'–6'. It is not clear whether the 400 tubes and 400 *buṭu*[...]s which are mentioned in the same list (lines r. i' 7'–8') have to be referred to the same *dappastu*.

[14] SAA 7, 66 r. i' 9'.

[15] AEAD 21a.

[16] CDA 398a. The term is generically intended as a cover or garment in AHw 1320b and CAD D 104b.

[17] Barnett 1976, pl. 65.

[18] Only one tassel is visible in the relief.

[19] StAT 3, 1 r.18–19.

[20] SAA 7, 115 i 11.

[21] ND 2758:6' (*Iraq* 23 [1961], 48, pl. XXVI); SAA 7, 96:3'; StAT 3, 1 r.18; TH 52:6.

[22] ND 2758:5' (*Iraq* 23 [1961], 48, pl. XXVI); StAT 3, 1 r.19.

[23] Kwasman suggests that it could also be referred to the right side of the *dappastu*. See Kwasman 2009, 118.

[24] Kwasman 2009, 114, K 6323+ ii 1–2.

Fig. 10.1: In the "picnic" in the royal garden, the king reclines on a couch with a decorated and tasselled bedcover covering his legs, while the queen is sitting on a chair. Both wear highly decorated garments, possibly enriched by golden appliqués of different shapes and sizes, such as encircled rosettes, discs, and stepped triangles (from Barnett 1976, pl. 63, detail).

that in the same list are mentioned beds among the grave goods for the royal dead.[25] Moreover, it represents a common item in enumerations of bedclothes in dowry lists of marriage contracts from Kalḫu.[26] *Dappastu*s for beds are listed in two triangular textile labels from Nineveh which possibly accompanied stocks of textiles.[27] The *dappastu* constituted one of the bedclothes which were used in Assyrian temples for the beds of the gods. From a text containing a memorandum on temple furnishings we learn that the *dappastu* was one of the bed textiles which were used as covering for the bed of the goddess Šērū'a in her shrine.[28] That this textile was strictly connected to beds in the daily life of the Assyrians is also evident from a private letter dealing with the adoption of

[25] Kwasman 2009, 114–115, K 6323+ ii 19', 32'.
[26] CTN 2, 1:3'; ND 2307 e.24 (*Iraq* 16 [1954], 37, pl. VI).
[27] SAA 7, 97:9'; SAA 7, 105:4'.
[28] SAA 7, 117 r.3–7.

a daughter. The text mentions what seem to be the basic household elements composing a bed: a wooden board (*lē'u*), blankets (*dappastu*), and a bedspread (*qarrāru*).[29] However, it seems that this textile could be used as covering for other pieces of the royal furniture as well; in a document from Kalḫu three talents of cloth of black (wool) and three talents of cloth of red wool for 12 *dappastus* are recorded, two of which were destined as covering of chairs.[30] This means that with six talents of wool cloth (*c.* 363.6 or 181.8kg) an Assyrian weaver could manufacture twelve of these textiles and that the quantity needed for one *dappastu* corresponded to half a talent (*c.* 30.3 or 15.1kg). Also this weight confirms that this type of textile could be very heavy; a possible explanation could be that with this term both blankets (or bedcovers) and large tapestries were designated.

As to the element indicated by the writing GAR-*nu*, it is possible that this form must be read as a *pirs* nominal form of the verb *šakānu*, "to place, set, install", i.e. as *šiknu*. This word is used in a Neo-Assyrian text to designate a textile. It occurs among various grave goods in a text concerning the royal funeral of a king;[31] in two passages of this text, *šiknu*-textiles are associated with mitres, leggings, and sleeves.[32] In connection to textiles, the *šiknu* also occurs in two texts of the second millennium BC. In an Old Assyrian document, two *kusītum*-garments with a *šiknum* are listed.[33] Interestingly, this item could also be associated with bedclothes; in fact, in a document from the city of Mari we are informed about a *ḫalû*-textile with a *šiknum* for a bed.[34] Also the *šiknu* of the first millennium BC, interpreted as designating a padding,[35] appears to have been used for both garments and bedclothes. However, in the case of our *dappastu*, it is difficult to think how a padding could be associated to four decorative metal objects. Perhaps, the most plausible solution is to consider *šiknum* as referring to the setting of the four metal items,[36] in other words, to the appearance or structure of the SIG.LU.KURs; the broken signs following the word in the two known attestations probably concerned the name of the material of this setting.[37]

Summing up, our *dappastu* represented a finely-executed blanket, perhaps destined to cover a bed of a goddess in an Assyrian temple. The exact function of the afore-mentioned metal decorative items escapes us, but it is plausible that the several hundreds of tubes (if this is the correct translation of the word *takkussu*) and of *buṭ*[*u*...]-elements, presumably consisting of very tiny and small pieces of metal, must have served to decorate the four bordering bands of this blanket, perhaps used in alternation or as single components of more elaborated designs. Other uses of these tubes are to be ruled out, in light of the fact that they were of precious metal. In fact, from the point of view of the textile technique, the use of metal tubes for the construction of tassels has been put forward in light of the two cylindrical tassels discovered in Tombs II and III at Nimrud, but the analysis of these tassels revealed that no bronze pin or tube was present inside them.[38] As for the four SIG.

[29] SAA 16, 53:8–10.

[30] ND 2758:5'–8' (*Iraq* 23 [1961], 48, pl. XXVI).

[31] Kwasman 2009, 116, K 6323+ r. i' 5', 18'.

[32] Kwasman 2009, 116, K 6323+ r. i' 4'–7', 15'–18'.

[33] StOr 46, 198:63 (Hecker *et al.* 1998, no. 429). See CAD Š/II 439a and Michel and Veenhof 2010, 242.

[34] RA 64, 33, no. 25:1. See CAD Š/II 439a. For the interpretation of the *ḫalû šiknu* as a "courtepointe", see Durand 2009, 40, 603.

[35] The interpretation of the *šiknu* as a pad or padding has been suggested by Kwasman in connection with the Neo-Assyrian occurrence of the word. See Kwasman 2009, 121, who, however, does not discuss the function of this textile in the light of the Old Assyrian and Mari attestations.

[36] For this meaning of the word, see CAD Š/II 436b–437a s.v. *šiknu* A 1a'.

[37] SAA 7, 64 r. i' 10'; 66 r. i' 2'.

[38] Crowfoot 1995, 114, 117.

LU.KUR GAR-*nu* (or *sik-lu-nat* GAR-*nu*), these items had probably to do with the decoration of the four angular tassels of the *dappastu*. Representations of garments worn by the king and other court members show that the tassels composing the fringed edge were clasped by elements. It is possible that in more elaborate textiles these elements were made of precious metal.[39] This is probably the case of the tassel of the bedcover of Assurbanipal's couch, although the representation of the juncture of the tassel to the bedcover's border in the relief does not seem to have been made with accuracy. If these considerations may be accepted, we may suppose that the *dappastu* was probably characterized by four angular tassels which were closed by small gold clasps.

The Findings in the Nimrud Tombs and Their Significance for the Identification of the First Millennium BC Dress Decorations in Iconography

Textile research on the Assyrian garments may greatly benefit from the combination of textual and iconographic materials in the identification of the items in question. A third type of evidence has been provided by the archaeological research on the burials of the eighth century capital of the Assyrian state, the city of Kalḫu (Nimrud). In particular, the discovery of Tomb II in the domestic quarter of the North-West Palace in Nimrud in 1989 by Iraqi archaeologists[40] revealed that, among various and precious grave goods which accompanied the skeletons of two women, one to be identified as Yabâ (wife of Tiglath-pileser III, 745–727 BC) and the other as Banitu (wife of Shalmaneser V, 726–722 BC) or Ataliya (wife of Sargon II, 721–705 BC),[41] there was a mass of blackened linen fabric which originally covered the bodies or was piled up over them.[42] More importantly, the tomb also contained a large variety of small objects, in part lying among the bones and in part in the folds of one layer of the solidified textile. This material included 700 tiny gold rosettes, star-shaped ornaments, circles, triangles, and banded agate studs with borders of gold granules. It has been assumed that all these tiny and finely made objects had been sewn onto the garments as decorative elements rather than being part of some broken piece of jewellery.[43] This material has been studied by Hussein and Suleiman, who published a catalogue with pictures and a brief description of the items.[44] Since then, no other in-depth studies have been carried out on the Nimrud treasures, if we except a summary panel description edited by Collon in a recent volume.[45] Given the fact that all the materials composing the Nimrud treasures are stored in an inaccessible place in Baghdad and are not available for study, the accurate study that the materials, especially

---

[39] On the use of jewelled tassels in the adornment of Assyrian costumes, see Houston 1954², 156–158 and pl. 8.2.

[40] Hussein and Suleiman 1999–2000, 87–133; Damerji 1999; 2008, 81–82.

[41] A new hypothesis has been put forward by Dalley, according to whom the two bodies contained in the sarcophagus of Tomb II at Nimrud belong to Yabâ and Ataliya. The name Banitu was probably the second name of Tiglath-pileser III's wife. See Dalley 2008, 171.

[42] On the textile remains found in the tombs, see Crowfoot 1995, 113–118. For pictures of them, see Crowfoot 1995, 116 fig. 5 and Hussein and Suleiman 1999–2000, 440 fig. 222. Analyses on the Nimrud textile fragments confirmed that flax had been used to fabricate the garments of the buried queens. See Crowfoot 1995, 117. Most recent analysis on these fragments also revealed the presence of cotton. See Toray 1996, 199.

[43] Crowfoot 1995, 113; Oates and Oates 2001, 83. A short description of the Tomb II dress decorations is given in Collon 2008, 114 with figs 14-q and 14-r. The pictures were reproduced from Hussein and Suleiman 1999–2000, 302 fig. 94, 306 fig. 98, and 307 fig. 99.

[44] Hussein and Suleiman 1999–2000.

[45] Collon 2008, 105–118.

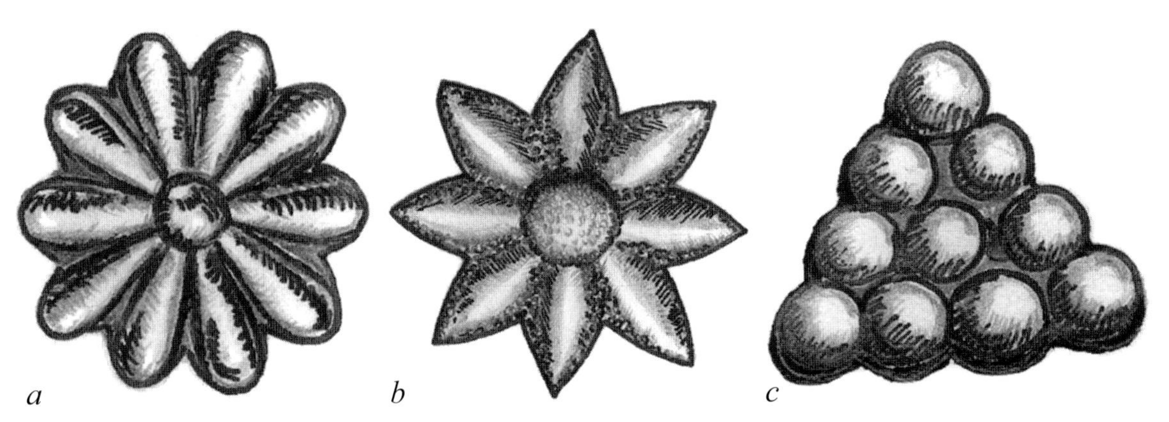

Fig. 10.2: Golden dress decorations from Tomb II in Nimrud: rosettes, stars, and triangles (drawing by the author from the pictures published in Hussein and Suleiman 1999–2000, figs 36, 98, 99).

the dress decorations, deserve is not possible at present.[46] Consequently, my observations will be limited to the available data.

Table 10.1 presents all the Nimrud items from Tombs I, II, and III (abbreviated in the table as respectively T1, T2, and T3) which may be interpreted as dress ornaments.

The Nimrud dress decorative elements belong to different typologies (Fig. 10.2). The most attested items are golden elements shaped as rosettes and stars, in all likelihood to identify with the *aiaru*s and the *kakkabtu*s which frequently occur in Neo-Assyrian administrative records. Star-shaped ornaments for textiles are well attested in other periods of the Mesopotamian history. Second millennium BC attestations of these items can be found in texts from the royal archives of Mari, from which we learn that *kakkabum*-ornaments were used to decorate both clothes and footwear.[47] Of a type of rosette among the decorative materials of the Tomb II were found 770 examples. To judge from the picture published by Hussein and Suleiman in their catalogue, this type is characterized by a ten-petalled structure[48] (Fig. 10.2a).

Other interesting golden dress decorations are represented by discs, wheels, hanging balls, domed studs, decorated and plain doughnuts. All of these were probably stitched on the garments of the two women. Some of these elements are pierced for pinning them to clothes, such as the rosettes (Nos. 8, 12, and 16) and the triangles[49] (No. 23), while others are provided with a suspension ring for fastening on garments, such as the buttons with globular protuberance in the middle (No. 14), the domed studs (No. 21), the eight-pointed stars (No. 22), and the doughnuts (Nos. 24 and 27). A number of pierced strips of gold sheet were found in Tombs II and III (Nos. 20 and 25). They are different in length, size, decoration, and number of holes. Unfortunately, no useful information

[46] In a personal communication (March 2012), Prof. John Curtis (British Museum) kindly informed me that a new publication of the Treasures will be carried out by Dr. Muzahim Mahmud and Prof. McGuire Gibson (University of Chicago). However, as Gibson explained to me, no detailed studies on the material are possible at present. I thank both Curtis and Gibson for this piece of information.
[47] Arkhipov 2012, 46.
[48] However, in Hussein and Suleiman 1999–2000, 241 fig. 36, it is said that the leaves of the rosettes are varying in number.
[49] It is possible that these "triangles" were actually a stylized representation of clusters of grapes. Note that *isḫunnatum*-shaped items are attested as ornaments in Mesopotamia. For these precious elements in Mari texts, see Arkhipov 2012, 79.

Table 10.1: the dress ornaments.

No.	Find (description)	Museum Number[1]	Details (weight, length)[2]	Literature
1	One fibula in the shape of a woman and a lion with an interwoven double chain and a carnelian seal (T1)	IM 108970, 108980, 108982	?	Hussein and Suleiman 1999–2000, fig. 12[3]
2	28 pieces consisting of nine grain-shaped beads sticking together longitudinally (T1)	IM 108969	109.95g (gross)	Hussein and Suleiman 1999–2000, fig. 16
3	Two items consisting of a central double palmette and chains ending with 16 smaller palmettes[4] (T2)	IM 105809–105810	24.2–23.83g	Hussein and Suleiman 1999–2000, fig. 27
4	Seven[5] agate eye stones set in disc-shaped gold frames with pomegranate-like protrusions (T2)	IM 105987	?	Hussein and Suleiman 1999–2000, fig. 29
5	11 small six- and globular-petalled rosettes with a hole in the middle (T2)	?[6]	?	Hussein and Suleiman 1999–2000, fig. 33
6	Nine buttons with convex surface (T2)	?	?	Hussein and Suleiman 1999–2000, fig. 33
7	Four wheel-shaped buttons (T2)	?	?	Hussein and Suleiman 1999–2000, fig. 33
8	Nine buttons in the shape of eight-petalled rosettes with a convex disc in the middle (T3)	IM 118084	23g	Hussein and Suleiman 1999–2000, fig. 33
9	Six spherical items connected to a short pipe (T3)	IM 118081	23.28g	Hussein and Suleiman 1999–2000, fig. 33
10	Four buttons with convex surfaces and decorated structure (T2)	?	?	Hussein and Suleiman 1999–2000, fig. 33
11	Four small wheel-shaped buttons with decorated structure (T2)	?	?	Hussein and Suleiman 1999–2000, fig. 33
12	Six buttons in the shape of eight-petalled rosettes with a concave circle in the middle, perhaps for holding precious stones (T3)	IM 118085	26.80g	Hussein and Suleiman 1999–2000, fig. 33
13	28 small disc-shaped buttons with convex surfaces (T3)	IM 118083	32.8g	Hussein and Suleiman 1999–2000, fig. 33
14	Ten buttons with globular protuberance in the middle (T2)	?	?	Hussein and Suleiman 1999–2000, fig. 33
15	Six six- and globular petalled rosettes with a hole in the middle (T2)	?	?	Hussein and Suleiman 1999–2000, fig. 33
16	Two eight-petalled rosettes with small pendants formed by chains and elements of various shapes (T3)	IM 118086–118087	3.8g–9.6g 3.2cm–4cm	Hussein and Suleiman 1999–2000, fig. 33
17	770 pieces in the shape of rosettes (T2)	IM 105983	½g–1g each	Hussein and Suleiman 1999–2000, fig. 36

18	Two fibulae in the shape of a woman and of a lion's head with interwoven wires and coloured stones hanging from them (T2)	IM 105959–105960	32.75g–39.75g	Hussein and Suleiman 1999–2000, fig. 78
19	Three fibulae ending with a hand-like element (T2)	IM 105892–105894	52g	Hussein and Suleiman 1999–2000, 92
20	Unspecified number of strips of different length and size, some of which decorated with recurring motifs (T2)	IM 105808	52.95g	Hussein and Suleiman 1999–2000, 93
21	1.160 domed studs (T2)	IM 105985	271g (gross)	Hussein and Suleiman 1999–2000, fig. 94
22	50 small eight-pointed stars with round centres (T2)	IM 105984	130g (gross)	Hussein and Suleiman 1999–2000, fig. 98
23	147 items in the shape of equilateral triangles whose external surface is decorated with globular elements (T2)	IM 105986	73.5g (gross)	Hussein and Suleiman 1999–2000, fig. 99
24	Two decorated doughnuts (T2)	IM 105872–105873	11.5g–12g	Hussein and Suleiman 1999–2000, fig. 102
25	Two pieces of a long strip decorated with recurrent intertwined leaves and branches forming rosettes (T3)	IM 115597	53.32g	Hussein and Suleiman 1999–2000, fig. 124
26	Two strap bands of golden wires ending with interwoven chains and decorated conical elements (T3)	IM 115506	295g 37cm (length of the main band) 4.9cm (length of the short band)	Hussein and Suleiman 1999–2000, fig. 131
27	Eight plain doughnuts[7] (T2)	IM 105864–105871	?[8]	Hussein and Suleiman 1999–2000, fig. 137

Notes

[1] Iraq Museum of Baghdad (IM).

[2] It is a pity that no systematic evaluation on the size of these dress decorations have been made by Hussein and Suleiman. The details contained in their study only refer to the weight of some the objects and, in very few cases, also to the length of them.

[3] See also Damerji 1999, fig. 14a–b. A carnelian seal, set in a golden frame with a suspension ring, presumably for hanging to a chain with a fibula, is illustrated in Hussein and Suleiman 1999–2000, 336 fig. 127.

[4] Note that the description in Hussein and Suleiman 1999–2000, 228 fig. 27 is erroneous. The decorative elements are described as "large rosettes".

[5] Not "two", as erroneously stated in Hussein and Suleiman 1999–2000, 231 fig. 29.

[6] Items nos. 5, 6, 7, 10, 11, 14, 15 of the table are neither described in detail nor specified as regards their museum numbers in Hussein and Suleiman 1999–2000, 236 fig. 33. The authors give the Museum nos. IM 105907, 105987, 105988, 110637, 118082, and 124997 without specifying the items to which they refer. In addition, note that the objects nos. 110637 and 124997 have been omitted in the lists of the finds in Hussein and Suleiman 1999–2000, 98–100, 104–111, 118–128.

[7] Note that in Hussein and Suleiman 1999–2000, 105 these items are erroneously described as "cylindrical pieces", while in Hussein and Suleiman 1999–2000, 346 as resembling "a round piece of cake".

[8] No specific details about the weight of the plain doughnuts are given in Hussein and Suleiman 1999–2000, 346 fig. 137, which only mention the total weight of the entire group of items, i.e. 73.90g.

is given in the catalogue of the Iraqi scholars about these important details.[50] From the published pictures, one may observe that the distance between the holes is quite regular in many of these pieces, even if in some cases more holes have been made in the same point of the strip (No. 20), presumably due to the necessity to adequately position the ornament to the area of the garment on which it was stitched. In the case of the long strip broken into two pieces (No. 25), the holes are very close each other and extend along the edge of all the four sides of the strip. The recurrent decorative designs of these strips are peculiar to Assyrian art; the one adorning the strips of Tomb II, constituted by a motif of two intertwining bands which form concentric circles, is also attested as a design on painted bricks and wall ornaments in the North-West Palace in Nimrud.[51] Disc-shaped buttons come in different typologies; among them, the ones with convex surface are the most numerous (Nos. 6, 10, and 13). The use of golden buttons as decorative elements of garments is attested at other sites of the Ancient Near East; those found in the royal necropolis of Ebla (*c.* 1750–1700 BC), for instance, show a motif constituted by concentric circles and four holes in the middle for pinning to the clothes.[52] Indeed, it is not clear to me whether some Nimrud pieces were actually fastened to clothes or used as jewellery. This is the case for the 28 pieces consisting of nine interlinked grain-shaped beads (No. 2); although Hussein and Suleiman qualify those as possible dress ornaments,[53] it is also possible that they were used as necklaces. Analogous pieces were found in the queens' tombs: one is composed of 58 elements formed by nine beads,[54] the other of 46 elements with the same number of beads.[55] Some of the Nimrud ornamental items are very elaborate, such as the fibulae (Nos. 1, 18) and the strap bands of golden wires and chains with pendants (No. 26). The last item is composed of two main bands (horizontal bands for the shoulder area) connected at their ends and two shorter bands (vertical bands for the neck area) attached at the middle of them; it was presumably used to decorate the neck and the shoulders of a robe of one of the buried queens.[56] The presence of seven agate eye stones set in gold frames (No. 4) among the dress decorations of Tomb II confirms that the adornment of luxury garments also made use of precious stones. This aspect is also documented in contemporary textual sources. An administrative document listing various textiles mentions two felted shawls or capes (*muklālu*) with the front part red and stones whose nature and number is not indicated.[57] Another text records a cloak (*kuzippu*) studded with (precious) stones.[58] The same qualification occurs in another text for a textile whose name, however, cannot be read on the tablet.[59]

The richness of this material witnesses the fine work of the Assyrian craftsmen as well as the aesthetics of the women belonging to the royal family in the eighth century BC. Given the huge number of the above-described golden elements, it is clear that the items in question had served to adorn various types of garments worn by the buried queens. Unfortunately, any possible reconstruction of the type of clothes and, especially, the specific place where each gold ornament

[50] Hussein and Suleiman 1999–2000, 301, 333.
[51] Layard 1853, I, figs 84, 86.
[52] Matthiae *et al.* 1995, 483 nos. 403, 404.
[53] Hussein and Suleiman 1999–2000, 218 fig. 16.
[54] Hussein and Suleiman 1999–2000, 290 fig. 83.
[55] Hussein and Suleiman 1999–2000, 291 fig. 84.
[56] Hussein and Suleiman 1999–2000, 340 fig. 131; Collon 2008, 114.
[57] SAA 7, 96:7'.
[58] SAA 7, 97:13'.
[59] SAA 7, 105:10'.

was pinned to the garments may only be based on a comparison with the extant iconographical evidence about Assyrian luxury garments. What is clear is that the management of all the precious materials which were supplied to the craftsmen working for the temple and the palace represented an important part of the activity of the state administrators, who compiled very detailed lists of precious objects with their weight. In a passage of a document issued by the state administration, unidentified items of gold to be used in association with clothes, presumably for wearing the gods' statues, are recorded with their weight.[60] It is not always clear whether the rosettes and the star-shaped elements recorded in these documents from Nineveh[61] refer to actual decorations for textiles or to ornamental items for other objects (parts of statues, temple furnishings, jewels, etc.). And the same can be said as regards the *takkusātu*, "tubes", which could also be used as parts of more elaborate pieces of jewellery.[62] We know, for example, that star-shaped ornaments could be used to adorn the base of quivers, bows, and bowcases.[63] A list enumerating items from Babylonian temples which were returned from Elam mentions rosettes and star-shaped ornaments, probably used as decorations for divine statues or for the garments which covered them. Among these items, there are rosettes of gold alloy (*aiarī sādāni*) associated to the Lady of Akkad[64] and 2/3 mina of gold for making four pure star-shaped ornaments (*kakkabāte ebbāte*) for the shoulder of the same goddess.[65] As dress decorations, *kakkabtu*-elements were placed on headbands (*kubšu*),[66] presumably used for gods' statues in temples. This item of clothing was worn by gods' statues and high officials of the king, and, more importantly, it constituted an important element of royal insignia.[67] Headgear worn by the Assyrian kings of the Neo-Assyrian period have the forms of a taller fez with conical top, diadem or upturned brim in front, and ribbons attached at the back of it. One or more horizontal bands decorating the royal fez are often characterized by rows of rosettes,[68] as documented in scenes illustrated on various Assyrian monuments, reliefs, and artefacts of this period. These rosette-shaped elements were probably not golden items attached to the fez, but woven fabric decorations of bright colour stitched on it. A white fez worn by Shalmaneser III depicted on glazed bricks shows a green six-petalled rosette on its front,[69] while the headgear worn by Sargon II could be white with three red bands adorned with white rosettes or red with white bands decorated with yellow rosettes.[70] One wonders whether other elements of cloth, metal or stone were used to decorate first millennium BC headgear; Mari texts, for example, show that turbans could be adorned with stone items in the shape of (heads of) pigs[71] or ducks.[72]

[60] SAA 7, 63 ii' 9–12.

[61] Rosettes are mentioned in SAA 7, 60 i 5. Stars occur in SAA 7, 60 ii 11, r. ii 6'; 63 ii' 1, 6; 64 i 2, 15; 67 i 4'; 68 ii' 3'; 74:2, 4; 89:10.

[62] Tubes are attested in SAA 7, 64 r. i' 8', 11'; 65 i 5'; 66 r. i' 3', 5', 7', 18', ii' 3'; 68 r. ii 1'; 72:1, 15'.

[63] SAA 7, 63 ii' 1–3, 6–8; 64 i 2; 89:10.

[64] SAA 7, 60 i 5–6.

[65] SAA 7, 60 ii 11–12.

[66] SAA 7, 74:3–4.

[67] See CAD K 485b–486a for references. For Neo-Assyrian attestations, see CTN 2, 155 r. v 14'; K 6323+ r. i' 4', 15' (Kwasman 2009, 116); PVA 271; SAA 3, 49 r.5'; SAA 7, 74:4; 96:8'; 105:11'; 120 ii' 16; SAA 10, 96 r.10, 16, 21; 184 r.6; SAA 11, 28:12.

[68] Reade 2009, 254, 256.

[69] Reade 2009, 250.

[70] Reade 2009, 256.

[71] Arkhipov 2012, 54 (*šaḫûm*).

[72] Arkhipov 2012, 56 (*ûsum*).

Table 10.2: the golden cylindrical pipes.

No.	Find	Museum Number	Details	Literature
1	57 cylindrical pipes (T2)	IM 105897	207.3g (gross)	Hussein and Suleiman 1999–2000, fig. 51
2	111 short cylindrical pipes (T2)	IM 105954	125.80g (gross)	Hussein and Suleiman 1999–2000, fig. 52
3	59 cylindrical pipes (T2)	IM 105955	128g (gross)	Hussein and Suleiman 1999–2000, fig. 71
4	Unspecified number of cylindrical pipes (T2)	IM 105971–105976	370.02g (gross)	Hussein and Suleiman 1999–2000, fig. 86

Among the precious items discovered in Nimrud a particular category of objects deserves to be considered. A large number of small pipes or tubes of gold was found among the jewellery and the dress decorations of Tomb II. In Table 10.2 the four groups of pipes are shown.

Hussein and Suleiman interpret the items Nos. 1 and 2 as "clothes hangers",[73] while no explanation of the function is given as regards the other two groups of pipes.[74] This interpretation does not seem to be convincing, since I do not understand how a garment could be hung by a chain formed by small and finely executed golden tubes. The alternative solution is that these pipes or tubes were used as elements of a necklace or as decorative elements for textiles. The first option seems to be confirmed by a comparison with a group of 84 golden tubes of 2.4cm each being different from the above-discussed ones; in this case, the tubes, whose total weight (gross weight) corresponds to 154.5g, have endings characterized by tiny granules.[75] All these elements were probably part of a series of necklaces which adorned the queen's neck. Golden cylindrical beads for necklaces have been discovered in other Near Eastern burial contexts. The ones found in Ebla, in the tomb of the "Signore dei Capridi", for instance, were of a golden typology of 1.6cm each.[76] However, the second possibility, i.e. that these tubes were used for adorning a textile, cannot be ruled out at all. As observed above, tubes occur in administrative records in the following quantities: 100,[77] 136,[78] 382,[79] and 404.[80] Only some of them are mentioned in association with textiles, as seen in the case of the textile called *dappastu*. If the comparison of the *takkussu*-elements occurring in the textual sources with the Nimrud tubes may be accepted, we may tentatively suggest that at least a part of the Nimrud golden tubes were used to decorate the garments which covered the queens' bodies.

The high number of some of the Nimrud dress decorations is astonishing. It reminds us of the hundreds of rosettes and *tenšûs* of the Neo-Babylonian textiles, as well as the hundreds of *takkussus* and *buṭu*[…]s which served to adorn the Neo-Assyrian *dappastus*. The quantities of certain objects, such as the 770 rosettes and the 1,160 domed studs, suggest that they were far from being isolated decorative elements. On the contrary, these items were diffusely stitched on the

[73] Hussein and Suleiman 1999–2000, 257–258 figs 51 and 52.
[74] Hussein and Suleiman 1999–2000, 278 fig. 71, 293 fig. 86.
[75] Hussein and Suleiman 1999–2000, 277 fig. 70.
[76] Matthiae *et al.* 1995, 471 no. 395.
[77] SAA 7, 64 r. i' 11'.
[78] SAA 7, 66 r. i' 3'.
[79] SAA 7, 64 r. i' 8'.
[80] SAA 7, 66 r. i' 5'.

whole surface of the garments, thus probably giving the queen's dress the appearance of a complete gold-made garment. This reminds us of what it is said in some letters of the royal correspondence from Mari concerning the fabrication of luxury clothes with appliqués.[81] In a letter dealing with instructions for the production of a cloth with appliqués, the sender (the king) asks his official that the decorated garment looks like a metal sheet.[82] In addition, the same letter informs us that the excessive weight of the appliqués could tear the garment in question.[83] These aspects may help us to a better understanding of the decorated luxury garments which are represented in the Assyrian palace reliefs. Garments worn by the Assyrian king show very elaborate patterns. In a relief slab from the Royal Palace of Dūr-Šarrukēn (modern Khorsabad), for example, King Sargon II wears a fringed shawl decorated with the motif of the double rosette within two concentric circles and an undergarment consisting of a long tunic having a square grid structure formed by squares containing small rosettes.[84] It is possible that, at least in the case of the undergarment, the rosettes were metal appliqués.[85] In light of the materials found in Tomb II, we may suggest that these decorative rosette-shaped elements were attached to the fabric-woven squares of the king's tunic. To do this, the palace tailors had probably at their disposal hundreds of these golden rosettes. An approximate estimate of the rosettes needed to adorn this type of royal tunic may be obtained by considering that the depicted row of squares containing rosettes in the lower part of the garment which is not covered by the shawl comprises thirteen of these elements. But this number refers to one side of the garment. This means that an entire row of decorations could comprise around twenty-five squares. Consequently, the whole surface of the royal tunic could comprise more than eight hundred of these decorative elements, a number not so far from that of the golden rosettes found in Tomb II and which reminds us of the several hundreds of metal tubes and *buṭu*[...]s mentioned in the above-discussed textual sources. Analogous observations may be made about the garments worn by Assurbanipal in the hunting scenes carved in the wall panels of the North Palace in Nineveh.[86] In the scene representing the king while hunting on horseback, the knee-length garment is completely covered by circled star-shaped ornaments, while the chest area is characterized by a rectangular panel bordered by bands with rows of rosettes, concentric circles, and other elements. Interestingly, the star-shaped decoration shows the same eight-pointed structure of the golden dress decorations from Nimrud. In all likelihood, all or part of the elements decorating Assurbanipal's garment were metal appliqués: the candidates seem to be the rosettes, the disc-shaped buttons, and the star-shaped ornaments of the typology documented in Nimrud.

Another example of possible link between the iconographical evidence and the dress decorations of Tomb II may be found in the case of the bronze friezes of the standards coming from the temple entrances of Sargon's palace at Khorsabad. The king is depicted in one of the friezes as wearing a

[81] Durand 1997, 271–278, nos. 133–139. The term used in these texts to indicate the appliqués is *taddêtum*. It seems that these ornaments also included embroideries of gold thread. See also 271. If really gold threads were used in Mesopotamia, this information may complete the analysis about the use of gold thread in antiquity given in Gleba 2008, 61–77.

[82] Durand 1997, 274, no. 136: "Il faut que cet habit, comme s'il était un habit de Tuttub, soit tissé et noué de façon soignée de chaîne et de trame et que son intérieur soit vraiment comme une feuille d'argent".

[83] Durand 1997, 274, no. 136: "Cet habit se verra mettre des orlets à la yamhadéenne et, comme une étoffe-*ḫuššûm*, du *ṣirpum* lui sera appliqué. Il ne faudrait pas que, lorsqu'on installera ensemble chaîne et trame, les ornaments ne soient (trop) lourds au moment où on les enfilera et que l'habit ne se déchire".

[84] Botta and Flandin 1849–50, pl. 12.

[85] Guralnick 2004, 226. The author also suggests that the rosettes could have been woven or embroidered, or that they could have consisted of fabric appliqués.

[86] Barnett 1976, pls. 5, 8, 10, 11, 12, 46, 47, 49, 50, 51, 52, 56, 59.

Fig. 10.3: Libbāli-šarrat's garment with rosette-shaped decorations in the Assur stele (from Andrae 1913, 7).

garment decorated by a vertical row of rosettes and a horizontal row of rosettes associated with a row of hanging triangles.[87] Interestingly, decorative dress elements in form of triangles were found among the precious objects of the queens' tomb. Analogous observations may be made about the motif of the circle or of the concentric circle, which appear on royal garments represented in various reliefs, such as, for instance, in the scene where Sennacherib is depicted as enthroned after the victory at Lachish; in this case, the garment worn by the king has concentric circles, some of which contain a central dot.[88] This fabric-woven decorative pattern could have been enriched by the addition of golden circles, discs, or wheels not so different from those which adorned the costumes of the two women of Tomb II. The garment worn by the Assyrian crown prince in the reign of Sennacherib, for example, shows a finely executed decoration on the bands which border the shawl as well as the sleeves, the shoulders, and the neck of the royal tunic. These bands are characterized by rows of rosettes or concentric circles.[89] In this case too, the small size of the rosettes and the circles suggests that these elements were metal appliqués, presumably of one of the types discovered in Nimrud.

If we now come to the description of the Assyrian queen's robe, we may see that some of the decorative patterns represented in mid-seventh century BC monumental art may be compared with the materials of the queens' tombs in Nimrud. In the well-known "banquet scene" of a wall panel from the North Palace in Nineveh,[90] Assurbanipal and his wife, Libbāli-šarrat, are depicted in a relaxed and feasting atmosphere in the royal garden, while enjoying the pleasures of wine and of some snacks served by female attendants. The queen is represented as enthroned and wearing a mural crown. Her fringed robe is constituted by an overcoat and a tunic showing the same decorative patterns, that is, an overall decoration of circles distributed throughout the garment with borders and sleeves enriched by outlined bands with rows of smaller circles, dots, and stepped triangles (Fig. 10.1). On a fragmentary stele from Assur (modern Qal'at Šerqāṭ) bearing a representation of the queen on the throne and an inscription which identify the woman as Libbāli-šarrat,[91] the queen's fringed overcoat has an overall decoration of rosettes and an outlined band with a row of smaller seven-petalled rosettes (Fig. 10.3).

There is no reason to think that the practice of decorating with metal items the luxury garments of the members of the king's family, as clearly documented in the eighth century queens' tombs

[87] Guralnick 2004, 226.
[88] Guralnick 2004, 228.
[89] Reproduced in Parpola and Watanabe 1988, 19.
[90] Barnett 1976, pl. 65.
[91] Andrae 1913, 6–8.

of Nimrud, stopped in Late Assyrian times. On the contrary, if we look at the representations of this period we may observe textiles with highly decorative patterning. Among the different materials found in the vaulted chambers of the queens' tomb there are some possible candidates for the dress decorations of Libbāli-šarrat's garments which are depicted in the above-discussed pictorial evidence. These are represented by the golden discs or the domed studs, the rosettes or the star-shaped items, and the triangle-shaped ornaments. All these items were probably attached to decorative bands which were previously woven as separate parts. Once prepared, these bands were then woven to the borders and to the sleeves of the garments. Additional elements were also stitched to the queen's robe, such as the decorated golden strips which were found in two Nimrud tombs. To judge from the decoration of Libbāli-šarrat's robe in the depicted scene, it is possible that a number of strips were stitched on the outlined bands adorning the neck, the sleeves, and the edge of both the overcoat and the tunic. The number of strips needed to decorate these parts varied according to the length of the single areas of the garment. Interestingly, ethnographic evidence from present-day manufacture of garments in Iraq attests to the continuity in the use of metal appliqués as dress decorations; in fact, gold rosettes of a type very similar to the eighth century BC Nimrud exemplars are still being stitched on garments in Mosul.[92]

Concerning the other golden elements adorning the queen's dress as depicted in the relief, we suppose that the decoration of the bands bordering the overcoat and the tunic was enriched by attaching small discs or domed studs of gold, while the triangle-decorations could have consisted in a variant of the golden triangle-shaped appliqués used by the Nimrud queens in the eighth century BC. The stepped structure of the triangles of Libbāli-šarrat's robe could have been inspired by the analogous structure of the Mesopotamian temple towers; this motif could have been chosen by the palace tailors in charge of the making of the queen's wardrobe for the special significance of the ziggurat as a symbol of Ištar, a goddess whose cult was strongly promoted by the Late Assyrian kings.[93] A second possibility is that also the overall circle-based decoration of the garment of Assurbanipal's wife could have been made by golden appliqués. Numerous discs and domed studs in origin decorated the robes of the Nimrud queens. Their large number, especially that of the domed studs, suggests that this second hypothesis cannot be ruled out at all. In all likelihood, the total number of the domed studs comprise items which adorned the garments of both the two buried queens of Tomb II; if so, a single garment could have been decorated in profusion with hundreds of these golden items, thus giving to the linen robe worn by the Assyrian queens the same brilliant appearance of the goddess' clothes.[94] Regarding the second example, in this case both the overall decoration and that of the band consist of rosettes, although of different size. Bracteates in the shape of rosettes of different size were found in the queens' tombs; presumably, they were applied to different areas of Libbāli-šarrat's robe. It is interesting to note that the seven-petalled rosette depicted on the band of the queen's overcoat resembles analogous golden elements of Tomb II at Nimrud, the unique difference being the number of petals, which in the Nimrud examples correspond to six, eight, as well as ten.

In light of the material discussed, we may assume that our queens, Yabâ and Banitu (or Ataliya), were accompanied in their last rest by tasselled overgarments and tunics decorated in profusion

[92] Damerji 2008, 82.

[93] See SAA 3, 7:9.

[94] In an Assurbanipal's hymn, Ištar of Nineveh is described as clothed with brilliance, with a crown gleaming like the stars, and with luminescent discs (*šanšānāti*) on her breasts shining like the sun. See SAA 3, 7:6–8.

Fig. 10.4: Stepped decoration on the Nimrud tassel and on female garments depicted in a palace wall panel from Nineveh (from Crowfoot 1995, 115 [tassel] and Layard 1853, II, 27 [relief]).

by a variety of golden appliqués, fibulae, and precious stones. Perhaps, they also wore a shawl decorated with stepped motifs both in the tassels and in the overall surface of the robe, as seems suggested by the tassels found in the burials (Fig. 10.4).

This study has shown the potential of combining sources of different nature to the end of reconstructing the peculiarities of the ancient garments. It is hoped that future research on the Nimrud treasures will take into due consideration the mine of information that the Assyrian queens have generously left to scholars of ancient textiles.

Acknowledgements

I wish to thank Marie-Louise Nosch, Mary Harlow, and Cécile Michel for reading the paper and for many comments and suggestions from which this contribution greatly benefitted.

Abbreviations

Abbreviations not included in this list follow those given in CAD.

AEAD Parpola, S. *et al.* 2007 *Assyrian-English-Assyrian Dictionary*. Helsinki.
AHw von Soden, W. 1958–81 *Akkadisches Handwörterbuch*, I–III. Wiesbaden.
CAD *The Assyrian Dictionary of the Oriental Institute of the University of Chicago*, Chicago 1956–2010.
CDA Black, J. *et al.* 2000² *A Concise Dictionary of Akkadian*, Santag, Arbeiten und Untersuchungen zur Keilschriftkunde 5, Wiesbaden.
CTN 2 Postgate, J. N. 1973 *The Governor's Palace Archive*, Cuneiform Texts from Nimrud 2, London.
PVA Landsberger, B. and Gurney, O. R. 1957/58 The Practical Vocabulary of Assur, *Archiv für Orientforschung* 18, 328–341.
SAA State Archives of Assyria, vols. 1–19, Helsinki 1987–.
StAT 3 Faist, B. 2007 *Alltagstexte aus neuassyrischen Archiven und Bibliotheken der Stadt Assur*, Studien zu den Assur-Texten 3, Wiesbaden.
TH Friedrich, J. *et al.* 1940 (1967 reprint) *Die Inschriften vom Tell Halaf. Keilschrifttexte und aramäische Urkunden aus einer assyrischen Provinzhauptstadt*, Archiv für Orientforschung, Beiheft 6, Berlin.

Bibliography

Andrae, W. 1913 *Die Stelenreihen in Assur*, Wissenschaftliche Veröffentlichungen der Deutschen Orient-Gesellschaft 24, Leipzig.

Arkhipov, I. 2012 *Le vocabulaire de la métallurgie et la nomenclature des objects en métal dans les textes de Mari*, Archives Royales de Mari 32/Matériaux pour le Dictionnaire de Babylonien de Paris, Tome III, Paris.

Barnett, R. D. 1976 *Sculptures from the North Palace of Ashurbanipal at Nineveh (668–627 B.C.)*, London.

Beaulieu, P.-A. 2003 *The Pantheon of Uruk During the Neo-Babylonian Period*, Cuneiform Monographs 23, Leiden/Boston.

Botta, P. E. and Flandin, E. 1849–50 *Monuments de Ninive, découvert et décrit par M. P. E. Botta, mesuré et dessiné par M. E. Flandin*, I–V, Paris.

Collon, D. 2008 Nimrud Treasures: Panel Discussion. In J. E. Curtis, H. McCall, D. Collon and L. al-Gailani Werr (eds), *New Light on Nimrud. Proceedings of the Nimrud Conference, 11th–13th March 2002*, London, 105–118.

Crowfoot, E. 1995 Textiles from Recent Excavations at Nimrud, *Iraq* 57, 113–118.

Dalley, S. 2008 The Identity of the Princesses in Tomb II and a New Analysis of Events in 701 BC. In J. E. Curtis, H. McCall, D. Collon and L. al-Gailani Werr (eds), *New Light on Nimrud. Proceedings of the Nimrud Conference, 11th–13th March 2002*, London, 171–175.

Damerji, M. S. B. 1999 *Gräber assyrischer Königinnen aus Nimrud*, Mainz.

Damerji, M. S. B. 2008 An Introduction to the Nimrud Tombs. In J. E. Curtis, H. McCall, D. Collon and L. al-Gailani Werr (eds), *New Light on Nimrud. Proceedings of the Nimrud Conference, 11th–13th March 2002*, London, 81–82.

Durand, J.-M. 1997 *Les documents épistolaires du palais de Mari, I*, Littératures anciennes du Proche-Orient 16, Paris.

Durand, J.-M. 2009 *La nomenclature des habits et des textiles dans les textes de Mari*, Archives Royales de Mari 30/Matériaux pour le Dictionnaire de Babylonien de Paris, Tome I, Paris.

Fales, F. M. and Postgate, J. N. 1992 *Imperial Administrative Records, Part I: Palace and Temple Administration*, State Archives of Assyria 7, Helsinki.

Gleba, M. 2008 *Auratae Vestes*: Gold Textiles in the Ancient Mediterranean. In C. Alfaro and L. Karali (eds), *Vestidos, textiles y tintes. Estudios sobre la producción de bienes de consumo en la Antigüedad. Actas del II Symposium Internacional sobre Textiles y tintes del Mediterráneo en el mundo antiguo (Atenas, 24 al 26 de noviembre, 2005)*, Purpureae vestes 2: Textiles and Dyes in Antiquity, Valencia, 61–77.

Guralnick, E. 2004 Neo-Assyrian Patterned Fabrics, *Iraq* 66, 221–232.

Hecker, K. *et al.* 1998 *Kappadokische Keilschrifttafeln aus den Sammlungen der Karlsuniversität Prag*, Praha.

Houston, M. G. 1954[2] *Ancient Egyptian, Mesopotamian and Persian Costume and Decoration*, A Technical History of Costume 1, London.

Hussein, M. M. and Suleiman A. 1999–2000, *Nimrud, A City of Golden Treasures*, Baghdad.

Kwasman, Th. 2009 A Neo-Assyrian Royal Funerary Text. In M. Luukko, S. Svärd and R. Mattila (eds), *Of God(s), Trees, Kings, and Scholars. Neo-Assyrian and Related Studies in Honour of Simo Parpola*, Studia Orientalia 106, Helsinki, 111–125.

Layard, A. H. 1853 *Monuments of Nineveh*, I–II, London.

Matthiae, P., Pinnock, F. and Scandone Matthiae, G. 1995 *Ebla. Alle origini della civiltà urbana. Trent'anni di scavi in Siria dell'Università di Roma "La Sapienza"*, Milano.

Michel, C. and Veenhof K. R. 2010 The Textiles Traded by the Assyrians in Anatolia (19th–18th centuries BC). In C. Michel and M.-L. Nosch (eds), *Textile Terminologies in the Ancient Near East and Mediterranean from the Third to the First Millennia BC*, Ancient Textiles Series 8, Oxford, 210–271.

Oates J. and Oates, D. 2001 *Nimrud. An Assyrian Imperial City Revealed*, London.

Oppenheim, A. L. 1949 The Golden Garments of the Gods, *Journal of Near Eastern Studies* 8, 172–193.

Parpola, S. and Watanabe, K. 1988 *Neo-Assyrian Treaties and Loyalty Oaths*, State Archives of Assyria 2, Helsinki.

Reade, J. 2009 Fez, Diadem, Turban, Chaplet: Power-Dressing at the Assyrian Court. In M. Luukko, S. Svärd, and R. Mattila (eds), *Of God(s), Trees, Kings, and Scholars. Neo-Assyrian and Related Studies in Honour of Simo Parpola*, Studia Orientalia 106, Helsinki, 239–264.

Roth, M. T. 1989–90 The Material Composition of the Neo-Babylonian Dowry, *Archiv für Orientforschung* 36/37, 1–55.

TORAY Industries, Inc., Fibers and Textiles Laboratories 1996 Report on the Analyses of Textiles Uncovered at the Nimrud Tomb-Chamber, Al-Rāfidān 17, 199–206.

Zawadzki, S. 2006 *Garments of the Gods: Studies on the Textile Industry and the Pantheon of Sippar According to the Texts from the Ebabbar Archive*, Orbis Biblicus et Orientalis 218, Fribourg/Göttingen.

11. *e-ri-ta*'s Dress: Contribution to the Study of the Mycenaean Priestesses' Attire

Tina Boloti

"It would seem that only in the sphere of religion had women achieved independent status"
Chadwick 1976, 115.

Among the eponymous women of the Linear B tablets from Pylos[1] *e-ri-ta*[2] stands out. Famous for her conflict with the Pylian *damos* over a land-holding,[3] she was the high priestess[4] of *pa-ki-ja-ne*,[5] a sanctuary site of preeminent importance to the Pylians, transliterated as *Sphagianes*.[6] She is attested in the tablets either by her name, plus her office,[7] or, alternatively, as *i-je-re-ja*[8] or *i-je-re-ja pa-ki-ja-na*,[9] as argued by Michel Lejeune.[10] Hence, she appears as the most frequently mentioned female religious functionary; she is followed by another priestess of *pa-ki-ja-ne*, known as *ka-ra-wi-po-ro*,[11] i.e. the *Keybearer*, second in the sacerdotal hierarchy, recorded nine times in total – in two cases in conjunction with her name *ka-pa-ti-ja*.[12] Apart from *e-ri-ta* and *ka-pa-ti-ja*, two more priestesses are attested by their name in the Pylian archive, *ka-wa-ra*[13] and *ke-i-ja*,[14] as well as a series of priests (*i-je-re-u*),[15] some of them also by their personal name.

Prominent priestly figures, like the aforementioned, who played undoubtedly a significant role within the complex palatial society, would be immediately identified in everyday life by their costumes, as the universal "dress code" semiotics (widely attested ethnographically) would entail

[1] Lindgren 1973.
[2] A name of ambiguous etymology, which should be read most probably as *Eritha*. See Aura Jorro 1985, 247, *s.v. e-ri-ta*.
[3] The conflict is recorded in the Pylos tablet Ep704.5–6 Cf. Witton 1960; also Αποστολάκης 1990, 116–148.
[4] Aura Jorro 1985, 273–274, *s.v. i-je-re-ja*.
[5] Aura Jorro 1993, 72–74, *s.v. pa-ki-ja-ne*. Cf. also Aura Jorro 1993, the toponym *pa-ki-ja-na*.
[6] Gérard-Rousseau 1968, 166–169, *s.v. pakijana*; Palaima 1995, 131 (e.g. PY Fr 1209).
[7] Cf. PY Ep 704.3/PY Eb 339.A, PY Ep 704.5–6.
[8] Cf. PY Eb 297.1, 317.1, 416.1; Ep 599.7.8.
[9] Cf. PY Eb 409.1, Eb 1176A, En 609.18, Eo 224.6.8.
[10] Lejeune 1960, 136–139; Gérard-Rousseau 1968, 108.
[11] Aura Jorro 1985, 324, *s.v. ka-ra-wi-po-ro*.
[12] Aura Jorro 1985, 316, *s.v. ka-pa-ti-ja*. Also Lindgren 1973, I, 60 (*s.v. ka-pa-ti-ja*); II, 72–72 (*s.v. ka-ra-wi-po-ro*). In the latter see also the variant spelling *Kapasija,* which is a dubious anthroponym according to Gérard-Rousseau 1968, 124.
[13] Gérard-Rousseau 1968, 108–109, *s.v. ijereja* [PY Qa 1289]; Aura Jorro 1985, 333, *s.v. ka-wa-ra*.
[14] Gérard-Rousseau 1968, 108–109, *s.v. ijereja* [PY Qa 1303]; Aura Jorro 1985, 336, *s.v. ke-i-ja*.
[15] Aura Jorro 1985, 274–275, *s.v. i-je-re-u*.

us to anticipate.[16] This would be assumed even more in rituals of the official religious calendar (i.e. processions, sacrifices etc.), in which they would participate as outstanding acting officials. Their formal attire would be complemented by characteristic accessories, such as headdresses, portable *insignia dignitatis* (e.g. different kinds of staffs and sceptres) as well as by jewellery of symbolic character, like necklaces, sealstones and signet rings. The latter, especially, except for their aesthetic value and significance as status symbols, would have been essential for the effective involvement of the religious personnel in administrative procedures, of which they were an active, integral part.[17]

Although Linear B tablets provide adequate evidence for certain types of dress (*we-a$_2$-no*, *ki-to*, *pa-we-a* etc.), raw materials and techniques,[18] since the textile industry constituted an important section of the palatial economy,[19] they offer virtually no direct information as far as priestly attire is concerned. Any attempt to correlate garments recorded in the tablets with those known from the contemporary iconography remains hypothetical or dubious.[20]

However, the lack of related textual evidence seems counterbalanced, at least to a certain extent, by the Late Bronze Age imagery, mostly of religious – even plausibly ceremonial – character. Religious functionaries, often female, participated undoubtedly as acting agents in rituals represented primarily in glyptic[21] and frescoes.[22] Their identification becomes possible either by their position in ceremonies and/or, at least in some cases, by their attire, divergent more or less by the current dress-types. Illuminating, in this respect, is the thoroughly discussed Hagia Triada sarcophagus from the middle 14th century BC (early LM IIIA2),[23] an epitome of contemporary religious iconography, in which traditional Minoan elements are combined with contemporary Mycenaean ones.[24] The ceremonies depicted on the long panels (A and B) of the sarcophagus,[25] where priestesses predominate, attest to the simultaneous use of various types of ritual garments, male and female, while on the side panels we gain, in all probability, ai insight to deities' attire. It is important to stress *ab initio* that the detected sartorial similarity between goddesses and priestesses is almost generally applied to the iconographic codes of that era, as attested by other examples, mostly in the glyptic imagery. As a consequence, their identification seems often ambiguous, especially in the case of fragmentary wall paintings.

[16] Wobst 1977; Wiessner 1989.

[17] Lupack 2008.

[18] Marinatos 1967, A18–A21; Τζαχίλη 1997; Del Freo *et al.* 2010.

[19] Killen 1984; Since 2000 at least nine related papers on this issue written by M.-L. Nosch, R. Firth, P. Militello, M.-E. Alberti, E. Luján, C. Varias and V. Petrakis, the most recent of which are concentrated in the *Kosmos* volume of the series *Aegaeum*, i.e. Nosch and Laffineur 2012.

[20] See for example the case of the so-called *we-a$_2$-no* garment in Marinatos 1967, A30 (where this name designates two distinctive types of long robe, i.e. with a central, vertical band and with diagonal bands), and in Jones 2009, 218–222 (where the same term has been attributed to the long robe with vertical band as well as to the garment of the women in the ivory trio from Mycenae).

[21] For a brief discussion see Niemeier 1989.

[22] Immerwahr 1990; Μπουλώτης 2005.

[23] For the dating of the sarcophagus to the early LM IIIA2 phase see DiVita 2000, 480; La Rosa 2000; Burke 2005, 403.

[24] The Hagia Triada sarcophagus, "the most important document of Minoan religion as well as the most difficult to interpret in its general significance" according to Nilsson 1950, 426, was originally published in Paribeni 1903. Nevertheless, the most thorough study of it is Long 1974. For a recent contextual analysis of the sarcophagus, providing a new aspect of this well-known Aegean artifact see Burke 2005.

[25] We follow here Militello 1998, 154–167, where the long and narrow sides of the sarcophagus are designated as A and B, C and D respectively.

Using the available iconographical evidence, we will attempt to "recreate" the wardrobe of the high priestess *e-ri-ta*, providing, at least hypothetically, a specific sartorial identity to her, making her emerge from her "aniconic" cadastral context of the Linear B tablets.

The dress: flounced skirt and long robe with vertical band

In order to dress the priestess *e-ri-ta* it seems methodologically correct to begin with the available iconographical data from Pylos itself, where this eminent woman lived and acted about the end of the 13th century BC.

One female figure,[26] at least, participates in the procession depicted in the wall-painting of Vestibule 5 (Fig. 11.1), dated to the last LH IIIB phase[27] of the Pylian palace.[28] The fresco, "a late reminiscence of the Knossian offering-bearers" according to Sarah Immerwahr,[29] represents almost exclusively men, *c.* 30cm. in height. They proceed to the left, arranged on two levels, with an oversize bull in the middle, the presence of which implies in all probability a sacrificial ritual. The majority of male participants wear long ceremonial bordered robes, while fewer are dressed in kilts.[30] The aforementioned woman, on the other hand, of the same size as the men, is clad in a flounced skirt, a typical garment of the Minoan and Mycenaean elites, which, in combination with an elaborately varied tight bodice, was also worn in ceremonial contexts as amply documented in different artistic media (wall-paintings, seal glyptic etc.). A number of female cult functionaries, and among them priestesses, apparently wore the same skirt – a type attested in the case of female divinities as well. Therefore, would just the presence of the Pylian woman in the procession justify her identification as a priestess?[31] This assumption would be quite plausible indeed, given the ritual character of the scene and its significant setting in the vestibule of the throne room. Nevertheless, the fragmentary state of the figure and the actual, bad preservation of the fresco, do not provide further distinctive features of her identity or elements of the role she played in this particular ritual.

A different, more illuminating aspect of the LH IIIB female priestly attire in Pylos is offered by a fragmentary fresco (Fig. 11.2) from the northwest plaster dump.[32] It depicts a half-size woman, walking to the right while her feet overlap a carved footstool, ivory in all probability, judging by its white colour.[33] This footstool,[34] with its closest parallel to the composition of the famous Tiryns

[26] Lang 1969, 68 (15H5/flounced skirt and feet).

[27] Despite the disagreements expressed as far as the correlation between absolute and relative chronologies of the Aegean Late Bronze Age, one should take into account Warren and Hankey 1989.

[28] Lang 1969, 64–68 (5H5–15H5), pl. 119. Apart from the abovementioned female figure, two more women, clad in flounced skirts and carrying *pyxides* and flowers, are depicted in the restored fresco (cf. pl. 119), albeit they are not included in the catalogue of the related fragments.

[29] Immerwahr 1990, 118.

[30] Immerwahr 1990, 117–118.

[31] See the opinion expressed in Whittaker 2007, 8.

[32] Although this fresco dump has been dated to LH IIIB, as noted in Lang 1969, 221, it is earlier than the ultimate decoration phase of the palace.

[33] For ivory footstools (*ta-ra-nu e-re-pa-te-jo*) in Linear B tablets from Pylos see Ventris and Chadwick 1973, 345–346 (Ta 722), where they are compared with the ivory decoration (with volute ends) of the fronts of two footstools found in the tombs 518 and 8, from Mycenae and Dendra respectively. See also the ideogram *220 denoting a footstool. For a similar footstool from the antechamber of the tholos tomb A at Archanes see Σακελλαράκης and Σαπουνά-Σακελλαράκη 1997, 165, 168, 721–729.

[34] As pointed out in Rehak 1995, 103 "Only four Aegean figures are represented using footstools, and all these are women."

Fig. 11.1: Palace of Pylos. Vestibule 5 wall sketch. After Lang 1969, plate 119.

Fig. 11.2: Pylos. The Priestess' feet. After Lang 1969, plate N.

Fig. 11.3: The Tiryns gold signet ring. After Μυλωνάς 1983, 211, fig. 166.

signet ring *CMS* I, no. 179 (Fig. 11.3), where a procession of Genii approaches an enthroned "goddess", designates respectively the Pylian woman as a leading processional figure – a high priestess, as argued reasonably by Mabel Lang.[35] The "priestess", however, does not wear the common flounced skirt but a long robe with a vertical central band, known otherwise as straight robe,[36] of which only the lower part has been here preserved. Linear and architectural motifs decorate the elaborate border of the dress, constituted by two horizontal bands: zigzags in the upper band, as well as in the band that goes up the side of the garment, and alternating blue and yellow beam-ends in the lower one. Without commenting on the dress-type, Lang rightly noted the structural similarity of the latter with a much earlier garment, that of the "goddess" in the LM II Knossian Procession fresco,[37] a fact that she attributed to common tradition.[38]

[35] Lang 1969, 85 (50Hnws/priestess' feet.)

[36] Long 1974, 38.

[37] LM II/IIIA according to Immerwahr 1990, 174–175 (KN No. 22).

[38] As noted in Lang 1969, 85: "the architectural border motifs which appeared first on the lower skirt border of the goddess in the Knossos Procession Fresco: i.e., tooth-ornament bands bordering friezes of pseudo-rosettes or beam-ends. Both the Knossian and Tirynthian examples seem somewhat coarser in execution than that which

The Knossian dress (Fig. 11.4), of which only the three decorative border zones have been preserved, the two outer with a row of half-rosettes and the middle with beam-ends (Fig. 11.5), was restored by Arthur Evans as a flounced skirt.[39] Evans also argued that the woman was a "goddess" holding, quite hypothetically, double-axes in both her hands.[40] Much later, Marc Cameron, commenting on Evans' restoration, suggested that the figure could be a "goddess" or a "priestess".[41] Her prominent position in the synthesis is indubitable,[42] as she is flanked by two groups of men, one of which offers her a piece of cloth, according to Christos Boulotis (Fig. 11.6).[43] However, her identity remains uncertain. The half-rosettes decorating her dress,[44] a divine symbol of the Hittites,[45] which also borders, as a kind of dado, the composition on the Tiryns gold signet (Fig. 11.3), would support her plausible interpretation as a goddess. Nevertheless, this motif would reasonably designate her mortal representative on Earth, a possibility supported by her equal size with the other processional figures and by her standing position.

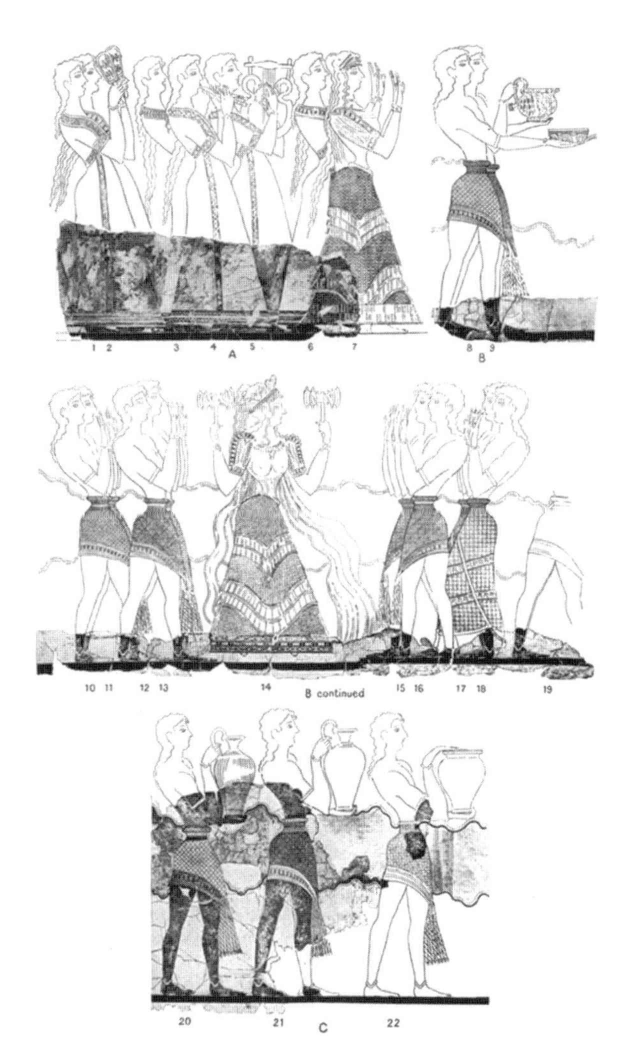

Fig. 11.4: Palace of Knossos. Successive groups of the "Procession fresco". In the centre of the Group B the so-called "goddess" (figure 14), clad in flounced skirt. Restored drawing by E. Gilliéron, fils. After Evans 1928, 723, fig. 450.

The evident similarity in the decorated border bands of the two dresses, the Pylian and the Knossian respectively, urged me to re-examine the latter. Thanks to a high resolution photograph (since the fresco is at the moment in the restoration laboratory of the Herakleion Archaeological Museum) I realized that we miss the part of the dress where the

appears here, but the tradition is indubitably the same".

[39] Evans 1928, 729, fig. 456a, where this decorative detail is depicted.

[40] Evans 1928, 721–724, fig. 450 (Group B).

[41] Cameron 1975, 139. Cameron also stresses the fact that Evans' restoration with double axes on her hands is arbitrary.

[42] Cameron 1975, 139.

[43] Boulotis 1987, 150, fig. 8.

[44] Half-rosettes are actually an unusual motif in the iconography of Aegean Bronze Age textiles, thus suitable to decorate the dress of an eminent figure. In real terms it seems more plausibly embroidered rather than woven.

[45] For the symbolic value of half-rosettes see Marinatos 2010, 135–139. For the presence of half-rosettes in the façade of a Minoan building depicted on the golden signet ring from Poros, Herakleion see Dimopoulou and Rethemiotakis 2003.

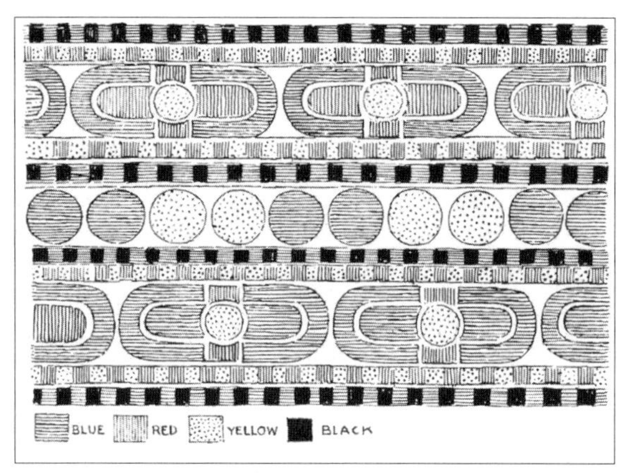

Fig. 11.5: Detail from the border of the "goddess' robe" in the "Procession fresco". After Evans 1928, 729, fig. 456a.

Fig. 11.6: Boulotis' restoration of the "goddess" from the "Procession fresco" receiving a piece of cloth. After Boulotis 1987, 154, fig. 8.

vertical band of the long robe type would be expected to end. Neither have I discerned, on the other hand, traces of a lateral curving band, indicative of a flounced skirt, as restored by Evans. Hence, it would be reasonable to argue that the dress of the Knossian "priestess" or "goddess" should be indeed a long robe with a vertical band, a suggestion further reinforced by the contemporary Aegean iconography: the border of flounced skirts is never decorated with horizontal bands. If my hypothesis is correct then from the Knossian Procession fresco we gain a significant element as far as the formal female priestly attire is concerned: the co-existence in ritual context of the two aforementioned types of dress, i.e. the long robe with vertical band and the flounced skirt. The latter is certainly worn in the same Knossian fresco by the leading female figure of the male processional group A (Fig. 11.4), a priestess perhaps, like the one in flounced skirt from the sacrificial procession of the Pylian Vestibule 5 (Fig. 11.1).

Long robes with vertical band, however, simpler than the abovementioned female garment, are also worn by six male processional figures (group A) of the Knossian Procession fresco, on the east wall of the Corridor (Fig. 11.4), and apparently by four more, on the west wall.[46] The former, preserved only in their lower half, were restored by Evans as musicians,[47] following the example of the lyre- and the flute-players (sides A and B respectively) on the Hagia Triada sarcophagus,[48] although the second one wears, it seems, a shorter version of this garment.[49]

The Hagia Triada sarcophagus, as already mentioned, constitutes a focal document in our discussion, since priestesses undoubtedly participate in the rituals depicted on its long sides. Apart from two women dressed in hide-skirts,

[46] Boulotis 1987, 148–150 (Group D), figs 3, 5. The adoption of the long robe by male cult functionaries was wide-spread from LM II–IIIA1 onwards as documented by some other wall paintings, like the Knossian Camp-stool fresco, with the latest example being that of the LH IIIB Pylian sacrificial procession of Vestibule 5, see *supra* Fig. 11.1.
[47] Evans 1928, 721–722, fig. 450.
[48] Militello 1998, 155–163.
[49] In Militello 1998, 291 this garment is described as plain.

the first in the libation scene of side A (Fig. 11.7)[50] and the second at the altar of side B (Fig. 11.8),[51] the other two indubitably priestly figures wear a long robe with vertical band. Among them, the "bucket-carrier" in the libation scene (Fig. 11.7), offers the most complete depiction of female priestly attire, since, in addition to the long robe, she also wears an elaborate headdress, a *polos*,[52] while one lentoid sealstone, at least, is discernible on her left wrist.[53] The same dress, but of different colour, is also worn by the fragmentary "priestess" of side B. The latter, with her hands towards the sacrificed bull, was restored wearing *polos* on her head in analogy to the priestess of side A.[54] Four out of five women following her are clad in straight robe as well, although more elaborate. However, due to their fragmentary state of preservation, their precise role in the ritual seems uncertain.

Although commonly classified as a Minoan artifact, Brendan Burke recently argued that the sarcophagus should be "connected to an emergent Mycenaean ideology". It is actually "a hybrid of Minoan and Mycenaean elements"[55] a view supported by the garments depicted. On the one hand, the indubitably Minoan hide-skirt, a peculiar type of ritual garment, used in Crete, in all probability from MM II onwards[56] and on the other hand the long robe with vertical band, which, in combination with *polos*, emerges as its Mycenaean counterpart. It is important to stress here that the latter appears in Crete after LM II, a period characterized by the Mycenaean presence on the island, with the earliest examples attested in the aforementioned Knossian Procession fresco.[57] However, if my hypothesis is correct, that the "goddess" or "priestess" from this particular fresco wore a straight robe, we cannot help wondering if she also wore *polos*, in analogy to the priestess on the Hagia Triada sarcophagus.

In Hagia Triada, again, long robes with vertical band are attested in two more frescoes, in the "Piccola"[58] and in the "Grande Processione".[59] In the "Grande Processione", which can be ascribed to the painter responsible for the sarcophagus,[60] this particular garment is worn by a partially preserved female figure (Fig. 11.9b), as well as by two men, a lyre-player and a bucket-carrier (Fig. 11.9a); the latter appears as a male counterpart of the similarly acting "priestess" on the sarcophagus' side A (Fig. 11.7).

---

[50] Militello 1998, 155–158. According to Long this is the front side of the sarcophagus cf. Long 1974, 35–53.

[51] Militello 1998, 160–163. That is the back side of the sarcophagus according to Long 1974, 61–71.

[52] I choose to use the Greek term *polos* (see Lidell-Scott 1436, *s.v. πόλος* esp. 5. crown of the head and 5.V. head-dress worn by goddesses) for this distinctive type of headdress, a term already used in Müller 1915.

[53] In Rehak 1994 cf. the case of a male priestly figure on the lentoid sealstone *CMS* I, no. 223, who wears a long robe with diagonal bands and a similar sealstone on his left wrist. For the use of the long robe with diagonal bands (known otherwise as "Syrian" robe) as typical dress of Aegean priests see Marinatos 1993, 127–128.

[54] Militello 1998, 161 (fig. 5). Photos of the sarcophagus before restoration are available in Nilsson 1950, 427, fig. 196. For the restoration of the sarcophagus in 1955 see Levi 1956.

[55] Burke 2005, 419.

[56] Σαπουνά-Σακελλαράκη 1971, 122–123; For a brief survey of the currently available iconographical evidence concerning the so-called hide-skirt see Boloti (forthcoming).

[57] As argued in Σακελλαράκης and Σαπουνά-Σακελλαράκη 1997, 616–617, fig. 654, the woman, buried during LM IIIA1 in the antechamber of the tholos A at Archanes, wore a long, priestly robe similar to those depicted in the Hagia Triada sarcophagus and the Knossos Procession fresco. The positions of the golden embroidered ornaments found within her burial clay sarcophagus seem to support this hypothesis.

[58] As noted in Militello 1998, 142–148, esp. 143, this garment is apparently worn by two, at least, out of seven or eight women in the lower frieze of the "Piccola Processione", which is stylistically dated to LM II–IIIA1.

[59] Militello 1998, 132–139.

[60] According to Chr. Boulotis (personal communication).

Fig. 11.7: The Hagia Triada sarcophagus. Side A. After Marinatos-Hirmer 1986, XXXII (above).

Fig. 11.8: The Hagia Triada sarcophagus. Side B. After Marinatos-Hirmer 1986, XXXI.

Quite exceptional for our theme is a small figure, depicted on a LH IIIB fresco fragment from the Cult Centre at Mycenae,[61] clad in a long robe with vertical central band (Fig. 11.10).[62] The figure,

[61] Κριτσέλη-Προβίδη 1982, 41–42 (B-2), table 6a. The fresco was found at the so-called Southwest Building.

[62] In Jones 2009, 318–321 has been suggested that the dress of the miniature figure is the *we-a$_2$-no* garment of the Linear B tablets, a kind of cloth made of either linen or wool cf. Nosch and Perna 2001, 472–473; Rougemont 2007, 47. Although this identification seems plausible it still remains dubious. Unlike Jones' suggestion of its Minoan origin we would stress that this type of dress, i.e. a long robe with a vertical, central band, appears in Crete only after LM II and it seems so closely associated to the Mycenaeans in the related iconography, as to support its Mycenaean origin instead.

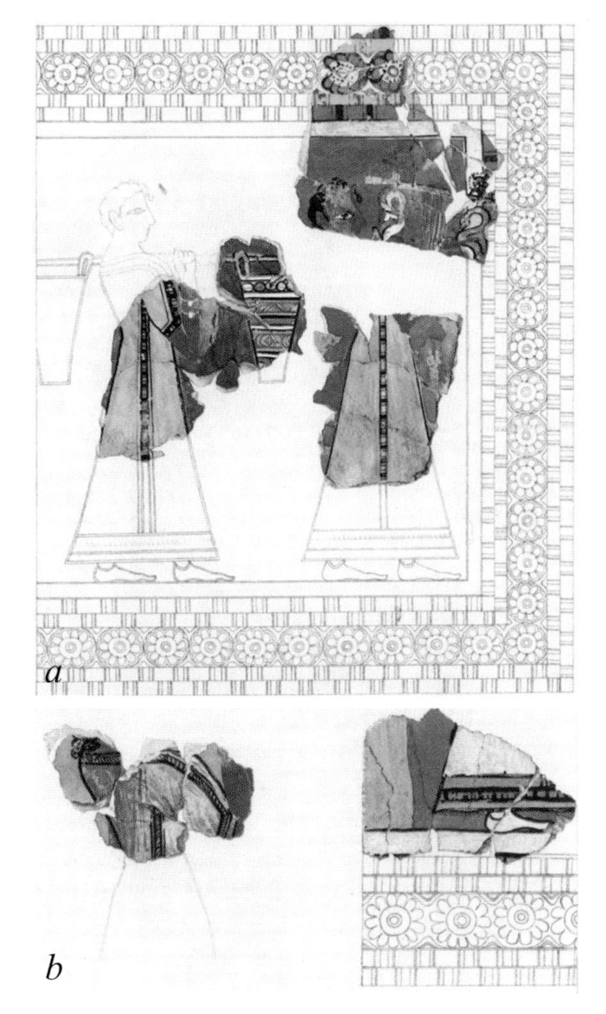

Fig. 11.9a and b: Hagia Triada. "La Grande Processione". After Militello 1998, pl. I.

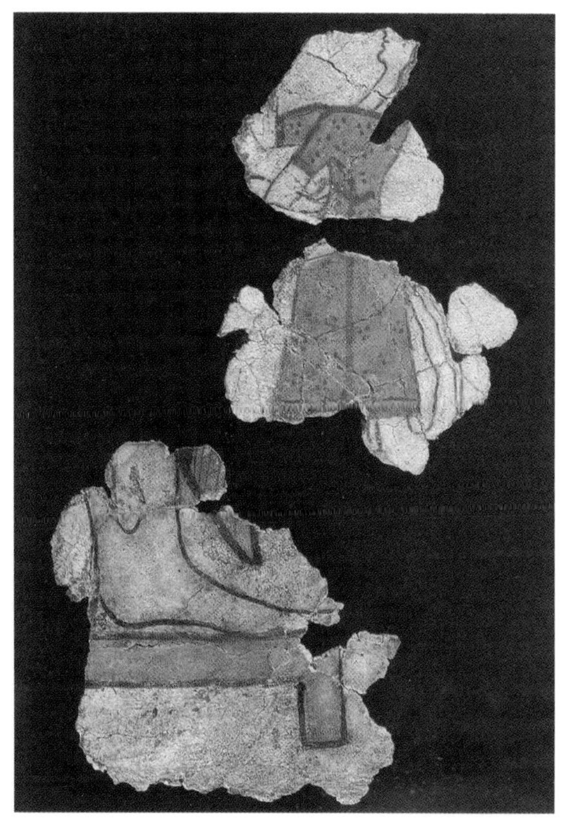

Fig. 11.10: Cult Centre at Mycenae. Fresco fragments depicting a half-life-size woman's hand holding a miniature female figure and woman's foot resting on a footstool. After The Mycenaean World 1988, 183 (152–153).

either held in the hand of a seated "goddess" alone or presented to her by a devotee, seems "a real little girl" rather than an idol according to Bernice Jones.[63] Jones also, comparing this fresco with the signet ring *CMS* I, no. 17 (Fig. 11.11), suggested that the figure possibly offers flowers to the "goddess". Nevertheless, despite its animated depiction, neither the way it is held nor its proportions, compared to the seated goddess or to the supposed standing devotee,

Fig. 11.11: The gold signet ring CMS I, 17 from Mycenae. After Μυλωνάς 1983, 187, fig. 141.

[63] Jones 2009, 317–318.

Fig. 11.12: The lentoid sealstone CMS I, 220 from Vaphio. After Σακελλαράκης 1972, pl. 95a.

support Jones' suggestion. Besides, figures/idols were undoubtedly among the offerings in the female Processions of the mainland Greece[64] – a cultic activity reflected in all probability in the Mycenaean festival *te-o-po-ri-ja*.[65] What does the long robe with vertical band reveal about the small figure on the fresco from Mycenae? Does it designate a divine figure, a priestess, or just a mere worshipper?

The headdress: diadem and *polos*

Universal semiotic codes of dress indicate that the head, as the most prominent part of the human body, would be reasonably adorned with a distinctive headdress, especially in the case of eminent individuals, like the Mycenaean priestesses we are discussing. However, what is the related evidence from the Late Bronze Age Aegean?

A peculiar type of headdress, designated as diadem, appears on the lentoid sealstone *CMS* I, no. 220 from the LH IIA tholos tomb of Vapheio in Laconia (Fig. 11.12).[66] Consisting of a row of projecting stems braced, in all probability, between two metal bands, it is worn by a female figure that carries an upright capricide and, seemingly, by the following woman, both clad in flounced skirts. According to Yannis Sakellarakis, who highlighted and restored this particular iconographical theme as an excerpt from a wider sacrificial procession, the first woman should be considered a high priestess leading the animal to the altar for sacrifice.[67] He also stressed that this type of headdress "has no parallel in the Creto-Mycenaean cycle"; nevertheless, he associated it with headdresses worn "by eminent women […] in formal occasions", like those on the Hagia Triada sarcophagus, with which "it might be compared only in its general elements".[68]

The simultaneous use of a specific headdress by priestesses and deities is undoubtedly attested once again on the Hagia Triada sarcophagus. There, the bucket-carrier "priestess" of the side A (Fig. 11.7) wears the so-called *polos*, while the "priestess" in front of the sacrificial table on side B was analogically restored wearing a similar headdress (Fig. 11.8). A *polos* is also worn by the two pairs of female "divinities" on the chariots of the sides C and D, drawn by griffins (Fig. 11.20) [69] and agrimia respectively (Fig. 11.21).[70]

---

[64] See Boulotis 1979, who restores a figurine carried by a processional woman from the palace of Tiryns, in conjunction with a piece of cloth.

[65] Gérard-Rousseau 1968, 211; Hiller 1984.

[66] The tradition of this kind of headdresses in mainland Greece seems to be attested by a diadem brought to light in 1989 during trial trenches conducted by G. Korres in the tholos tomb 1 at Myrsinochori/Routsi, see *Ergon* 1989, 28–30. More details about this diadem in Κορρές 1996, 56–57.

[67] Σακελλαράκης 1972.

[68] Σακελλαράκης 1972, 252. We note that on the lentoid sealstone *CMS* I, no. 221, also from Vapheio, the same scene is depicted, though divergent in details. As far as it can be detected, the "priestess", discernible behind the animals' head has an elaborate coiffure adorned with headband, which seems to be a reminiscent of the goddess from the Xeste 3 at Akrotiri.

[69] Long 1974, 29–32. Cf. also the same scene depicted on the gold signet ring *CMS* V.1B, no. 137 from the tholos tomb at Antheia, dated to LH IIA–IIB. Apart from the *polos* with a plume all the other elements related to the figures represented are dubious, due to their vague rendering.

[70] Long 1974, 54–57.

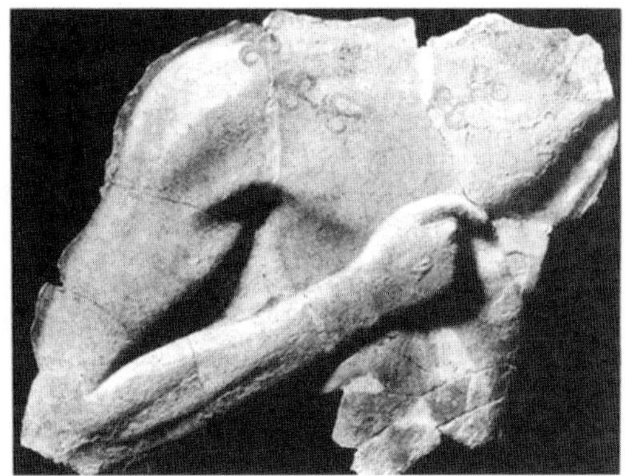

Fig 11.13: Palace of Knossos. The relief fresco fragments restored afterwards as the so-called "Priest-King". After Evans 1928, 776 (fig. 504B) and 780 (fig. 508).

The *polos*, an elaborate, flat, cylindrical hat, covered atop, with an attached plume or flower (often a lily), firstly appears in Neopalatial Crete and is commonly worn by sphinxes.[71] The motif of a *polos*-wearing sphinx, usual in the LM I seal glyptic,[72] was adopted afterwards by the Mycenaeans and featured in various media[73] until the end of the Late Bronze Age in mainland Greece;[74] there, it became an emblematic headdress, apparently made by reinforced cloth or leather, in a variety of colours (brown, yellow or blue), worn also, as it seems, by divinities and priestesses.

A quite elaborate precursor of *polos* has been attested in the MM III/LM I relief fresco of the "Priest-King" (Fig. 11.13).[75] Evans argued that this sophisticated headdress, with beaded decoration round its lower borders, and waz-lilies rising above, was an *insigne dignitatis* denoting the double authority of the Knossian figure. However the attribution of this lily-crown to a man was reasonably disputed by Wolf-Dietrich Niemeier. He preferred to put it instead on the head of a sphinx, restored hypothetically next to a male figure.[76] The issue still remains open: in an extensive article, Maria Shaw recently tried to once again associate the lily-crown with the preserved male torso.[77] Despite

[71] For a possible association between sphinxes and the kingship cf. the comment of Poursat,1973, 114, related to a MM II terracotta plaque of a sphinx from the Quartier Mu at Malia. Poursat argues that the latter might represent Kingship' s authority, as in the case of Ancient Egypt.

[72] See for example *CMS* II.3, no. 118, *CMS* II.7, nos. 83–84.

[73] For a selection of ivory artifacts, such as *pyxides* and plaques, with *polos*-wearing sphinxes cf. Poursat 1977, 43–45, 81, 92–93, 113, 148–149, 153–154, 156–159, 169–170.

[74] For sphinxes depicted wearing a *polos* on LH IIIB sarcophagi from Tanagra or on LH IIIC Mycenaean pictorial vases see Αραβαντινός 2010, 120–124 and Vermeule and Karageorghis 1982, 144, 224 (XI.91) respectively.

[75] Evans 1928, 775–795, pl. XIV (as restored).

[76] For the revised reconstruction of the relief in the 1980s see Niemeier 1987; Niemeier 1988. Apart from Niemeier, doubts on Evans' restoration of the relief fresco were expressed also in Coulomb 1979 and Coulomb 1990. Niemeier and Coulomb argued that these fresco fragments could have been attributed to more than one figures, especially in the case of the "crown", since it was normally a typical female headdress. Although, the original reconstruction of Evans seems further supported by the ring impression *CMS* II.8.1, no. 248 from Knossos, where a male figure, flanked by two huge dogs, wears a headdress which resembles that associated with the "Priest-King", we cannot overlook that "due to the incomplete state of preservation of the sealing, the drawing of the ring is not 100 percent certain" as stressed in Marinatos 2007, 272.

[77] Shaw 2004.

her attempt, approved also by Nanno Marinatos,[78] we cannot ignore that these fresco fragments are problematic, both stratigraphically[79] and chronologically[80] and even if Shaw is right, the identification of the male figure as "a king or a god",[81] or even a religious (?) functionary, is due primarily to this particular headdress.

In mainland Greece, on the other hand, *polos* seem to have been adopted by the 15th century BC judging by its earliest representation on the head of the enthroned "goddess" on the Tiryns gold signet *CMS* I, no. 179,[82] as well as by another seated female figure on the, more or less, contemporary lentoid sealstone *CMS* VII, no. 118 from Mycenae, now in the British Museum.[83] This woman, the divine nature of whom is ascertained by the fact that she is seated on a lion's head throne and is flanked by two lions, wears *polos*, in combination with a long garment, although it is doubtful whether it is a robe, like that of the "goddess" on the Tiryns ring, or a cloak, since her upper limbs are not discernible.[84]

The transference of *polos* from the divine iconography to the priestly one, from the head of the divinities to the head of their earthly representatives, the priestesses,[85] took place evidently before the beginning of LM IIIA2, a *terminus ante quem* provided by the Hagia Triada sarcophagus. Despite the ambiguous nature of female figures depicted with *polos*, often due to their fragmentary state of preservation, this headdress seems established also on the head of mortal women[86] in mainland Greece from the 14th century BC onwards, i.e. more or less simultaneously with the Hagia Triada sarcophagus; at least, four fresco examples from Mycenaean palatial centres provide evidence for it.

A *polos*-crowned woman has recently been identified by Boulotis[87] among the participants in the Theban female procession (Fig. 11.14),[88] generally agreed to be the earliest on the Greek mainland, dated to the 14th century BC.[89] This woman, the only one so far among the processional figures with a headdress of this kind, could plausibly be assigned as an eminent individual, a "priestess" in all probability.

---

78 Marinatos 2007.

79 Although it may seem convenient to attribute all the plaster pieces to a single figure we cannot ignore the fact that they were found scattered in a dumping fill at a depth of about 2 m. and they could belong to a composition which "may have involved many figures, possibly a procession heading toward the Central Court" as admitted in Shaw 2004, 76.

80 In Shaw 2004, 77 the proposed dates range from MM IIIB to LM IB, or occasionally even later.

81 View supported in Marinatos 2007, 271.

82 As pointed out in Renfrew 1985, 24 "In the interpretation of early religious iconography 'Cherchez le monstre' can be a useful first step".

83 According to Evans 1935, 402, fig. 333, it is "said to have been found at Mycenae".

84 An iconographic variation of this theme offers the sealstone *CMS* VI, no. 313, in the Ashmolean Museum, probably also originating from Mycenae, with a standing, woman flanked by a pair of lions. The woman, described as a goddess, wears a long robe and a peculiar headdress (*polos*?), while at its right a floating sacral knot is depicted. Cf. Evans 1935, 402, fig. 334.

85 In Long 1974, 37 it has been argued that this type of headdress is not "restricted to women and sphinxes" as attested by "an ivory *pyxis* from Tsountas' excavations at Mycenae (tomb 49) showing two men wearing a headdress of this type and leading a sphinx" [a view reproduced also in Lenuzza 2012, footnote 13]. For the *pyxis* see Poursat 1977, 92 (297/2476. *Pyxis* avec homme et sphinx), pl. XXVIII.

86 It is important to note that Holland 1929 already suggested that the gold and glass paste plaques found in some Mycenaean tombs belonged to headdresses of this kind, while in the Tomb 3 of the Kladeos cemetery near Olympia a row of glass paste plaques was found encircling a skull cf. Yalouris 1967; the latter was designated as a diadem in Long 1974, 37.

87 Μπουλώτης 2000, 1116–1117, footnote 91.

88 Immerwahr 1990, 200–201 (Th No. 1). Nine to twelve life-size women, found by Keramopoulos in 1909 in Room N of the "House of Kadmos", studied and restored by H. Reusch (1948–1949). Albeit Reusch dated the fresco fragments to LH II, Immerwahr proposes a LH IIIA chronology.

89 Immerwahr 1990, 115–117.

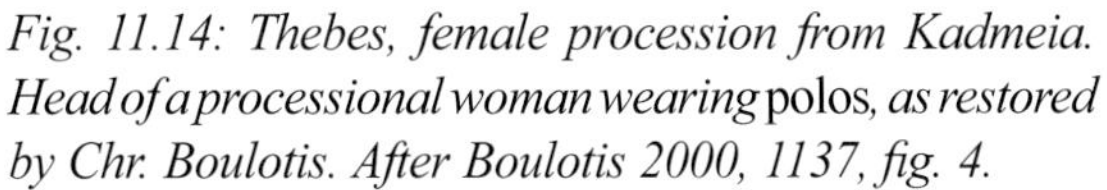

Fig. 11.14: Thebes, female procession from Kadmeia. Head of a processional woman wearing polos, *as restored by Chr. Boulotis. After Boulotis 2000, 1137, fig. 4.*

Fig. 11.15: After Rodenwaldt 1921, 50, fig. 26.

The *polos*-crowned female figure[90] restored by Gerhard Rodenwaldt from three fresco fragments found near the megaron of Mycenae could also be a processional woman.[91] It consists of a fragmentary *polos* and part of a woman's neck and shoulder (Fig. 11.15), dated to LH IIIA/B1, a middle phase of the palace's decoration according to Immerwahr.[92] Thanks to the sacral knot attached on her neck, this woman has been connected directly to the well-known fresco image "La Parisienne" from the Knossian palace[93] and identified as a female priestly figure, a possibility stressed by Rodenwaldt: "Das Tragen der Schleife, die auch gesondert als Kultsymbol erscheint, kann wohl nur als Abzeichen einer priesterlichen Funktion der Dargestellten aufgefasst werden".[94] Rodenwaldt, however, did not comment on the *polos* issue, although the latter would additionally support the priestly identity of the figure. On the contrary, he identified as a "goddess" another *polos*-crowned figure from Mycenae, the famous Plastered Head, dated to the 13th century BC (Fig. 11.16).[95] The latter (16.8 cm in height) found during Tsountas' excavations within the citadel,[96] was possibly joined to a torso, as attested by a vertical cavity at the bottom (for a wooden peg in all

[90] Immerwahr 1990, 117, 191 (MY No. 2). The above-mentioned fresco fragments, found in 1886 by Tsountas outside West Portal ("Pithos Area"), belong to the image of a female procession.

[91] Rodenwaldt 1921, 50, fig. 26.

[92] Immerwahr 1990, 191.

[93] Actually there are two "Parisiennes" according to Cameron's restoration of the well-known "Campstool fresco" from the palace of Knossos. For the three different restorations of this fresco proposed so far see Lenuzza 2012. In any case, "La Parisienne" – and her female companion – appears to be by far the most important figures in the fresco. Lenuzza 2012, 256, as earlier Cameron 1975, 60, has suggested that she should be considered an important religious figure, possibly a high-priestess.

[94] Rodenwaldt 1921, 51.

[95] Rodenwaldt 1912a, 31, footnote 3, addendum.

[96] The plastered head was found in 1896 within the citadel of Mycenae in the debris of a building near the west side of the fortification wall see Τσούντας 1902.

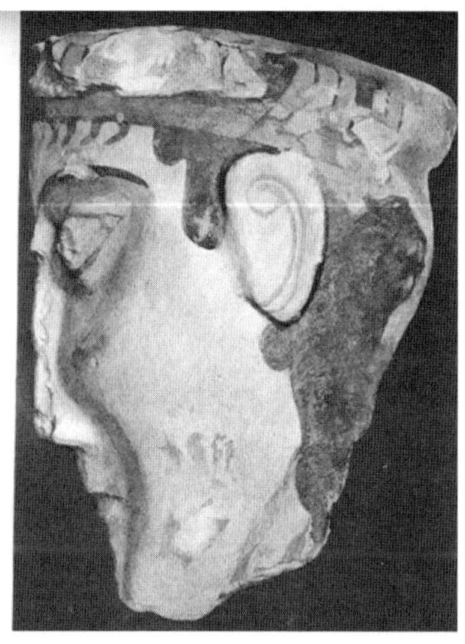

Fig. 11.16: Mycenae, acropolis. The female plastered head. After Τσούντας 1902, pl. 1.

Fig. 11.17: Pylos. The so-called "White goddess". After Lang 1969, pl. 127.

probability), forming the statue of a goddess or a sphinx,[97] depending on the preferred interpretation.

From Pylos, where our discussion of Mycenaean female priestly attire began, comes another *polos*-crowned woman in a LH IIIB fragmentary fresco. The figure, known as the "White Goddess" (Fig. 11.17)[98] was found in the same context with the abovementioned "priestess" wearing the straight robe (Fig. 11.2),[99] i.e. in the fresco dump on the northwestern slope. Actually, only the head of this life-size figure, facing left, has been preserved and she is identified as a "goddess" in contrast to the half-life size "priestess", facing right.[100] Using elements from the composition on the Tiryns gold signet *CMS* I, no. 179 (Fig. 11.3),[101] Lang argued that the "goddess" would be enthroned and approached by the "priestess", who overlaps with her feet the footstool of the throne (Fig. 11.2). She also supported

[97] As Lang points out: "It is this cap [note: *polos*] which is largely responsible for the belief that the head belonged to a sphinx, but the connection is made more tenuous by the parallel with 49 H nws, who is not a sphinx but wears a spiral crown" Lang 1969, 57. For the suggestion that this Plastered Head belonged to a sphinx cf. Müller 1915, 17–18.
[98] Lang 1969, 83–85 (49Hnws/White Goddess).
[99] Lang 1969, 85 (50Hnws/priestess' feet).
[100] Lang 1969, 84 noted that the height of the 'priestess' "might be as much as 0.90m; the seated goddess might be as little as 1.10m high".
[101] Lang 1969, 84.

Fig. 11.18: Cult Centre at Mycenae, the Shrine of the Fresco. "Goddess" holding sheaths of grain. After Μυλωνάς 1983, 144, fig. 113.

the idea that the difference detected in the background colour between the two figures "is not an objection to this association since it is only natural that the background color should change at least once in a scene of this size, and that if it is blue for the sky above, it can as well be red below".[102]

Despite Lang's argumentation the afore-mentioned differences in size and background cannot be lightly ignored.[103] Actually, the so-called "White Goddess" could easily be disassociated from the "priestess" and be treated independently as a mortal, processional figure,[104] walking towards the left. It is noteworthy that from the same fresco dump comes a fragmentary fresco of life-size processional women in flounced skirts, two of whom carry wild roses.[105] Furthermore, the *polos* she wears is appropriate not only to divine figures, as we have seen above, but also to mortal women with priestly status.

The identity of the *polos*-crowned woman, dated to the mid-13th century BC, from the Shrine of the Fresco at the Cult Centre of Mycenae also seems ambiguous (Fig. 11.18). The female figure, of whom only the upper part and one foot with fringed hem survive, wears a garment knotted over her right shoulder (the latter overlaps an underdress with short sleeves) while on her head she wears *polos* with a plume. Albeit "both the knotted, fringed garment and the plumed hat suggest that she is a priestess",[106] the tail of an animal behind her, clearly leonine in its tuft, creates ambivalence over the interpretation. Since the rest of the animal is missing, save for two clawed paws towards the lower right of the picture, it has been interpreted either as a griffin or as a lion.[107] Hence, the presence of a griffin or even a lion in such a pose would more reasonably suggest a goddess rather than a priestess.

[102] Lang 1969, 84.
[103] The same skepticism is expressed in Immerwahr 1990, 118.
[104] Actually Lang 1969, 84 pointed out that the closest parallel to the so called "White goddess" is "a very similar head from the Theban Procession Fresco".
[105] Lang 1969, 86–89 (51 H nws); Immerwahr 1990, 118.
[106] Morgan 2005, 167.
[107] Marinatos restored it as a griffin while Rehak as a lion, see Morgan 2005, 168 and footnote 37.

Fig. 11.19: Mycenae, acropolis. The "Palladion". After Μυλωνάς 1983, 208, fig. 164.

Nevertheless she has, like many of the terracotta figures,[108] raised arms, and in each hand she holds sheaths of grain.[109] The gesture can be either votive or divine, but as here it is directed towards the sacred platform she probably "represents the priestess as goddess impersonator".[110]

The identity of a mortal female wearing *polos* on the plaster plaque from the Cult Centre at Mycenae, known as the "Palladion" seems more certain (Fig. 11.19).[111] Unfortunately due to its bad state of preservation, we cannot discern more details. The fact is that the woman painted on the left certainly wears a yellow *polos* on her head but, unlike the woman on the right clad in a flounced skirt, her dress is unspecified, although we could suppose the same attire for both due to the heraldic scheme of the composition. Nevertheless, thanks to the *polos*, it would be reasonable to suggest that she is a priestess paying tribute to the central figure, actually a figure-of-eight shield.

Could the association of the *polos* with Mycenaean deities,[112] at least from the 15th century, or with Mycenaean priestesses, at least from the 14th century BC onwards, be useful for a new approach to the Mycenaean terracotta figurines, especially those with *polos* on their head?[113] The latter, produced in quantity from the LH IIIA, could be affordable representations of female deities,[114]

---

[108] See, for example, Pliastika 2012, 611, referring to basic morphological characteristics of Mycenaean terracotta figures of type A.

[109] In Burke 2012, 175–176 this long-held identification is really challenged since it is argued that the woman holds *Pinna nobilis* fan shells instead.

[110] Morgan 2005, 168.

[111] Rodenwaldt 1912b.

[112] The establishment of *polos* on the head of "goddesses" has been attested also by some terracotta female figures of the 13th century BC, designated as deities cf. for example, the figure found by G. Welter on Mt. Oros, Aigina in the 1930s, see Pilafidis-Williams 1995, figs 1–3; also, the figure from the acropolis of Midea (West Gate area, room VI), dated to the end of the 13th century BC in Δημακοπούλου and Διβάρη-Βαλάκου 2010, 28, figs 41–43.

[113] As noted in French 1971, 118, *polos* is a late criterion on small figurines, of type T and Psi (late LH IIIA2 and LH IIIB); *ibid.* 146–147.

[114] See the noteworthy comment on the female figure from Mt Oros on Aigina made in Pilafidis-Williams 1995, 231: "On the whole, the Oros figure seems to have been largely influenced by small figurines".

Fig. 11.20: The Hagia Triada sarcophagus. Side C. After Marinatos-Hirmer 1986, XXXIII.

Fig. 11.21: The Hagia Triada sarcophagus. Lower part of side D. After Marinatos-Hirmer 1986, XXXII (below).

used as "amulets" or as votive offerings, an equivalent of the modern Christian crosses or of cheap paper icons? Just suggestions open to discussion.

Deities and priestesses: the sartorial similarity

The ambiguity concerning priestesses and goddesses in the case of *polos* could be extended to their attire in general. The Hagia Triada sarcophagus testifies indubitably to the sartorial similarity between the two groups: the garments worn by the "goddesses" on the narrow sides of the sarcophagus, those riding the griffin-drawn chariot (Fig. 11.20), considered to be Mycenaean,[115] (side C) and the agrimi-drawn chariot (Fig. 11.21), considered to be Minoan (side D), respectively,[116] are identical to the garments of the "priestesses" on the long sides (A and B, see *supra*). Hence, the charioteer on the griffin-drawn chariot[117] and the passenger on the agrimi-drawn one[118] wear long robes with vertical band while the passenger of the chariot on side C and the charioteer on side D wear, as it seems, another type of long robe, that with diagonal bands (otherwise known as "Syrian robe"), a typical "stole" for priests, according to Marinatos.[119] All four "goddesses", however, wear *poloi* on their head.

In fact, the sartorial similarity between

[115] Cf. the same scene depicted on the signet ring *CMS* V.1B, no. 137 from Antheia.

[116] Since "the primary function of both griffin and agrimi is to identify the goddesses ridding", as correctly pointed out in Long 1974, 57 "the griffin-goddesses might be Mycenaean deities in contrast to the Minoan agrimi-goddesses."

[117] The east side of the sarcophagus according to Long 1974, 29–34.

[118] The west side of the sarcophagus according to Long 1974, 54–60. For the identification of these animals as agrimi see Long 1974, 55–57.

[119] See Marinatos 1993, 127–130. The Syrian origin and the use of this garment as priestly attire were suggested in Evans 1935, 397–419. For a different view expressed by Rehak who argued that "the diagonally-banded robed men [...] are middle administrators rather than priests" see Rehak 1995, 111.

goddesses and priestesses is not restricted to the Hagia Triada sarcophagus. It is amply detected in the seal imagery of religious/ritual character, mostly on golden signet rings of the early Late Bronze Age in mainland Greece (LH II-LH IIIA1).[120] The female figures depicted there, both goddesses[121] and priestesses[122] wear the well known flounced skirt. The golden signet ring *CMS* I, no. 17 from Mycenae in indicative in this case (Fig. 11.11). On this ring, stylistically dated to LBA I–II, both the leading figure, the priestess in all probability, and the seated woman, designated as a deity receiving flowers, are clad in a similar flounced skirt. Three other female figures in this scene, two of which are smaller in size share similarities in hair style; hence, the heads of the "goddess" and the "priestess" are adorned with hair bands and flower pins – the latter are attributes, in all probability, of the goddess and, consequently, of her worshippers.

Dressing up the priestess *e-ri-ta*

After the iconographic research above, we return to our initial question: what would constitute the official attire of the famous high-priestess *e-ri-ta*? According to the acknowledged semiotic codes of dress, it seems more than reasonable that she would have been dressed in a distinctive way as to be immediately identified either during the performance of her duties, in the sanctuary site of *pa-ki-ja-ne*, or within Pylian palatial society in general.

Given the prominent anatomical position of the head and its related semiotic connotations, we should firstly imagine her wearing a *polos*, simple or composite, with a plume atop or, perhaps, a floral motif. This typical Mycenaean headdress, of apparently Minoan origin or inspiration, is attested in LH IIIB Pylos thanks to the fragmentary fresco of the so-called "White Goddess" (Fig. 11.17), which could be equally interpreted as a high priestess acting in a palatial ceremony. Less probably *e-ri-ta* would have worn a diadem like that depicted on the LH IIA sealstone *CMS* I, no. 220 from Vapheio (Fig. 11.12), judging by the lack of iconographical parallels[123] and by the chronological distance of about two and a half centuries between the latter and the lifetime of *e-ri-ta*.

Polos, as a dressing accessory, could have been combined with a long robe with vertical band or a flounced skirt,[124] but also with less ordinary dress types, like the one worn by the goddess or priestess with sheaths of grain from the Cult Centre at Mycenae (Fig. 11.18).[125] *Polos* and flounced skirts are apparently coupled in the case of the two attendants, probably priestesses, flanking the figure-of-eight shield on the "Palladion" from the Cult Centre at Mycenae, without excluding other fragmentary Mycenaean wall-paintings (Thebes, Mycenae, Pylos). Nevertheless, the first combination, i.e. *polos* and long robe, are much better attested, so far, for priestesses and goddesses thanks to the evidence provided by the Hagia Triada sarcophagus and the Tiryns gold signet (Fig. 11.3). Hence, if we accept that *e-ri-ta* would have worn a kind of *polos* on her head, it would seem more probable that it was combined with a long robe, of the type with a vertical, central band, like the one worn by the standing "priestess" from the fragmentary Pylian fresco (Fig. 11.2). In an sophisticated version of the female priestly attire we would add, cautiously, a sacral

[120] See Niemeier 1989, where he has collected all the available evidence related to religious scenes.
[121] See Niemeier 1989, 169–174 (Groups 2–4) and 181–183 (Group 6).
[122] See Niemeier 1989, 167–169 (Group 1) and 174–181 (Group 5).
[123] The exact use of so-called golden crowns from the Grave Circle A at Mycenae remains dubious. Cf. for example the finds from the Shaft grave IV in Karo 1930, 71–72 (229–230), pl. XLI.
[124] This possibility is indirectly attested in the case of the Theban Procession fresco.
[125] Morgan 2005, 167–168 argued that "Both the knotted, fringed garment and the plumed hat suggest that she is a priestess".

knot behind her neck, as attested by "La Parisienne" and by the processional female figure from Mycenae, according to Rodenwaldt's restoration (Fig. 11.15). An additional hint to this case might be supplied, however, by the female figure from the Cult Centre at Mycenae, who holds sheaths of grain (Fig. 11.18). Apart from *polos*, she wears a garment knotted over one shoulder that would recall slightly a moderate version of sacral knot.

The formal attire of *e-ri-ta*, whose clothing would have been fabricated in the textile workshops attached to the sanctuaries,[126] would have been complemented in all probability with precious jewellery. The latter, apart from being status symbols of the high priestess of *pa-ki-ja-ne*, with symbolisms well established on the contemporary religious codes and beliefs, would visualize emphatically her particular connection with certain divinities, on behalf of which she exerted her authority. Hence, a necklace with beads in the form of a figure-of-eight shields, for example, would have suggested a functional connection with the divinity, the emblem/attribute of which would have been this particular symbol.[127] Furthermore, we may plausibly argue that jewellery of apparently symbolic character, found *in corpore* within selected Mycenaean tombs,[128] would indicate the possible priestly identity of their owners. The symbolic use of jewellery, established in Minoan Crete, as attested by some early representations of high status figures, like the "Priest-King" from Knossos[129] was adopted by the Mycenaeans, as early as the Shaft Graves period, to be continued, with a gradual decline of lavishness, until the end of the Mycenaean palatial system, i.e. *e-ri-ta*'s era.

Following the same interpretative *modus*, in the formal attire of *e-ri-ta* we would also include some *insignia dignitatis*: a kind of sceptre, in the simpler version of a staff[130] with or without elaborate finial, according to a well attested Minoan tradition, or, occasionally, some other meaningful emblems, like the double axe, adopted by the Mycenaeans together with other sacred symbols.[131] Sceptres made of precious materials (gold, ivory etc.), which sporadically accompanied burials in the Mycenaean mainland,[132] would apparently indicate a kind of political and/or religious

[126] It is reasonable to suppose that the garments of the religious functionaries were produced, as a rule, by workers in the sanctuary textile workshops, e.g. like those attested at Thebes, Lupack 2008, 105–110, esp. 105–106. Cf. the case of the *wanax*, who, according to Palaima 1997, 412 "should have his own craft specialists to attend to the needs of his person and functions". Apart from a *ke-ra-me-u*, "potter", as *wa-na-ka-te-ro*, i.e. related to the *wanax*, is characterized a *ka-na-pe-u*, "fuller", responsible, as argued, for the cloth finishing processes. Would these pieces of cloth have been used for the *wanax*'s official attire?

[127] Μπουλώτης 1999, 21–34 (general discussion on Mycenaean jewellery), especially 32–34 and 47–48 (LH II–IIIA figure-of-eight shield pendant from the so called "Treasure from Thebes").

[128] Cf. Ξενάκη-Σακελλαρίου 1985, 192–196, table 84 (gold pendant –X 2946– in the form of a woman carrying a *pyxis*-like rectangular object, who wears a necklace and a kind of headdress, from the chamber tomb 68 at Mycenae); Μπουλώτης 1999, 47–48 (references to gold figure-of-eight shields pendants from LH II tombs in mainland Greece, i.e. from a tholos tomb in Pylos and a chamber tomb in Prosymna).

[129] Two more characteristic examples from the early Late Bronze Age Aegean should be referred to here: the so-called "prince" on the "Chieftain cup" from Hagia Triada and the goddess with the griffin in the fresco of the Crocus-gatherers from Akrotiri (Xeste 3). The latter wears two necklaces, one with ducks and the other with dragon-flies, in all probability attributes of her divine nature.

[130] Cf. in Hallager 1985 the staff depicted on the well-known "Master Impression".

[131] For the adoption of the double-axe by the Mycenaeans see Rodenwaldt 1912a, 157–158 (227. Fragmente einer Kultdarstellung), pl. XVI.6 (double axes in conjunction with flowers); Lambrinudakis 1981, 62, figs 10, 12 (votive bronze double axes from the Mycenaean period in the sanctuary of Apollon Maleatas).

[132] Karo 1930, 84 (308+309), fig. 20, pl. XVIII (gold staff-sceptre in two pieces, ~78.5 cm. long, from Grave IV of the Circle A at Mycenae, the only one of this kind preserved *in corpore* in the prehistoric Aegean); for an ivory staff ending in the head of a griffin (sceptre head?) from Kadmeia, Thebes, dated to the 14th–13th centuries BC, see *Mycenaean*

Fig. 11.22:Cult Centre at Mycenae. Reconstructed drawing of the paintings in the Shrine of the Fresco. After Morgan 2005, 167, fig. 10.5.

authority during the lifetime of the deceased. In any case, it is indubitable that during *e-ri-ta*'s era, female figures used *insignia dignitatis*, denoting high sacerdotal or divine identity.[133] Relevant evidence is provided by the fresco from the homonymous Shrine at the Cult Centre of Mycenae: a standing woman in flounced skirt (priestess or goddess) holds in her extended hand a staff (pole or spear?) which apparently constitutes the equivalent of the large sword kept vertically by the woman, in a straight fringed garment, opposite her, possibly as her divine emblem (Fig. 11.22).[134]

World 1988, 252 (no. 272).

[133] As noted in Morgan 2005, 164–165, four of the idols from the homonymous Shrine in the Cult Centre of Mycenae, with both arms across the chest or with one arm raised and one across the chest, i.e. those with the basic poses 2 and 3 according to Andrew Moore, were supposed to carry axe-hammer, as indicated by the preservation of shafts in their hands.

[134] As noted in Morgan 2005, 168: "The clothing does not permit us to speculate on the divine versus mortal status of these women. Both types can be worn by cult functionaries or goddesses. They serve here to distinguish the two, but given their balanced position in relation to columns and platform, it is perhaps more likely that both figures belong to the divine sphere."

The well documented economic/productive activities of the Mycenaean sanctuaries,[135] lead us finally to the quite plausible assumption that *e-ri-ta*, the high-priestess in the main sanctuary site of the Pylian territory, possessed some sphragistic media, like all the palatial officials engaged in these. Through this prism it seems reasonable to assume that Late Bronze Age Aegean seals and signet rings with religious scenes would not have been merely prestigious accessories; they could have functioned equally as administrative devices in the hands of priests and priestesses. Especially rings depicting rituals with exclusively female participants, such as the signet *CMS* I, no. 17 (Fig. 11.11), found in the vicinity of the Cult Centre at Mycenae, or three signets from Aidonia,[136] which would have been owned by priestesses. The female burial[137] from the tholos A at Archanes is revealing in this respect:[138] three golden signet rings with religious themes[139] placed next to her chest indicate her possible priestly identity.[140]

The aforementioned hypothesis seems to be supported by a small LH IIIB fresco fragment from Pylos. Found in the same fresco dump as the "priestess" with the long robe (Fig. 11.2) and the *polos*-crowned "White Goddess" (Fig. 11.17), it belongs, in all probability, to a female wrist, with two perforated lentoid sealstones attached on it by four threads or wires.[141] The co-existence of these three iconographic elements (*polos*, long robe with vertical band and sealstones on the wrist), as components probably of the same, fragmentary composition but not necessarily of the same figure, recalls the "priestess" of the Hagia Triada sarcophagus (Fig. 11.7); her priestly appearance includes all these distinctive features. Additional evidence, this time from the late 13th century, i.e. *e-ri-ta*'s era, is the female figure with sheaths of grain from the Cult Centre at Mycenae (Fig. 11.18).[142] The woman, ambiguously identified as priestess or goddess, wears a lentoid sealstone on her right wrist as well as a *polos* on her head. Despite the fact that "goddesses" are also depicted wearing sealstones on their wrists,[143] a reflection, evidently, of the priestly attire in the divine

---

[135] Lupack 2008.

[136] *CMS* V, Suppl. 1B, nos. 113–115.

[137] As noted in Σακελλαράκης and Σαπουνά-Σακελλαράκη 1997, 168 "for the person buried [...] we have information only by the study of the movable finds. Due to the limited sceletological material we do not gain evidence for the gender or the age of the deceased. The finds albeit are very illuminating. The lack of weapons, the abundance of domestic pottery and the richness of jewellery indicate a female burial."

[138] The tholos A at Archanes, dated to the LM IIIA1 period, is the best preserved tomb of this type in Crete, as noted in Σακελλαράκης and Σαπουνά-Σακελλαράκη 1997, 654–661, fig. 721. Furthermore, it has the same form as the tholos tomb of Atreus in Mycenae and Minyas in Orchomenos. For its excavation and the woman buried in the antechamber see *ibid.* 158–168.

[139] The first one with a scene of tree worship, the other two with figure-of-eight shields – in one combined with sacral knots. Two more gold signet rings with figure-of-eight shields on their bezel were found in the SW corner of the antechamber together with beads of glass paste and gold, placed at first in a wooden *pyxis* probably, as recorded in Σακελλαράκης and Σαπουνά-Σακελλαράκη 1997, 654–661.

[140] We underline here that in Σακελλαράκης and Σαπουνά-Σακελλαράκη 1997, 167 it was argued that «the person in the burial chest wore a long priestly robe adorned with gold».

[141] According to Lang 1969, 184 (13 M nws), two round stones are depicted, held in place by four curving lines on white ground. For a less elaborate bracelet on a white arm against white ground see also Lang 1969, 86–89 (51 H nws). For a comparable bracelet on the wrist of the Cup-bearer from the Knossian Procession fresco see Evans 1927, 705, fig. 441, pl. XII.

[142] Morgan 2005, pl. 24b.

[143] See, for example, in Figs 11.20 and 11.21 of this paper, the female "divinities"/drivers of the chariots in the narrow panels C and D of the Hagia Triada sarcophagus, wearing lentoid sealstones on their wrists. Moreover, the lentoid sealstone depicted on both wrists of the so-called "Dove Goddess", a terracotta figure from the Postpalatial Shrine of the Double Axes at Knossos, LM IIIB according to Evans 1928, 335–340, fig. 193a1 and a2, or LM IIIA1/2 according to Ρεθεμιωτάκης 1998, 67–68. Four oversized rings or amygdaloid seals as bracelets and armlets are also worn on the

sphere, it seems more reasonable to attribute the hand of the Pylian fresco fragment to a woman of the local elite, participant in a ritual, probably a priestess, since the precious sealstones on her hand indicate her interference in administrative and economic activities.

The choice of the Pylian high priestess *e-ri-ta* as a study case for the Mycenaean female priestly attire needs no further justification; she is undoubtedly the most eminent female priestly figure in the whole Late Bronze Age Aegean. Whatever her apparel would be, the same could also be applied, *mutatis mutandis*, to other priestesses of the Mycenaean palatial centres of mainland Greece as well as Crete during its Mycenaean phase. Despite the predominance of a visual, coded language in the expression of religious attitudes it would be reasonable to assume local variants of established priestly attire, on a diachronic as well as on a synchronic level, even within the same community. The related iconographic evidence from Hagia Triada (sarcophagus and frescoes), Knossos (Procession fresco) and mainland Greece (frescoes from the palatial centres), as we have seen, give a hint of a relative multiplicity. Moreover, the latter could be attributed to the ranking of religious functionaries and duties, as attested in the Linear B tablets.[144]

The priestess *e-ri-ta* stands at the end of a Mycenaean palatial tradition of female priestly attire, a tradition detectable from the 2nd half of the 15th century until the end of the 13th century BC. Although her dress and accompanying accessories would have made her immediately recognizable within Pylian society, we cannot stop wondering whether, and to what degree, she was allowed to express her personal tastes through her apparel. Could she have dictated her own personal sartorial choices? Would have she differed from her contemporary *ka-ra-wi-po-ro ka-pa-ti-ja*, who also performed her duties in *pa-ki-ja-ne*? However, these and other similar questions will remain unanswered.

Acknowledgements

My warmest thanks are offered to Dr. Chr. Boulotis and to Prof. Emerita I. Tzachili for fruitful discussions on various aspects of this issue, as well as to V. Petrakis for useful comments on the text. I am also indebted to Dr. G. Rethemiotakis, Director of the Archaeological Museum of Herakleion, for a high-resolution photograph that allowed me to re-examine the garment of the so-called "goddess" from the Knossian Procession fresco.

Abbreviations

AA	*Archäologischer Anzeiger*
AJA	*American Journal of Archaeology*. The Journal of the Archaeological Institute of America
AJP	*American Journal of Philology*
AM	*Mitteilungen des Deutschen Archäologischen Instituts, Athenische Abteilung*

hands of a terracotta figure, conventionally named as "Lady M", from the House M quarter in the Mycenae citadel, a new area with indications of cult activity contemporary to the shrines of the Cult Centre, as referred to in Pliatsika 2012, 617–618, pl. CLa–c. Similarly decorated, and probably made by the same person, is another terracotta figure, named as the "Brussels' Lady", now in the Musées Royaux in Brussels, of the same origin in all probability.

[144] Olivier 1960.

ArchEph	*Archaiologike Ephemeris*
ArchKorrBl	*Archaeologisches Korrespondenzblatt*
ASAtene	*Annuario della Scuola archeologica di Atene e delle Missioni italiane in Oriente*
BSA	*Annual of the British School at Athens*
CMS	*Corpus der Minoischen und Mykenischen Siegel*
Ergon	*To Ergon tes Archaiologikes Etaireias*
Kadmos	*Kadmos*. Zeitschrift für vor- und frühgriechische Epigraphik
Mycenaean World 1988	*The Mycenaean World, Five Centuries of early Greek culture, 1600–1100 BC.* Catalogue of the exhibition held in the National Archaeological Museum, 15 December 1988–13 March 1989, Athens 1988.

Bibliography

Αποστολάκης, Ι. Μ. 1990 *Η Δικαιοσύνη στη Μυκηναϊκή Εποχή. Λειτουργία και Απονομή* Αθήνα-Κομοτηνή
Αραβαντινός, Β. 2010 *Το Αρχαιολογικό Μουσείο Θηβών*. Αθήνα
Aura Jorro, F. 1985 *Diccionario Micénico*, Vol. I. Madrid
Aura Jorro, F. 1993 *Diccionario Micénico*, Vol. II. Madrid
Boloti, T. (forthcoming) Minoan "hide skirt" in the Mycenaean Mainland: reality or artistic tradition? *In the honorary volume for Prof. I. Tzachili.*
Boulotis, Chr. 1979 Zur Deutung des Freskofragmentes Nr. 103 aus der tirynther Frauenprozession. *ArchKorrBl* 9, 59–67.
Boulotis, Chr. 1987 Nochmals zum Prozessionsfresko von Knossos: Palast und Darbringung von Prestige-Objekten. In Hägg and R. Marinatos, N. (eds), *The Function of the Minoan Palaces, Proceedings of the Fourth International Symposium at the Swedish Institute in Athens, 10–16 June 1984*. 145–155.
Μπουλώτης, Χρ. 1999 Το μυκηναϊκό κόσμημα. In *Ελληνικά κοσμήματα. Από τις συλλογές του Μουσείου Μπενάκη*. Αθήνα.
Μπουλώτης, Χρ. 2000 Η τέχνη των τοιχογραφιών στη μυκηναϊκή Βοιωτία, *Επετηρίς της Εταιρείας Βοιωτικών Μελετών*, Τόμος Γ', Τεύχος α', 1095–1149.
Μπουλώτης, Χρ. 2005 Πτυχές θρησκευτικής έκφρασης στο Ακρωτήρι, *Αλς* 3, 20–75.
Burke, B. 2005 Materialization of Mycenaean Ideology and the Ayia Triada Sarcophagus. *AJA* 109, 403–422.
Burke, B. 2012 Looking for Sea-silk in the Bronze Age Aegean. In M.-L. Nosch and R. Laffineur (eds). *Kosmos. Jewellery, Adornment and Textiles in the Aegean Bronze Age, Proceedings of the 13th International Aegean Conference, University of Copenhagen, Danish National Research Foundation's Centre for Textile Research, 21–26 April 2010*. Leuven/Liege, 171–178.
Cameron, M. A. S. 1975 *A General Study of Minoan Frescoes with Particular Reference to Unpublished Wall Paintings from Knossos,* 4 vols. (diss. University of Newcastle upon Tyne).
Chadwick, J. 1976 *The Mycenaean World*. Cambridge.
Coulomb, J. 1979 Le 'Prince aux Lis' de Knossos reconsideré, *BCH* 103, 29–50.
Coulomb, J. 1990 Quartier sud de Knossos: Divinité ou athlète? *Cretan Studies* 2, 99–110.
Del Freo, M. *et al.* 2010 The Terminology of Textiles in the Linear B Tablets, including some Considerations on Linear A Logograms and Abbreviations. In C. Michel and M.-L. Nosch (eds), *Textile Terminologies in the Ancient Near East and Mediterranean from the third to the first millennia BC*, Ancient Textiles Series 8. Oxford 338–373.
Δημακοπούλου Κ. and Διβάρη-Βαλάκου, Ν. 2010 *Η Μυκηναϊκή Ακρόπολη της Μιδέας*. Αθήνα
Dimopoulou N. and Rethemiotakis, G. 2003 The "Sacred Mansion" Ring from Poros, Herakleion. *AA* 118, 1–22.
Di Vita, A. 2000 Atti della Scuola: 1996–1997. *ASAtene* 74–75, 467–586.
Evans, A. 1928 *The Palace of Minos at Knossos* II. London
Evans, A. 1935 *The Palace of Minos at Knossos* IV. London
French, E. 1971 The Development of Mycenaean Terracotta Figurines. *BSA* 66, 102–187.

Gérard-Rousseau, M. 1968 *Les mentions religieuses dans les tablettes mycéniennes*. Roma.

Hallager, E. 1985 *The Master Impression: A Clay Sealing from the Greek-Swedish Excavations at Kastelli, Khania*, SIMA LXIX. Göteborg

Hiller, S. 1984 Te-o-po-ri-ja. In Glotz G. (ed.) *Aux origines de l'Hellénisme. La Crète et la Grèce. Hommage à Henri van Effenterre.* Paris, 139–150.

Holland, L. B. 1929 Mycenaean Plumes. *AJA* 33, 173–205.

Immerwahr, S. A. 1990 *Aegean painting in the Bronze Age*. University Park, Pennsylvania.

Jones, R. B. 2009 New Reconstructions of the "Mykenaia" and a Seated Woman from Mycenae. *AJA* 113.3, 309–337.

Karo, G. 1930 *Die Schachtgräber von Mykenai.* München

Killen, J. T. 1984 The Textile Industries at Pylos and Knossos. In C. Shelmerdine and Th. Palaima, (eds) *Pylos Comes Alive: Industry and Administration in a Mycenaean Palace*. New York, 49–63.

Κορρές, Γ. 1996 Η συμπληρωματική ανασκαφή εις Ρούτση Μυρσινοχωρίου Πυλίας και τα στέμματα του Πρωτομυκηναϊκού κόσμου. In *Πρώτο Επιστημονικό Συμπόσιο «Ανασκαφή και Μελέτη: Πρόσφατες εξελίξεις της έρευνας σε θέματα Αρχαιολογίας και Ιστορίας της Τέχνης στο Πανεπιστήμιο Αθηνών», Αθήνα, 3 και 4 Απριλίου 1996, Αίθουσα Εκδηλώσεων (A*ula) Φιλοσοφικής Σχολής. Πρόγραμμα και Περιλήψεις Ανακοινώσεων, 56–57.

Κριτσέλη-Προβίδη, Ι. 1982 *Τοιχογραφίες του Θρησκευτικού Κέντρου των Μυκηνών.* Αθήνα.

Lambrinudakis, V. 1981 Remains of the Mycenaean Period in the Sanctuary of Apollon Maleatas. In R. Hägg and N. Marinatos, *Sanctuaries and Cults in the Aegean Bronze Age. Proceedings of the First International Symposium at the Swedish Institute in Athens, 12–13 May 1980*. Stockholm, 59–65.

Lang, M. 1969 *The palace of Nestor at Pylos in Western Messenia, Vol. II, The Frescoes*. Princeton.

La Rosa, V. 2000 The Painted Sarcophagus: Determining the Chronology. In S. Sherrat (ed.), *The Wall Paintings of Thera. Proceedings of the First International Symposium, Petros M. Nomikos Conference Centre, Thera, Hellas, 30 August–4 September 1997*, vol. II. Athens, 996–997.

Lejeune, M. 1960 Prêtres et prêtresses dans les documents mycéniens. In *Hommages à Georges Dumézil*, Latomus 45. Paris, 129–139.

Lejeune, M. 1971 Prêtres et prêtresses dans les documents mycéniens. In *Mémoires de la philologie mycénienne II*. Roma, 85–93.

Lenuzza, V. 2012 Dressing Priestly Shoulders: suggestions from the Campstool fresco. In M.-L. Nosch and R. Laffineur (eds), *Kosmos: Jewellery, Adornment and Textiles in the Aegean Bronze Age. Proceedings of the 13th International Aegean Conference/13e Rencontre égéenne internationale, University of Copenhagen, Danish National Research Foundation's Centre for Textile Research, 21–26 April 2010*. Leuven/Liege 255–263.

Levi, D. 1956 The Sarcophagus of Haghia Triada restored, *Archaeology* 9, 192–199.

Lindgren, M. 1973 *The People of Pylos. Prosopographical and Methodological Studies in the Pylos archives*, BOREAS 3:I. Uppsala.

Long, Ch. 1974 *The Ayia Triadha Sarcophagus. A Study of Late Minoan and Mycenaean Funerary Practices and Beliefs*. Göteborg.

Lupack, S. 2008 *The Role of the Religious Sector in the Economy of Late Bronze Age Mycenaean Greece*. Oxford.

Marinatos, Sp. 1967 Kleidung, Haar- und Barttracht, *Archaeologia Homerica*. Göttingen.

Marinatos, N. 1993 *Minoan Religion. Ritual, Image and Symbol*. Columbia.

Marinatos, N. 2007 The Lily Crown and Sacred Kingship in Minoan Crete. In P. Betancourt *et al.* (eds), *Krinoi kai Limenes: Studies in Honor of Joseph and Maria Shaw*, 271–276.

Marinatos, N. 2010 *Minoan Kingship and the Solar Goddess. A Near Eastern Koine*. Urbana/Chicago.

Militello, P. 1998 *Haghia Triada I, Gli Affreschi*. Padova.

Morgan, L. 2005 The Cult Centre at Mycenae and the duality of life and death. In L. Morgan (ed.), *Aegean Wall Painting. A Tribute to Marc Cameron*, British School at Athens Studies 13, 159–171.

Müller, V. 1915 *Der Polos, die griechische Götterkrone*. Berlin.

Niemeier, W.-D. 1987 Das Stuckrelief des Prinzen mit der Federkrone aus Knossos und minoische Gottheiten. *AM* 102, 65–98.

Niemeier, W.-D. 1988 The Priest-King Fresco from Knossos: A New Reconstruction and Interpretation. In E. B. French and K. A. Wardle (eds), *Problems in Greek Prehistory, Papers Presented at the Centenary Conference of the British School of Archaeaology at Athens, Manchester April 1986*. Bristol, 235–244.

Niemeier, W.-D. 1989 Zur Ikonographie von Gottheiten und Adoranten in den Kultszenen auf Minoischen und Mykenischen Siegeln. In *Fragen und Probleme der bronzezeitlichen ägäischen Glyptik, Beiträge zum 3. Internationalen Marburger Siegel-Symposium, 5.–7. September 1985*, *CMS* Beiheft 3. Berlin, 163–184.

Nilsson, M. 1950 *The Minoan-Mycenaean Religion and its Survival in Greek Religion*. Lund, 2nd edition.

Nosch, M.-L. and Perna, M. 2001 Cloth in the Cult. In R. Laffineur and R. Hägg (eds), *Potnia, Deities and Religion in the Aegean Bronze Age. Proceedings of the 8th International Aegean Conference/8e Rencontre* égéenne *internationale, Göteborg, Göteborg University, 12–15 April 2000*, Aegaeum 22. Liege, 471–477.

Nosch, M.-L. and Laffineur, R. (eds) 2012 *Kosmos: Jewellery, Adornment and Textiles in the Aegean Bronze Age. Proceedings of the 13th International Aegean Conference/13e Rencontre égéenne internationale, University of Copenhagen, Danish National Research Foundation's Centre for Textile Research, 21–26 April 2010,* Aegaeum 33. Leuven/Liège.

Olivier, J.-P. 1960 *A propos d' une "liste" de desservants de sanctuaire dans les documents en Lineaire B de Pylos*. Bruxelles.

Palaima, Th. 1995 The Nature of the Mycenaean *Wanax*: Non-Indo-European Origins and Priestly Functions. In P. Rehak (ed.), *The Role of the Ruler in the Prehistoric Aegean: Proceedings of a Panel Discussion Presented at the Annual Meeting of the Archaeological Institute of America, New Orleans, Louisiana, 28 December 1992*, Aegaeum 11, 119–139.

Palaima, Th. 1997 Potter and Fuller: The Royal Craftsmen. In R. Laffineur and P. Betancourt (eds), *TEXNH: Craftsmen, Craftswomen and Craftsmanship in the Aegean Bronze Age. Proceedings of the 6th International Aegean Conference, Philadelphia, Temple University, 18–21 April 1996,* Aegaeum 16. Liège/Austin, 407–412.

Paribeni, R. 1903 Lavori esequiti dalla Missione Archeologica Italiana nel palazzo e nella necropoli di Haghia Triada dal 23 febbraio al 15 luglio 1903, *RendLinc* ser. 5.12, 340–351.

Pilafidis-Williams, K. 1995 A Mycenaean terracotta figure from Mount Oros on Aigina. In C. Morris (ed.), *Klados. Essays in honour of J. N. Coldstream*. London, 229–234.

Pliatsika, V. 2012 Simply Divine: the Jewellery, Dress and Body Adornment of the Mycenaean Clay Female Figures in Light of New Evidence from Mycenae. In M.-L. Nosch and R. Laffineur (eds), *Kosmos*, 609–626.

Poursat, J.-Cl. 1973 Le sphinx minoen: Un nouveau document. In *Antichità cretesi: Studi in onore di Doro Levi*. Catania, 111–114.

Poursat, J.-Cl. 1977 *Catalogue des ivoires Mycéniens du Musée National d'Athènes*. Paris.

Rehak, P. 1995 The Aegean 'Priest' on *CMS* I 223. *Kadmos* 33, 76–84.

Renfrew, C. 1985 *The Archaeology of Cult, The Sanctuary at Phylakopi*. London.

Ρεθεμιωτάκης, Γ. 1998 *Ανθρωπόμορφη πηλοπλαστική στην Κρήτη. Από τη Νεοανακτορική έως την Υπομινωική περίοδο*, Βιβλιοθήκη της Εν Αθήναις Αρχαιολογικής Εταιρείας αριθ. 174. Αθήνα.

Rodenwaldt, G. 1912a *Tiryns II, Die Fresken des Palastes*. Athens.

Rodenwaldt, G. 1912b Votivpinax aus Mykenai. *AM* 37, 129–140, pl. VIII.

Rodenwaldt, G. 1921 *Der Fries des Megarons von Mykenai*. Halle.

Rougemont, F. 2007 Flax and Linen Textiles in the Mycenaean Palatial Economy. In C. Gillis and M.-L. Nosch (eds), *Ancient Textiles. Production, Craft and Society. Proceedings of the First International Conference on Ancient Textiles, held at Lund, Sweden, and Copenhagen, Denmark, on March 19–23, 2003*. Oxford, 46–49.

Σακελλαράκης, Ι. 1972 Το θέμα της φερούσης ζώον γυναικός εις την κρητομυκηναϊκήν σφραγιδογλυφίαν. A*rch*E*ph*, 245–258.

Σακελλαράκης, Γ. and Σαπουνά-Σακελλαράκη Ε. 1997 *Αρχάνες. Μια νέα ματιά στη μινωική Κρήτη*. Αθήνα.

Σαπουνά-Σακελλαράκη, Ε. 1971 *Μινωικόν Ζώμα*, Βιβλιοθήκη της Εν Αθήναις Αρχαιολογικής Εταιρείας αριθ. 71. Αθήνα.

Shaw, M. C. 2004 The 'Priest-King' Fresco from Knossos: Man, Woman, Priest, King, or Someone Else. In A. P. Chapin (ed.), ΧΑΡΙΣ*: Essays in honor of Sara A. Immerwahr*, 65–84.

Τζαχίλη, Ι. 1997 *Υφαντική και υφάντρες στο προϊστορικό Αιγαίο, 2000–1000 π.Χ. Ηράκλειο.*

Τσούντας, Χρ. 1902 Κεφαλή εκ Μυκηνών. *Arch*E*ph*, 1–10.

Ventris, M. and Chadwick, J. 1973 *Documents in Mycenaean Greek*. Cambridge. 2nd edition.

Vermeule, E. and Karageorghis, V. 1982 *Mycenaean Pictorial Vase Painting*. Cambridge, Mass.

Warren, P. and Hankey, V. 1989 *Aegean Absolute Chronology*. Bristol.

Whittaker, H. 2007 Reflections on the Social Status of Mycenaean Women. In L. Larsson Lovén and A. Strömberg (eds), *Public roles and personal status, men and women in Antiquity. Proceedings of the second Nordic Symposium on Gender and Women's History in Antiquity, Copenhagen 3–5 October 2003*. Sävedalen, 3–18.

Wiessner, P. 1989 Style and Changing Relations between the Individual and Society. In I. Hodder (ed.), *The Meanings of Things. Material culture and symbolic expression*, One World Archaeology 6. London, 56–63.

Witton, W. 1960 The Priestess Eritha. *AJP* 81, n. 4, 415–421.

Wobst, H. M. 1977 Stylistic Behavior and Information Exchange. In C. Cleland (ed.), *Papers for the Director: research essays in honor of James B. Griffin*, Anthropological Papers of the University of Michigan 61. Ann Arbor, 317–342.

12. Flax and Linen in the First Millennium Babylonia BC: The Origins, Craft Industry and Uses of a Remarkable Textile

Louise Quillien

> "Utu: Young lady, the green flax is full of loveliness; Inanna, the green flax is full of loveliness, like barley in the furrow, in loveliness and charm; sister, a grand length of linen does take one's fancy; Inanna, a grand length of linen does take one's fancy; let me grub it up for you, and give it to you green, young lady, let me bring you green flax! Inanna, let me bring you green flax!"[1]

This Sumerian poem, sung by the god Utu to the goddess Inanna, lists all the stages of linen craftwork, from the uprooting of the plant to the manufacturing of a fabric for the nuptial bed of the goddess. Flax is known and used for textiles in Mesopotamia from Neolithic times, even if its transformation required elaborate techniques and precise knowledge. The *Linum usitatissimum L.* is one of the oldest plant species domesticated in Mesopotamia,[2] and it is still cultivated today in Iraq.[3] Of the many varieties of flax, this species is the best suited for textile manufacturing because it can yield long fibres. Flax can also produce oil, but in Mesopotamia, sesame is preferred to flax for this purpose.[4]

The oldest fragments of textiles found in Near East are made of linen. They come from the Nahal Hemar cave in Judea and date from the Neolithic period.[5] But during the 4th millennium BC, wool becomes the most frequently used textile fibre in Mesopotamia. During the three following millennia, however, the knowledge required for preparing, spinning and weaving flax is not forgotten. In 1st millennium BC Babylonia, we can observe, in some contexts, a renewed interest in the use of flax. At that time, the word referring to flax in its various forms (plant, fibre and manufactured object), is the Sumerian "gada", with the Akkadian equivalent "*kitû*". The cuneiform texts dealing with the use of flax in Babylonia during the 1st millennium BC mostly come from Uruk and Sippar's archives and date from the "long sixth century BC".[6] These administrative texts concern the

[1] Sumerian poem *The Bridal Sheet*, 3rd millennium BC, translated by Jacobsen 1987, 13–15.
[2] Forbes 1965, 27.
[3] Al Rawi 1980, 276–288.
[4] Reculeau 2009, 13–37.
[5] Schick 1988.
[6] The cuneiform texts dealing with linen in these archives cover a period from the 7th year of Nabopolassar, 619 BC (BM 77276 = ZA 4 137) to the 31st year of Darius I, 490 BC (BM 65133). The temples archives in general are concentrated on the "long 6th century", according to M. Jursa's expression (Jursa 2010, 5). The use of flax in the temples presents a remarkable continuity throughout the period. The political change after the conquest of Babylonia by Cyrus I has

temples' organisation and control of the textile craft industry to manufacture the rich garments regularly offered to the deities. In the temples' archives, several texts mention linen: a ritual text,[7] a judicial inquiry[8] and various inventories of goods and properties.[9] They show that, apart from divine garments, linen clothes are worn by the priests, the temples' officials and the craftsmen. Another group of texts comes from private archives, including marriage contracts,[10] a letter,[11] and lease contracts.[12] These texts are less numerous, but they present a broader spatial and chronological distribution. Furthermore, they give information about the use of linen outside the temples. The archaeological contemporary textile remains are scarce in Babylonia. But other disciplines like archaeobotany[13] and art history[14] can shed light on the texts. Furthermore, the situation in Babylonia can be compared with other countries and periods of Antiquity.[15] The Babylonian sources are rich enough to reconstruct the "chaîne opératoire" from flax to linen textiles in 1st millennium BC Babylonia, in order to discuss the value of this textile and its place in society. With this purpose, we will analyse successively the production of flax, the specialised craft of linen within the temples, and the use of linen textiles in Babylonian society.

The origin and the price of flax

The cultivation of flax

Places of cultivation

According to the text MMA 86.11.210, dated to the 14th year of Nabopolassar, craftsmen of the Ebabbar temple, the great temple of Sippar, are sent to the countryside in order to provide flax for the temple. They travel to a place called Bēl-iqbi, which hosts the largest complex of palm grove belonging to the Ebabbar's properties.[16] Bēl-iqbi is located far from Sippar, near Borsippa, on the bank of the Euphrates.[17] The text reads as follows: "Flax that the weavers have carried from the hands of Bēl-iqbi's gardeners, 2000 hands of flax that Ilû-rabû-nā'id gave to Šamaš-aḫ-iddin, including: 500 (taken) as the *šibšu*-tax; 1000 (bought) for ten shekels of silver; 500 exchanged for three gur of dates which was at their disposal. (...)".[18]

no consequences in the worship of the gods. Furthermore, the end of Uruk's archives during the reign of Darius I and the end of Sippar's archives under Xerxes I, does not mean that linen was no longer used in temples after this period.

[7] UVB 15 40, Uruk.

[8] CT 2, 2, Sippar, 19th year of Darius I's texts reign (503–502 BC).

[9] For example, the text Nrg 28, from Sippar, dated from the 1st year of Neriglissar's reign (559 BC), is an inventory of goods coming from Babylon and given to the temple of Sippar by an official of the temple.

[10] Cyr 183 (Roth n°19), Sippar, 4th year of Cyrus' reign (535–534 BC); CT 49 165 (Roth n°38), Babylon, reign of Antiochus ; BM 76968 (Roth n°42), Borsippa, 108th year of Seleucid era (203 BC).

[11] TCL 9 117.

[12] BE 09, 65; BE 09, 86a; EE 14/CBS 4999; EE 19/CBS 12861; IMT 16/Ni. 507; IMT 18/Ni. 528. These texts are land rentals pertaining to the Murašu's archives. They come from Nippur, and are dated to Artaxerxes I's reign, (464–424 BC). I thank G. Tolini for pointing me to these texts.

[13] H. Reculeau 2009 13–37, summarizes the archaeobotanical studies of flax in Mesopotamia.

[14] The Assyrian bas-reliefs depict garments. They do not represent the garments mentioned in Babylonian sources, but they can help to make hypotheses about the techniques known in Mesopotamia and about how the clothes are worn.

[15] For example, the studies by F. Médard (unpublished) on flax production in Neolithic Switzerland; Vogelsang-Eastwood 1992 on linen in Egypt; F. Rougemont on Mycenaean linen craft industry.

[16] These lands are the most ancient properties of the Ebabbar temple; they were bought at the beginning of Nabopolassar's reign (Jursa 2010, 346–347).

[17] Jursa 2010, 346–347.

[18] Extract of the text MMA 86.11.210, dated Nbp 26.II.14, published by Petschow RIDA III/1 167: "(Line 1) gada *šá*

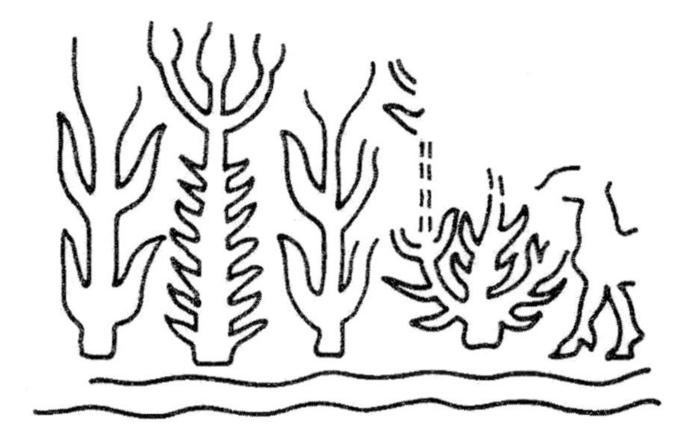

Fig. 12.1: The plant with three stalks depicted on an Uruk vase dated to the 3rd millennium BC might be flax, according to Crawford 1985, 74.

Fig. 12.2: This cylinder seal from Susa (from Rova 1994 n°419) might depict flax harvest according to Breniquet 2008, 273.

Another text, YBC 9273, dated to Nabuchodnezzar's reign, indicates that two men received money from the Eanna temple of Uruk, to collect flax in the steppe, or the neighbourhood of the city, where crops are growing:[19] "Twenty shekels of silver to buy flax, which were carried by Ina-tēšī-eṭir, son of Tabnīa, grandson of Nūr-Sîn, and Nanaia-iddin, the bleacher, to the steppe (…)".[20]

Six lease contracts, dated to the reign of Artaxerxes I, belonging to the Murašu archive, indicate that flax has to be paid as rent by the tenants,[21] along with cereals, vegetables and onions. The lands are irrigated and often rented with their canals.

Babylonian lands are suitable for the cultivation of flax and palm trees; they grow in the same lands. Flax requires a rich soil, a significant water supply and meticulous care. Its cultivation is a gardening work, as is palm cultivation. Flax can be grown in small irrigated fields, with crop rotation because the plants deplete the soil,[22] or on the wet banks of rivers unsuitable for other crops. Today, flax is still grown in gardens and floodplains in Iraq.[23] The scarcity of texts dealing with the cultivation of flax, in comparison with date palm, shows that it was a rare plant, on the fringes of other production. But during the Achaemenid period, according to the Murašu archive, its cultivation was encouraged by landowners.

---

^{lú}uš-bar gada *ina* šu^{II} ^{lú}nu-giš-kiri$_6$-meš (2) *šá* ^{d}en-*iq-bi*^{ki} *iš-šu-ú* dub (3) 2-lim šu^{II} *šá* gada ^{Id}gal-ní-tug *ina* igi ^{Id}utu-šeš-mu (4) *ina lìb-bi* 5-me *šib-šú* 1-lim *a-na* 10 gín kù-babbar (5) 5-me *ku-mu* 3 gur zú-lum-ma *e-šu-ú-m*[*a*] (6) *šá ina* igi-*šú-nu i-te-ṭer* (…).

[19] Coquerillat 1968 explains how during the Neo-Babylonian period, the area around Uruk was devoted to cereal crops or palm trees in irrigated lands belonging mainly to the Eanna.

[20] Extract of the text YBC 9273, date : Nbk 25, edition Payne 2007, 109: "(Line 1) 1/3 gín kù-babbar *šá a-na* ki-l[am] (2) *šá* gada$^{\text{hi-a}}$ *ina* šu^{II} ^{I}*ina*-sùḫ-[sur] (3) a-*šú šá* ^{I}*tab-né-e-a* a ^{I}zalag$_2$-⌜^{d}30⌝ (4) *ù* ^{Id}*na-na-a*-mu (5) ^{lú}*pu-ṣa-a-a a-na* edin (6) *šu-bu-lu* (…)".

[21] BE 09,65 (rent: 500 hands of flax) ; BE 09,86a (rent: 2500 hands of flax); EE 14 = CBS 4999 (rent: 500 hands of flax) ; EE 19 = CBS 12861 (rent: 200 hands of flax); IMT 16 = Ni 507 (rent 500 hands of flax) ; IMT 18 = Ni 528 (rent: 300 hands of flax).

[22] According to Latin authors, flax impoverishes soils. As Pliny writes: "flax burns fields and damages the ground" (*Natural History* XIX, V). Crop rotation in flax fields is also practiced in Egypt, see Kemp and Vogelsang-Eastwood 2001, 27.

[23] Al-Rawi 1980, 276–288.

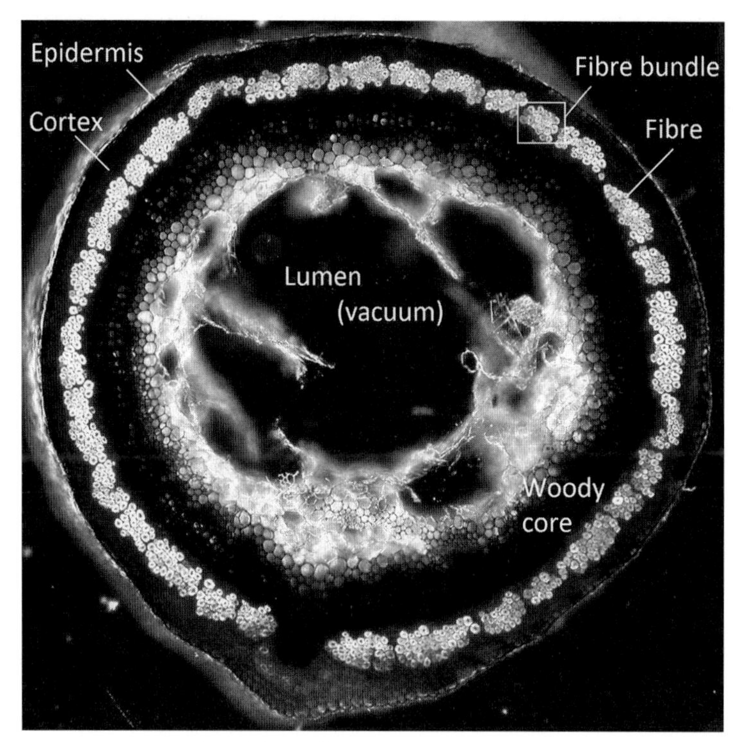

Fig. 12.3: Flax stem cross-section.

Cultivation techniques
In hot climate countries, flax is a winter crop, sown in autumn and harvested in spring.[24] The text MMA 86.11.210, concerning the collection of flax from the gardeners by two craftsmen, dates from May (26 Aiaru), the time of harvest.

Techniques of fibre extraction are not described in cuneiform tablets. The Sumerian poem *The Bridal Sheet*[25] and archaeological discoveries[26] indicate that the main stages of this work are the same in various places and periods in Antiquity,[27] even if there are differences related to cultural and environmental contexts.

Flax is harvested by hand. The stalks are pulled out to preserve the entire length of fibres. The seeds are removed, perhaps with a comb.[28] Then the flax is retted. Each flax stem is made of a woody core, with a central cavity. Fibres are located around the stalk, just behind the bark. They are agglomerated in bundles with pectin. The retting dissolves the pectin of the stalks,

[24] Today, in Iraq, flax is sown in October, irrigated during the growth and harvested in May (Renfrew 1985, 63).
[25] Jacobsen 1987, 13–15.
[26] Breniquet 2006, 167–176 proposes that the bone combs found in Neolithic levels at Ramad, Syria, were used for combing flax, not for carding wool.
[27] F. Medard shows how the steps of linen work were already known in Neolithic times in Switzerland (Médard 2006). These steps are also depicted on Egyptian mural paintings (Vogelsang-Eastwood 1992), and they are well described by Latin authors (Pliny, *Natural History* XIX, I).
[28] Breniquet 2006, 173.

frees the fibres and isolates them from the woody elements of the stem. Retting is well known in Mesopotamia, as evoked by the Sumerian poem *The Bridal Sheet*.[29] In dry climate countries, flax bundles are immersed in a pond or in a small water stream for a few days before being removed at the appropriate time. Then they are scutched: the stalks are beaten with a wooden tool to remove any woody elements without damaging the fibres. Finally, the linen fibres are combed and sorted according to their quality.

YBC 9273, which concerns money given by the Eanna temple to receive flax from the Uruk countryside, is dated to the month Ululu, which means September. Unlike MMA 86.11.210, it does not specify in what locality the flax had to be purchased. By this time of the year, the flax is probably already retted, scutched, combed and collected in handfuls.

Flax collection

The temples are the best documented flax producers and consumers in cuneiform sources. MMA 86.11.210 shows three ways for temple agents to acquire the flax grown in temple fields. Five hundred hands of flax[30] are collected as a *šibšu* tax. This tax usually concerns agricultural production such as cereals, dates and vegetables. Therefore, flax could be cultivated in fields devoted to this produce. Five hundred hands of flax are exchanged for dates. Dates are to be considered here as a means of payment. One hundred hands of flax are purchased with silver, evidence for the growing monetisation of the Babylonian economy during the 1st millennium BC.[31]

According to YBC 9273, the personnel responsible for supplying flax for the temples comes from two social groups, the city elite and the craftsmen. Ina-tēšī-eṭir is a member of the city elite, because his family name is mentioned. He may be the guarantor of silver entrusted by the temple for the purchase. Nanaia-iddin is a bleacher craftsman; his presence is needed to choose good quality flax.

In most of the administrative texts produced by temples, craftsmen receive silver themselves to buy flax for the temples. Their expertise is therefore necessarily required to ensure the quality of the materials. They can buy it in local markets where landowners sell their agricultural production. Purchase seems to be the most common way for the temple to acquire flax. The cultivation of flax in Babylonia remains rare, although temples and individuals – mentioned in the Murašu archive – have encouraged flax production. In addition to local production, Babylonia must import flax from other countries, especially Egypt.

The importation of flax from the Levant to Babylonia.

Flax is one of the precious goods imported from the West during the 1st millennium BC. For example, two shipment inventories of the merchant Nādin-aḫi, TCL 12, 84 and YOS 6, 168, record 153 minas of linen thread together with metals, precious stones, dyed wool and other goods imported from the Levant. The Eanna temple has given money to the merchant for this purchase. The two texts are dated from the 5th year of Nabonidus' reign.[32] Great business operations in the

[29] "Brother, when you have brought me green flax, who will ret it for me? Who will ret it for me? Who will ret its fibers for me?" Jacobsen 1987, 13.

[30] In Akkadian sources, unwrought flax is counted in "hands", perhaps because flax is worked handful by handful at each stage of its transformation from plant to fibres ready for spinning.

[31] See Jursa 2010, 469–753 for the monetization during 1st millennium BC Babylonia.

[32] These texts have been discussed by Oppenheim 1967, 239. According to Joannès 1999, 15–16 before shipping, the temple defines its needs. Then the merchant uses his experience to choose products according to their quality, cost and availability. At the time of final delivery, the temple's administration evaluates products, verifies the purchase prices and

Levant like this one are rare. The goods are carried by caravans from the West to Babylon, and then shipped on the Euphrates.[33] At Babylon, they follow the canals all the way to local centres.[34]

The origin of the imported goods is sometimes mentioned: copper from Ionia, iron from Egypt, alum from Lebanon. Linen surely comes from these regions as well, probably from Egypt. The high quality of Egyptian linen was well known during Antiquity.[35] CT 2, 2 demonstrates that a piece of linen fabric from Egypt has a great value in Babylonian temples. This tablet relates how a craftsman stole a strip of linen pertaining to the god's wardrobe. Responsible for finding the missing linen piece, he tried to replace it with a linen fabric from Egypt taken from another craftsman, Ubalissu-Gula. But Ubalissu-Gula proved that he had purchased the piece of linen from an Egyptian man at Babylon. Thus, the capital of Babylonia is a centre of redistribution of Egyptian linen. The text CT2, 2 reads:

"Guzānu, the priest *šangu* of Sippar and the *ērib bīti* of Šamaš said:

> This linen *šupallitu* fabric does not belong to Šamaš ! It's Bēl-ittannu who took the linen strip which was in [the menders] workshop.

They also declared:

> Šamaš's linen fabric is not lost; [it had to be] in [Bēl-ittannu's] workshop; and this linen *šupallitu* fabric is not belonging to Šamaš!

They also questioned Ubalissu-Gula, saying:

> This linen *šupallitu* fabric which had been given, this one, from whom did you receive it?
>
> This linen *šupallitu* fabric, in the presence of Erībaia, son of Šum-libši-Marduk, of Šumaia, son of Nāṣir, of Šum-iddin, son of Bēl-apla-iddin, of Širktu, Šamaš oblat, I got it from the hands of an Egyptian, for flour and dates".[36]

Two other texts (YOS 6 115 and TBER 68/69) mention respectively 15 and 5 linen tunics *salḫu*[37] purchased by Eanna's agents in Babylon, with gold, silver and aromatics. They probably have gone through trading channels and they must have a special value justifying their purchase in Babylon, because the temple craftsmen can also make *salḫu* tunics themselves.

Imported linen also arrives in Babylonia through war and tribute. This is well attested in Neo-Assyrian sources. In a Sippar text, FLP 1595, a royal donation dated from the 13th year of Nabonidus' reign, the king offers golden plates and bowls, cedar, juniper and linen tunics *salḫu* to the Ebabbar temple as part of the spoils.

---

clears the accounts. These two texts are the account of one unique commercial shipment. See also Graslin 2009, 38–42.

[33] Oppenheim 1967, 239.

[34] Graslin 2009, 205 presents a map of the commercial roads used to import textiles in Mesopotamia.

[35] Herodotus, II, 105 praises the high quality of Egyptian linen, which is also reflected in the archaeological remains (Schrenk 2006).

[36] CT 2 2; date: Dar 19; edition: Joannès 1992, 181–184; "(…) (10) ^{I}*gu-za-nu* ^{lú}šangu *sip-par*^{ki} *ù* ^{lú}ku$_4$-é ^{d}utu (11) ^{Id}en-*it-tan-nu i-šá-il-ma iq-bu-ú um-ma šu-mal-li-tu$_4$* gada [*a-ga-a*] (12) *ul šá* ^{d}utu *ši-i* ^{Id}en-*it-tan-nu ši-iš-ṭu šá* gada *šá ina* é-šu^{II} *šá* ^{lú}túg-kal-kal (13) *ik-ta-šad iq-bu-ú um-ma* gada *šá* ^{d}utu *ul ḫal-liq ina* é šu^{II} [^{Id}en-*it-tan-nu ši-i*] (14) *šu-pal-li-tu$_4$ a-ga-a ul šá* ^{d}utu *ši-i* ^{I}din-*su*-^{d}*gu-la* [*i-šá-il-ma iq-bu-ú*] (15) *um-ma šu-pal-li-tu$_4$ a-ga-a šá nad-nu ša-i ul-tu a*-[*a-im-ma*] (16) [*ta-ra-a*]*š-ši áš-šú* ^{I}din-*su*-^{d}*gu-la iq-bu um-ma* šu-*pal-li*-[*tu$_4$ a-ga-a*] (17) [*ina*] *gub-zu šá* ^{I}su-*a* a-*šú šá* ^{I}mu-*lib-ši*-^{d}amar-utu ^{I}mu-*a* a-*šú šá* ^{I}*na-ṣir* ^{I}mu-mu a-*šú šá* (18) ^{Id}en-a-mu ^{I}*ši-rik-ti* ^{lú}*ši-rik* ^{d}utu *a-na qí-me ù* [zú-lum-ma] (19) *ina* šu^{II} ^{lú}*mi-ṣir-a-a an-da-ḫar-šú* (…)".

[37] The reading *salḫu* is better than *šalḫu* (CAD Š/I, 242) accordind to S. Zawadzki, 2006, 105.

Fig 12.4: Price of hands of flax.

Text	Date	Place	Flax (hand)	Price (shekels of silver)	Shekles/100 hands
MMA 86.11.210	Nbp26.II.14 (612 BC)	Sippar	1000 1225	10 12	1 1 (+25 handful as a remission)
Nbn 164	Nbn21.VI.4 (552 BC)	Sippar	11600	164	1.4
Nbn 163	Nbn21.VI.4 (552 BC)	Sippar	2700	18	0.6
Nbn 370	Nbn12.IX.9 (546 BC)	Sippar	300	2	0.6

Fig. 12.5: Price of weighed linen.

Text (gada/*kitû)*	Date	Place	Weight (mina)	Price (shekel of silver)	shekel/mina
YBC 9273	Nbk12?.VI.x	Uruk	140	18.633	0.1331
Nbn 163	Nbn21.VI.04 (552 BC)	Sippar	120	32	0.2667
GCCI 1,278	Nbn16.IV.08 (548 BC)	Uruk	294.5	22.63	0.5749
BIN 1 161	X 15.II.2 (554 BC)	Uruk	4	2	0.5

But imported linen remains rare in Babylonian sources. Local linen arrives in the temples in handfuls of fibres, while imported linen is always thread or fabric. The cuneiform texts do not tell to what extent Babylonian linen was local or imported.

The price of flax and linen

Unwrought flax is counted in 'hands' in Babylonian texts, while wool is always weighed. This can be explained by the fact that bundles of flax are hand-held during all the stages of the transformation from stalks to fibres ready for spinning.[38] The same way of counting flax exists in Mycenaean Palatial administration.[39]

With one shekel of silver, it is possible to buy between 100 and 200 hands of flax. It is the quantity necessary to make a linen *salḫu* tunic, according to Nbn 163. One shekel of silver is the price of four minas of wool at Uruk under Nabonidus' reign, and five or six minas of wools are required to make a túg-kur-ra, the standard woollen cloth worn by workers.[40] According to these data, the price of flax is not too expensive compared to wool. However, temples can decide by themselves the price of the flax they buy when it comes from their own fields.[41] The text MMA

[38] F. Médard and C. Jespersen have used experimental archaeology to reconstruct linen work in Switzerland during Neolithic times. They evaluate the weight of one handful of raw flax: 130 grams, which can give 15 grams of high quality combed flax (Médard and Jespersen unpublished, 1–3).

[39] According to Rougemont 2009, the Mycenaean sign *SA*, could mean the same, "handful" of flax. The word always means unwrought flax. The *SA* are always counted, never weighed.

[40] Jursa 2010, 620.

[41] According to MMA 86.11.210, part of the harvest of the temple's fields is acquired by purchase or by exchange whereas

86.11.210 indicates that 500 hands of flax are worth 540 litres of dates. As a comparison, the monthly ration of dates which can be given to the temple's craftsmen at Sippar according to BM 64124 is 180 or 360 litres.[42]

Some texts indicate the price of weighed linen. But we do not know whether it is raw flax, linen thread, or linen fabric, because the Sumerian word gada means linen in all these different stages.[43] It is therefore normal that the price indicated varies widely.

Imported linen is always in the form of thread or fabric. According to YOS 6 168, linen thread costs 0.732 shekels of silver for one mina (50 gram), during the sixth year of Nabonidus' reign, (550 BC). The price of a linen tunic *salḫu* is four shekels according to TBER 68/69 (first year of Nabonidus, 555–554 BC), and 4.6 shekels according to YOS 6 115 (seventh year of Nabonidus, 549 BC). The three texts come from Uruk. According to these texts the linen tunic is more expensive when it is imported.

Mentions of prices are too scarce to determine the value of linen. The price can vary according to its origin, the circumstances of purchase, and the moment of the transaction. Linen is expensive, because the sums involved are always significant.[44] Linen is imported together with luxury goods. When flax is produced locally, prices are not exceptionally high compared to wool. Flax is a valuable product, available in sufficient quantities to meet the temples' needs.

A specialised linen craft industry in the temples of Uruk and Sippar

The profession 'linen weaver' (*išpar kite*) appears during the Neo-Babylonian period, for the first time in Mesopotamian history. The majority of the texts which give details about techniques and organisation of the textile craft industry come from temple archives. Linen textiles were also produced outside the religious sphere because linen clothes for urban elite are known, but this craft industry is not well documented. As Oppenheim explains, the "care and feeding of the gods" is central in Mesopotamian religion. The offering of new clothes and the regular changing of the wardrobe of the god's statues are among the most important parts of the worship of the deities.[45] A large personnel of craftsmen, under temple control, is in charge of the making and caring of these precious fabrics and garments. Temples scribes carefully record the materials and silver given to the craftsmen for their work, and also the final products delivered by the craftsmen to the temples when their work is completed. These texts do not describe the craftsmen's work, but they give a lot of detail about its organisation and about the different professions and tasks.[46]

---

another part is taken as a tax without retribution to the farmer.

[42] Text dated to the 11th year of Nabonidus (Bongenaar 1997, 302–303).

[43] For example, in the text GCCI 2 381, the word gada means non-spun flax (line 2 : "5 ma-na 1/3 ˹5˺ gín gada[ḫi-a] *ḫal-ṣu*", "5 mina 5 shekles 1/3 of combed flax"), while in the text Nbn 163, the word gada means linen fabric (line 2–5: "2 gú-un gada *kab-ba-ri* ki-lá 4-ta *šid-da-nu* 1 [gada]*bu-lu-ú šá* [d]*a-a* 1 [gada]*bu-lu-ú šá* [d]*bu-ne-ne*", "two talents of thick linen including four curtains *šiddû*, a curtain *bulû* of Aia, a curtain *bulû* of Bunene".

[44] As a comparison, the average wage of a craftsman varies between one and five shekels of silver per month during the reign of Nabonidus and Cyrus according to Jursa 2010, 679.

[45] Oppenheim 1964, 183–196, and Beaulieu 2003, Zawadzki 2006.

[46] Zawadzki 2006, 3–22 gives a typology of the texts concerning textile craft industry at Sippar.

The administration of linen craft industry in the temples

Professions

The organisation of textile craft industry in the temples has already been well studied.[47] There is a major difference between the prebendary craftsmen,[48] who have a higher status and a share in the offerings, and non-prebendary craftsmen, who have a lower status and who are paid with rations.

In the Ebabbar temple of Sippar, there is a specialised profession explicitly named 'linen weaver' (*išpar kite*). These craftsmen do not have a prebend. They form a group clearly separated from the 'weavers of coloured wool' (*išpar birmi*). The example of Šulā, a linen weaver in Sippar, shows the range of their functions. Šulā is responsible with another craftsman for the temple's silver given for the purchase of flax during at least four years (Nbn 163, Nbn 164). He is responsible for all the stages of production of linen fabrics: spinning (BM 62100), weaving of linen tunics *salḫu* (Cyr 326) and curtains (Nbn 502). Linen weavers often work in teams. Šulā is the head of one of these teams (Cyr 326).

At Sippar, linen weavers can also work as 'bleacher' (*puṣāiā*). They have to bleach the fabrics: *puṣu* (Nbn 492, Camb 415), and to wash them: *zukkû* (CT 55 439). They may also do repairs because they receive raw flax and linen thread. There are only a few bleachers among the linen weavers, and they often head a team of craftsmen. The specialisation of bleacher certainly requires a deeper technical knowledge and gives them a higher status and more responsibilities. Bleachers often receive materials: alkali (CT 55 439), linen thread (Nbn 805), thick linen (Nbn 117). They also receive silver for buying materials (Nbn 370, BM 75708). Another category of bleacher exists only at Sippar, the *mupaṣṣu*.[49] Unlike the *pūṣāia,* who belongs consistently to the linen weavers, the *mupaṣṣu* Ardīa is not a linen weaver. He bleaches fabrics (Nbn 115), but he works with menders, not with linen weavers (Nbn 115, BM 64941, BM 64007[50]).

At Uruk, the profession linen weaver does not exist. However, there are many *pūṣāia* bleachers. This group of craftsmen is in fact the equivalent of the group named linen weavers at Sippar, because they perform the same activities. They are responsible for both supplying raw flax to the temple (YBC 9273), and receiving combed flax (UCP 9/I 68). They deliver linen to the temple, once spun (YBC 9385) or woven (YOS 6 74). They wash linen fabrics (Eames 527).[51] They often work in teams, and they can be involved in contracts for the manufacturing of special pieces of fabric (GCCI 1 412).

The menders receive fabrics to repair. They do not make clothes. They work with both wool and linen textiles. Some of them are specialised in linen mending like Arrabi, who works on linen fabrics (Nbn 1090), but also receives dyed wool to make repairs (Nbn 415). The mending can be a specialisation of the linen weavers, the bleachers and the coloured wool weavers. All these professions are paid in rations.

Other craftsmen have a prebend: the weavers *išparu* /lú uš-bar and the launderer *ašlāku*/lú túg-babbar. They work with both wool and linen. Their function is to prepare garments and fabrics for

[47] See Bongenaar 1997; Da Riva 2002, and Zawadzki 2006 for Sippar; Kleber 2008; Payne 2007 for Uruk.

[48] The word 'prebend' comes from medieval vocabulary. It has been chosen by the historians to translate the Akkadian word *isqu*. Prebendary income, in Mesopotamian temples, is a wage given to perform a function in the temple. This function is hereditary, and implies a contact with the divine. The wage often includes a part of the offerings presented to the gods. On the prebendaries, see Waerzeggers 2010.

[49] The word *mupaṣṣu* comes from *peṣû*: 'white', it means 'bleacher, washerman' according to the CAD M/II 209 and 'washerman' according to the AHw/III, 674.

[50] Bertin 1877–8.

[51] Payne 2007, 119.

religious ceremonies. They receive linen fabrics from the linen weaver. For example, Nergal-iddin, replacing the prebendary weaver Kutû, receives linen garments in the text Nbn 696.

The prebendary launderers *ašlāku* also receive linen and woollen garments before the ceremonies.[52] They have to wash: *ana zikûti* (YOS 17 251) and to prepare them before religious ceremonies.[53] At Sippar, they are always responsible for the same garments recorded in standardised lists.[54] Uruk texts indicate that they receive aromatics, perhaps to perfume the garments or to please the gods while working. They also sew the golden sequins which adorned garments.[55]

Linen weavers are in charge of the manufacture of linen fabrics. They are organised in teams and pass their knowledge from father to son or by apprenticeship. They may be specialised in bleaching or mending. Bleaching is a specialisation of linen weavers, while dyeing is the work of coloured wool weavers. There is another clear division between the craftsmen who manufacture, bleach and mend linen clothes and those who prepare them for the ceremonies. Only the first ones are specialised in linen fabrics. Only the second ones have prebendaries and participate more closely in the worship of the gods.

Storage of linen

At Sippar, linen fabrics are stored in specific containers. The *nakmaru* is the most frequently used and can contain fabrics.[56] S. Zawadzki thinks that the *nakmaru* is a basket (Nbn 660). It is large enough to contain 18 garments (Nbn 252). The *nakmaru* may be a wicker trunk used for fabrics only made of linen. At Uruk, the *nakmaru* is used to store the golden stars and rosettes sewn on the woollen garments *kusītu* (NCBT 1008).[57]

Two texts from Sippar indicate that linen fabrics can also be stored in a *šaddu* container. In the Neo-Babylonian period, it corresponds to a chest where jewellery and precious stones are kept.[58] It can have a catch (CT 55 429). This chest is smaller than the *nakmaru*: it contains no more than three fabrics (Nbn 1090 and Nbn 1121). At Uruk, the *šaddu* is used to keep only gold. It could mean that at Sippar, some linen clothes are valuable enough to be stored in the same boxes used for jewellery. The woollen fabrics are never stored in the *nakmaru* nor in the *šaddu*. Linen fabrics are rare, they have a high value and they need care. This is perhaps the reason why they are stored in such special chests. Furthermore, linen and wool have different properties. Linen is less attacked by moths than wool, but it needs to be protected from moisture which turns the fabric yellow. They also must be stored flat or rolled to avoid fold marks.

Places of work

Texts do not refer to specific places of work for linen weavers. Temple craftsmen work most of the time outside the temple because their activities are often dirty. Some texts could indicate that a linen craft industry exists which is not controlled by the temples. Waerzeggers has proved that in

[52] The word *ašlāku* is traditionally translated «fuller», but Lackenbacher has noticed that the *ašlāku*'s work included also the finishing, the washing and the repairing of clothes. The term 'launderer' seems more appropriate for these different tasks.
[53] Zawadzki 2006, 66.
[54] These lists are called "*miḫṣu tenû*" and "*ana tabê*" in the typology of Stefan Zawadzki 2006, 3–21.
[55] Payne 2007, 87.
[56] "A storage container made of reed" CAD N/I, 188 ; "A wicker basket (Ein Tragkorb)" AHw/III, 722.
[57] The text NCBT 1008 indicates that the golden stars, rosettes and other sequins are stored in a *nakmaru* basket. These sequins come from the *kusītu* garment, which is made of wool.
[58] According to the CAD Š/I : 42.

various cities, such as Uruk, Babylon, and Borsippa, bleachers are working for the urban elite.[59] The separation between temple workforce and these urban craftsmen is not clear.[60] The urban bleachers are probably specialists of linen work, as temple bleachers. Woollen clothes can be washed in the houses, but the bleaching of linen requires complex knowledge and specialised craftsmen. The texts concerning urban craft industry are rare. To study the stages of linen work, it is necessary to turn back to the temple archives.

Stages of linen work in the temples of Uruk and Sippar

Manufacture

The temple craftsmen receive raw flax. They are therefore responsible for the whole manufacturing process from the spinning to the weaving of linen clothes. The temple scribes have to control the entry and exit of temple materials and garments. The scribes do not tell the craftsmen how to do their work. Nevertheless, the cuneiform tablets contain some information about the techniques used by linen weavers.

Spinning is carried out by giving a twist to fibres to intertwine them and produce a continuous and solid thread. Because of moisture, flax wraps spontaneously in an S direction.[61] The impulsion of rotation can be given between the fingers, on the thigh, or with a whorl. Whorls found in Mesopotamia date back to the 3rd millennium.[62] The Sumerian poem *The Bridal Sheet* mentions, after spinning, the doubling of the thread.[63] It is possible to spin very fine thread with flax.[64] Three words mean thread in Akkadian. *Ṭīmu* means woollen or linen thread,[65] as does *ṭimītu,* attested at Sippar only.[66] In contrast, the *ṭumānu* means only linen thread.[67] This word appears during the Neo-Babylonian period. The craftsmen sometimes buy the thread (Nbn 805) or spin flax themselves. The quantities of linen thread delivered by craftsmen to the temple are small, between five and 75 shekels (40 to 625 grams).[68] When thread is imported, the amounts are more significant, up to 153 minas (76.5 kilograms) in the text TCL 12 84.[69] This is also the case in the work contracts between the temple and a team of craftsmen for the delivery of thread and fabrics after receipt

[59] Waerzeggers 2006.

[60] It is not clear whether the temples had a full-time dedicated workforce or if the same men worked as urban craftsmen and occasionally for the temple. Elizabeth Payne suggests that one of the craftsmen mentioned in private bleaching contracts at Uruk (YOS 16 68 and CTMMA 3103), Libluṭ son of Nabû-šumu-ukīn, may be the same man who appears in temple archives as a linen weaver and bleacher (An Or 9, 9 III), Payne 2007, 178. See also Jursa 2005, 145–146.

[61] Breniquet 2008, 110.

[62] Breniquet 2008, 116–121.

[63] "Inanna : Brother, when you have brought it to me already spun, who will double up for me ? Who will double up for me? Who will double its thread for me?" Jacobsen 1987, 14.

[64] The fineness of Egyptian linen is famous. At Susa, remains of linen fabrics on copper axes, in a prehistoric tomb, are made of very fine thread. According to the experts, the fineness of the thread was more significant than modern thread made with machines. Al Jadir 1972, 59.

[65] CAD T, 112; AHw/IV, 1394; Beaulieu 2003, 16 and Zawadzki 2006, 31.

[66] CAD Ṭ, 112; AHw/IV, 1392; Zawadzki 2006, 31; Oppenheim 1967, 247–248.

[67] According the CAD Ṭ, 125, it means "a fine thread or fabric" and for the AHw/IV, 1394 it is "linen canvas (eine Leinwand)". Beaulieu 2003, 16 considers that it is woven linen. But Zawadzki 2003, 31 observes that at Sippar, *ṭumānu* is often weighed, whereas linen fabrics are counted. He chooses the translation "linen thread". Oppenheim 1967, 247–248 notices that the *ṭumānu* do not appear in the same contexts as the two other terms for thread in Sippar: the *ṭumānu* is the only one which is imported from the Levant. But it can also be made locally by temple craftsmen.

[68] Quantities of linen *ṭumānu* delivered: 10 shekels (UCP 9/I 20); 75 shekels (NBC 4859); 18.5 shekels (NCBT 702); 40 shekels (YBC 9385).

[69] Also 37.14 minas (18.5 kilograms) of linen thread imported in NCBT 632.

of raw materials. In the contract BM 62100, four minas (two kilograms) of linen thread *ṭumānu* are delivered, six minas (three kilograms) in BM 72810. In the last text, the thread is made by a woman, named Muranātu, the only woman linen spinner known in the archives. We do not know if the craftsmen spin flax themselves or if they have a spinning team, maybe women, working for them.[70] The text Nbn 164 gives an idea of the productivity of flax spinning:

> "(1) Balance of accounts (made) with linen weavers, which (goes from) the first year (of) Nabonidus, king of Babylon, until the month Ulûlu 21th day 4th year of Nabonidus, king of Babylon.
> (5) [x] minas one shekel of silver from the 1st year of Nabonidus, king of Babylon, [x mines] two shekels of silver from the second year, [x mi]nas from the 3rd year, total two minas 2/3 (mina) four shekels of silver had been given to Šulā, Uššāia and their workers [... fo]r 11 600 hands of flax.
> (10) From which:
>
> – ten *salḫu* of *kibsu* were delivered for 1800 hands of flax, on the month Aiaru, 2nd year
> – one talent seven minas of thick (linen) were delivered for 2700 hands, (for) nine *salḫu* of *kibsu*;
> (15) one *ḫullānu* for 1650 hands [.......] on the month Aiaru, 3rd year;
> – [......] *salḫu* of *kibsu*, for 450 hands [............] 3rd year were delivered;
> – [..................] for 2700 hands of flax were delivered on the month Ulûlu, 21th day
> (20) 4th year; four minas 17 shekels of linen thread were delivered (for) 200 hands.
> 2000 hands for 18 *salḫu* are at the disposal [of] Šulā and his workers, the remainder.
> (25) Month Ulûlu, 4th year of Nabonidus, king of Babylon, accounts are settled."[71]

Two thousand and fifty seven shekels (21.42 kilograms) of linen thread are spun with 200 hands of flax. According to the same text, one hand of linen weight 1.5 shekels (12.5 grams). Therefore, 200 hands of flax, weighting 300 shekels, give 257 shekels of linen thread. The loss of weight during spinning is low: 14.3%. The raw flax given to the craftsmen for spinning is already combed and selected for its length and quality.[72] The thread is then allocated to craftsmen or workshops as is suggested in the Uruk text YOS 6 113:

> "Linen thread which have been given for the weaver, the 7th year of Nabonidus king of Babylon: 5/6 mina (for) the weav[er x] shekels for the cella[73] month Ulûlu 1st day; ½ mina (for) the weaver ten shekels for [the *cella*] month Ulûlu 16th day (etc.)".[74]

The weaving of linen fabrics is not described in cuneiform texts. But the tablets give information

[70] In many Mesopotamian palaces and temples, spinning and weaving workshops existed. In Lagash, teams of women and children worked in workshops in exchange for rations (Lambert 1961). At Mari, the women weavers šal-uš-bar were more numerous than the male weavers (J. Bottéro, Archives Royales de Mari VII, 274). Finally, the texts called "the slave documents" may indicate that women worked in spinning teams in the Babylonian palace of Dûr Yakin (Durand, 1979).

[71] Nbn 164, date: Nbn 21.VI.04 "(1) *e-peš* níg-ka$_9$ *šá it-ti*$^{!}$ ^{lú}uš-bar gada (2) *šá* ta mu 1-kam ^{Id}nà-i lugal é^{ki} (3) *a-di* iti kin u$_4$ 21-kam mu 4-kam (4) ^{I}nà-i lugal tin-tir^{ki} (5) [x] ⸢ma⸣-na 1 gín kù-babbar *šá* mu 1-kam ^{I}nà-i lugal é^{ki} (6) [x ma-na] 3 gín kù-babbar *šá* mu 2-kam (7) [x ma]-na *šá* mu 3-kam pap 2 ma-na 2/3 4 gín kù-babbar (8) [*šá* ma]-na 21 lim 6 me šu^{II} *šá* gada *a-na* (9) $^{⸢\text{I}⸣}$*šu-la-a* <*u*> ^{I}*uš-šá-a-a u* ^{lú}érin-meš-*šú-nu* šum-*in* (10) ⸢*ina*⸣ *lib-bi* 10 ^{gada}*sal-ḫu šá kib-su a-na* (11) 1 lim 8 me šu^{II} *šá* gada iti gu$_4$ mu 2-kam *it-tan-nu* (12) 1 gú-un 7 ma-na *kab-ba-ru a-na* (13) 2 lim 7 me šu^{II} *šá* gada 9 *sal-ḫu* (14) 1-*en ḫu-ul-la-nu a-na* 1 lim 6 me 50 ⸢šu$^{\text{II!}}$⸣ (15) [*šá* gada] iti gu$_4$ mu-3-kam *it-tan-na* (16) [......] *sal-ḫi* ⸢*šá*$^{?}$⸣ *kib-su a-na* 4 me 50 šu^{II} (17) [............] mu 3-kam *it-tan-na* (18) [..........] 1-en PU DA [..........] (19) *a-na* 2 lim 7 me šu^{II} *šá* gada *ina* iti kin (20) u$_4$ 21-kam mu 4-kam *it-tan-nu* (21) 4 ma-na 17 gín *tu-ma-na-a-a-ti* (22) 2-me šu^{II} *it-tan-nu* (23) 2 lim šu^{II} *a-na* 18 *sal-ḫi ina* igi [......] (24) ^{I}*šu-la-a u* ^{lú}érin-meš-*šú re-*⸢*hi*⸣ (25) iti kin u$_4$ 21-kam mu 4-kam ^{I}nà-i ⸢lugal⸣ (26) lugal tin-tir^{ki} nì-ka$_9$ ki *šú-nu ep-šú*."

[72] Otherwise, the loss during the spinning would have been more significant. According to Médard, unpublished, 130 grams of unwrought linen yields 35 grams of fibres after retting and scutching. Only 15 grams of these fibres are suitable for spinning.

[73] Akkadian "*papaḫu*" (CAD P, 104).

[74] YOS 6 113; date: Nbn 08; Salonen no 233: "(1) ^{gada}*ṭi-mu šá a-na* ^{lú}uš-bar sì-*na* mu 7-kám ^{Id}nà-i lugal tin-tir$^{\text{[ki]}}$ (2) 5/6 ma-na 5 gín ^{lú}uš-[bar x] gín *a-na* é-*pa-pa-ḫi* iti ⸢kin⸣ u$_4$ 1-kam (3) 1 ½ ma-na ^{lú}uš-bar 10 gín *a-na* [.........] iti kin u$_4$ 16-kam (...)".

about the yield and the time of work. Every year, the temple administration determines its needs and orders new garments from the craftsmen. The temple furnishes the raw materials and linen weavers had to weave fabrics within the year. Nbn 163 and Nbn 164, both coming from Sippar and dated to the same day, are examples of these orders. Nbn 164 summarises the number of hands of flax given to the weavers from the first to the third year of Nabonidus reign and the fabrics made by the weavers with this flax. At the end of the text, the remaining flax is given for the work of the fourth year. Nbn 163 assigns a new quantity of flax for the next year. The linen fabrics are very frequently used for worship at Sippar. A system of yearly commands with strict control is organised by the temple's administration. Thus, the temples are regularly supplying new linen fabrics and garments. At Uruk, these orders for linen fabrics are formal contracts between the temple and the craftsmen, and they concern the linen curtains *gildû* frequently used in the gods' cella (PTS 3053, GCCI 1 412, YBC 3715). Occasionally, the temples hire specialised craftsmen who come from outside the city to weave exceptional pieces of linen fabric. Peek 2 is an hiring contract sealed the 9 Šabattu, 14th year of Nabopolassar, at Babylon, under the authority of the *šangu*, the priest of Sippar who may have come to Babylon for the New Year's celebrations. The text says:

> "750 hands of flax, property of Šamaš treasure, in charge of Madānu-aḫ-iddin. Madānu-aḫ-iddin will deliver during the month Aiaru two pieces of fabric[75] 12 cubits long, four cubits wide, work of the 14th year. Marduk-nadin-aḫi and Arad-Nabî his son, are the guarantors. Madānu-aḫ-iddin will deliver to Šamaš one piece of fabric, 12 cubits long, four cubits wide, during the month Šabattu, work of the 13th year".[76]

This contract involves a craftsman with a Babylonian name, Madānu-aḫ-iddin,[77] who probably does not belong to the Ebabbar's personnel. The temple of Sippar hires him for his specialised knowledge. He has to weave two large linen fabrics called *kīpu* each year with flax given by the temple. Madānu-aḫi-iddin had not yet delivered the work of the 13th year. The temple's administration concludes a new contract, dated to year 14, to oblige him to respect his engagements.

The organisation of the work of linen weavers is known, and weaving techniques are less well documented. There are a few indications of the size of linen fabrics. But, according to Zawadzki, the dimensions might have been noted only when they were exceptional.[78] The most frequently mentioned linen fabrics are small (two metres by two metres).[79] They may have been woven on a horizontal loom. This loom is the most suitable for linen, according to Breniquet, because the tension of the threads is moderated and the weft is beaten horizontally.[80] But the larger fabrics, for example the *kipû* measuring two metres by six metres may have been made with another type of loom. The warp weighed loom can be used with linen. Another hypothesis is the use of a vertical

[75] *Kipānu* : plural of *kīpu*. This text contains the only occurrence of this word according to the CAD K, 401.

[76] Extract of the text Peek 2, date: Nbp 09.XI.14, edition: Theo G. Pinches, *Inscribed Babylonian Tablets in the possession of Sir Henry Peek*, London, 1888: "(Line 1) 7 me 50 šu^{II} *šá* gada (2) níg-ga ^{d}utu *ina* ugu (3) ^{Id}di-ku$_{5}$-šeš-sì-*na* (4) 2 *ki-pa-a-nu šá* 12 <kùš> àm uš (5) 4-kùš sag-ki *iš-ka-ri* (6) *šá* mu 14-kam ^{Id}di-ku$_{5}$-šeš-mu (7) *ina* iti gu$_{4}$ *i-nam-din* (8) ^{Id}amar-utu-mu-šeš *ù* ^{I}ìr-nà (9) a-*šú pu-ut na-šu-ú* 1-*en* (10) ^{gada}*ki-i-pi* 12-kùš uš (11) *ù* 4-kùš sag-ki *ina* iti šu (12) *iš-ka-ri šá* mu 13-kam (13) ^{Id}di-ku$_{5}$-šeš-mu *a-na* ^{d}utu (14) *i-nam-din* (…)".

[77] Madānu is the chamberlain of Marduk, the great god of Babylon.

[78] Zawadzki 2006, 104.

[79] According to the text Peek 2; 750 hands of flax are used to make two fabrics of 12m^{2}, so 31.25 hands of flax gives 1m^{2} of linen fabric. According to this equivalence, the linen *salḫu* fabrics of the text Nbn 164 have the following dimensions: 3.5m^{2}, 5.76m^{2}, 9.6m^{2} and 3.5m^{2}. This calculation is an approximation: the weight of one hand of flax can vary and the *salḫu* can be of different qualities.

[80] Breniquet 2008, 136.

loom with two beams. This loom is used to weave flax in Egypt and is known in Mesopotamia during 1st millennium BC.[81]

After weaving, fabrics are sometimes sewn together. The texts indicate that some fabrics were used to make another one. The expression is, in Akkadian: "fabric one for fabric two" (*ana*) or "fabric two from fabric one" (*ša*). The technical process behind these expressions is not mentioned. The first fabric may have been tailored, sewn, decorated or arranged differently to make the second one. For example, two linen *salḫu* fabrics are used to make one linen curtain for a canopy called *dallat šamê*, at Sippar, according to the text BM 64591.[82] The Uruk text FLP 1613 perhaps refers to the sewing of a linen curtain; it mentions thread given to the craftsmen for 'stitching' the linen curtains *gildû*.[83]

Decoration

At Uruk and Sippar, the statues of the gods wear not only rich garments but also ornaments in gold and precious stones.[84] Golden sequins of various shapes are sewn onto the garments. According to Uruk documentation, four clothes are adorned with sequins.[85] Only one of them, the *pišannu*, may be made of linen, although its meaning remains unclear.[86] However, linen was used as wire for the necklace of goddesses, because of its strength and resistance. The text YOS 6 216 describes: "a necklace (for Ištar) of 88 beads, grenade shape, in striped agate (with) a gold frame (and) 88 golden lions, carnelian beads and a turquoise bead in the middle, held between two golden buttons on a linen wire".[87]

Usually linen fabrics used for worship are white. In the temples of Uruk and Sippar, dyeing is the work of the wool weaver, whereas linen weavers are specialised in bleaching. But some linen garments could also be colourful. One text mentioned a coloured linen tunic *salḫu* (BM 61025). Linen is not easy to dye, the colour is pale and is not fast. However, linen was sometimes dyed in Mesopotamia, according to the poem *The Bridal Sheet.*[88] In Egypt, linen can be dyed too.[89] Linen weavers occasionally receive dyeing materials, as in the text YOS 6 74: "64 linen fabrics [...] and 15 minas of *ḫurātu* dye,[90] one *pi* (36 litres) of *uqnātu* dye,[91] offering from Šamaš-mukīn-aḫḫi the *ša-reš*i officer, have been delivered by Šamaš-iddin, bleacher".[92] Here, the quantity of dye is very

[81] "The only clear representation of a Mesopotamian loom is an horizontal loom and one must wait the 1st millennium to see the mentions of "superior" and "inferior" beams", Breniquet 2008, 179.
[82] Text edited by Zawadzki 2006, 136–137.
[83] "*A-na ta-ki-pi gíd-da-la-né-é*".
[84] See Oppenheim 1949, Joannès 1992 and Beaulieu 2003.
[85] Oppenheim 1949, 179.
[86] The *pišannu* is preceded by the determinative « gada » (linen) in the texts BM 63912, Nbn 213, BIN 2 126, CT 56 388 and BIN 1 145. But instead of being a garment, it could also means a basket adorned with linen and coloured wool, which was used to store jewels.
[87] YOS 6 216, date: Nbn 14.VI.10; edition: Beaulieu 2003, 146: "(1) 1 gú 88 na$_4$ *nu-úr-mu-ú* babbar-dil (2) *man-di-tu*$_4$ kù-gi 60+28 *kur-ṣu-ú* kù-gi (3) na$_4$ gug na$_4$ aš-gì-gì-šá *bi-rit* (4) *ina* 2 *pi-in-gu* kù-gi *ina* dur gada-ḫa *ṣa-bit*".
[88] "Inanna: Brother, when you have brought it to me already doubled up, who will dye for me? Who will dye for me? Who will dye its thread for me?" Jacobsen 1987, 14.
[89] Goyon 1996.
[90] The *hurātu* dye may be gallnut (Joannès 1984, 143) or madder (Stol 1983, 533).
[91] The *uqnātu* might derive from *unqû,* a word meaning lapis-lazuli. It could be a dye of the same colour.
[92] YOS 6 74 ; date : Nbn 15.XI.06; " (1) 1+šú 4 gada$^{\text{ḫi-a}}$ *la-⸢x⸣ ù* (2) 15 ma-na ^{giš}ḫab (3) 1 pi *ú-qu-na-a-ta* (4) *ir-bi šá* ^{Id}utu-du-šeš (5) ^{lú}sag ^{Id}utu-mu (6) ^{lú}*pu-ṣa-a-a* (7) igi-*ir*".

large. Alum is used as a mordant.[93] But even if linen weavers can receive dye, it does not mean that linen was dyed with it.

Another technique can produce a coloured tunic *salḫu*, the embroidery of linen with coloured wool. Some clothes are made of wool and linen. For example, in the text GCCI 2 381, Amêl-Nanaia, the bleacher, received blue purple wool and combed flax to make the *šiddu* curtains of the goddess Nanaia. The text Nbn 349 indicates that blue purple wool is given to weavers for making the *mutattu* of a linen *kibsu*. The *mutattu* means, during the Neo-Babylonian period, a headband of dyed wool,[94] or a headdress.[95] It also may be a woven strip of coloured wool sewn onto the linen fabric, or embroidery, for example a braided trim.[96] Embroidery is well known in ancient Mesopotamia.[97] It is an easy technique for creating patterns on fabrics. A rare text, Cyr 232, shows that garments of the gods can be decorated with complex patterns, line 25: "a cloth (made of) red wool (with) a lion pattern" (1 túg$^{\text{hi-a}}$ ^{síg}ḫé-me-da *ur-maḫ*). Most of the time, linen fabrics are white in the temples of Uruk and Sippar. But some special linen garments may have been decorated with coloured wool embroidery.

Taking care of the linen garments and fabrics

Linen garments and fabrics have to be presented in perfect condition at the time of ceremonies. Linen needs to be regularly bleached. The bleaching process decolorises all the elements remaining on the linen cellulosic fibre, without damaging the fibre itself. Bleaching comes after weaving in the Sumerian poem *The Bridal Sheet*.[98] It is difficult to bleach linen thread without altering it,[99] and it is easier to bleach an entire fabric. Bleaching can go wrong if the craftsmen do not take enough precautions. It is a long process that requires a lot of practice. An apprentice bleacher has to learn the art of bleaching with a master over six years, according to the text Cyr 131:[100] "Nabû-šum-iddin son of Ardīa son [of …….. and] ^{f}Ina-Esagil-bêlet daughter [of] Šamaš-ilû, his wife, have given to Libluṭ, son of Uššāia, Nidintu, [their slave], for six years, for (teaching him) the profession of [bleacher]. **(6)** He will teach him the complete bleacher work."[101]

Cuneiform texts indicate that alkali and oil (Nbn 502) are used for bleaching.[102] The alkali is derived from tamarix, a tree whose ashes give soda. Alkali mixed with sesame oil gives soap. Juniper resin can be added to improve the smell.[103] The bleaching process is not described in the texts. In preindustrial Europe, linen fabrics are soaked in a bath of fermented water with germinated barley

[93] Cardon 2003, 333–334.

[94] CAD M/II, 312 "a headband" (meaning two); AHw/III, 689 "Half (Hälfte)".

[95] "An (elaborate) headdress", Zawadzki 2006, 132–133.

[96] Breniquet identifies in Mesopotamian iconography some bands or strips which may have been manufactured with tablet weaving. This technique allows to make woollen braid trims with elaborate patterns. (Breniquet 2008, 186–192).

[97] For example, fragments of linen fabrics found in royal tombs at Nimrud, dated to the second half of the 8th century BC, are made of linen tabby and embroidery playing with the natural colour of linen (beige and brown) to create patterns. (Crowfoot 1995, 113)

[98] Jacobsen 1987, 14.

[99] Baines 1989, 157–160.

[100] The apprenticeship contracts in 1st millennium Babylonia are analysed by J. Hackl in Jursa 2010, 694–725.

[101] Extract of the text Cyr 313, date: Cyr 25.V.08; edition Petschow, RLA 6 557 and 560: "(Line 1) ^{Id}nà-mu-mu a-*šú šá* ^{I}ìr-*ia* a [………………] (2) ^{mí}*ina*-é-sag-íl-*be-let* dumu-mí[-*šú šá*] (3) ^{Id}utu-dingir-*ú-a* dam-*šú* ^{I}*ni-din-ti* [lú *qal-la-šu-nu*] (4) *a-di* 6-ta mu-an-na-meš *a-na* ^{lú}⸢*pu*!⸣-[*ṣa-am-mu-ú-tu*] (5) *a-na* ^{I}*lib-luṭ* a-*šú šá* ^{I}*uš-šá-a-a id-din-nu* (6) ^{lú}*pu-ṣa-am-mu-ú-tu qa-tu-ú* (7) *ú-lam-mad-su* (…)".

[102] The word used in the texts is the Sumerian giš-naga. According to Zawadzki 2006 63–64, it is not the equivalent of the Akkadian *uḫulu* (alkali) but it had to be read gad-šu-naga, equivalent of the Akkadian *bīnu* (tamarix).

[103] According to Zawadzki, text BM 83647 (Zawadzki 2006, 65).

for two days. Then, fabrics are put in a vat with boiling water, ash and oil. They are washed, while water is regularly thrown on them. Fabrics are finally hung out in the sun and dampened, over several days. The process must be repeated many times to obtain a shade more and more white.[104]

Linen textiles also need to be washed. It is a less complex and shorter process than bleaching. The process is described in a Sumerian humoristic story "At the fuller".[105] The fabric is placed in water, beaten, plucked and washed with soap. Then it is 'tumbled': rinsed and wrung thoroughly. The clean linen fabrics are dried with special care of the edges.[106] This process is repeated several times. The same technique is attested in Egypt and in pre-industrial Europe.[107] The activities of the washer are not described in the cuneiform texts, but the materials they receive are precisely recorded. They have a higher status than linen weavers and they prepare the garments for worship. At Uruk, washers received precious aromatics.[108] They may use them to perfume the clothes. Indeed, the smell is part of the god's radiance in Babylonian religion.

After being worn, linen garments need to be repaired by the menders. The condition of the garments is often specified in the texts. New (*eššu*) garments are given to the most important gods and goddesses. Old ones (*labīru*) are given to minor deities. 'Old' garments do not mean that they are in bad condition. Fabrics can be called 'open' (*peṭu*), a term used only for linen fabrics. These fabrics are not torn because a new fabric can be called 'open' (BM 60307:6). An open fabric might be a cloth voluntarily split, to make a tunic, or a type of very loose weave which let the light pass between the threads, as is often the case with linen. The menders have to repair the garments and to finish them, because they also receive new garments. Washing and bleaching alter the fabric, the threads must be tightened, and holes must be mended. The linen garments are precious, so they are reused several times. The existence of a specialised craft industry of linen in the temples of Uruk and Sippar shows the value of the material.

Uses of linen textiles in 1st millennium BC Babylonia

Linen fabrics of the gods

Statues of the gods, in Babylonian temples are dressed with rich garments. The furnitures for worship is also decorated with precious fabrics. Linen has a special place in the worship.

Linen garments and fabrics

At Sippar, three standard linen fabrics are regularly offered to the gods: the *salḫu*, the *kibsu* and the *ḫullānu*. The three words are preceded by the determinative gada meaning linen. The *salḫu*[109] is very common. Most of the time, it is white, but it can also be coloured (BM 61025). In this case, it may have been dyed, as the text NCBT 1069 indicates line 20: "one *salḫu* which is given for dye". The size of the *salḫu* varies from 3.5m² to 5.76m² according to Zawadzki, and it weights

[104] Baines 1989, 162–163. She describes the bleaching as it was done in the region of Harlem, in Holland, during the 18th century.

[105] Forster 1993, 89–90 (UET 6/2 414).

[106] Waerzeggers 2006, 94.

[107] The same process is described by Baines 1989, 161 for 18th century Europe. The important steps are also depicted on Egyptian wall painting, in an idealized form (Vogelsang-Eastwood 1992).

[108] Payne 2007, 92–93.

[109] "A piece of linen fabric" CAD Š, 242; "linen cloth for deities (ein Leinengewand für Götterbillder)" AHw/IV, 1147.

between 1.4 and 5.5 kilograms.[110] It is woven by linen weavers of the temple (Nbn 164 and 164). But its production and circulation are not limited to the sanctuaries. A *salḫu* can be purchased on the market (CT 57 259) at Babylon (TBER Pl 68–69) or imported from the Levant (YOS 6, 115). The word *salḫu* is never preceded by the determinative túg, so it is not a complex garment. The *salḫu* is probably not cut, assembled and sewn, but it is worn draped around the body.[111] Indeed, linen is a lightweight fabric suitable for underwear. All the statues of the gods of Sippar and some of the gods of Uruk including Ištar (PTS 2094[112]), are wearing a *salḫu*. It can also be used as fabric, during the processions (Nbk 312), and to cover furniture for the worship: a throne (BM 63909),[113] a canopy (BM 64591),[114] a carriage (CT 55 815). In the text NCBT 1069: the *salḫu* is used as a sail for a processional boat and as an altar cover. The word *salḫu* applies then to a linen fabric of standard size and rectangular shape, suitable for various uses.

The *kibsu*[115] is closely linked to the *salḫu*. The expression *salḫu ana kibsu*, (one *salḫu* for one *kibsu*) or *salḫu* ša *kibsu* (one *salḫu* from one *kibsu*) are common (Nbk 312, Nbn 164). The *kibsu* is a fabric more advanced in the production stages than the *salḫu*.[116] The *salḫu* may be cut to make the *kibsu*; the text BM 64591[117] mentions that the *kibsu* is smaller. The *kibsu* may also be more decorated. The text Nbn 349 indicates that the *kibsu* can be trimed with a *muttatu*. The *kibsu* is given to numerous deities at Sippar. It is used also for covering altars of divine symbols (Nbk 312:20), chariots (CT 55 815) or thrones (Nbk 312:26). The word is rare at Uruk and appears in three texts, TCL 12 109 and PTS 2687 with the writing *ki-ba-su*[118] and NCBT 1069 where it is associated with the *salḫu*. The *kibsu* is a fabric made with the *salḫu*, decorated and mostly used as a furniture fabric.

The meaning of *ḫullānu* has been discussed elsewhere.[119] According to text Nbn 164, 1650 hands of linen are required to weave one *ḫullānu* which means that it weighs 20.6 kilograms. The *ḫullānu* is made by the linen weavers in smaller quantities than the *salḫu* and the *kibsu*. It appears in lists of garments at Sippar, but it is only worn by the two main deities: Šamaš and Bunene. It is also associated with bedspreads in the texts Nbn 115:12–13 and Nbn 252. At Uruk, the *ḫullānu* is preceded by the determinative túg and given to numerous deities according to PTS 2094. This heavy fabric is used as a garment for the main deities at Sippar and Uruk. It can also be transformed in a bedspread. We do not know if the *ḫullānu* is made of linen at Uruk.

---

110 Zawadzki 2006, 105–109.
111 Zawadzki 2006, 105–109.
112 Partly edited by Beaulieu 2003, 53, 180, 202, 220, 244, 258, 277, 284.
113 Bertin 1292.
114 Edited by Zawadzki 2006, 136–137.
115 "A piece of linen fabric" CAD K, 339; "a cloth (ein Kleidungsstück)" AHw/II, 472.
116 The *kibsu* never appears in the working order for the manufacturing of linen fabrics. It is not manufactured directly by the craftsmen but made from a *salḫu*.
117 Edited by Zawadzki 2006, 136–137.
118 Most often, the word is written *kib-su*. The writing *ki-ba-su* appears in this two Uruk texts and in BM 63909, a text from Sippar.
119 "A blanket or a wrap of linen or wool" CAD H, 229 and Beaulieu 2003, 15; "blanket (Decke)" AHw/II, 354; "coverlet or shirt", Zawadzki 2006, 109–111.

Linen curtains
In the temples of Sippar and Uruk, curtains for the deities' *cella* are always made of linen.[120] The linen *šiddu* is the most common curtain in the documentation.[121] The weight of the *šiddu* varies between 20 minas or ten kilograms (Nbn 502, Camb 36) and 35 minas or 17 kilograms (BM 84054).[122] Their size must be important. This curtain is made by the linen weavers (Nbn 163) but it can bear some decoration of wool. The text GCCI 2 381 says that a *šiddu* is made of 305.3 shekels of combed flax and 31 shekels of blue purple wool. The text CT 4 27:14 mentioned that the *šiddu* can have a woollen braided cord (*nīri*). According to the CAD, the *šiddu* curtain masks the offering during the rituals (RAcc 22, 4), or surrounds the statues during the new-year festival (RAcc 115 r.6).

The *gildû* curtain exists at Uruk and Sippar.[123] It can be made of other linen curtains: two (or two pairs) of *šiddu* curtains are sent to the city of Baṣ for making one *gildû* (CT 56, 10). According to NBC 8350, 30 minas (15 kilograms) of combed flax are used to make one *gildû*. At Uruk, their manufacture is controlled by contracts between the temple and the craftsmen (YBC 3715, GCCI 1 412, PTS 3053).[124] According to these texts, the *gildû* is employed as a veil to close the doors of god and goddesses' *cella*.

The giš-ig an-*e* or *dalat šamê* is a canopy with linen curtains. Some texts mentioned the *kitû* ša *dalat šamê*, the linen of the canopy (BM 72810:14, Nbn 1121:12 and Camb 415:9). In the *cella*, the gods' statues stand under linen curtains. According to Zawadzki, the size of this curtain varies between height and 13 cubits length.[125] The *dalat šamê* is not woven by the craftsmen but it is made of another linen fabric: the *salḫu* (BM 64591:11–14[126] and BM 66166:11).

Sippar texts show the circulation of linen fabrics, in two directions. Firstly, they circulate from major deities to minor ones. Secondly, linen fabrics can change their function and their aspect during their life. A linen tunic *salḫu* can become a curtain or a cover. The life of a linen fabric is long because of its solidity. This process of recycling can be explained by the price of the materials. Even in the temples, nothing precious must be wasted.

Textiles made of linen or wool.
Some textiles are made of different materials. The *sūnu* is a strip.[127] At Sippar, the word is not preceded by a determinative. The *sūnu* can be made of linen fabrics as the *kibsu* or the *salḫu* (BM 63503:22, Camb 412:5, Nbn 848:12). It also can be woven by the linen weavers (BM 65592:12). But the text YOS 17 254 indicates that half a mina of purple wool is used to make one *sūnu*. At

[120] The different curtains are all preceded by the determinative gada meaning linen.
[121] The CAD Š II, 407 propose the translation "cloth, curtain", but the word is never mentioned as a garment in the Neo-Babylonian texts. AHw:IV 1230 "side, edge, curtain (Seite, Rand, Vorhang)".
[122] Edited by Zawadzki 2006, 61–63.
[123] According to the CAD, the *gidlu* is "a string (of garlic)"; according to the AHw/II, 287, it is a "twisted cord (gedrehte Schnur)". The writing varies a lot (*gi-da-li-e* BM 84054 and CT 4 28, *gi-di-il-'* CT 56, 10, *gi-da-lu-ú* CT 55 439, *gi-da-lú-ú* NBC 8350, *gi-da-la-né-e* YBC 3715 and FLP 1613, gada-lá GCCI 1 412 and PTS 3053).
[124] See Payne 2007, 107.
[125] See the discussion about the size of the *dallat šamê* of Zawadzki 2006, 135–136.
[126] Edited by Zawadzki 2006, 136–137.
[127] "A piece of clothing or part thereof" CAD S, 388 and Beaulieu 2003, 15; "II cloth, bandage (Tuch od Binde)" AHw/III, 1059; "the *sūnu* might have been a head covering" (…) "in lists of Šamaš garments it could have been a kind of a belt" (…) it could have "a loincloth function" even if its "function change in response to specific circumstances" Zawadzki, 2006, 102–105.

Sippar, the *sūnu* is made of linen or/and wool. At Uruk, the word *sūnu* is always preceded by the determinative túg (cloth) and nothing proves that it is made of linen.

The *taḫapšu* is a tablecloth or a blanket.[128] At Sippar, even if the word is never preceded by the determinative gada (linen), the *taḫapšu* is always associated with the linen *salḫu*, *kibsu*, and *ḫullānu*. The *taḫapšu* can be made of a *salḫu* or a *kibsu* (Nbn 696:11–12). But wool is also used to manufacture the *taḫapšu* according to the texts Nbk 240, Nbn 948 and Nbn 494.

The *guḫalṣu* is a scarf or a braid.[129] At Uruk the word is often preceded by the determinative gada (linen) and followed by the expression "of purple wool" (YOS 17 301:17–18, YOS 19 270:13, GCCI 2 121:16). The text YOS 7 183:7 mentioned "a *guḫalṣu* of black fabric and thread" (*guḫalṣu ša* mud *u ṭimu*). The *guḫalṣu* could be a strip of linen with woollen embroidery, or a braid made of linen and wool. Numerous linen fabrics and garments have some elements made of coloured wool. The material of a garment can vary depending on its uses and on the city's traditions.

Rare linen garments.

In the temple archives, the linen fabrics and garments offered regularly to the gods are always the same. They are less numerous than woollen ones. In some rare texts other linen clothes appear. For example, the text BM 91002[130] from Sippar mentions a *mezēḫu ša* gada. It may be a scarf[131] occasionally made of linen, especially when the cloth is given to Šamaš. The gada-*ṣuppatu*[132] appears in three texts from Sippar, BM 61731, Nbn 731, CT 55 792. The *ṣuppatu* is made with one mina of linen according to BM 61731. But the word can also be preceded by the determinative síg for wool. In CT 55 792, it is delivered by a dyer and weighs 5.1 minas (2.5 kilograms).

Some linen garments are mentioned only once, such as the gada-*kīpu* (Peek 2) and the gada-*laripe*.[133] The gada-*buṣu*[134] (NCBT 597) may refer to byssus. At Sippar and at Uruk, all the deities wear a linen tunic *salḫu*. At Sippar, Šamaš and Bunene wear a linen coat *ḫullānu*.

At Sippar, linen is suitable for Šamaš because the god Sun wears only white garments,[135] and linen is whiter than wool. With the exception of *salḫu* tunics, linen is more often used for furniture than for the garments. The *salḫu* and *kibsu* at Sippar are often transformed into blankets, altar cloths, or curtains. Linen curtains are very important in worship, especially at Uruk, according to the documentation. They are spread at the doors, on the canopy or in front of the statues of the gods and goddesses. In the Neo-Babylonian *mis pî* ritual, a linen curtain must be placed in front of the statue of the god.[136] Linen is a perfect material to make curtains because of its solidity, its fineness and transparency. Behind a linen curtain, the statue is at the same time hidden and visible. It represents the real but supernatural presence of the god in the *cella*.

[128] CAD T, 40 "a woollen or linen blanket or stole"; "blanket" Beaulieu 2003, " a blanket for horse (eine Decke für Pferde)" AHw/IV, 1301; "a blanket or a coverlet", Zawadzki 2006, 134–135.

[129] "A special type of garment, perhaps a scarf, also a kind of coloured thread or a braid" CAD G, 123; "a scarf or a braid"; "wire, braid (Draht, Borte)" AHw/II, 296; Beaulieu, 2003, 15. Zawadzki differentiates the *guḫalṣu* and the *guḫalṣētu* (Zawadzki, 2006, 111–114).

[130] Zawadzki 2009

[131] "A scarf or a belt" CAD M II, 46 ; "sash (Schärpe)" AHw/III, 650; Beaulieu 2004, 16, Zawadzki 2006 , 151–152.

[132] CAD Ṣ, 249 "a strip of carded wool", AHw/IV, 1112 "spread carded wool (Lage gekämmte Wolle)".

[133] Not mentioned in the CAD.

[134] AHw/I, 143; CAD B, 350 *buṣu* meaning D.

[135] Zawadzki 2006.

[136] Dick 1999, BM 43749: 38.

Linen in Babylonian society

The garments worn by the statues of the gods are archaic in their fashion and do not correspond to the garments of contemporary peoples. The letters, marriage contracts, rituals, and lists shed light on the use of linen in several spheres of Babylonian society: clergy, soldiers and notables.

Priests occasionally wear linen cloth. According to the ritual UVB 15 40, linen is reserved for some of them. The consecrated lamentation priest (gala) and the chief of lamentation priest (gala-maḫ) wear the linen garment *ḫalpu*. A woollen *ḫalpu* is worn by another priest. Therefore, the use of linen depends on the function of the priest, not on the kind of cloth. The *ḫalpu* is a garment attested only in this ritual. The king also wears a linen garment during this ceremony, the *naḫlaptu gada*. The *naḫlaptu* is a shawl or a coat and belongs to the god's wardrobe.[137] The two terms come from the same root: the verb *ḫalāpum* which mean "to cover, clothe".[138] White linen has a sacred value, helping the king to be in contact with the gods during the ritual.

Linen clothes appear in lists of soldiers' equipment. The *karballatu*,[139] a headgear, can be made of linen. The *karballatu* is often mentioned with the *šir'am*, and pertains to the basic military dress. Usually, the word is not preceded by a determinative. But in TCL 9 117, the *karballatu* is said to be 'of linen' (*ša* gada). This text is a long list of supplies the author has sent to his lord. The lord is an administrator of the temple who bears the title of *bēl piqitti*. We do not know if the purpose of this equipment is military or civil. The *karballatu* is written with the determinative gada in one marriage contract (Cyr 183). But most of the times, the word is preceded by the determinative túg for 'cloth', and could be made of another material. According to the attestations, only notables or officials wear a linen *karballatu*. The *šir'am* cloth is another very common cloth worn by the soldiers and by workers.[140] Usually, it must be in wool or in leather. Only two texts mention linen *šir'am*. The same letter TCL 9, 117, where it is associated with the linen *karballatu*, and Nrg 28, a list of all the effects of an official travelling to Babylon. The list includes two *šir'am* for men, and one *šir'am* for women. Only the second one is "in linen" (*ša* gada). In the text TCL 9, 117, a *qablu ša* gada is listed with the *karballatu* and the *šir'am*. The word *qablu* means 'the hips' or 'the middle'. This cloth is probably a linen belt. According to these texts, some elements of military dress can be made in linen, but it seems exceptional and reserved for persons of high distinction.

The marriage contracts (*nudunnû*) list the property given by the family of the wife to the husband. The dowry includes cash, real estate, and personal items. The wife brings with her objects for the comfort of the household, including clothes.[141] However, the items listed in these contracts are only the valuable properties, belonging to the heritage of the family. The objects of everyday use and low value are not mentioned. Several contracts detail the garments and fabrics contained in the dowry. Three of them include linen clothes. In the marriage contract CT 49 165 from Babylon, dated to the Seleucid Era, many garments are listed. Most of them are made of wool, but there is one gada*ṣiprētu*, a dyed linen fabric.[142] The contract Cyr 183, from Sippar, dated to Cyrus' reign,

[137] "Wrap, outer garment" CAD H, 48; "robe, coat (Gewand, Mantel)" AHw/III, 715; "a kind of decorative shirt or blouse, an outer garment". Zawadzki, 2006, 114–116.

[138] CAD Ḫ, 34, *ḫalāpu* A meaning (2).

[139] "A piece of linen headgear for soldiers" CAD K, 215; "(pointed) cap – (Spitz) Mütze" AHw/II, 449.

[140] "One: Leather coat, two: a garment" CAD S, 312–313; "(Shirt of) armour – Panzer(hemd)" AHW/III, 1029; "a long tunic worn near the body" S. Zawadzki, 2009, 414; "a cardigan" F. Joannès 2010, 406; "a tunic, sometimes a chainmail" L. Oppenheim, JCS IV 1950, 192.

[141] M. Roth 1989, 1–2.

[142] The word *ṣiprētu* is attested in another text during the Neo-Babylonian period, according to the CAD Ṣ, 204 in the

Fig. 12.6: Attestations of linen garments for profane uses.

Text/Edition	Date/Place	Description	Linen fabric or garment
BM 76968/72 M. Roth 1989 b. n°42	Ant 23.XI.108 (204 BC) Borsippa	Marriage contract	3 linen fabrics (3 gada$^{ḫi\text{-}a}$) 1 ^{gada}*a-mur-sak-ku* 1 linen cloth for the head (*kitû šá muḫḫi qaqqadu*) 2 ^{gada}*šārsili*
CT 49 165 M. Roth 1989 b n°38	Ant 20.XII.[x] Babylon	Marriage contract	dyed linen (^{gada}*ṣiprēti*)
Cyr 183 M. Roth 1989 b. n° 19	Cyr 10. ?. 04 (535-534 BC) Sippar	Marriage contract	2 linen headgear (2 ^{gada}*karballutu*)
BM 76136:6	? Sippar	Inventory of objects for worship (garments and vases)	2 ^{gada}*a-mur-sak-ku*
Nrg 28	Nrg 16.IX.01 (559 BC) Sippar	Inventory of the properties of an official of Sippar travelling to Babylon	1 gada-túg? 1 linen *šir'am* for women (1 ^{túg}*šir'am ša* gada *amīltu*)
TCL 9 117	? Uruk	Letter to a lord, inventory of the goods sent to him.	1 *ša qablu* gada 1 linen *šir'am* (1 ^{túg}*šir'am ša* gada) 2 linen headgear (2 ^{túg}*karbalātu ša* gada)
YBC 3941	Nbk 22.IV.42 (562 BC) Uruk	Inventory of objects of a house	1 túg-gada *bāqqu*

mentions two linen *karballutu* in the dowry. This linen military garment is recorded in the list after silver, objects in bronze and iron, and furniture. It does not occupy the first place in the list. On the contrary, in BM 76968, linen garments are recorded first. In this marriage contract from Borsippa, dated to the Seleucid Era, the clothes are listed before silver, bronze objects and furniture. Of thirteen garments, seven are in linen: three linen fabrics (gada$^{ḫi\text{-}a}$), a ^{gada}*a-mur-sak-ku*,[143] a linen cloth for the head, perhaps a turban, a veil or a headdress (gada *ša muḫḫi qaqqadi*), and a ^{gada}*šarsilu*. The meaning of some of these garments is difficult to determinate without other attestations.

Some inventories mentioned linen garments. The text BM 76136,[144] dated to the Seleucid Era, is an inventory of garments and vases, probably for the cult, and mentions the linen *a-mur-sak-ku*. In the text YBC 3941, an inventory of commodities of a house from Uruk, dated to Nebuchadnezzar's reign, appears a túg gada *bāqqu*. The meaning of this word remains obscure. In the text BM 61494,[145] a goldsmith receives several clothes including a linen *maššanu*. But the attestations of craftsmen's clothes are rare. There is no other mention of these two last words. According to the sources at our disposal, linen garments in Babylonian society seem to be scarce and reserved for specific uses.

expression *ṣiprētu ša ṣupātu* (Camb 235), the word may be derived from *ṣirpu*, "dyed wool". The *ṣiprētu* could also be the plural of *ṣipru*: "comet tail" and mean a kind of train, maybe doubled. Thanks to F. Joannès for this interpretation.

[143] See Zawadzki 2010, 415.

[144] Zawadzki 2010, 409–429. This text comes from Babylon or Borsippa.

[145] Zawadzki 2010, 423.

In 1st millennium BC Babylonia, wool is still the most common textile material. But linen has a special place in Babylonian society. Flax is grown in the vicinity of cities like Uruk, Sippar and Nippur, but its cultivation is rarely mentioned in cuneiform documentation. Flax is also imported from the Levant and from Egypt. It is one of the goods exchanged in Babylon. The relatively important use of flax and linen in Babylonian sanctuaries can be explained by the wealth of the temples, their vast land holdings, and also by the growth of trade and long distance import channels to Babylonia. At the end of the Achaemenid period, the Murašu archive show that linen cultivation is promoted by private entrepreneurs.

The production of linen textiles is known in the context of the Babylonian temples of Sippar and Uruk. Some fabrics and garments of the gods are made of linen. Specialised craftsmen are in charge of their manufacture, decoration and care. The technical study reveals the peculiarities of their work: the making, the bleaching, the storage of linen are specific. Despite the technical constraints, the linen weavers are able to produce very large and heavy linen fabrics. They know also how to decorate them with coloured wool. They have a very advanced technical knowledge. The appearance of the profession 'linen weaver' at Sippar in the 1st millennium BC shows that this craft industry was of special importance during this period.

Linen fabrics could have been appreciated for their strength, their fineness, and their whiteness once bleached. Linen fabrics for worship are diverse in their aspect and use. Linen seems to be preferred for undergarments, for furniture, but also for prestigious garments for Šamaš at Sippar, and for the king and priests during rituals. Outside the temples, linen clothes are mentioned in the dowries and in various inventories. If wool is always the major textile fibre, linen occupies a significant place. Linen is present in almost all spheres of society to which texts give us access.

Aknowledgements

I present my warmest thanks to Michael Jursa, Elizabeth Payne and Stefan Zawadzki for having pointed me to pertinent texts and for having sent to me numerous transcriptions of unpublished cuneiform tablets. The transliterations of the NCBT, NBC and YBC texts from Yale and Princeton were provided by Elizabeth Payne. The texts BM 65133; BM 75708; BM 64941; BM 62100; BM 72810; BM 61025; BM 72810; BM 63503; BM 65592; BM 61731; BM 60307; and BM 60135 are cited by courtesy of Stefan Zawadzki who shared his transcriptions before their publication in *Garments of the Gods*, band 2. Oxford, (2013).

List of abreviations.

AHw	Von Soden, W. 1965–1981 *Akkadisches Handwörterbuch*. Göttingen.
BE	Clay, A. T. 1908 *Legal and Commercial Transactions, The Babylonian Expedition of the University of the Pennsylvania, Series A: Cuneiform Texts, VIII/I*. Pennsylvania.
Bertin	Bertin, G. 1883–4 *Copies of Babylonian Terra-cotta Dated Tablets, Principally Contracts*.
BIN1	Keiser, C. E. 1917 *Letters and Contracts from Erech Written in the Neo-Babylonian Period*. New Haven
BM	British Museum.

Camb — Strassmaier, J. N. 1890 *Inschriften von Cambyse, König von Babylon (529–521), Babylonische Texte Heft VIII–IX*. Leipzig.

CAD — Gelb, I. J., Jacobsen, T., Landsberger, B. and Oppenheim, A. L. (eds) 1956–2010 *The Assyrian Dictionary of the Oriental Institute of the University of Chicago*. Chicago.

Cyr — Strassmaier, J. N. 1880 *Inschriften von Cyrus, König von Babylon, (538–529 v. Chr.), Babylonische Texte Heft VII*. Leipzig.

CT 2 — Pinches, T. G. 1896 *Cuneiform Texts from Babylonian Tablets in the British Museum, part II*. London.

CT 49 — Kennedy, D. A. 1968, *Late Babylonian Economic Texts, Cuneiform Texts from Babylonian Tablets in the British Museum, part. ILXIX*. London.

CT 55, 56, 57 — Pinches, T. G. and Finkel, I. L. 1981 *Neo-Babylonian and Achaemenid Economic Texts, Cuneiform Texts from Babylonian Tablets in the British Museum, part. LV, LVI and LVII*. London.

Dar — Strassmaier, J. N. 1897 *Inschriften von Darius, König von Babylon (521–485 v. Chr), Babylonische Texte Heft X–XII*. Leipzig.

EE — Stolper, M. W. 1985 *Entrepreneurs and Empire, The Murašu Archive, the Murašu Firm, and Persian Rule in Babylonia, PIHANS 54*. Leiden.

GCCI — Dougherty, R. P. 1923–1933 *Goucher College Cuneiform Inscriptions, Archives from Erech, time of Nabuchadrezzar and Nabonidus (band 1)*, New Heaven, Yale University Press. New Haven.

GCCII — Dougherty, R. P. 1923–1933 *Goucher College Cuneiform Inscriptions, Archives from Erech, Neo Babylonian and Persian Periods (band 2),* New Heaven, Yale University Press. New Haven.

IMT — Donbaz, V. and Stolper, M. W. 1997 *Istanbul Murašu Texts, PIHANS 79*. Leiden.

MMA — Nöldenke, A. B. 1977 *Cuneiform Texts in the Metropolitan Museum of Art (New York)*. Paris.

NBC — Tablets in the Nies Babylonian Collection, Yale University.

NCBT — Newell Collection of Babylonian Tablets, Yale University.

Nbk — Strassmaier, J. N. 1889 *Inschriften von Nabuchodonosor, König von Babylon, (604–561 v. Chr), Babylonische Texte Heft V–VI*. Leipzig.

Nbn — Strassmaier, J. N. 1889 *Inschriften von Nabonide, König von Babylon (555–538*), *Babylonische Texte Heft I–IV*. Leipzig.

Nrg — Evetts, B. T. A. 1892 *Inscriptions of the Reign of Evil-Merodach, Neriglissar and Naborosoarchos*. Leipzig.

Peek — Pinches, T. G. 1988 *Inscribed Babylonian Tablets in the Possession of Sir Henry Peek*. London.

PTS — Tablets in the Princeton Theological Seminary.

TCL XII — Conteneau, P. 1927 *Contrats néo-babyloniens*, de *Téglath-Phalasar à Nabonide, Textes Cunéiformes du Louvre XII*. Paris.

TCL XIII — Conteneau, P. 1929 *Contrats néo-babyloniens II; achéménides et séleucides, Textes Cunéiformes du Louvre XIII*. Paris.

UCP 9/1 and 2 — Lutz, H. F. 1927 *Neo-Babylonian Administrative Documents from Erech, Part I and Part II*. New Haven/London.

UVB	Jordan, J. 1930 *Erster Vorläufiger Bericht* über *die von der Notgemeinschaft der Deutschen Wissenschaft in Uruk-Warka unternommen Ausgrabungen*. Berlin.
VS 6, 20	Königlichen Museen zu Berlin (ed.) 2010 *Vorderasiatische Schriftdenkmäler der Königlichen Museen zu Berlin.* Berlin.
YBC	Tablets in the Babylonian Collection, Yale University.
YOS 6	Dougerthy, R. P. 1920 *Records from Erech, time of Nabonidus*, *Yale Oriental Series VI*, *Babylonian Text*. New Haven/London
YOS 7	Tremayne, A. 1925 *Records from Erech, time of Cyrus and Cambyse*, *Yale Oriental Series VII, Babylonian Text*. New Haven and London.
YOS 17	Weisberg, D. 1980 *Texts from the Time of Nebuchadnezzar, Yale Oriental Series XVII, Babylonian Text*. New Haven/London.
YOS 19	Beaulieu, P. A. 2000 *Legal and Administrative Texts from the Reign of Nabonidus*, *Yale Oriental Series XIX, Babylonian Text*. New Haven/London.
ZA	Zeitschrift für Assyriologie.

Bibliography

Al-Jadir, W. 1972 Le métier des tisserands à l'époque assyrienne, filage et tissage. In *Sumer* 28, 53–74.

Al-Rawi, F. N. H. 1980 *Flora of Iraq, Band 4*. Baghdad.

Baines, P. 1989 *Linen, Hand Spinning and Weaving*. London.

Barber, E. J. W. 1991 *Prehistoric Textiles, The Development of Cloth in the Neolitic and Bronze Ages with Special Reference to the Aegean*. Princeton.

Beaulieu, P.-A. 2003 *The Pantheon of Uruk During the Neo-Babylonian Period, Cuneiform Monographs 23*. Leiden/Boston.

Bottéro, J. 1957 *Textes économiques et administratifs*, *Archives Royales de Mari VII*. Paris.

Breniquet, C. 2006 Ce lin, qui me le peignera? Enquête sur la fonction des peignes en os du néolithique précéramique levantin. *Syria 83*, 167–176.

Breniquet, C. 2008 *Essai sur le tissage en Mésopotamie, des premières communautés sédentaires au milieu du IIIe millénaire avant J.-C.*, *Travaux de la maison René-Ginouvès 5*. Paris.

Bongenaar, A. C. V. M. 1997 *The Neo-Babylonian Ebabbar Temple at Sippar: Its Administration and its Prosopography*, *PIHANS 80*. Leiden.

Cardon, D. 2003 *Le monde des teintures naturelles*. Paris.

Cocquerillat, D. 1968 *Palmeraies et cultures de l'Eanna d'Uruk (559–520)*, *Ausgrabungen der Deutschen Forschungsgemeinschaft in Uruk-Warka 8*. Berlin.

Crawford, H. 1985 A Note on the Vegetation of the Uruk Vase. *Bulletin of Sumerian Agriculture* 2, 73–76.

Crowfoot, E. 1995 Textiles from Recent Excavations at Nimrud. *Iraq* 57, 113-118.

Da Riva, R. 2002 *Der Ebabbar-Tempel von Sippar in frühneubabylonischer Zeit (640–580 v. Chr.), Alter Orient Altes Testament 291*. Münster.

Dick, M. B. (ed.) 1999 *Born in Heaven, Made on Earth: The Making of the Cult Image in the Ancient Near East*. Winona Lake, Indiana.

Durand, J.-M. 1979, Les « slave documents » de Merodach-Baladan. *Journal Asiatique*, 245–260.

Durand, J.-M. 2009 *La nomenclature des habits et des textiles dans les textes de Mari, Archives Royales de Mari 30*. Paris.

Forbes, R. J. 1965 *Studies in Ancient Technology*, *Band 3*. Leiden.

Forster, B. R. 1993 *Before the Muses, An Anthology of Akkadian Litterature*. Bethesda, Maryland.

Goyon, J.-Cl. 1996 Le lin et sa teinture en Egypte. Des procédés ancestraux aux pratiques importées (VIIe siècle av. J.-C. à l'époque récente). In *Aspects de l'artisanat du textile dans le monde méditerranéen (Egypte, Grèce, Monde Romain), Coll. de l'Institut d'Archéologie et d'Histoire de l'Antiquité*.

Graslin, L. 2009 *Les échanges à longue distance en Babylonie au Ier millénaire, une approche économique, Orient et Méditerranée 5*. Paris.

Hrouda, B. 1965 *Die kulturgeschichte des assyrischen flachbildes, Band 2*. Bonn.

Jacobsen, T. 1987 *The Harps That Once, Sumerian Poetry in Translation*. New-Haven/London.

Joannès, F. 1992 Les temples de Sippar et leurs trésors à l'époque Néo-Babylonienne. *Revue d'Assyriologie* 86, 159–184.

Joannès, F. 1999 Structures et opérations commerciales en Babylonie à l'époque néo babylonienne. In J. G. Dercksen (ed.), *Trade and Finance in Ancient Mesopotamia*. PIHANS 24.

Joannès, F. 2010 Textile Terminology in the Neo-Babylonian Documentation. In C. Michel and M.-L. Nosch, *Textile Terminologies in the Ancient Near East and Mediterranean from the Third to the First Millennnia BC*, 400–408.

Jursa, M. 2005 *Neo-babylonian Legal and Administrative Documents, Typology, Contents and Archives, Guide to the Mesopotamian Textual Record 1*. Münster.

Jursa, M. 2010 *Aspects of the Economic History of Babylonia in the First Millennium BC: Economic Geography, Economic Mentalities, Agriculture, the Use of Money and Problem of Economic Growth. Alter Orient und Altes Testament 337*. Münster.

Kleber, K. 2008 *Tempel und Palast. Die Beziehungen zwischen dem König und dem Eanna-Tempel im spätbabylonischen Uruk, Alter Orient Altes Testament 358*. Münster.

Kemp, B. J. and Vogelsang-Eastwood, G. M. 2001 *The Ancient Textile Industry at Amarna, Sixty-eighth Excavation Memoir, Egypt Exploration Society*. London.

Lackenbacher, S. 1982 Un texte vieux-babylonien sur la finition des textiles. *Syria* 59, 129–149.

Lambert, M. 1961 Recherches sur la vie ouvrière, les ateliers du tissage de Lagaš. *Archiv Orientální 29*, 422–443.

Levey, M. 1960 Chemistry and Chemical Technology in Ancient Mesopotamia, *J. Chem. Educ. 37*. Netherland.

Matsushima, E. 1993 Divine Statues in Ancient Mesopotamia: Their Fashioning and Clothing and Their Interaction with the Society. In E. Matsushima (ed.) *Official Cult and Popular Religion in the Ancient Near East*, 207–217. Heidelberg.

McCorriston, J. 1997 The Fiber Revolution, Textile Extensification, Alienation and Social Stratification in Ancient Mesopotamia. *Current Anthropology 38–4*, 517–549.

Médard, F. 2006 *Les activités du filage au Néolithique sur le Plateau Suisse, Analyse technique, économique et sociale*. Paris.

Médard, F. and Jespersen, C. (unpublished, personnal communication) *Le filage du lin, question de rendement, Journées d'archéologiques expérimentale, Bilan no 2.*

Michel, C. 2006 Femmes et production textile à Aššur au début du IIe millénaire av. J.-C. *Techniques et culture 46*, 281–297.

Oppenheim, A. L. 1949 The Golden Garments of the Gods. *Journal of Near Eastern Studies 8–3*, 172–193.

Oppenheim, A. L. 1964 The Care and Feeding of the Gods. *Ancient Mesopotamia*, 183–198.

Oppenheim, A. L. 1967 Essay on Overland Trade in the First Millenium BC. *Journal of Cuneiform Studies 21*, 236–254.

Parker, R. A. and Dubberstein, V. H. 1956 *Babylonian Chronology 626 BC–AD 75*. Providence.

Payne, E. E. 2007 *The Craftsmen of the Neo-Babylonian Period: A Study of the Textile and Metal Workers of the Eanna Temple* (provided by the author).

Reculeau, H. 2009 Le point sur la 'plante à huile': réflexions sur la culture du sésame en Syrie-Mésopotamie. In *Journal des Médecines cuneiforms 13*, 13–37.

Renfrew, J. M. 1985 Finds of Sesame and Lineseed in Ancient Iraq. *Bulletin of Sumerian Agriculture 2*, 63.

Roth, M. T. 1989 The Material Composition of the Neobabylonian Dowry. *Archiv für Orientforschung 36/37*, 1–55.

Roth, M. T. 1989 *Babylonian Marriage Agreements 7th–3rd Centuries BC, Alter Orient und Altes Testament 222*. Neukirchen-Vluyn.

Rougemont, F. 2007 Flax and Linen Textiles in the Mycenaean Palatial Economy. In Gillis C. and Nosch M.-L. (eds) *Ancient Textiles: Production, Craft and Society. Proceeding of the Conference held in Lund/Falsterbo, Sweden and Copenhagen, Denmark, on 19–23 March 2003*, 46–49.

Rougemont, F. 2009 Les phases de la production textile à l'époque mycénienne, de la fibre au produit fini: bref aperçu des données épigraphiques. *Cahier des thèmes transversaux ARSCAN 9*, 2007–2008, 37–47.

Rova, E. 1994 Ricerce sui sigilli a cylindro vicino-orientali del period di Uruk/Jemdet Nasr, *Instituto per l'Oriente C.A. Nallino*. Rome.
Salonen, E. 1980 *Neubabylonische Urkunden verschiedenen Inhalts III*. Helsinki.
Schick, T. 1988 Nahal Hemar Cave: Cordage, Bastetry and Fabrics, *Atigot* 18, 31–42.
Schrenk, S. 2006 *Textiles in Situ: Their Find Spot in Egypt and Neighbouring Countries in the First Millenium CE, Riggisberger Berichte 13*. Abegg-Stiftung, Switherland.
Stol, M. 1983 Leder(industrie). In *RLA 6,* 527–543.
Vogelsang-Eastwood, G. 1992 *The Production of Linen in Pharaonic Egypt, Stichting textile research*. Leiden.
Waerzeggers, C. 2006 Neo-babylonian laundry. *Revue d'Assyriologie 1/2006, band 100*, 83–96.
Waerzeggers, C. 2010 *The Ezida Temple of Borsippa, Priesthood, Cult, Archives*. Leiden.
Waetzoldt, H. 1972 *Untersuchungen zur neusumerischen Textilindustrie*. Rome.
Zawadzki, S. 2006 *Garments of the Gods. Studies on the Textile Industry and the Pantheon of Sippar according to the Texts from the Ebabbar Archive, Orbis Biblicus et Orientalis 218*. Fribourg.
Zawadzki, S. 2010 Garments in non-cultic context. In C. Michel and M.-L. Nosch, *Textile Terminologies in the Ancient Near East and Mediterranean from the Third to the First Millennnia BC*, 409–429.

13. Two Special Traditions in Jewish Garments and the Rarity of Mixing Wool and Linen Threads in the Land of Israel

Orit Shamir

Thousands of Roman textiles were discovered in the Land of Israel. However none of the textiles studied from Jewish sites show specifically Jewish characteristics. Hebrew texts make clear the extent of foreign influence on clothing: the Talmud Yerushalmi[1] and the Babylonian Talmud[2] both list eighteen essential garments for men and the terms are almost entirely based on Greek or Latin words.[3]

Thus, the basic items of clothing worn by Jews did not differ significantly from those worn by other inhabitants of the Graeco-Roman world.[4] Although the basic items of clothing are the same, there are two traditions in Jewish garments that are distinctive:

a. The laws of *sha'atnez* – Jewish law forbids the weaving of woollen threads together with linen: This is mentioned twice in the Bible:

   It is written in Leviticus 19:19 ...Nor shall you wear a garment of cloth made of two kinds of material". The prohibition of hybrid material is mentioned in the context of other hybrids such as cattle breeding and planting of different species together in a single field. *Sha'atnez* garments are mentioned but the materials are not listed.

   In Deuteronomy 22:11 it is written – "You shall not wear cloth of wool and linen mixed together". *Sha'atnez* applies only to sheep's wool and linen. However, any other combination of textiles does not create *sha'atnez* such as the combinations of materials like cotton, silk, camel wool and mohair.

b. *Tzitzit* – tassels at each corner of the mantle. The Torah states: "Speak to the children of Israel, and say to them, that they shall make themselves fringes on the corners of their garments throughout their generations, and they shall put on the corner fringe a blue (*tekhelet*) thread".[5]

---

[1] Shabbath 16.5.
[2] Shabbath 120a.
[3] Sheffer and Granger-Taylor 1994, 241.
[4] Roussin 1994, 183, 188.
[5] Numbers 15, 38; Babylonian Talmud, Menahoth 38–52; Yadin 1963, 182–187.

Sha'atnez

Although thousands of textiles in the Land of Israel were examined by the author,[6] not one piece of *sha'atnez* was found at Roman Jewish sites. This stands in contrast to other sites like in Syria e.g. Dura Europos and Palmyra,[7] and in Coptic Egypt which yielded great quantities of textiles made of mixed linen and wool.[8]

The few examples of mixed wool and linen (*sha'atnez*) textiles dated before the Byzantine period in the Land of Israel include:

1. Kuntillat 'Ajrud is located on an isolated hill, on the border between the southern Negev and Sinai Peninsula, nowadays in Egypt, near the junction of ancient roads traversing the Sinai desert from the first half of 8th century BC (Iron Age II). The linen textiles in general and the *sha'atnez* in particular, reflect the religious function of Ajrud, as a site inhabited by priests and probably these *sha'atnez* textiles belonged to them. Two of the *sha'atnez* textiles are undyed and undecorated, and made of linen warp and wool weft. The third textile is made of linen ornamented with selfbands and red wool (madder) and blue (indigo) linen threads in the warp. This textile was explained by A. Sheffer and A. Tidhar[9] as a type considered by Bible commentators as reserved for the high priest (Fig. 13.1) (see discussion below).
2. Wadi ed-Dâliyeh is located 15 km north of Jericho, consists of caves. 56 textiles catalogued including: linen (36), wool (12), linen mixed with wool (3), camel hair (3), camel hair mixed with wool (1), linen braids (3). These three *sha'atnez* textiles belonged to Samaritan refugees are dated to 365–335 BCE.[10] They have warps of linen and wefts of wool bands. One of them (No. 2) is a woman's scarf or veil, with the remains of a fringe at one end. It is decorated with shaded bands in grey-green, purple, blue and red. No. 3 has blue-green bands, No. 4 has a red band. The Samaritans are a religious sect based in the Nablus area, and the Torah – the Five Books of Moses – is their holy book. They still live in Israel today and keep the same religious rules. Their temple was at Mount Gerizim near Nablus. The Samaritans rebelled against Greek rule, but were suppressed with great cruelty. Some of their leaders fled to caves in Wadi ed-Dâliyeh and were killed there. It is possible that among them were their priests who were allowed to wear *sha'atnez*.
3. In Nabataean burials at `En Tamar dated to Roman period 2nd–3th centuries CE, there is a small group of linen textiles decorated with wool red bands (Fig. 13.2).[11] The Nabateans controlled the Spice routes joining Petra and Gaza northwards to Syria and westwards to the Mediterranean.

There are few examples of linen sewing threads on a wool textile: Cave of Letters (No. 45)[12] and Masada (two textiles). The Cave of Letters located at Nahal Hever, Judean Desert (Fig. 13.3) dated to the Roman period – Jewish Bar Kokhba revolt (132–135 CE) against Rome. It is a highly inaccessible cave and was used as refuge and not for dwelling.

[6] Shamir 2007
[7] Pfister and Bellinger 1945, 25, No. 256; Pfister 1934,13; 1937, Pls. 2:C, 4:F.
[8] Baginski and Tidhar 1980; Zemer 2010.
[9] Sheffer and Tidhar 2012.
[10] Crowfoot 1974, 60, 63.
[11] Shamir 2003, 37.
[12] Yadin 1963. 92 were published. An additional 254 textiles were found during excavations conducted in 2000–1 at the Cave of Letters under the direction of R. Freund and R. Arav, Vandenabeele *et. al.* 2006

Fig. 13.1: Kuntillat 'Ajrud. Linen textile decorated with wool bands. Photo by Havraham H.

The priest's girdle (belt)

In view of the biblical prohibition against wearing mixed wool and linen garments, it seems surprising to find these remnants of *sha'atnez* at Kuntillat 'Ajrud.[13] The strong northern-Israelite influence is reflected in the finds, showing that 'Ajrud was actually an Israelite and not Judean site.[14] According to the excavator, Prof. Ze'ev Meshel, it was established in order to demonstrate the control of Israel and the God of Israel over the road leading to the Red Sea and over the kingdom of Judea, just like the border-temples of other periods and along other borders demonstrate this dominance. The Israeli king (Joash) settled a group of priests and Levites from Israel there, who would fulfil both his commands as well as commandments of God: "...for every matter pertaining to God and affairs of the King".[15] It was the priests and Levites who gave the site its national and religious character and were responsible for at least some of the inscriptions mention various deities which were discovered at the site.[16]

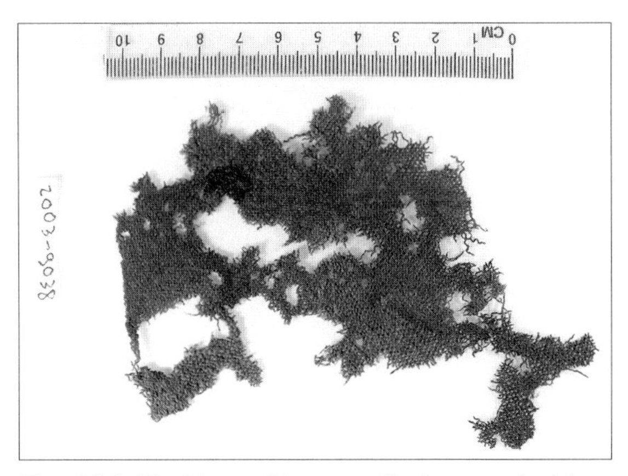

Fig. 13.2: `En Tamar. Linen textile decorated with wool bands (IAA. No. 2003-9038).

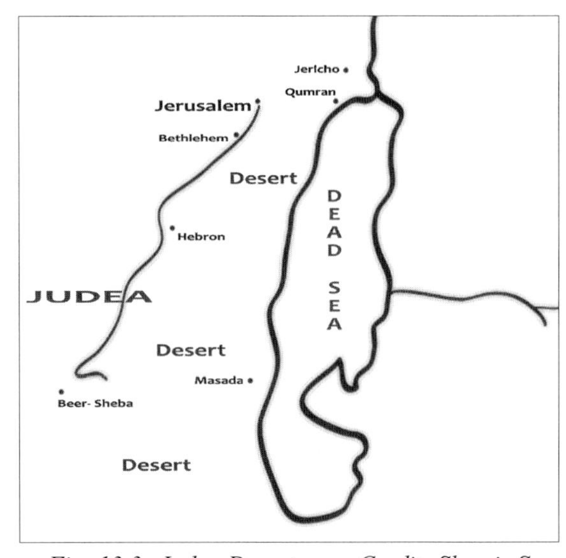

Fig. 13.3: Judea Desert map. Credit: Shamir S.

[13] Sheffer and Tidhar 2012, 307.
[14] During the Iron Age, two kingdoms of Hebrews emerged as important local powers in the ancient Levant: Israel in the north emerged in the 9th century BCE and Judah emerged in the 8th century BCE in the south.
[15] 1 Chr. 26, 32.
[16] Meshel 2012, 69.

The Bible does not explain why it is forbidden to mix these two fibres together, but ancient and modern interpreters gave different explanations. One explanation is connected with the priests' garments. The priests were allowed to wear *sha'atnez*. Why is *sha'atnez* need in the Temple? Why were the priests obligated to dress in clothes made of wool and linen together?

In order to distinguish between the priests' worship and the Jewish public worship, *sha'atnez* was forbidden only for the public – the prohibition was designed to separate priestly from public practice. Additionally, the prohibition was a way of setting aside this fibre blend only for holy purposes.

Josephus Flavius (Joseph ben Matityahu, 37–100 CE) wrote in *Antiquities of the Jews* that wearing *sha'atnez* was prohibited and was reserved for the priests of Israel only.

Although the High Priest's garments were different from the ordinary priest's garments, all of them wore *sha'atnez*. Ordinary priests wore *sha'atnez* only at their girdle.

Sheffer and Tidhar[17] noted that the priests were required to wear linen in the Temple, never wool: "It shall be that when they enter at the gates of the inner court, they shall be clothed with linen garments; and wool shall not be on them while they are ministering in the gates of the inner court and in the house"[18] In Mesopotamia, where the dominant fibre was wool, the priests were dressed also in linen. But The Bible instructs that the High Priest's vestment should be highly decorated and coloured, for honour and for beauty: "And you shall make sacred garments for Aaron your brother, for honor and for beauty".[19]

The Bible describes the priest's girdle: "And the sash of fine twisted linen, and blue and purple and scarlet material, the work of the weaver, just as the Lord had commanded Moses".[20] Rabbinic Judaism maintains that *sha'atnez* was permitted in the case of the priest's girdle, in which linen was woven with purple, blue, and scarlet yarn. According to the Rabbis (Judaic studies teacher, religious authority in Judaism), the purple, blue, and scarlet was made from wool.

Although the High Priest's garments were different from the ordinary priest's garments, all of them wore *sha'atnez*. In order to distinguish between the priestly worship and Jewish public worship *sha'atnez* was forbidden for the public.

Besides, the High Priest is only one person and probably his location and worship was at the Kingdom of Israel which had two central temples: Dan and Bethel.

Missing weft threads (spaces)

A few textiles were found with missing weft threads:

a. On one linen textile (a tunic sheet or headscarf) from the Cave of Letters there are two thin 'bare bands' where weft threads are missing. Yadin thought that they had been removed[21] (Fig. 13.4). We re-examined the textile and did not find any remains of fibres. The 'bare bands' have remains of one single thread all along the width of the textile at each of the bands. Parallel to the 'bare bands', *c.* 15cm from them there are two self-bands made of linen. I think the missing threads are linen self-bands that were taken apart to be used for sewing or another purpose.

[17] Sheffer and Tidhar 2012, 307.
[18] Ezekiel 44, 17–18.
[19] Exodus 28:2.
[20] Exodus 28:6.
[21] Yadin 1963, 261–262, No. 74.

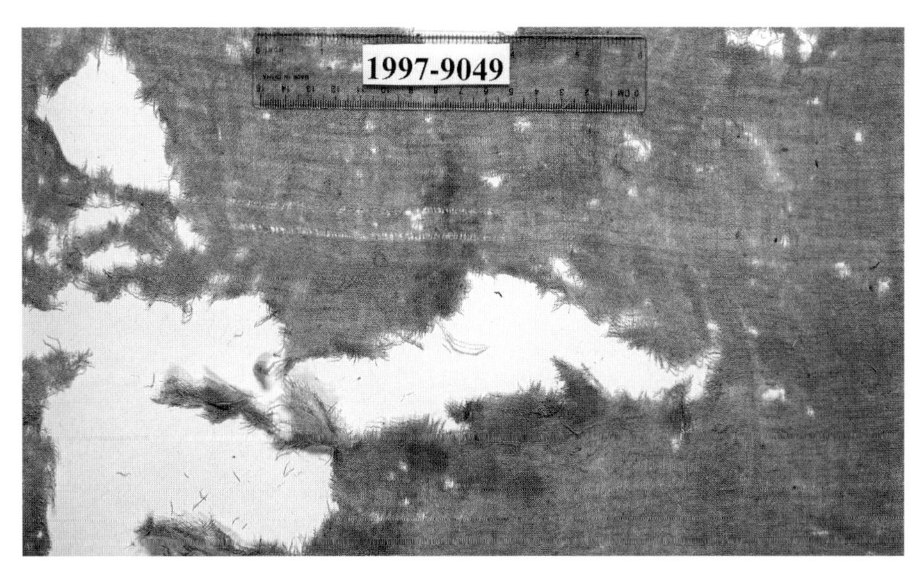

Fig. 13.4: Cave of Letters, missing weft threads (IAA. No. 1997-9049).

b. Several textiles with elaborate fringes at Qumran (Nos. 2, 3, 17, 31, 57) have an open space of missing weft threads, then a woven strip, and long fringe ends. No trace of wool or another filling was found in the space.
c. A linen textile from Avior Cave near Jericho has remains of red wool in the area of the missing weft threads.[22]

There are a number of explanations for this phenomenon:

a. Religious explanation:
   Yadin, Precker and Sheffer[23] assumed that wool threads had been intentionally removed from linen textiles. They think that these textiles had probably been bought by Jews from the Romans, and in order to avoid *sa'atnez*, they took out the threads.
b. Technological explanation:
   This space e.g. at Qumran could have accommodated the upper rod of the warp-weighted loom, to help start the weaving with a tight, straight, well-arranged warp (Fig. 13.5).[24]
c. Aesthetic explanation:
   It was a decoration[25] when the open bands are deliberately left empty as at Palmyra.[26] This practice of leaving an open space at both ends of a cloth was observed in Syria, such spaces are incorporated in fringes on headcloths for men and women.[27]

[22] Sheffer 1998.
[23] Yadin 1963, 262; Precker 1992, 170; Sheffer 1998.
[24] Crowfoot 1951, 31, Cat. No. 1.
[25] Crowfoot 1951, 31.
[26] J. P. Wild, pers. comm.
[27] Crowfoot 1955, 20.

Fig. 13.5: Wrapper from Qumran with space (IAA. No. 351297).

d. Preservation explanation:
   At Palmyra some linen textiles have open bands near fringes, where the wool has disintegrated.[28] In this case we usually will find remains of fibres such as in the textiles of Kasr al-Yahud near Jericho[29] and 'En Tamar.[30]

Sha'atnez conclusions

The concern to avoid *sha'atnez* during the Roman period, despite the hardship of war against the Roman army and the certain temptation to buy these textiles from non-Jews at the markets is impressive and caused technical weaving problems.

Stitching wool textiles with linen threads or vice verse is also forbidden in *sha'atnez*. Their presence in the Cave of the Letters can be explained by the harsh siege conditions of the Roman army.

Another important fact is the almost complete absence of mixed wool and linen (*sha'atnez*) textiles at non-Jewish sites, except in a few cases in the Roman period in a Nabatean burial at 'En Tamar[31] suggesting that most of the textiles in Israel during the Roman period were produced by Jews and purchased by the non-Jewish population. There is a great resemblance between the Nabatean and Jewish textiles (1st–2nd centuries CE), including shaded bands and the number of threads per cm.

The linen textiles in general and the *sha'atnez* in particular, reflect the religious function of Ajrud, as a site inhabited by priests and probably these *sha'atnez* textiles belonged to them.

---

[28] Crowfoot 1951, 30–31
[29] Shamir 2005.
[30] Shamir 2006.
[31] Shamir 2006.

Fig. 13.6: A man wrapped with tallit (mantle) and Tzitzit are attached to the four corners. Courtesy of "The Galilee Experience".

Fig. 13.7: Modern Tzitzit wikipedia, photographer Drosenbach.

Tzitzit–Ritual Tassels

In the following I will present the material findings.

What is *Tzitzit*?

The Hebrew noun *tzitzit* is the name for specially knotted ritual fringes worn by observant Jews. *Tzitzit* (Fig. 13.6) are attached to the four corners of the *tallit* (mantle). Wearing the *tzitzit* is commanded in Deuteronomy:[32] "You shall make yourself twisted threads, on the four corners of your garment with which you cover yourself." According to the Torah, the purpose of wearing *tzitzit* is to remind Jews of their religious obligations. In addition, it serves as a reminder of the Exodus from Egypt.[33] The tassel (*tzitzit*) on each corner is made of four strands bearing knots (Fig. 13.7). There are different interpretations concerning the number of strands but they are beyond the scope of this work. Women were exempt from this commandment.

[32] Deuteronomy 22:12.

[33] Numbers 15:40.

Blue (*tekhelet*) Dye

Determining what exactly *tekhelet* would have looked like has been the subject of conjecture and curiosity among rabbis, religious commentators and scientists for centuries.[34] The story of the search for the source for the dye *tekhelet* – Biblical blue – is one of intrigue, deception, and deduction. It weaves together clues from Torah scholarship, archaeology, and chemistry, and its major players include the great Chasidic Rebbe, a former Chief Rabbi of Israel, archaeologists, marine biologists and chemists.[35]

As mentioned above, *tzitzit* or part of it was dyed blue (*tekhelet*): "Rabbi Meir said: Whoever observes the mitzva of *tzitzit* is considered as if he greeted the Divine Presence, for *tekhelet* resembles the sea, and the sea resembles the sky, and the sky resembles God's holy throne."[36] There are also other interpretations of the meaning of *tekhelet* but they are beyond the scope of this work.

This explains why among the commentators, who base their comments in part on those images, there is no consensus about the blue tint, and colour ranges from blue to green and black.[37] The dye originated from sea snails as required by Jewish law.

Textiles Found in Israel dyed with Tyrian purple

Although thousands of textiles have been examined by the author[38] and others, up to the present time the true dye (*tekhelet*) from a breed of *murex trunculus* was found in only two textiles from Masada. One is blue and the other is purple.[39] Three more were found recently at Wadi Muraba'at.[40] Most scholars assume that the *tekhelet* was produced from *murex trunculus*.

The Finds

A single detached fringe was found at Kuntillat 'Ajrud (among scraps of fabric and threads). It was knotted from few undyed linen threads. It could be a *tzizith* or a regular tassel at the edge of a garment.[41]

But what is the meaning of the tassels discovered at the Cave of Letters? Three separate tassels, identified by the excavator Prof. Y. Yadin[42] as *tzizith* (Fig. 13.8), were found in the Cave of the Letters. They were found with a bundle of dyed unspun wool fibres (Fig. 13.9) in the Letters-skin contained bundle of Bar-Kokhba letters. They were wrapped in a piece of woollen mantle decorated with H-shaped design (No. 38) and a linen cloth (No. 80).[43]

The tassels, unconnected to any garment, are made of undyed linen threads, S-spun, tied before dyeing to purple unspun wool fibres.[44] The fibres beneath the tying, near the fold, had not been dyed.

---

[34] Kraft 2011.
[35] http://www.tekhelet.com/brochure.htm
[36] Sifre, Shelach, 15, 39.
[37] Amar Z, http://www.shaalvim.co.il/torah/maayan-article.asp?id=580 (in Hebrew).
[38] Shamir 2007.
[39] Examined by Prof. Zvi C. Koren, Koren 1997; Kraft 2011.
[40] Sukenik, Iluz, Shamir and Amar 2013.
[41] Sheffer and Tidhar 2012:299, Fig. 9:12.
[42] Yadin 1963, 182–187.
[43] Yadin 1963, 237, 263.
[44] Yadin 1963, 183 describes the knots in details.

As a bundle the tassels were double-dyed with madder and indigo.[45] Double dyeing is complicated because it is difficult to determine the exact amount of each dye colour. It is not possible to know which dye was used first, blue or red.[46] These fibres are not dyed with blue dye which originated from murex as required by Jewish law. This dye was very expensive, and according to Talmudic sources often imitated with the help of less expensive plant dyes.[47]

The absence of tassels on the mantles was explained by Yadin in the following way: that some of the mantles were used as shrouds and the Jews used to take the tassels of the deceased's mantle before burial. However, not all the mantles found at the Cave of the Letters were used as shrouds and none of them had *tzizith*.[48]

The Absence of *tzizith* at Qumran

Tzizith have not come to light at Qumran. If they had been used they would probably have survived like other organic materials.[49]

The sectarians, dwellers of Qumran, wore only linen garments.[50] *Tzizith* made of wool tied to a linen mantle is *sha'atnez*.[51]

Another explanation is that the textiles at the Qumran caves are in secondary use and perhaps the Jews removed the tassels when the mantle went out of use.

Tzizith Conclusions

Over the years the Rabbis, scientists and others visited the storeroom of the Israel Antiquities Authority in Jerusalem where all these artefacts were kept and we had many discussions about them.

Fig. 13.8: The Cave of the Letters, tzizith, (IAA. No. 1993-205).

[45] Koren 2005, 199; re-examined by Sukenik, pers. comm.
[46] Sukenik pers. comm.
[47] Yadin 1963:185.
[48] See discussion in Magness 2011:112.
[49] E.g. textiles and phylactery cases, Shamir 2008.
[50] Shamir and Sukenik 2010, 2011.
[51] Babylon Talmud Menachot 40, 1, but see discussion below.

Tzizith have not been found at any other Jewish site and this is not accidental and raises doubts about Yadin's interpretation of the findings.

Tzizith is spun and plied but at the Cave of Letters they are not. Yadin[52] thought this was because they were not yet finished, but after they have been tied to the linen threads it is impossible to spin and ply them.

Yadin also relied on the dye analyses made by Abrahams and Edelstain[53] who identified the dyes as indigo and kermes. As kermes dye was not found in any of the woollen textiles of the Cave of the Letters, Yadin[54] came to the conclusion that these tassels are *tzizith*. But Koren[55] and then Sukenik[56] later identified it as madder. Besides, kermes has been found only in three textiles in Israel, all from a Nabatean site.[57]

Yadin[58] also noted that according to the Talmudic sources, the dye for the *tzizith* must be solely for this specific purpose. As described above, the tassels were found with a bundle of the same fibres and dye.

Linen does not absorb any dye (except of blue produced from indigo) and Sukenik recently made an experiment of dyeing linen with different sources of dyes and arrived at this conclusion.[59] It can be perhaps assumed that it was used for 'checking' the dye in order to see what dye was obtained, and the linen, which does not absorb any dye, used in contrast to the dyed fibres.

Yadin[60] also considered the question and wrote: "Could they have served for 'checking' the dye, prior to dyeing the entire bundle?...". But he developed this idea in another direction to explain why *tzizith* could be made of wool and linen *(sha'atnez)*.

Or these tassels were samples that the dyer wanted to present, demonstrating his wares and his dyeing abilities, as mentioned in the Jewish sources.[61]

The dyer may deliberately combine linen thread and wool fibres because the fibres are designed to be used as examples only, or attempts to dye, but not to be worn and therefore there is no fear of *sha'atnez*: There is only a prohibition of *kilayim* regarding thread that has been spun and woven, as it says "You shall not wear *Sha'atnez*".[62]

It might be assumed that the woman from the Cave of the Letters who was found with fleece among her belongings (Fig. 13.9), bought it in market with the samples from the dyer.

Josephus and Philo do not mention *tzizith*[63] and maybe the majority of Jews didn't wear it. The Rabbinic sources criticize ordinary Jews for not observing these commandments.[64]

[52] Yadin 1963, 186.
[53] Yadin 1963, 182.
[54] Yadin 1963, 183.
[55] Koren 2005.
[56] Koren 2005 and then N. Sukenik pers. comm.
[57] 'En Rahel, 1th century CE, Shamir 1999; 2003; Koren 1999.
[58] Yadin 1963, 185.
[59] N. Sukenik, pers. comm.
[60] Yadin 1963, 183.
[61] Tosefta, Shabbat 1, 5.
[62] Mishnah Kilayim 9, 8.
[63] Magness 2011, 237.
[64] Magness 2011, 236.

Fig. 13.9: The Cave of the Letters, fleece of dyed unspun wool fibers (IAA. No. 1993-205).

Acknowledgements

My thanks are due to: Nahum Ben-Yehuda, Prof. Zvi C. Koren, Prof. Zohar Amar, Dr. Yitzhak Lifshitz, Dr. Benjamin Orbach, Prof. Menahem Kahana, Effie Meir and Dr. Naama Sukenik. Despite their help the interpretations are my responsibility. All photos except 1, 3, 6, 7 are: Courtesy of the Israel Antiquities Authority. Photos by Clara Amit.

Bibliography

Baginski, A. and Tidhar, A. 1980 *Textiles from Egypt 4th–13th Centuries CE*. Jerusalem.

Bellinger, L. 1962 Textiles. In H. D. Colt (ed.), *Excavations at Nessana*. Princeton, 91–105.

Crowfoot, G. M. 1951 Linen Textiles from the Cave of Ain Feshkha in the Jordan Valley. *Palestine Exploration Quarterly* 83, 5–31.

Crowfoot, G. M. 1955 The Linen Textiles. In D. Barthelemy and J. T. Milik (eds) *Discoveries in the Judaean Desert* I*: Qumran Cave*. Oxford, 18–38.

Crowfoot, E. 1974 Textiles. In P. W Lapp and N. L. Lapp (eds). *Discoveries in the Wâdi ed-Dâliyeh, AASOR* 41. Cambridge, Mass., 60–77.

Freund, R. A. and Arav R. 2001 Return to the Cave of Letters – What Still Lies Buried? *BAR* 27, 25–39.

Koren, Z. C. 1997 The Unprecedented Discovery Of The Royal Purple Dye On The Two Thousand Year-Old Royal Masada Textile. *American Institute for Conservation, The Textile Specialty Group Postprints* 7, 23–34.

Koren, Z. C. 2005 Chromatographic Analyses of Selected Historic Dyeings from Ancient Israel. In R. Janaway and P. Wyeth (eds), *Scientific Analysis of Ancient and Historic Textiles: Informing, Preservation, Display and Interpretation*. London, 194–201.

Kraft, D. 2011 Rediscovered, Ancient Color is Reclaiming Israeli. *The New York Times* 27.2.2012. http://www.nytimes.com/2011/02/28/world/middleeast/28blue.html

Magness, J. 2011 *Stone and Dung, Oil and Spit: Jewish Daily Life in the Time of Jesus*. Grand Rapids.
Meshel, Z. 2012 The nature of the Site and its Biblical Background. In Z. Meshel *et al.* (eds), *Kuntillet 'Ajrud. An Iron Age II Religious Site on the Judeah-Sinai Border*. Jerusalem, 60–69.
Milik, J. T 1977 Qumran Grotto 4. *DJD* 6, 35, VI:8.
Pfister, R. 1934 *Textiles de Palmyre*, Paris.
Pfister, R. and Bellinger, L. 1945 *Excavations at Dura Europos* IV, 2. *The Textiles*. New Haven.
Precker, R. 1992 *Vegetable Dyes and the Dyeing Industry in the Eastern Mediterranean during the Hellenistic, Roman and Byzantine Period.* MA thesis, Bar-Ilan University. Ramat Gan (Hebrew).
Roussin L. A. 1994 Costume in Roman Palestine: Archaeological Remains and the Evidence from the Mishnah. In J. L. Sebesta and L. Bonfante (eds), *The World of Roman Costume*. Madison, WI, 182–190.
Shamir, O. 2003 Textiles, Basketry and Cordage from Nabatean Sites along the Spice Route between Petra and Gaza. In R. Rosental-Heginbottom (ed.), *The Nabateans in the Negev*. Haifa, 35–38.
Shamir, O. 2005 Tunics from Kasr al-Yahud. In L. Cleland, M. Harlow and L. Llewellyn-Jones (eds). *The Clothed Body in the Ancient World*. Oxford, 162–168.
Shamir, O. 2006 Textiles, Basketry and Cordage and Fruits from 'En Tamar: Preliminary Report. In P. Bienkowski and K. Galor. *Crossing the Rift Valley*. Oxford, 191–194.
Shamir, O. 2007 "*Textiles in the Land of Israel from the Roman Period till the Early Islamic Period in the Light of the Archaeological Finds*". Institute of Archaeology, The Hebrew University. PhD dissertation. Supervised by: Prof. Gideon Foerster (Institute of Archaeology – The Hebrew University) and Dr. John Peter-Wild (Department of Art, History and Archaeology – University of Manchester).
Shamir, O. 2008 (2nd edition). Organic Materials. In D. T. Ariel *et al.* (eds), *The Dead Sea Scrolls*. Jerusalem, 116–134.
Shamir, O. and Sukenik, N. 2011 Qumran Textiles and the Garments of Qumran's Inhabitants. *Dead Sea Discoveries* 18, 206–225.
Shamir, O. and Sukenik, N. 2010 The Christmas Cave Textiles Compared to Qumran Textiles. *Archaeological Textiles Newsletter* 51, 26–30.
Sheffer, A. 1998 Bar Kokhba Period Textiles from Avior Cave. In H. Eshel and D. Amit (eds), *Refuge Caves of the Bar Kokhba Revolt*. Tel Aviv, 169–188. (Hebrew).
Sheffer, A. and Granger-Taylor, H. 1994 Textiles from Masada: A Preliminary Selection. In Y. Aviram, G. Foerster, and E. Netzer (eds), *Masada* IV. Jerusalem, 153–256.
Sheffer, A. and Tidhar, A. 2012 Textiles and Basketry at Kuntillat `Ajrud. In Z. Meshel *et al.* (eds), *Kuntillet 'Ajrud. An Iron Age II Religious Site on the Judeah-Sinai Border*, Jerusalem, 289–312. *(after* 1991 *`Atiqot* 20;1–26 with comment by the editor).
Sheffer, A. and Tidhar, A. 1991 The Textiles from the `En-Boqeq Excavation in Israel. *Textile History* 22, 3–46.
Sukenik, N., Iluz, D., Shamir, O., Varvak, A. and Amar, Z. 2013 Purple-Dyed Textiles From Wadi Murabba'at – Historical, Archaeological and Chemical Aspects. *Archaelogical Textiles Review (ATR)* 55, 46–54.
Vandenabeele, P., Adwards, H. G. M., Shamir O., Gunneweg, J. and Moens, L. 2006 Raman Spectroscopic Study of Archaeological Textile Samples from the 'Cave of Letters'. In J. Gunneweg, C. Greenblatt and A. Adriens (eds). *Bio- and Material Cultures at Qumran*. Stuttgart. 131–138.
Yadin, Y. 1963 *Finds from the Bar Kokhba Period in the Cave of Letters*. Jerusalem.
Zemer, A. 2010 *Coptic Textiles from Egypt in Ancient Times*. Haifa.